Reliability of Large and Complex Systems

Reliability of Large and Complex Systems

Second Edition

Krzysztof Kołowrocki
Department of Mathematics
Gdynia Maritime University
Gdynia, Poland

ELSEVIER

AMSTERDAM • BOSTON • HEIDELBERG • LONDON • NEW YORK • OXFORD
PARIS • SAN DIEGO • SAN FRANCISCO • SINGAPORE • SYDNEY • TOKYO

Elsevier
32 Jamestown Road, London NW1 7BY
225 Wyman Street, Waltham, MA 02451, USA

Second edition 2014

British Library Cataloguing-in-Publication Data
A catalogue record for this book is available from the British Library

Library of Congress Cataloging-in-Publication Data
A catalog record for this book is available from the Library of Congress

ISBN: 978-0-08-099949-4

For information on all Elsevier publications
visit our website at store.elsevier.com

This book has been manufactured using Print On Demand technology. Each copy is produced to order and is limited to black ink. The online version of this book will show color figures where appropriate.

Contents

List of Figures

List of Tables

Notations

E_i	components of series, parallel and 'm out of n' systems
E_{ij}	components of series–parallel and parallel–series systems
T_i	component lifetimes of two-state series, parallel and 'm out of n' systems
T_{ij}	component lifetimes of two-state series–parallel and parallel–series systems
T	a two-state system lifetime
$R(t)$	a component reliability function of a two-state homogeneous system
$F(t)$	a component lifetime distribution function of a two-state homogeneous system
$R^{(i)}(t)$	component reliability functions of two-state non-homogeneous series, parallel and 'm out of n' systems
$F^{(i)}(t)$	component lifetime distribution functions of two-state non-homogeneous series, parallel and 'm out of n' systems
$R^{(i,j)}(t)$	component reliability functions of two-state non-homogeneous series–parallel and parallel–series systems
$F^{(i,j)}(t)$	component lifetime distribution functions of two-state non-homogeneous series–parallel and parallel–series systems
$\overline{\boldsymbol{R}}_n(t)$	a reliability function of a two-state homogeneous series system
$\overline{\boldsymbol{R}}'_n(t)$	a reliability function of a two-state non-homogeneous series system
$\boldsymbol{R}_n(t)$	a reliability function of a two-state homogeneous parallel system

$\boldsymbol{R}'_n(t)$	a reliability function of a two-state non-homogeneous parallel system
$\boldsymbol{R}_n^{(m)}(t)$	a reliability function of a two-state homogeneous 'm out of n' system
$\boldsymbol{R}'_n(m)(t)$	a reliability function of a two-state non-homogeneous 'm out of n' system
$\overline{\boldsymbol{R}}_{k_n l_n}(t)$	a reliability function of a two-state homogeneous parallel–series system
$\overline{\boldsymbol{R}}'_{k_n l_n}(t)$	a reliability function of a two-state non-homogeneous parallel–series system
$\boldsymbol{R}_{k_n l_n}(t)$	a reliability function of a two-state homogeneous series–parallel system
$\boldsymbol{R}'_{k_n l_n}(t)$	a reliability function of a two-state non-homogeneous series–parallel system
$\overline{\mathfrak{R}}(t)$	a limit reliability function of two-state homogeneous series and parallel–series systems
$\overline{\mathfrak{R}}'(t)$	a limit reliability function of two-state non-homogeneous series and parallel–series systems
$\mathfrak{R}(t)$	a limit reliability function of two-state homogeneous parallel and series–parallel systems
$\mathfrak{R}'(t)$	a limit reliability function of two-state non-homogeneous parallel and series–parallel systems
$\mathfrak{R}^{(0)}(t)$	a limit reliability function of a two-state homogeneous 'm out of n' system
$\mathfrak{R}^{(\mu)}(t)$	a limit reliability function of a two-state homogeneous 'm out of n' system
$\overline{\mathfrak{R}}^{(1)}(t,\cdot)$	a limit reliability function of a two-state homogeneous 'm out of n' system
$E(T)$	a mean lifetime of a two-state system
$\sigma(T)$	a lifetime standard deviation of a two-state system
z	a number of reliability states of a multi-state component and a multi-state system
$T_i(u)$	multi-state component lifetimes of series, parallel and 'm out of n' systems in a state subset

$T_{ij}(u)$	multi-state component lifetimes of series–parallel and parallel–series systems in a state subset
$T(u)$	a multi-state system lifetime in a state subset
$R(t,\cdot)$	a multi-state component reliability function of a homogeneous system
$F(t,\cdot)$	a multi-state component lifetime distribution function of a homogeneous system in a state subset
$R^{(i)}(t,\cdot)$	multi-state component reliability functions of homogeneous series, parallel and '*m* out of *n*' systems
$F^{(i)}(t,\cdot)$	multi-state component lifetime distribution functions of homogeneous series, parallel and '*m* out of *n*' systems in a state subset
$R^{(i,j)}(t,\cdot)$	multi-state component reliability functions of homogeneous series–parallel and parallel–series systems
$F^{(i,j)}(t,\cdot)$	multi-state component lifetime distribution functions of homogeneous series–parallel and parallel–series systems in a state subset
$\overline{\boldsymbol{R}}_n(t,\cdot)$	a reliability function of a multi-state homogeneous series system
$\overline{\boldsymbol{R}}'_n(t,\cdot)$	a reliability function of a multi-state non-homogeneous series system
$\boldsymbol{R}_n(t,\cdot)$	a reliability function of a multi-state homogeneous parallel system
$\boldsymbol{R}'_n(t,\cdot)$	a reliability function of a multi-state non-homogeneous parallel system
$\boldsymbol{R}_n^{(m)}(t,\cdot)$	a reliability function of a multi-state homogeneous '*m* out of *n*' system
$\overline{\boldsymbol{R}}_n^{(\overline{m})}(t,\cdot)$	a reliability function of a multi-state homogeneous '*m* out of *n*' system
$\boldsymbol{R}_n^{\prime(m)}(t,\cdot)$	a reliability function of a multi-state non-homogeneous '*m* out of *n*' system
$\overline{\boldsymbol{R}}_n^{\prime(\overline{m})}(t,\cdot)$	a reliability function of a multi-state non-homogeneous '*m* out of *n*' system

$\overline{\boldsymbol{R}}_{k_n,l_n}(t,\cdot)$	a reliability function of a multi-state homogeneous parallel–series system
$\overline{\boldsymbol{R}}'_{k_n,l_n}(t,\cdot)$	a reliability function of a multi-state non-homogeneous parallel–series system
$\boldsymbol{R}_{k_n,l_n}(t,\cdot)$	a reliability function of a multi-state homogeneous series–parallel system
$\boldsymbol{R}'_{k_n,l_n}(t,\cdot)$	a reliability function of a multi-state non-homogeneous series–parallel system
$\overline{\mathfrak{R}}(t,\cdot)$	a limit reliability function of multi-state homogeneous series and parallel–series systems
$\overline{\mathfrak{R}}'(t,\cdot)$	a limit reliability function of multi-state non-homogeneous series and parallel–series systems
$\mathfrak{R}(t,\cdot)$	a limit reliability function of multi-state homogeneous parallel and series–parallel systems
$\mathfrak{R}'(t,\cdot)$	a limit reliability function of multi-state non-homogeneous parallel and series–parallel systems
$\mathfrak{R}^{(0)}(t,\cdot)$	a limit reliability function of a multi-state homogeneous '*m* out of *n*' system
$\mathfrak{R}^{(\mu)}(t,\cdot)$	a limit reliability function of a multi-state homogeneous '*m* out of *n*' system
$\overline{\mathfrak{R}}^{(1)}(t,\cdot)$	a limit reliability function of a multi-state homogeneous '*m* out of *n*' system
r	a critical reliability state of a system
$\mathbf{r}(t)$	a risk function of a multi-state system
$M_i(u)$	a multi-state component mean lifetime in a state subset
$\sigma_i(u)$	a multi-state component lifetime standard deviation in a state subset
$\overline{M}_i(u)$	a multi-state component mean lifetime in a state
$M(u)$	a multi-state system mean lifetime in a state subset
$\sigma(u)$	a multi-state system lifetime standard deviation in a state subset
$\overline{M}(u)$	a multi-state system mean lifetime in a state

δ — a permitted level of a multi-state system risk function

τ — a moment of exceeding a permitted multi-state system risk level

$D_{\overline{\Re}_l}$ — domains of attraction of limit reliability functions $\overline{\Re}_l(t)$ of two-state homogeneous series system

$\boldsymbol{R}^{(m)}_{k_n,l_1,l_2,\ldots,l_{k_n}}(t)$ — a reliability function of a homogeneous two-state series–'m out of k_n' system

$\overline{\boldsymbol{R}}^{(\overline{m})}_{k_n,l_1,l_2,\ldots,l_{k_n}}(t)$ — a reliability function of a homogeneous two-state series–'m out of k_n' system

$\boldsymbol{R}^{(m)}_{k_n,l_n}(t)$ — a reliability function of a homogeneous and regular two-state series–'m out of k_n' system

$\overline{\boldsymbol{R}}^{(\overline{m})}_{k_n,l_n}(t)$ — a reliability function of a homogeneous and regular two-state series–'m out of k_n' system

$\overline{\boldsymbol{R}^{(m_1,m_2,\ldots,m_{k_n})}_{k_n,l_1,l_2,\ldots,l_{k_n}}}(t)$ — a reliability function of a two-state 'm_i out of l_i'–series system

$\overline{\overline{\boldsymbol{R}}^{(\overline{m}_1,\overline{m}_2,\ldots,\overline{m}_{k_n})}_{k_n,l_1,l_2,\ldots,l_{k_n}}}(t)$ — a reliability function of a two-state 'm_i out of l_i'–series system

$\overline{\boldsymbol{R}^{(m)}_{k_n,l_n}}(t)$ — a reliability function of a homogeneous and regular two-state 'm out of k_n'–series system

$\overline{\overline{\boldsymbol{R}}^{(\overline{m})}_{k_n,l_n}}(t)$ — a reliability function of a homogeneous and regular two-state 'm out of k_n'–series system

$\Re^{(m)}_i(t)$ — a limit reliability function of a homogeneous and regular two-state series–'m out of k_n' system

$\Re^{(\overline{m})}_i(t)$ — a limit reliability function of a homogeneous and regular two-state series–'m out of k_n' system

$\overline{\Re^{(m)}_i}(t)$ — a limit reliability function of a homogeneous and regular two-state 'm out of k_n'–series system

$\overline{\overline{\Re}^{(\overline{m})}_i}(t)$ — a limit reliability function of a homogeneous and regular two-state 'm out of k_n'–series system

$\boldsymbol{R}_{r,k_n,l_n}(t)$ — a reliability function of a two-state series–parallel system of order r

$\mathfrak{R}_i(t)$	a limit reliability function of a two-state series–parallel system of order r
$\overline{\boldsymbol{R}}_{r,k_n,l_n}(t)$	a reliability function of a two-state parallel–series system of order r
$\overline{\mathfrak{R}}_l(t)$	a limit reliability function of a two-state parallel–series system of order r
ρ	a factor reducing a component failure rate
$\boldsymbol{R}_n^{(1)}(t)$	a reliability function of a two-state series system with components improved by reducing their failure rates by a factor ρ
$\boldsymbol{R}_n^{(2)}(t)$	a reliability function of a two-state series system with a single hot reservation of its components
$\boldsymbol{R}_n^{(3)}(t)$	a reliability function of a two-state series system with a single cold reservation of its components
$\boldsymbol{R}_n^{(4)}(t)$	a reliability function of a two-state series system with a single mixed reservation of its components
$\boldsymbol{R}_n^{(5)}(t)$	a reliability function of a two-state series system with its single hot reservation
$\boldsymbol{R}_n^{(6)}(t)$	a reliability function of a two-state series system with its single cold reservation
$\mathfrak{R}^{(1)}(t)$	a limit reliability function of a two-state series system with components improved by reducing their failure rates by a factor ρ
$\mathfrak{R}^{(2)}(t)$	a limit reliability function of a two-state series system with a single hot reservation of its components
$\mathfrak{R}^{(3)}(t)$	a limit reliability function of a two-state series system with a single cold reservation of its components
$\mathfrak{R}^{(4)}(t)$	a limit reliability function of a two-state series system with a single mixed reservation of its components
$\mathfrak{R}^{(5)}(t)$	a limit reliability function of a two-state series system with its single hot reservation

$\Re^{(6)}(t)$	a limit reliability function of a two-state series system with its single cold reservation
$\boldsymbol{T}^{(1)}$	a lifetime mean value of a two-state series system with components improved by reducing their failure rates by a factor ρ
$\boldsymbol{T}^{(2)}$	a lifetime mean value of a two-state series system with a single hot reservation of its components
$\boldsymbol{T}^{(3)}$	a lifetime mean value of a two-state series system with a single cold reservation of its components
$\boldsymbol{T}^{(4)}$	a lifetime mean value of a two-state series system with a single mixed reservation of its components
$\boldsymbol{T}^{(5)}$	a lifetime mean value of a two-state series system with its single hot reservation
$\boldsymbol{T}^{(6)}$	a lifetime mean value of a two-state series system with its single cold reservation
$Z(t)$	a complex system operation process
z_b	a complex system operational state
$[p_b(0)]_{1\times\nu}$	a vector of probabilities of an operation process $Z(t)$ at initial states z_b
$[p_{bl}]_{\nu\times\nu}$	a matrix of probabilities of transitions of an operation process $Z(t)$ between operation states z_b and z_l
$p_b(t)$	a transient probability of process $Z(t)$ at an operation state z_b at a moment t
p_b	a limit value of a transient probability $p_b(t)$
θ_{bl}	a conditional sojourn time of a process $Z(t)$ at operational states
$[H_{bl}(t)]_{\nu\times\nu}$	a matrix of conditional distribution functions of sojourn times θ_{bl}
m_{bl}	a mean values of a sojourn time θ_{bl}
θ_b	an unconditional sojourn time of process $Z(t)$ at an operation state z_b
$H_b(t)$	an unconditional distribution function of a sojourn time θ_b at an operation state z_b

m_b	a mean value of an unconditional sojourn time θ_b at an operation state z_b
$\hat{\theta}_b$	an operation process total sojourn time $\hat{\theta}_b$ at the particular operation state z_b during the fixed system opetation time
$\widehat{m}_b$	an expected value of an operation process total sojourn time $\hat{\theta}_b$ at the particular operation state z_b during the fixed system opetation time
$T_{ij}^{(b)}(u)$	a conditional lifetime in a subset of reliability state of a component E_{ij} of a multi-state complex system at an operation states z_b
$[R_{ij}(t,\cdot)]^{(b)}$	a conditional reliability function of a component E_{ij} of a multi-state complex system at an operation states z_b
$[R_{ij}(t,u)]^{(b)}$	a coordinate of a conditional reliability function of a component E_{ij} of a multi-state complex system at an operation states z_b
$T^{(b)}(u)$	a conditional lifetime in a subset of reliability states of a multi-state complex system at operational states z_b
$[\boldsymbol{R}(t,\cdot)]^{(b)}$	a conditional reliability function of a multi-state complex system at operational states z_b
$[\boldsymbol{R}(t,u)]^{(b)}$	a coordinate of a conditional reliability function of a multi-state complex system at operational states z_b
$M_b(u)$	a multi-state complex system conditional mean lifetime in a subset of reliability states
$T(u)$	an unconditional lifetime in a subset of reliability states of a non-homogeneous multi-state complex system
$\boldsymbol{R}(t,\cdot)$	an unconditional reliability function of a multi-state complex system
$\boldsymbol{R}(t,u)$	a coordinate of an unconditional reliability function of a multi-state complex system
$\boldsymbol{r}(t)$	a risk function of a multi-state complex system
$M(u)$	a multi-state complex system mean lifetime in a subset of reliability states

$\sigma(u)$	a multi-state complex system lifetime standard deviation in a subset of reliability states
$\overline{M}(u)$	a multi-state complex system mean lifetime in a particular reliability state
δ	a permitted level of a multi-state complex system risk function
τ	a moment of exceeding a permitted multi-state complex system risk level δ
$\dot{p}_b$	an optimal limit value of a transient probability $p_b(t)$
$\dot{m}_{bl}$	an optimal mean values of a sojourn time θ_{bl}
$\dot{m}_b$	an optimal mean value of an unconditional sojourn time θ_b at an operation state z_b
$\dot{\hat{m}}_b$	an optimal expected value of an operation process total sojourn time $\hat{\theta}_b$ at the particular operation state z_b during the fixed system opetation time
$\dot{\boldsymbol{R}}(t,\cdot)$	an optimal and unconditional reliability function of a multi-state complex system
$\dot{\boldsymbol{r}}(t)$	an optimal risk function of a multi-state complex system
$\dot{M}(u)$	an optimal multi-state complex system mean lifetime in a subset of reliability states
$\dot{\sigma}(u)$	an optimal multi-state complex system lifetime standard deviation in a subset of reliability states
$\dot{\overline{M}}(u)$	an optimal multi-state complex system mean lifetime in a particular reliability state
$\dot{\tau}$	an optimal moment of exceeding a permitted multi-state complex system risk level δ
$[\Re(t,\cdot)]^{(b)}$	a conditional limit reliability function of a multi-state complex system at operational states z_b
$[\Re(t,u)]^{(b)}$	a coordinate of a conditional limit reliability function of multi-state complex system at operational states z_b
$\Re(t,\cdot)$	an unconditional limit reliability function of a multi-state complex system
$\Re(t,u)$	a coordinate of an unconditional limit reliability function of multi-state complex system

Preface

The book is concerned with the application of limit reliability functions to the reliability evaluation of large systems. Two-state and multi-state large systems composed of independent components are considered. The main emphasis is on multi-state systems with degrading (ageing) components because of the importance of such an approach in safety analysis, assessment and prediction, and analysing the effectiveness of operation processes of real technical systems.

Many technical systems belong to the class of complex systems as a result of the large number of components they are built of and their complicated operating processes. This complexity very often causes evaluation of system reliability and safety to become difficult. As a rule these are series systems composed of large numbers of components. Sometimes the series systems have either components or subsystems reserved and then they become parallel–series or series–parallel reliability structures. We meet large series systems, for instance, in piping transportation of water, gas, oil and various chemical substances. Large systems of these kinds are also used in electrical energy distribution. A city bus transportation system composed of a number of communication lines, each serviced by one bus, may be a model series system, if we treat it as not failed, when all its lines are able to transport passengers. If the communication lines have at their disposal several buses we may consider it as either a parallel–series system or an '*m* out of *n*' system. The simplest example of a parallel system or an '*m* out of *n*' system may be an electrical cable composed of a number of wires, which are its basic components, whereas the transmitting electrical network may be either a parallel–series system or an '*m* out of *n*'–series system. Large systems of these types are also used in telecommunication, in rope transportation and in transport using belt conveyers and elevators. Rope transportation systems like port elevators and ship-rope elevators used in shipyards during ship docking and undocking are model examples of series–parallel and parallel–series systems.

Taking into account the importance of the safety and operating process effectiveness of such systems, it seems reasonable to expand the two-state approach to multi-state approach in their reliability analysis. The assumption that the systems are composed of multi-state components with reliability states degrading in time without repair gives the possibility for more precise analysis of their reliability, safety and operational processes' effectiveness. This assumption allows us to distinguish a system reliability critical state to exceed which is either dangerous for the environment or does not assure the necessary level of its operational process effectiveness. Then, an important system reliability characteristic is the time to the moment of exceeding the system reliability critical state and its distribution, which is called the system risk function. This distribution is strictly related to the system multi-state reliability function that is a basic characteristic of the multi-state system.

In the case of large systems, the determination of the exact reliability functions of the systems and the system risk functions leads us to very complicated formulae that are often useless for reliability practitioners. One of the important techniques in this situation is the asymptotic approach to system reliability evaluation. In this approach, instead of the preliminary complex formula for the system reliability function, after assuming that the number of system components tends to infinity and finding the limit reliability of the system, we obtain its simplified form.

The mathematical methods used in the asymptotic approach to the system reliability analysis of large systems are based on limit theorems on order statistics distributions considered in very wide literature [3,9–11,15,22,23,28–31,37–41,44,45, 125,143,149,151,169,172]. These theorems have generated the investigation concerned with limit reliability functions of the systems composed of two-state components [5,7,25–29,48–50,58–74,93,94,119–125,135,136,148,152,167,175]. The main and fundamental results on this subject that determine the three-element classes of limit reliability functions for homogeneous series systems and for homogeneous parallel systems have been established by Gniedenko [41]. These results are also presented, sometimes with different proofs, for instance in subsequent works [7,15,23,30,65,93]. The generalizations of these results for homogeneous '*m* out of *n*' systems have been formulated and proved by Smirnow [151], where the seven-element class of possible limit reliability functions for these systems has been fixed. Some partial results obtained by Smirnow may be found in Ref. [93] and additionally with the solution of the speed of convergence problem in Ref. [31]. As has been done for homogeneous series and parallel systems, classes of limit reliability functions have been fixed by Chernoff and Teicher [23] for homogeneous series–parallel and parallel–series systems. Their results were concerned with the so-called 'quadratic' systems only. They have fixed limit reliability functions for the homogeneous series–parallel systems with the number of series subsystems equal to the number of components in these subsystems, and for the homogeneous parallel–series systems with the number of parallel subsystems equal to the number of components in these subsystems. These results may also be found for instance in later works [7,65].

All the results so far described have been obtained under the linear normalization of the system lifetimes. Of course, there is the possibility of looking for limit reliability functions of large systems under other than linear standardization of their lifetimes. In this context, the results obtained by Pantcheva [143] and Cichocki [27] are exemplary. Pantcheva [143] has fixed the seven-element classes of limit reliability functions of homogeneous series and parallel systems under power standardization for their lifetimes. Cichocki [27] has generalized Pantcheva's results to hierarchical series–parallel and parallel–series systems of any order.

The book contains the results described above and their newest generalizations for large two-state systems and their developments for multi-state systems' asymptotic reliability analysis under the linear standardization of the system lifetimes and the system lifetimes in the state subsets, respectively.

Generalizations presented here of the results on limit reliability functions of two-state homogeneous series, and parallel systems for these systems in case they are non-homogeneous, are mostly taken from Refs. [81,93]. A more general problem is concerned with fixing the classes of possible limit reliability functions for the so-called 'rectangular' series–parallel and parallel–series systems. This problem for homogeneous series–parallel and parallel–series systems of any shape, with different numbers of subsystems and numbers of components in these subsystems, has been progressively solved in Refs. [62–65,68,70]. The main and new result of these works was the determination of seven new limit reliability functions for homogeneous series–parallel systems as well as for parallel–series systems. This way, new 10-element classes of all possible limit reliability functions for these systems have been fixed. Moreover, in these works it has been pointed out that the type of system limit reliability function strongly depends on the system shape. These results allow us to evaluate reliability characteristics of homogeneous series–parallel and parallel–series systems with regular reliability structures, i.e. systems composed of subsystems having the same numbers of components. The extensions of these results for non-homogeneous series–parallel and parallel–series systems have been formulated and proved successively in Refs. [65,69–72,81]. These generalizations additionally allow us to evaluate reliability characteristics of the series–parallel and parallel–series systems with non-regular structures, i.e. systems with subsystems having different numbers of components. In some of the cited works, as well as the theoretical considerations and solutions, numerous practical applications of the asymptotic approach to real technical system reliability evaluation may also be found [29,48–50,60,66,71,73, 93,95,152,167,173].

More general and practically important complex systems composed of multi-state and degrading in time components are considered among others in Refs. [1,2,4–6,8,12,16–21,32,34–35,43,53–57,73–84,93,116,117,122,126–132,134, 137–139,142,144,147,150,174–177]. An especially important role they play in the evaluation of technical systems reliability and safety and their operating process effectiveness is defined in the book for large multi-state systems with degrading components. The most important results regarding generalizations of the results on limit reliability functions of two-state systems are dependent on transferring them to series, parallel, '*m* out of *n*', series–parallel and parallel–series multi-state systems with degrading components are given in Refs. [74–84]. Some of these publications also contain practical applications of the asymptotic approach to the reliability evaluation of various technical systems [74–79,81–84,93].

The results concerned with the asymptotic approach to system reliability analysis have become the basis for the investigation concerned with domains of attraction for the limit reliability functions of the considered systems [25,93,95,119,120]. In a natural way they have led to investigation of the speed of convergence of the system reliability function sequences to their limit reliability functions [93]. These results have also initiated the investigation of limit reliability functions of '*m* out of

n'–series, series–'m out of n' systems [25,135,136] and systems with hierarchical reliability structures [25–27], as well as investigations on the problems of system reliability improvement and optimization [121,122].

The aim of the book is to deliver a complete elaboration of the state of art on the method of asymptotic approach to reliability evaluation for as wide as possible a range of large systems included mainly in Ref. [85]. Pointing out the possibility of this method's extensive practical application in the operating processes of these systems is also an important reason for this book. The book contains complete current theoretical results of the asymptotic approach to reliability evaluation of large two-state and multi-state series, parallel, 'm out of n', series–parallel and parallel–series systems together with their practical applications to the reliability evaluation of a wide range of technical systems. Some partial results on the asymptotic approach to reliability evaluation of 'm out of n'–series, series–'m out of n' and hierarchical systems, and their application to large systems reliability improvement are also presented in the book.

Additionally, the book delivers a very wide discussion on the reliability modelling, prediction and optimization of complex multi-state systems operating at variable conditions with changing at times of their reliability structures and their components reliability parameters. The discussion is based on the results included in Refs. [13,14,46–47,86–93,96–115,123,124,157–159,161,163–166,171]. Moreover, the asymptotic approach to the large complex systems at varying operation conditions is presented and illustrated with some applications [154–156,160,162].

The following construction of the book has been assumed. In chapters concerned with two-state systems the results and theorems are presented without the proofs but with exact reference to the literature where their proofs may be found. Moreover, the procedures of the results' practical applications are described and applied to the model two-state systems reliability evaluation. In chapters concerned with multi-state systems, the recent theorems about their multi-state limit reliability functions are formulated and briefly justified. Next, the procedures of the result applications are presented and applied to real technical systems reliability and risk evaluation. Moreover, the possibility of computer-aided reliability evaluation of these systems is suggested and its use is presented.

The book contains complete actual solutions of the formulated problems for the considered large systems reliability evaluation in the case of any reliability functions of the system components.

The book consists of preface, ten chapters, summary and bibliography.

In Chapter 1, some basic notions necessary for further considerations are introduced. The asymptotic approach to the system reliability investigation and the system limit reliability function is defined.

In Chapter 2, two-state homogeneous and non-homogeneous series, parallel, 'm out of n', series–parallel and parallel–series systems are defined. Their exact reliability functions are also determined.

Basic notions of the system multi-state reliability analysis are introduced in Chapter 3. Further the multi-state homogeneous and non-homogeneous series, parallel, 'm out of n', series–parallel and parallel–series systems with degrading components are defined and their exact reliability functions are determined. Moreover, the notions of the multi-state limit reliability function of the system, its risk function and other multi-state system reliability characteristics are introduced.

Chapter 4 is concerned with limit reliability functions of two-state systems. Three-element classes of limit reliability functions for homogeneous and non-homogeneous series systems are fixed. Some auxiliary theorems that allow us to justify facts on the methods of those systems' reliability evaluation are formulated and proved. The chapter also contains the application of one of the proven facts to the reliability evaluation of a non-homogeneous gas pipeline that is composed of components with Weibull reliability functions. The accuracy of this evaluation is also illustrated. Three-element classes of possible limit reliability functions for homogeneous and non-homogeneous parallel systems are fixed as well. Some auxiliary theorems that allow us to justify facts on the methods of these systems' reliability evaluation are formulated and proved. The chapter also contains the application of one proved fact to the reliability evaluation of a homogeneous energetic cable used in the overhead electrical energy distribution that is composed of components with Weibull reliability functions. The accuracy of this evaluation is illustrated in a table and a figure. The class of limit reliability functions for a homogeneous 'm out of n' system is fixed and the '16 out of 35' lighting reliability is evaluated in this chapter. This chapter also contains the results of investigations on limit reliability functions of two-state homogeneous and non-homogeneous series–parallel systems. Apart from formulated and proved auxiliary theorems that allow us to justify facts on the methods of these systems' reliability evaluation, their 10-element classes of possible limit reliability functions are fixed. In this chapter, in the part concerned with applications there are two formulated and proved facts that determine limit reliability functions of series–parallel systems in the cases where they are composed of components having the same and different Weibull reliability functions. On the basis of those facts the reliability characteristics of a homogeneous gas pipeline composed of two lines of pipe segments and a non-homogeneous water supply system composed of three lines of pipe segments are evaluated. The results of investigations on limit reliability functions of two-state homogeneous and non-homogeneous parallel–series systems are given in this chapter as well. Theorems, which determine 10-element classes of possible limit reliability functions for these systems in the cases where they are composed of identical and different components, are formulated and justified. Moreover, some auxiliary theorems that are necessary in practical reliability evaluation of real technical systems are formulated and proved. In the part concerned with applications one fact is formulated and proved and then applied to evaluation of the reliability of a model homogeneous parallel–series system. Moreover, the procedure of the reliability evaluation of more complicated large systems that structures are compositions of large two-state systems considered in this chapter's previous sections is presented. This procedure application is illustrated in the exact and approximate reliability evaluation of the two-state non-homogeneous system composed of a large series–parallel system and a large series–'m out of k' system linked in series and the two-state homogeneous system composed of three large '*m* out of *l*' systems linked in series.

Generalizations of the results of Chapter 4 on limit reliability functions of two-state systems, consisting of their transferring to multi-state series, parallel, '*m* out of *n*', series–parallel and parallel–series systems, are done in Chapter 5. The classes of all possible limit reliability functions for these systems in cases when they are composed of identical and different (in the reliability sense) components are fixed. The newest theorems that allow us to evaluate the reliability of large technical systems of those kinds are formulated and proved in Chapter 5 as well. Apart from the main theorems fixing the classes of multi-state limit reliability functions of the considered system, some auxiliary theorems and corollaries allowing their direct applications to reliability evaluation of real technical objects are also formulated and proved. Moreover, in this chapter there are wide applications depending on the results applying to the evaluation of reliability characteristics and risk functions of different multi-state transportation systems. The results concerned with multi-state series systems are applied to the reliability evaluation and risk function determination of homogeneous and non-homogeneous pipeline transportation systems, a homogeneous model telecommunication network and a homogeneous bus transportation system. The results concerned with multi-state parallel systems are applied to reliability evaluation and risk function determination of an energetic cable used in an overhead electrical energy distribution network and to reliability and durability evaluation of a three-level steel rope used in rope transport. Results on limit reliability functions of a homogeneous multi-state '*m* out of *n*' system are applied to durability evaluation of a steel rope. A model homogeneous series–parallel system and homogeneous and non-homogeneous series–parallel pipeline systems composed of several lines of pipe segments are estimated as well. Moreover, the reliability evaluation of the model homogeneous parallel–series electrical energy distribution system is performed. This chapter is completed by presentation of the procedure of the reliability evaluation of more complicated large systems whose structures are compositions of large multi-state systems considered in the previous sections. This procedure application is illustrated in the exact and approximate reliability evaluation of the non-homogeneous multi-state system composed of a large series–parallel system and a large series–'*m* out of *k*' system linked in series and the homogeneous multi-state system composed of three large '*m* out of *l*' systems linked in series.

Chapter 6 is devoted to the multi-state asymptotic reliability analysis of port and shipyard transportation systems. Theoretical results of this chapter and Chapter 5 are applied to the reliability evaluation and the risk functions determination of some selected port transportation systems. The results of the asymptotic approach to reliability evaluation of non-homogeneous multi-state series–parallel systems are applied to the transportation system used in the Baltic Grain Terminal of the Port of Gdynia for transporting grain from its elevator to the rail carriages. The results of the asymptotic approach to the reliability evaluation of the non-homogeneous multi-state series–parallel systems are applied to the piping transportation system used in the Oil Terminal in Dębogorze. This transportation system is set up to take the oil from the tankers that deliver it to the unloading pier located at the breakwater of the Port of Gdynia. The results of the asymptotic approach to reliability evaluation of non-homogeneous multi-state series–parallel and series systems are applied to the transportation system used in the Baltic Bulk Terminal of the Port of Gdynia for loading bulk cargo on the ships. The results of this chapter and Chapter 5 are also applied to reliability evaluation and risk function determination of the shipyard transportation system. Namely, the results of the asymptotic approach to reliability evaluation of homogeneous multi-state parallel–series systems are applied to the ship-rope transportation system used in the Naval Shipyard of Gdynia for docking ships coming for repair. The reliability analysis performed on the considered systems in this chapter is based on the data concerned with the operation processes and reliability of their components coming from experts, from component technical norms and from their producers' certificates.

In Chapter 7, the classes of possible limit reliability functions are fixed for the considered systems in the case where their components have exponential reliability functions. Theoretical results are represented in the form of a very useful guide containing algorithms placed in tables and giving sequential steps for proceeding in the reliability evaluation in each of the possible cases of the considered system shapes. The application of these algorithms for reliability evaluation of the multi-state non-homogeneous series transportation system, the multi-state model homogeneous series–parallel, the multi-state non-homogeneous series–parallel pipeline transportation system and the multi-state non-homogeneous parallel–series bus transportation system is illustrated. The evaluation of reliability functions, risk functions, mean values of sojourn times in subsets of states and mean values of sojourn times in particular states for these systems is carried out. The calculations are performed using a computer program based on the algorithms, thus allowing automatic evaluation of the reliability of large real technical systems.

In Chapter 8, the open problems related to the topics considered in the book are presented. The domains of attraction for previously fixed limit reliability functions of the series, parallel, '*m* out of *n*', series–parallel and parallel–series systems are introduced. More exactly, there are formulated theorems giving conditions which reliability functions of the components of the system have to satisfy in order that the system limit reliability function is one of the functions from the system class of all limit reliability functions. Some examples of the result application for series systems are also illustrated. The very important practical problem of the speed of convergence of system reliability function sequences to their limit reliability functions is investigated as well. An exemplary theorem is presented, which allows the differences between the system limit reliability functions and the members of their reliability function sequences to be estimated. Next, an example of the speed of convergence evaluations of reliability function sequences for a homogeneous series–parallel system is given. Partial results of the investigation on the asymptotic approach to reliability evaluation of '*m* out of *n*'–series, series–'*m* out of *n*' and hierarchical systems and on system reliability improvement are presented. These result applications are illustrated graphically as well.

Chapter 9 is devoted to modelling, prediction and optimization of the operation and reliability of technical systems related to their operation processes. The emphasis is on multi-state systems composed of ageing components and on changing their reliability structures and their components' reliability parameters during their operation processes that are called the complex systems. The integrated general model of complex systems' reliability, linking their reliability models and their operation processes models and considering variables at different operation states, their reliability structures and their components reliability parameters, is constructed. This system reliability general model, together with the linear programming, is proposed for operation and reliability optimization of complex technical systems. Those theoretical tools are applied to modelling, prediction and optimization of the operation processes and reliability characteristics of the multi-state non-homogeneous system composed of a series–parallel and a series–'*m* out of *k*' subsystems linked in series, the multi-state homogenous '*m* out of *l*'–series system and the port grain transportation system composed of three multi-state non-homogeneous series–parallel subsystems linked in series changing their reliability structures and their components reliability parameters at variable operation conditions.

In Chapter 10, the combination of the results on the reliability of complex multi-state systems related to their operation processes and the results concerning the limit reliability functions of the multi-state systems is proposed, to obtain the results on the asymptotic approach to the evaluation of the large complex multi-state systems reliability at the variable operation conditions. The asymptotic approach to the large complex system reliability evaluation and the large complex system limit reliability function is defined. Limit reliability functions of selected large complex systems composed of components having exponential reliability functions are fixed. The method of using these results is illustrated by their application to the approximate evaluation of reliability characteristics of the large multi-state non-homogeneous system composed of a series–parallel and a series–'*m* out of *k*' subsystems linked in series, the large multi-state homogenous '*m* out of *l*'–series system and the port grain transportation system composed of three large multi-state non-homogeneous series–parallel subsystems linked in series changing their reliability structures and their components reliability parameters at variable operation conditions. The accuracy of the approximation of the exact reliability characteristics fixed for those systems in Chapter 9 by their approximate evaluation obtained in this chapter is discussed as well.

The book is completed by the summary, which contains the evaluation of the presented results, the formulation of open problems concerned with large and complex systems' reliability and the perspective of further investigations on the considered problems.

1 Basic Notions

Considering the reliability of two-state systems we assume that the distributions of the component and the system lifetimes T do not necessarily have to be concentrated in the interval $<0, \infty)$. It means that a reliability function

$$R(t) = P(T > t), \quad t \in (-\infty, \infty)$$

does not have to satisfy the usually demanded condition

$$R(t) = 1 \quad \text{for } t \in (-\infty, 0)$$

This is a generalization of the normally used concept of a reliability function. This generalization is convenient in the theoretical considerations. At the same time, from the achieved results on the generalized reliability functions, for particular cases, the same properties of the normally used reliability functions appear.

From that assumption it follows that between a reliability function $R(t)$ and a distribution function

$$F(t) = P(T \leq t), \quad t \in (-\infty, \infty)$$

there exists a relationship given by

$$R(t) = 1 - F(t) \quad \text{for } t \in (-\infty, \infty)$$

Thus, the following corollary is obvious.

Corollary 1.1
A reliability function $R(t)$ is non-increasing, right-continuous and moreover

$$R(-\infty) = 1, \quad R(+\infty) = 0$$

Definition 1.1
A reliability function $R(t)$ is called degenerate if there exists $t_0 \in (-\infty, \infty)$, such that

$$R(t) = \begin{cases} 1, & t < t_0 \\ 0, & t \geq t_0 \end{cases}$$

Reliability of Large Systems. DOI: http://dx.doi.org/10.1016/B978-0-08-099949-4.00001-5

Corollary 1.2
A function

$$R(t) = 1 - \exp[-V(t)], \quad t \in (-\infty, \infty)$$

is a reliability function if and only if a function $V(t)$ is non-negative, non-increasing, right-continuous,

$$V(-\infty) = \infty, \quad V(+\infty) = 0$$

and moreover $V(t)$ can be identically equal to ∞ in an interval.

Corollary 1.3
A function

$$\overline{R}(t) = \exp[-\overline{V}(t)], \quad t \in (-\infty, \infty)$$

is a reliability function if and only if a function $\overline{V}(t)$ is non-negative, non-decreasing, right-continuous,

$$\overline{V}(-\infty) = 0, \quad \overline{V}(+\infty) = \infty$$

and moreover $\overline{V}(t)$ can be identically equal to ∞ in an interval.

Corollary 1.4
A function

$$R^{(0)}(t) = 1 - \sum_{i=0}^{m-1} \frac{[V(t)]^i}{i!} \exp[-V(t)], \quad t \in (-\infty, \infty), \quad m \in N$$

is a reliability function if and only if a function $V(t)$ is non-negative, non-increasing, right-continuous,

$$V(-\infty) = \infty, \quad V(+\infty) = 0$$

and moreover $V(t)$ can be identically equal to ∞ in an interval.

Corollary 1.5
A function

$$R^{(\mu)}(t) = 1 - \frac{1}{\sqrt{2\pi}} \int_{-\infty}^{-v(t)} e^{-x^2/2}\,dx, \quad t \in (-\infty, \infty), \quad 0 < \mu < 1$$

is a reliability function if and only if a function $\nu(t)$ is non-increasing, right-continuous,

$$\nu(-\infty) = +\infty, \quad \nu(+\infty) = -\infty$$

and moreover $\nu(t)$ can be identically equal to $-\infty$ or equal to $+\infty$ in an interval.

Corollary 1.6
A function

$$\overline{R}^{(1)}(t) = \sum_{i=0}^{\overline{m}} \frac{[\overline{V}(t)]^i}{i!} \exp\left[-\overline{V}(t)\right], \quad t \in (-\infty, \infty), \ \overline{m} \in N$$

is a reliability function if and only if a function $\overline{V}(t)$ is non-negative, non-decreasing, right-continuous,

$$\overline{V}(-\infty) = 0, \ \overline{V}(+\infty) = \infty$$

and moreover $\overline{V}(t)$ can be identically equal to ∞ in an interval.

Agreement 1.1
In further considerations if we use symbols $V(t)$ and $\overline{V}(t)$, we always mean functions of properties given in Corollaries 1.2–1.6. If $V(t)$ and $\overline{V}(t)$ are identically equal to ∞ we assume that

$$\exp[-V(t)] = 0, \ \exp[-\overline{V}(t)] = 0, \ [V(t)]^i \exp[-V(t)] = 0 \text{ and } [\overline{V}(t)]^i \exp[\overline{V}(t)] = 0$$

If we say that $V(t)$, $\nu(t)$ and $\overline{V}(t)$ are non-negative, non-increasing or non-decreasing and right-continuous we mean the intervals where

$$V(t) \neq \infty, \ \nu(t) \neq \infty \text{ and } \overline{V}(t) \neq \infty$$

Moreover, we denote the set of continuity points of a reliability function $R(t)$ by C_R and the set of continuity points of a function $V(t)$ and points such that $V(t) = \infty$ by C_V. We denote the set of continuity points of a reliability function $\overline{R}(t)$ by $C_{\overline{R}}$ and the set of continuity points of a function $\overline{V}(t)$ and points such that $\overline{V}(t) = \infty$ by $C_{\overline{V}}$. Similarly, we denote the set of continuity points of reliability functions $R^{(0)}(t)$, $R^{(\mu)}(t)$ and $\overline{R}^{(1)}(t)$ by $C_{R^{(0)}}$, $C_{R^{(\mu)}}$ and $C_{\overline{R}}(1)$ respectively and the set of continuity points of a function $\nu(t)$ and points such that $\nu(t) = -\infty$ or points such that $\nu(t) = +\infty$ by C_ν.

According to Definition 1.1, Corollaries 1.2–1.6 and Agreement 1.1, we assume the next definitions.

Definition 1.2
A function $V(t)$ defined for $t \in (-\infty, \infty)$, non-negative, non-increasing, right-continuous and such that

$$V(-\infty) = \infty, \quad V(+\infty) = 0$$

is called degenerate if there exists $t_0 \in (-\infty, \infty)$ such that

$$V(t) = \begin{cases} \infty, & t < t_0 \\ 0, & t \geq t_0 \end{cases}$$

Definition 1.3
A function $\nu(t)$ defined for $t \in (-\infty, \infty)$, non-increasing, right-continuous and such that

$$\nu(-\infty) = +\infty, \quad \nu(+\infty) = -\infty$$

is called degenerate if there exists $t_0 \in (-\infty, \infty)$ such that

$$\nu(t) = \begin{cases} +\infty, & t < t_0 \\ -\infty, & t \geq t_0 \end{cases}$$

Definition 1.4
A function $\overline{V}(t)$ defined for $t \in (-\infty, \infty)$, non-negative, non-decreasing, right-continuous and such that

$$\overline{V}(-\infty) = 0, \quad \overline{V}(\infty) = \infty$$

is called degenerate if there exists $t_0 \in (-\infty, \infty)$ such that

$$\overline{V}(t) = \begin{cases} 0, & t < t_0 \\ \infty, & t \geq t_0 \end{cases}$$

Under those definitions the following corollaries are clear.

Corollary 1.7
A reliability function

$$R(t) = 1 - \exp[-V(t)], \quad t \in (-\infty, \infty)$$

is degenerate if and only if a function $V(t)$ is degenerate.

Corollary 1.8
A reliability function

$$\overline{R}(t) = \exp[-\overline{V}(t)], \quad t \in (-\infty, \infty)$$

is degenerate if and only if a function $\overline{V}(t)$ is degenerate.

Corollary 1.9
A reliability function

$$R^{(0)}(t) = 1 - \sum_{i=0}^{m-1} \frac{[V(t)]^i}{i!} \exp[-V(t)], \quad t \in (-\infty, \infty), \quad m \in N$$

is degenerate if and only if a function $V(t)$ is degenerate.

Corollary 1.10
A reliability function

$$R^{(\mu)}(t) = 1 - \frac{1}{\sqrt{2\pi}} \int_{-\infty}^{-v(t)} e^{-x^2/2}\, dx, \quad t \in (-\infty, \infty), \quad 0 < \mu < 1$$

is degenerate if and only if a function $\nu(t)$ is degenerate.

Corollary 1.11
A reliability function

$$\overline{R}^{(1)}(t) = \sum_{i=0}^{\overline{m}} \frac{[\overline{V}(t)]^i}{i!} \exp\left[-\overline{V}(t)\right], \quad t \in (-\infty, \infty), \quad \overline{m} \in N$$

is degenerate if and only if a function $\overline{V}(t)$ is degenerate.

The asymptotic approach to the reliability of two-state systems depends on the investigation of limit distributions of a standardized random variable

$$(T - b_n)/a_n$$

where T is the lifetime of a system and $a_n > 0$, $b_n \in (-\infty, \infty)$ are suitably chosen numbers called normalizing constants.

Since

$$P((T - b_n)/a_n > t) = P(T > a_n t + b_n) = \boldsymbol{R}_n(a_n t + b_n)$$

where $R_n(t)$ is a reliability function of a system composed of n components, then the following definition becomes natural.

Definition 1.5

A reliability function $\mathfrak{R}(t)$ is called a limit reliability function or an asymptotic reliability function of a system having a reliability function $\boldsymbol{R}_n(t)$ if there exist normalizing constants $a_n > 0$, $b_n \in (-\infty, \infty)$ such that

$$\lim_{n\to\infty} \boldsymbol{R}_n(a_n t + b_n) = \mathfrak{R}(t) \quad \text{for } t \in C_{\mathfrak{R}}$$

Thus, if the asymptotic reliability function $\mathfrak{R}(t)$ of a system is known, then for sufficiently large n, the approximate formula

$$\boldsymbol{R}_n(t) \cong \mathfrak{R}((t-b_n)/a_n), \quad t \in (-\infty, \infty) \tag{1.1}$$

may be used instead of the system exact reliability function $\boldsymbol{R}_n$(t).

From the condition

$$\lim_{n\to\infty} \boldsymbol{R}_n(a_n t + b_n) = \mathfrak{R}(t) \quad \text{for } t \in C_{\mathfrak{R}}$$

it follows that setting

$$a_n = a a_n, \quad \beta_n = b a_n + b_n$$

where $a > 0$ and $b \in (-\infty, \infty)$, we get

$$\lim_{n\to\infty} \boldsymbol{R}_n(a_n t + \beta_n) = \lim_{n\to\infty} \boldsymbol{R}_n(a_n(at+b) + b_n) = \mathfrak{R}(at+b) \quad \text{for } t \in C_{\mathfrak{R}}$$

Hence, if $\mathfrak{R}(t)$ is the limit reliability function of a system, then $\mathfrak{R}(at+b)$ with arbitrary $a > 0$ and $b \in (-\infty, \infty)$ is also its limit reliability function. That fact, in a natural way, yields the concept of a type of limit reliability function.

Definition 1.6

The limit reliability functions $\mathfrak{R}_0(t)$ and $\mathfrak{R}(t)$ are said to be of the same type if there exist numbers $a > 0$ and $b \in (-\infty, \infty)$ such that

$$\mathfrak{R}_0(t) = \mathfrak{R}(at+b) \quad \text{for } t \in (-\infty, \infty)$$

Agreement 1.2

In further considerations we assume the following notations:

$x(n) \ll y(n)$ or $x(n) = o(y(n))$, where $x(n)$ and $y(n)$ are positive functions, means that $x(n)$ is of order much less than $y(n)$ in a sense

$$\lim_{n\to\infty} x(n)/y(n) = 0$$

$x(n) \approx y(n)$ or $x(n) = r(y(n))$, where $x(n)$ and $y(n)$ are either positive or negative functions, means that $x(n)$ is of order $y(n)$ in a sense

$$\lim_{n \to \infty} x(n)/y(n) = 1$$

$x(n) \gg y(n)$ or $x(n) = O(y(n))$, where $x(n)$ and $y(n)$ are positive functions, means that $x(n)$ is of order much greater than $y(n)$ in a sense

$$\lim_{n \to \infty} x(n)/y(n) = \infty$$

2 Two-State Systems

We assume that

$$E_i, \quad i = 1, 2, \ldots, n, \quad n \in N$$

are two-state components of the system having reliability functions

$$R_i(t) = P(T_i > t), \quad t \in (-\infty, \infty)$$

where

$$T_i, \quad i = 1, 2, \ldots, n$$

are independent random variables representing the lifetimes of components E_i with distribution functions

$$F_i(t) = P(T_i \leq t), \quad t \in (-\infty, \infty)$$

The simplest two-state reliability structures are series and parallel systems. We define these first.

Definition 2.1
A two-state system is called series if its lifetime T is given by

$$T = \min_{1 \leq i \leq n} \{T_i\}$$

The scheme of a series system is given in Figure 2.1.

The above definition means that the series system is not failed if and only if all its components are not failed, and therefore its reliability function is given by

$$\overline{\boldsymbol{R}}_n(t) = \prod_{i=1}^{n} R_i(t), \quad t \in (-\infty, \infty)$$

Reliability of Large Systems. DOI: http://dx.doi.org/10.1016/B978-0-08-099949-4.00002-7

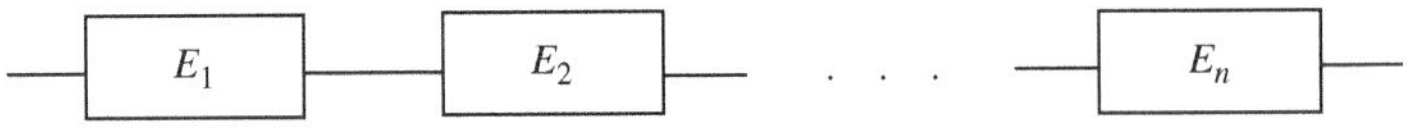

Figure 2.1 The scheme of a series system.

Definition 2.2
A two-state series system is called homogeneous if its component lifetimes T_i have an identical distribution function

$$F(t) = P(T_i \le t), \quad t \in (-\infty, \infty), \; i = 1, 2, \ldots, n$$

that is if its components E_i have the same reliability function

$$R(t) = 1 - F(t), \quad t \in (-\infty, \infty)$$

The above definition results in the following simplified formula:

$$\overline{\boldsymbol{R}}_n(t) = [R(t)]^n, \quad t \in (-\infty, \infty) \tag{2.1}$$

for the reliability function of the homogeneous two-state series system.

Definition 2.3
A two-state system is called parallel if its lifetime T is given by

$$T = \max_{1 \le i \le n}\{T_i\}$$

The above definition means that the parallel system is failed if and only if all its components are failed and therefore its reliability function is given by

$$\boldsymbol{R}_n(t) = 1 - \prod_{i=1}^{n} F_i(t), \quad t \in (-\infty, \infty)$$

The scheme of a parallel system is given in Figure 2.2.

Definition 2.4
A two-state parallel system is called homogeneous if its component lifetimes T_i have an identical distribution function

$$F(t) = P(T_i \le t), \quad t \in (-\infty, \infty), \; i = 1, 2, \ldots, n$$

that is if its components E_i have the same reliability function

$$R(t) = 1 - F(t), \quad t \in (-\infty, \infty)$$

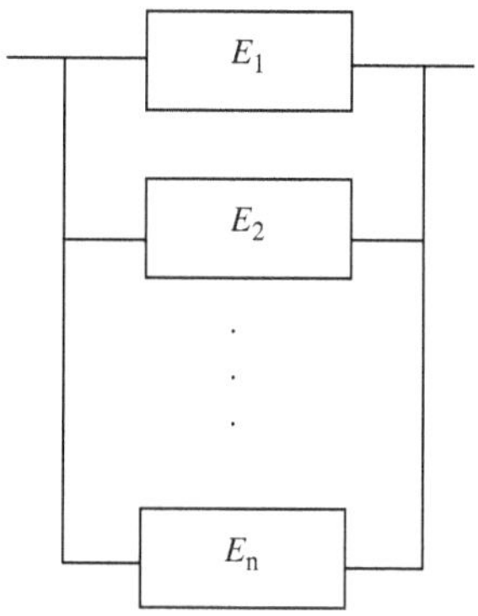

Figure 2.2 The scheme of a parallel system.

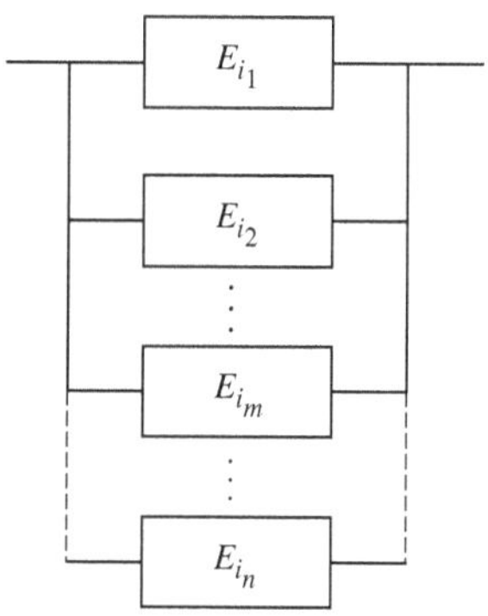

Figure 2.3 The scheme of an '*m* out of *n*' system.

Under this definition, we get the following formula:

$$\boldsymbol{R}_n(t) = 1 - [F(t)]^n, \quad t \in (-\infty, \infty) \tag{2.2}$$

for the reliability function of the homogeneous two-state parallel system.

Definition 2.5

A two-state system is called an '*m* out of *n*' system if its lifetime T is given by

$$T = T_{(n-m+1)}, \quad m = 1, 2, \ldots, n$$

where $T_{(n-m+1)}$ is the mth maximal order statistic in the sequence of component lifetimes $T_1, T_2, \ldots, T_n$.

The scheme of an '*m* out of *n*' system is given in Figure 2.3, where $i_1, i_2, \ldots, i_n \in \{1, 2, \ldots, n\}$ and $i_j \neq i_k$ for $j \neq k$.

The above definition means that the two-state '*m* out of *n*' system is not failed if and only if at least m out of its n components are not failed. The two-state '*m* out of *n*' system becomes a parallel system if $m = 1$, whereas it becomes a series system if $m = n$. The reliability function of the two-state 'm out of n' system is given either by

$$\boldsymbol{R}_n^{(m)}(t) = 1 - \sum_{\substack{r_1, r_2, \ldots, r_n = 0 \\ r_1 + r_2 + \cdots + r_n \leq m-1}}^{1} \prod_{i=1}^{n} [R_i(t)]^{r_i} [F_i(t)]^{1-r_i}, \quad t \in (-\infty, \infty)$$

or by

$$\overline{\boldsymbol{R}}_n^{(\overline{m})}(t) = \sum_{\substack{r_1, r_2, \ldots, r_n = 0 \\ r_1 + r_2 + \cdots + r_n \leq \overline{m}}}^{1} \prod_{i=1}^{n} [F_i(t)]^{r_i} [R_i(t)]^{1-r_i}, \quad t \in (-\infty, \infty), \ \overline{m} = n - m$$

Definition 2.6
A two-state 'm out of n' system is called homogeneous if its component lifetimes T_i have an identical distribution function

$$F(t) = P(T_i \leq t), \quad t \in (-\infty, \infty), \ i = 1, 2, \ldots, n$$

that is if its components E_i have the same reliability function

$$R(t) = 1 - F(t), \quad t \in (-\infty, \infty)$$

The reliability function of the homogeneous two-state 'm out of n' system is given either by

$$\boldsymbol{R}_n^{(m)}(t) = 1 - \sum_{i=0}^{m-1} \binom{n}{i} [R(t)]^i [F(t)]^{n-i}, \quad t \in (-\infty, \infty) \tag{2.3}$$

or by

$$\overline{\boldsymbol{R}}_n^{(\overline{m})}(t) = \sum_{i=0\overline{m}} \binom{n}{i} [F(t)]^i [R(t)]^{n-i}, \quad t \in (-\infty, \infty), \ \overline{m} = n - m \tag{2.4}$$

Other basic, slightly more complex, two-state reliability structures are series–parallel and parallel–series systems. To define them, we assume that

$$E_{ij}, \quad i = 1, 2, \ldots, k_n, \ j = 1, 2, \ldots, l_i, \ k_n, \ l_1, l_2, \ldots, l_{k_n} \in N$$

are two-state components of the system having reliability functions

$$R_{ij}(t) = P(T_{ij} > t), \quad t \in (-\infty, \infty)$$

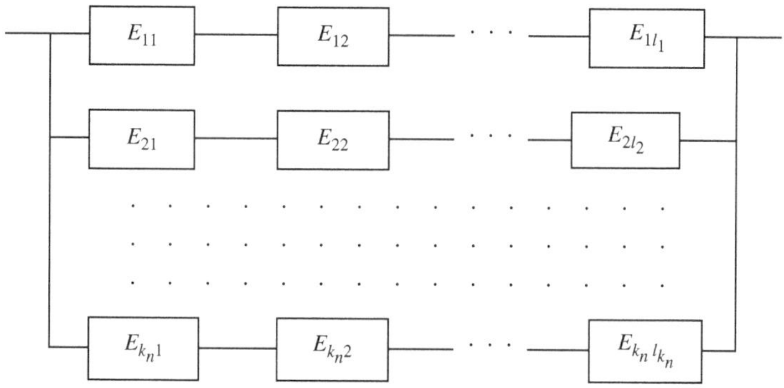

Figure 2.4 The scheme of a series–parallel system.

where

$$T_{ij}, \quad i = 1, 2, \ldots, k_n, \; j = 1, 2, \ldots, l_i$$

are independent random variables representing the lifetimes of components E_{ij} with distribution functions

$$F_{ij}(t) = P(T_{ij} \le t), \quad t \in (-\infty, \infty)$$

Definition 2.7

A two-state system is called series–parallel if its lifetime T is given by

$$T = \max_{1 \le i \le k_n} \{ \min_{1 \le j \le l_i} \{T_{ij}\}\}$$

The scheme of a series–parallel system is given in Figure 2.4.

By joining the formulae for the reliability functions of two-state series and parallel systems it is easy to conclude that the reliability function of the two-state series–parallel system is given by

$$\boldsymbol{R}_{k_n, l_1, l_2, \ldots, l_{k_n}}(t) = 1 - \prod_{i=1}^{k_n} \left[1 - \prod_{j=1}^{l_i} R_{ij}(t) \right], \quad t \in (-\infty, \infty)$$

where k_n is the number of series subsystems linked in parallel and l_i are the numbers of components in the series subsystems.

Definition 2.8

A two-state series–parallel system is called homogeneous if its component lifetimes T_{ij} have an identical distribution function

$$F(t) = P(T_{ij} \le t), \quad i = 1, 2, \ldots, k_n, \; j = 1, 2, \ldots, l_i$$

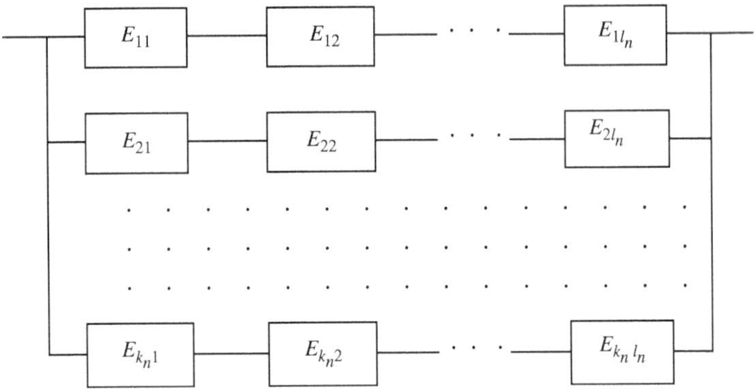

Figure 2.5 The scheme of a regular series–parallel system.

that is if its components E_{ij} have the same reliability function

$$R(t) = 1 - F(t), \quad t \in (-\infty, \infty)$$

Definition 2.9
A two-state series–parallel system is called regular if

$$l_1 = l_2 = \cdots = l_{k_n} = l_n, \quad l_n \in N$$

that is if the numbers of components in its series subsystems are equal.

The scheme of a regular series–parallel system is given in Figure 2.5.

The reliability function of the homogeneous regular two-state series–parallel system is given by

$$\boldsymbol{R}_{k_n,l_n}(t) = 1 - [1 - [R(t)]^{l_n}]^{k_n}, \quad t \in (-\infty, \infty) \tag{2.5}$$

where k_n is the number of series subsystems linked in parallel and l_n is the number of components in the series subsystems.

Definition 2.10
A two-state system is called parallel–series if its lifetime T is given by

$$T = \min_{1 \le i \le k_n} \{ \max_{1 \le j \le l_i} \{T_{ij}\}\}$$

The scheme of a parallel–series system is given in Figure 2.6.

By superposition of the formulae for the reliability functions of two-state parallel and series systems it is easy to conclude that the reliability function of the two-state parallel–series system is given by

$$\overline{\boldsymbol{R}}_{k_n,l_1,l_2,\ldots,l_{k_n}}(t) = \prod_{i=1}^{k_n} \left[1 - \prod_{j=1}^{l_i} F_{ij}(t) \right], \quad t \in (-\infty, \infty)$$

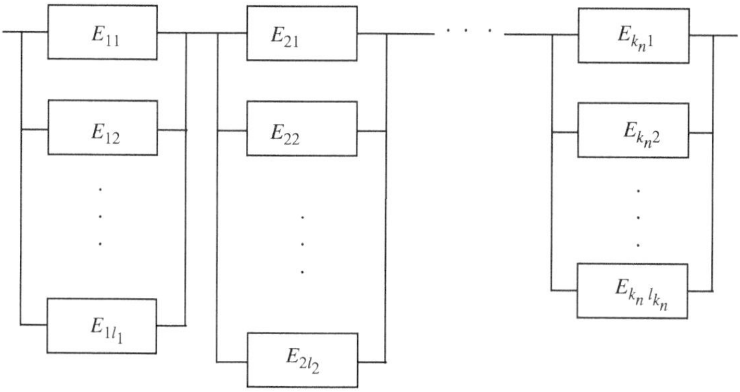

Figure 2.6 The scheme of a parallel–series system.

where k_n is the number of parallel subsystems linked in series and l_i are the numbers of components in the parallel subsystems.

Definition 2.11

A two-state parallel–series system is called homogeneous if its component lifetimes T_{ij} have an identical distribution function

$$F(t) = P(T_{ij} \le t), \quad i = 1, 2, \ldots, k_n, \ j = 1, 2, \ldots, l_i$$

that is if its components E_{ij} have the same reliability function

$$R(t) = 1 - F(t), \quad t \in (-\infty, \infty)$$

Definition 2.12

A two-state parallel–series system is called regular if

$$l_1 = l_2 = \cdots = l_{k_n} = l_n, \quad l_n \in N$$

that is if the numbers of components in its parallel subsystems are equal.

The scheme of a regular parallel–series system is given in Figure 2.7.

The reliability function of the homogeneous regular two-state parallel–series system is given by

$$\overline{\boldsymbol{R}}_{k_n,l_n}(t) = [1 - [F(t)]^{l_n}]^{k_n}, \quad t \in (-\infty, \infty) \tag{2.6}$$

where k_n is the number of parallel subsystems linked in series and l_n is the number of components in the parallel subsystems.

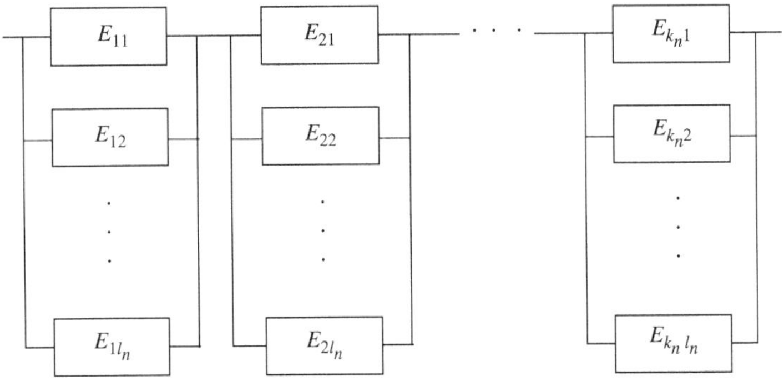

Figure 2.7 The scheme of a regular parallel−series system.

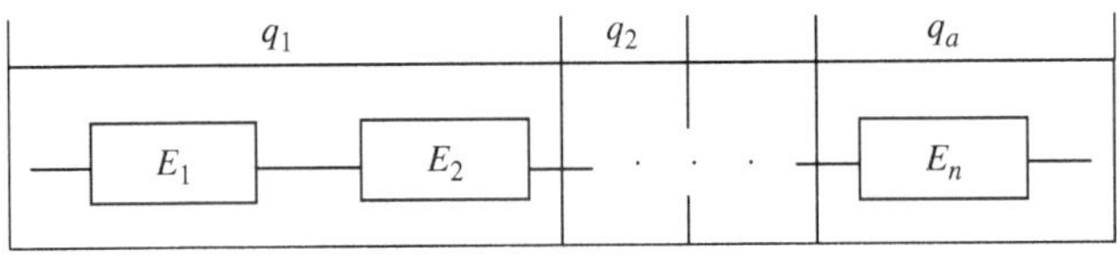

Figure 2.8 The scheme of a non-homogeneous series system.

Definition 2.13

A two-state series system is called non-homogeneous if it is composed of a, $1 \le a \le n$, different types of components and the fraction of the ith type component in the system is equal to q_i, where $q_i > 0$, $\sum_{i=1}^{a} q_i = 1$. Moreover,

$$R^{(i)}(t) = 1 - F^{(i)}(t), \quad t \in (-\infty, \infty), \; i = 1, 2, \ldots, a \tag{2.7}$$

is the reliability function of the ith type component.

The scheme of a non-homogeneous series system is given in Figure 2.8.

It is easy to show that the reliability function of the non-homogeneous two-state series system is given by

$$\overline{\boldsymbol{R}}'_{k_n, l_n}(t) = \prod_{i=1}^{a} (R^{(i)}(t))^{q_i n}, \quad t \in (-\infty, \infty) \tag{2.8}$$

Definition 2.14

A two-state parallel system is called non-homogeneous if it is composed of a, $1 \le a \le n$, different types of components and the fraction of the ith type component in the system is equal to q_i, where $q_i > 0$, $\sum_{i=1}^{a} q_i = 1$. Moreover,

$$R^{(i)}(t) = 1 - F^{(i)}(t), \quad t \in (-\infty, \infty), \; i = 1, 2, \ldots, a \tag{2.9}$$

is the reliability function of the ith type component.

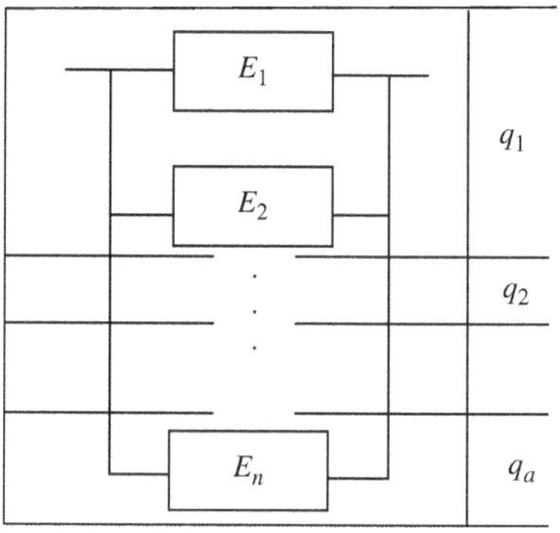

Figure 2.9 The scheme of a non-homogeneous parallel system.

The scheme of a non-homogeneous parallel system is given in Figure 2.9.

It is possible to work out that the reliability function of the non-homogeneous two-state parallel system is given by

$$R'_n(t) = 1 - \prod_{i=1}^{a} (F^{(i)}(t))^{q_i n}, \quad t \in (-\infty, \infty) \tag{2.10}$$

Definition 2.15

A two-state 'm out of n' system is called non-homogeneous if it is composed of a, $1 \le a \le n$, different types of components and the fraction of the ith type component in the system is equal to q_i, where $q_i > 0$, $\sum_{i=1}^{a} q_i = 1$. Moreover,

$$R^{(i)}(t) = 1 - F^{(i)}(t), \quad t \in (-\infty, \infty), \; i = 1, 2, \ldots, a \tag{2.11}$$

The scheme of a non-homogeneous 'm out of n' system is given in Figure 2.10, where $i_1, i_2, \ldots, i_n \in \{1, 2, \ldots, n\}$ and $i_j \neq i_k$ for $j \neq k$.

The reliability function of the non-homogeneous two-state 'm out of n' system is given either by

$$\boldsymbol{R}'^{(m)}_n(t) = 1 - \sum_{\substack{0 \le r_i \le q_i n \\ r_1 + r_2 + \cdots + r_a \le m-1}} \prod_{i=1}^{a} \binom{q_i n}{r_i} [R^{(i)}(t)]^{r_i} [F^{(i)}(t)]^{q_i n - r_i}, \; t \in (-\infty, \infty) \tag{2.12}$$

or by

$$\overline{\boldsymbol{R}}'_n(m)(t) = \sum_{\substack{0 \le r_i \le q_i n \\ r_1 + r_2 + \cdots + r_a \le \overline{m}}} \prod_{i=1}^{a} \binom{q_i n}{r_i} [F^{(i)}(t)]^{r_i} [R^{(i)}(t)]^{q_i n - r_i}, \quad t \in (-\infty, \infty) \tag{2.13}$$

where $\overline{m} = n - m$.

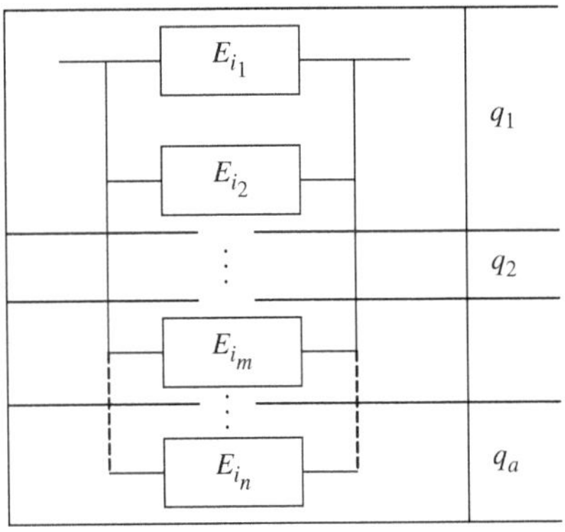

Figure 2.10 The scheme of a non-homogeneous 'm out of n' system.

Definition 2.16

A two-state regular series–parallel system is called non-homogeneous if it is composed of a, $1 \le a \le k_n$, $k_n \in N$, different types of series subsystems and the fraction of the ith type series subsystem is equal to q_i, where $q_i > 0$, $\sum_{i=1}^{a} q_i = 1$. Moreover, the ith type series subsystem consists of e_i, $1 \le e_i \le l_n$, $l_n \in N$, types of components with reliability functions

$$R^{(i,j)}(t) = 1 - F^{(i,j)}(t), \quad t \in (-\infty, \infty), \ j = 1, 2, \ldots, e_i$$

and the fraction of the jth type component in this subsystem is equal to p_{ij}, where $p_{ij} > 0$ and $\sum_{j=1}^{e_i} p_{ij} = 1$.

The scheme of a regular non-homogeneous series–parallel system is shown in Figure 2.11.

The reliability function of the regular non-homogeneous two-state series– parallel system is given by

$$\boldsymbol{R}'_{k_n,l_n}(t) = 1 - \prod_{i=1}^{a} [1 - (R^{(i)}(t))^{l_n}]^{q_i k_n}, \quad t \in (-\infty, \infty) \tag{2.14}$$

where

$$R^{(i)}(t) = \prod_{j=1}^{e_i} (R^{(i,j)}(t))^{p_{ij}}, \quad i = 1, 2, \ldots, a \tag{2.15}$$

Definition 2.17

A two-state regular parallel–series system is called non-homogeneous if it is composed of a, $1 \le a \le k_n$, $k_n \in N$, different types of parallel subsystems and the fraction of the ith type parallel subsystem is equal to q_i, where $q_i > 0$, $\sum_{i=1}^{a} q_i = 1$. Moreover, the ith type parallel subsystem consists of e_i, $1 \le e_i \le l_n$, $l_n \in N$, types of components with reliability functions

$$R^{(i,j)}(t) = 1 - F^{(i,j)}(t), \quad t \in (-\infty, \infty), \ j = 1, 2, \ldots, e_i$$

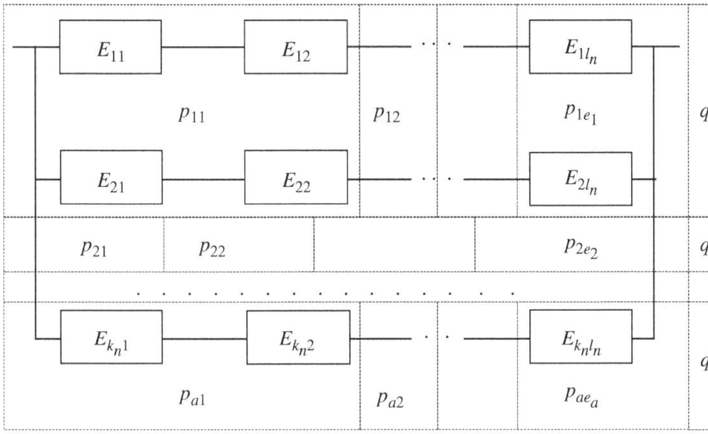

Figure 2.11 The scheme of a regular non-homogeneous series–parallel system.

and the fraction of the jth type component in this subsystem is equal to p_{ij}, where $p_{ij} > 0$ and $\sum_{j=1}^{e_i} p_{ij} = 1$.

The scheme of a regular non-homogeneous parallel–series system is shown in Figure 2.12.

The reliability function of the regular non-homogeneous two-state parallel–series system is given by

$$\overline{\boldsymbol{R}}'_{k_n,l_n}(t) = \prod_{i=1}^{a} [1-(F^{(i)}(t))^{l_n}]^{q_i k_n}, \quad t\in(-\infty, \infty) \tag{2.16}$$

where

$$F^{(i)}(t) = \prod_{j=1}^{e_i} (F^{(i,j)}(t))^{p_{ij}}, \quad i = 1,2,\ldots,a \tag{2.17}$$

Remark 2.1

In our further considerations, we suppose that n, k_n and l_n are positive real numbers and we investigate the families of the reliability functions $\overline{\boldsymbol{R}}_n(t)$, $\overline{\boldsymbol{R}}'_n(t)$, $\boldsymbol{R}_n(t)$, $\boldsymbol{R}'_n(t)$, $\boldsymbol{R}_n^{(m)}(t)$, $\boldsymbol{R}'^{(m)}_n(t)$, $\overline{\boldsymbol{R}}^{(\overline{m})}_{k_n,l_n}(t)$, $\boldsymbol{R}'^{\overline{m}}_{\overline{n}}(t)$ for $n\in(0, \infty)$ and the families of the reliability functions $\boldsymbol{R}_{k_n,l_n}(t)$, $\boldsymbol{R}'_{k_n,l_n}(t)$, $\overline{\boldsymbol{R}}_{k_n,l_n}(t)$, $\overline{\boldsymbol{R}}'_{k_n,l_n}(t)$ corresponding to the pair (k_n, l_n), where $k_n \in (0, \infty)$, $l_n \in (0, \infty)$. This assumption is necessary in proving theorems that are cited in the next parts of the book. However, from the practical point of view it is important that n, k_n and l_n be natural numbers. The return to natural numbers is trivial since the positive real number can be represented by the sum of its natural part and its real remaining part r. Then the expression of the form $[R(t)]^r$ exists in the formulae for the reliability functions of the considered systems. Since this

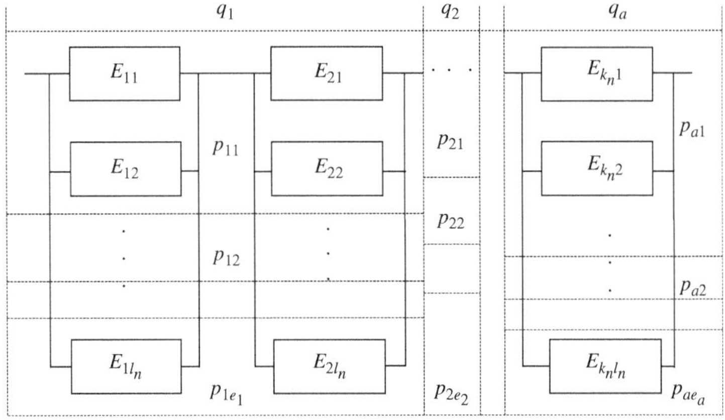

Figure 2.12 The scheme of a regular non-homogeneous parallel–series system.

expression is a reliability function, it means that the system or subsystem is composed of one component that has a reliability function different from the reliability function $R(t)$ of the remaining components. Such single components do not have a significant influence on the limit reliability functions of the considered systems, which are composed of large numbers of components.

3 Multi-State Systems

In the multi-state reliability analysis to define systems with degrading (ageing) components we assume that:

- E_i, $i = 1, 2, \ldots, n$, are components of a system;
- all components and a system under consideration have the state set $\{0, 1, \ldots, z\}$, $z \geq 1$;
- the state indexes are ordered, the state 0 is the worst and the state z is the best;
- $T_i(u)$, $i = 1, 2, \ldots, n$, are independent random variables representing the lifetimes of components E_i in the state subset $\{u, u+1, \ldots, z\}$, while they were in the state z at the moment $t = 0$;
- $T(u)$ is a random variable representing the lifetime of a system in the state subset $\{u, u+1, \ldots, z\}$ while it was in the state z at the moment $t = 0$;
- the system state degrades with time t without repair;
- $e_i(t)$ is a component E_i state at the moment t, $t \in (-\infty, \infty)$, given that it was in the state z at the moment $t = 0$;
- $s(t)$ is a system state at the moment t, $t \in (-\infty, \infty)$, given that it was in the state z at the moment $t = 0$.

The above assumptions mean that the states of the system with degrading components may be changed in time only from better to worse. The way in which the components and the system states change is illustrated in Figure 3.1.

Definition 3.1
A vector

$$R_i(t, \cdot) = [R_i(t, 0), R_i(t, 1), \ldots, R_i(t, z)], \quad t \in (-\infty, \infty), \;\; i = 1, 2, \ldots, n$$

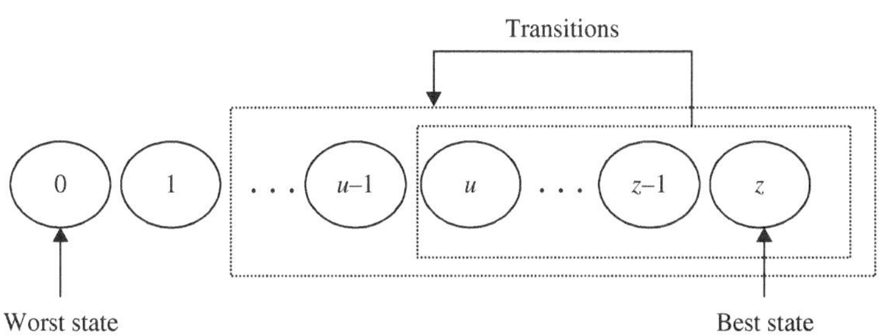

Figure 3.1 Illustration of states changing in system with ageing components.

Reliability of Large Systems. DOI: http://dx.doi.org/10.1016/B978-0-08-099949-4.00003-9

where

$$R_i(t,u) = P(e_i(t) \geq u | e_i(0) = z) = P(T_i(u) > t), \quad t \in (-\infty, \infty), \quad u = 0, 1, \ldots, z \tag{3.1}$$

is the probability that the component E_i is in the state subset $\{u, u+1, \ldots, z\}$ at the moment t, $t \in (-\infty, \infty)$, while it was in the state z at the moment $t = 0$, is called the multi-state reliability function of a component E_i.

Under this definition we have

$$R_i(t,0) \geq R_i(t,1) \geq \cdots \geq R_i(t,z), \quad t \in (-\infty, \infty), \quad i = 1, 2, \ldots, n$$

Further, if we denote by

$$p_i(t,u) = P(e_i(t) = u | e_i(0) = z), \quad t \in (-\infty, \infty), \quad u = 0, 1, \ldots, z$$

the probability that the component E_i is in the state u at the moment t, while it was in the state z at the moment $t = 0$, then by Eq. (3.1)

$$R_i(t,0) = 1, \quad R_i(t,z) = p_i(t,z), \quad t \in (-\infty, \infty), \quad i = 1, 2, \ldots, n \tag{3.2}$$

and

$$p_i(t,u) = R_i(t,u) - R_i(t,u+1), \quad u = 0, 1, \ldots, z-1, \quad t \in (-\infty, \infty), \quad i = 1, 2, \ldots, n \tag{3.3}$$

Moreover, if

$$R_i(t,u) = 1 \quad \text{for } t \leq 0, \quad u = 1, 2, \ldots, z, \quad i = 1, 2, \ldots, n$$

then

$$M_i(u) = \int_0^\infty R_i(t,u)\mathrm{d}t, \quad u = 1, 2, \ldots, z, \quad i = 1, 2, \ldots, n \tag{3.4}$$

is the mean lifetime of the component E_i in the state subset $\{u, u+1, \ldots, z\}$,

$$\sigma_i(u) = \sqrt{N_i(u) - [M_i(u)]^2}, \quad u = 1, 2, \ldots, z, \quad i = 1, 2, \ldots, n \tag{3.5}$$

where

$$N_i(u) = 2\int_0^\infty tR_i(t,u)\mathrm{d}t, \quad u = 1, 2, \ldots, z, \quad i = 1, 2, \ldots, n \tag{3.6}$$

is the standard deviation of the component E_i lifetime in the state subset $\{u, u+1, \ldots, z\}$ and

$$\overline{M}_i(u) = \int_0^{\infty} p_i(t,u)\mathrm{d}t, \quad u = 1, 2, \ldots, z, \ i = 1, 2, \ldots, n \tag{3.7}$$

is the mean lifetime of the component E_i in the state u, in the case when the integrals defined by Eqs. (3.4), (3.6) and (3.7) are convergent.

Next, according to Eqs. (3.2)–(3.4) and (3.7), we have

$$\overline{M}_i(u) = M_i(u) - M_i(u+1), \quad u = 0, 1, \ldots, z-1; \ \overline{M}_i(z) = M_i(z), \ i = 1, 2, \ldots, n \tag{3.8}$$

Definition 3.2

A vector

$$\boldsymbol{R}_n(t, \cdot) = [\boldsymbol{R}_n(t,0), \boldsymbol{R}_n(t,1), \ldots, \boldsymbol{R}_n(t,z)], \quad t \in (-\infty, \infty)$$

where

$$\boldsymbol{R}_n(t,u) = P(s(t) \geq u | s(0) = z) = P(T(u) > t), \quad t \in (-\infty, \infty), \ u = 0, 1, \ldots, z \tag{3.9}$$

is the probability that the system is in the state subset $\{u, u+1, \ldots, z\}$ at the moment t, $t \in (-\infty, \infty)$, while it was in the state z at the moment $t = 0$, is called the multi-state reliability function of a system.

Under this definition, we have

$$\boldsymbol{R}_n(t,0) \geq \boldsymbol{R}_n(t,1) \geq \cdots \geq \boldsymbol{R}_n(t,z), \quad t \in (-\infty, \infty)$$

and if

$$p(t,u) = P(s(t) = u | s(0) = z), \quad t \in (-\infty, \infty), \ u = 0, 1, \ldots, z \tag{3.10}$$

is the probability that the system is in the state u at the moment t, $t \in (-\infty, \infty)$, while it was in the state z at the moment $t = 0$, then

$$\boldsymbol{R}_n(t,0) = 1, \ \boldsymbol{R}_n(t,z) = p(t,z), \quad t \in (-\infty, \infty) \tag{3.11}$$

and

$$p(t,u) = \boldsymbol{R}_n(t,u) - \boldsymbol{R}_n(t,u+1), \quad u = 0, 1, \ldots, z-1, \ t \in (-\infty, \infty) \tag{3.12}$$

Moreover, if

$$\boldsymbol{R}_n(t,u) = 1 \ \text{ for } \ t \le 0, \quad u = 1, 2, \ldots, z$$

then

$$M(u) = \int_0^\infty \boldsymbol{R}_n(t,u)\mathrm{d}t, \quad u = 1, 2, \ldots, z \tag{3.13}$$

is the mean lifetime of the system in the state subset $\{u, u+1, \ldots, z\}$,

$$\sigma(u) = \sqrt{N(u) - [M(u)]^2}, \quad u = 1, 2, \ldots, z \tag{3.14}$$

where

$$N(u) = 2\int_0^\infty t\,\boldsymbol{R}_n(t,u)\mathrm{d}t, \quad u = 1, 2, \ldots, z \tag{3.15}$$

is the standard deviation of the system sojourn time in the state subset $\{u, u+1, \ldots, z\}$ and moreover

$$\overline{M}(u) = \int_0^\infty p(t,u)\mathrm{d}t, \quad u = 1, 2, \ldots, z \tag{3.16}$$

is the mean lifetime of the system in the state u while the integrals (3.13)–(3.15) are convergent.

Additionally, according to Eqs. (3.11)–(3.13) and (3.16), we get the following relationship:

$$\overline{M}(u) = M(u) - M(u+1), \quad u = 0, 1, \ldots, z-1; \ \ \overline{M}(z) = M(z) \tag{3.17}$$

Definition 3.3

A probability

$$r(t) = P(s(t) < r | s(0) = z) = P(T(r) \le t), \quad t \in (-\infty, \infty)$$

that the system is in the subset of states worse than the critical state r, $r \in \{1, \ldots, z\}$ while it was in the state z at the moment $t = 0$ is called a risk function of the multi-state system or, in short, a risk.

Under this definition, from Eq. (3.1), we have

$$\boldsymbol{r}(t) = 1 - P(s(t) \ge r | s(0) = z) = 1 - \boldsymbol{R}_n(t, r), \quad t \in (-\infty, \infty) \tag{3.18}$$

and if τ is the moment when the risk exceeds a permitted level δ, then

$$\tau = \boldsymbol{r}^{-1}(\delta) \tag{3.19}$$

where $\boldsymbol{r}^{-1}(t)$, if it exists, is the inverse function of the risk function $\boldsymbol{r}(t)$.

Definition 3.4

A multi-state system is called series if its lifetime $T(u)$ in the state subset $\{u, u+1, \ldots, z\}$ is given by

$$T(u) = \min_{1 \le i \le n} \{T_i(u)\}, \quad u = 1, 2, \ldots, z$$

The above definition means that a multi-state series system is in the state subset $\{u, u+1, \ldots, z\}$ if and only if all its components are in this subset of states.

It is easy to work out that the reliability function of the multi-state series system is given by

$$\overline{\boldsymbol{R}}_n(t, \cdot\,) = [1, \overline{\boldsymbol{R}}_n(t, 1), \ldots, \overline{\boldsymbol{R}}_n(t, z)]$$

where

$$\overline{\boldsymbol{R}}_n(t, u) = \prod_{i=1}^{n} R_i(t, u), \quad t \in (-\infty, \infty), \;\; u = 1, 2, \ldots, z$$

Definition 3.5

A multi-state series system is called homogeneous if its component lifetimes $T_i(u)$ in the state subsets have an identical distribution function

$$F_i(t, u) = F(t, u), \quad u = 1, 2, \ldots, z, \;\; t \in (-\infty, \infty), \;\; i = 1, 2, \ldots, n$$

that is if its components E_i have the same reliability function

$$R_i(t, u) = R(t, u) = 1 - F(t, u), \quad u = 1, 2, \ldots, z, \;\; t \in (-\infty, \infty), \;\; i = 1, 2, \ldots, n$$

The reliability function of the homogeneous multi-state series system is given by

$$\overline{\boldsymbol{R}}_n(t, \cdot\,) = [1, \overline{\boldsymbol{R}}_n(t, 1), \ldots, \overline{\boldsymbol{R}}_n(t, z)] \tag{3.20}$$

where

$$\overline{\boldsymbol{R}}_n(t, u) = [R(t, u)]^n, \quad t \in (-\infty, \infty), \;\; u = 1, 2, \ldots, z \tag{3.21}$$

Definition 3.6
A multi-state system is called parallel if its lifetime $T(u)$ in the state subset $\{u, u+1, \ldots, z\}$ is given by

$$T(u) = \max_{1 \le i \le n} \{T_i(u)\}, \quad u = 1, 2, \ldots, z$$

The above definition means that the multi-state parallel system is in the state subset $\{u, u+1, \ldots, z\}$ if and only if at least one of its components is in this subset of states.

The reliability function of the multi-state parallel system is given by

$$\boldsymbol{R}_n(t, \cdot) = [1, \boldsymbol{R}_n(t, 1), \ldots, \boldsymbol{R}_n(t, z)]$$

where

$$\boldsymbol{R}_n(t, u) = 1 - \prod_{i=1}^{n} F_i(t, u), \quad t \in (-\infty, \infty), \; u = 1, 2, \ldots, z$$

Definition 3.7
A multi-state parallel system is called homogeneous if its component lifetimes $T_i(u)$ in the state subsets have an identical distribution function

$$F_i(t, u) = F(t, u), \quad u = 1, 2, \ldots, z, \; t \in (-\infty, \infty), \; i = 1, 2, \ldots, n$$

that is if its components E_i have the same reliability function

$$R_i(t, u) = R(t, u) = 1 - F(t, u), \quad u = 1, 2, \ldots, z, \; t \in (-\infty, \infty), \; i = 1, 2, \ldots, n$$

The reliability function of the homogeneous multi-state parallel system is given by

$$\boldsymbol{R}_n(t, \cdot) = [1, \boldsymbol{R}_n(t, 1), \ldots, \boldsymbol{R}_n(t, z)] \tag{3.22}$$

where

$$\boldsymbol{R}_n(t, u) = 1 - [F(t, u)]^n, \quad t \in (-\infty, \infty), \; u = 1, 2, \ldots, z \tag{3.23}$$

Definition 3.8
A multi-state system is called an 'm out of n' system if its lifetime $T(u)$ in the state subset $\{u, u+1, \ldots, z\}$ is given by

$$T(u) = T_{(n-m+1)}(u), \quad m = 1, 2, \ldots, n, \; u = 1, 2, \ldots, z$$

where $T_{(n-m+1)}(u)$ is the mth maximal order statistic in the sequence of the component lifetimes $T_1(u)$, $T_2(u)$, ..., $T_n(u)$.

The above definition means that the multi-state 'm out of n' system is in the state subset $\{u, u+1, \ldots, z\}$ if and only if at least m out of its n components are in this state subset; and it is a multi-state parallel system if $m = 1$ and it is a multi-state series system if $m = n$.

It can be simply shown that the reliability function of the multi-state 'm out of n' system is given either by

$$\boldsymbol{R}_n^{(m)}(t,\cdot) = [1, \boldsymbol{R}_n^{(m)}(t,1), \ldots, \boldsymbol{R}_n^{(m)}(t,z)]$$

where

$$\boldsymbol{R}_n^{(m)}(t,u) = 1 - \sum_{\substack{r_1,r_2,\ldots,r_n=0\\ r_1+r_2+\cdots+r_n \le m-1}}^{1} [R_i(t,u)]^{r_i}[F_i(t,u)]^{1-r_i}, \; t\in(-\infty,\infty),\; u=1,2,\ldots,z$$

or by

$$\overline{\boldsymbol{R}}_n^{(\overline{m})}(t,\cdot) = \left[1, \overline{\boldsymbol{R}}_n^{(\overline{m})}(t,1), \ldots, \overline{\boldsymbol{R}}_n^{(\overline{m})}(t,z)\right]$$

where

$$\overline{\boldsymbol{R}}_n^{(\overline{m})}(t,u) = \sum_{\substack{r_1,r_2,\ldots,r_n=0\\ r_1+r_2+\cdots+r_n \le \overline{m}}}^{1} [F_i(t,u)]^{r_i}[R_i(t,u)]^{1-r_i}, \; t\in(-\infty,\infty),\; \overline{m}=n-m,\; u=1,2,\ldots,z$$

Definition 3.9

A multi-state 'm out of n' system is called homogeneous if its component lifetimes $T_i(u)$ in the state subsets have an identical distribution function

$$F_i(t,u) = F(t,u), \quad u = 1,2,\ldots,z, \; t\in(-\infty,\infty), \; i = 1,2,\ldots,n$$

that is if its components E_i have the same reliability function

$$R_i(t,u) = R(t,u) = 1 - F(t,u), \quad u = 1,2,\ldots,z, \; t\in(-\infty,\infty), \; i = 1,2,\ldots,n$$

The reliability function of the homogeneous multi-state 'm out of n' system is given either by

$$\boldsymbol{R}_n^{(m)}(t,\cdot) = \left[1, \boldsymbol{R}_n^{(m)}(t,1), \ldots, \boldsymbol{R}_n^{(m)}(t,z)\right] \tag{3.24}$$

where

$$\boldsymbol{R}_n^{(m)}(t,u) = 1 - \sum_{i=0}^{m-1} \binom{n}{i} [R(t,u)]^i [F(t,u)]^{n-i}, \quad t \in (-\infty, \infty), \quad u = 1, 2, \ldots, z \tag{3.25}$$

or by

$$\overline{\boldsymbol{R}}_n^{(\overline{m})}(t,\cdot) = \left[1, \overline{\boldsymbol{R}}_n^{(\overline{m})}(t,1), \ldots, \overline{\boldsymbol{R}}_n^{(\overline{m})}(t,z)\right] \tag{3.26}$$

where

$$\overline{\boldsymbol{R}}_n^{(\overline{m})}(t,u) = \sum_{i=0\overline{m}} \binom{n}{i} [F(t,u)]^i [R(t,u)]^{n-i}, \; t \in (-\infty, \infty), \; \overline{m} = n - m, \; u = 1, 2, \ldots, z \tag{3.27}$$

Other basic multi-state reliability structures with components degrading in time are series–parallel and parallel–series systems. To define them, we assume that

- E_{ij}, $i = 1, 2, \ldots, k_n$, $j = 1, 2, \ldots, l_i$, $k_n, l_1, l_2, \ldots, l_{k_n} \in N$, are components of a system;
- all components E_{ij} have the same state set as before $\{0, 1, \ldots, z\}$;
- $T_{ij}(u)$, $i = 1, 2, \ldots, k_n$, $j = 1, 2, \ldots, l_i$, $k_n, l_1, l_2, \ldots, l_{k_n} \in N$, are independent random variables representing the lifetimes of components E_{ij} in the state subset $\{u, u+1, \ldots, z\}$, while they were in the state z at the moment $t = 0$;
- $e_{ij}(t)$ is a component E_{ij} state at the moment t, $t \in (-\infty, \infty)$, while they were in the state z at the moment $t = 0$.

Definition 3.10
A vector

$$R_{ij}(t,\cdot) = [R_{ij}(t,0), R_{ij}(t,1), \ldots, R_{ij}(t,z)], \; t \in (-\infty, \infty), \; i = 1, 2, \ldots, k_n, \; j = 1, 2, \ldots, l_i$$

where

$$R_{ij}(t,u) = P(e_{ij}(t) \geq u \mid e_{ij}(0) = z) = P(T_{ij}(u) > t), \quad t \in (-\infty, \infty), \quad u = 0, 1, \ldots, z$$

is the probability that the component E_{ij} is in the state subset $\{u, u+1, \ldots, z\}$ at the moment t, $t \in (-\infty, \infty)$, while it was in the state z at the moment $t = 0$, is called the multi-state reliability function of a component E_{ij}.

Definition 3.11
A multi-state system is called series–parallel if its lifetime $T(u)$ in the state subset $\{u, u+1, \ldots, z\}$ is given by

$$T(u) = \max_{1 \leq i \leq k_n} \{ \min_{1 \leq j \leq l_i} \{T_{ij}(u)\}\}, \quad u = 1, 2, \ldots, z$$

The reliability function of the multi-state series–parallel system is given by

$$\boldsymbol{R}_{k_n,l_1,l_2,\ldots,l_n}(t,\cdot)=[1,\boldsymbol{R}_{k_n,l_1,l_2,\ldots,l_n}(t,1),\ldots,\boldsymbol{R}_{k_n,l_1,l_2,\ldots,l_n}(t,z)]$$

and

$$\boldsymbol{R}_{k_n,l_1,l_2,\ldots,l_{k_n}}(t,u)=1-\prod_{i=1}^{k}\left[1-\prod_{j=1}^{l_i}R_{ij}(t,u)\right],\quad t\in(-\infty,\infty),\ u=1,2,\ldots,z$$

where k_n is the number of series subsystems linked in parallel and l_i are the numbers of components in the series subsystems.

Definition 3.12

A multi-state series–parallel system is called homogeneous if its lifetimes $T_{ij}(u)$ in the state subset have an identical distribution function

$$F_{ij}(t,u)=F(t,u),\quad u=1,2,\ldots,z,\ t\in(-\infty,\infty),\ i=1,2,\ldots,k_n,\ j=1,2,\ldots,l_i$$

that is if its components E_{ij} have the same reliability function

$$R_{ij}(t,u)=R(t,u)=1-F(t,u),\ u=1,2,\ldots,z,\ t\in(-\infty,\infty),\ i=1,2,\ldots,k_n,\ j=1,2,\ldots,l_i$$

Definition 3.13

A multi-state series–parallel system is called regular if

$$l_1=l_2=\cdots=l_{k_n}=l_n,\quad l_n\in N$$

The reliability function of the multi-state homogeneous regular series–parallel system is given by

$$\boldsymbol{R}_{k_n,l_n}(t,\cdot)=[1,\boldsymbol{R}_{k_n,l_n}(t,1),\ldots,\boldsymbol{R}_{k_n,l_n}(t,z)] \tag{3.28}$$

and

$$\boldsymbol{R}_{k_n,l_n}(t,u)=1-[1-[R(t,u)]^{l_n}]^{k_n},\quad t\in(-\infty,\infty),\ u=1,2,\ldots,z \tag{3.29}$$

where k_n is the number of series subsystems linked in parallel and l_n is the number of components in the series subsystems.

Definition 3.14

A multi-state system is called parallel–series if its lifetime $T(u)$ in the state subset $\{u,u+1,\ldots,z\}$ is given by

$$T(u)=\min_{1\le i\le k_n}\{\max_{1\le j\le l_i}\{T_{ij}(u)\}\},\quad u=1,2,\ldots,z$$

The reliability function of the multi-state parallel–series system is given by

$$\overline{\boldsymbol{R}}_{k_n,l_1,l_2,\ldots,l_{k_n}}(t,\cdot\,) = [1, \overline{\boldsymbol{R}}_{k_n,l_1,l_2,\ldots,l_{k_n}}(t,1), \ldots, \overline{\boldsymbol{R}}_{k_n,l_1,l_2,\ldots,l_{k_n}}(t,z)]$$

and

$$\overline{\boldsymbol{R}}_{k_n,l_1,l_2,\ldots,l_{k_n}}(t,u) = \prod_{i=1}^{k_n}\left[1 - \prod_{j=1}^{l_i} F_{ij}(t,u)\right], \quad t \in (-\infty,\infty), \ \ u = 1,2,\ldots,z$$

where k_n is the number of its parallel subsystems linked in series and l_i are the numbers of components in the parallel subsystems.

Definition 3.15
A multi-state parallel–series system is called homogeneous if its lifetimes $T_{ij}(u)$ in the state subset have an identical distribution function

$$F_{ij}(t,u) = F(t,u), \quad u = 1,2,\ldots,z, \ \ t \in (-\infty,\infty), \ \ i = 1,2,\ldots,k_n, \ \ j = 1,2,\ldots,l_i$$

that is if its components E_{ij} have the same reliability function

$$R_{ij}(t,u) = R(t,u) = 1 - F(t,u), \ \ u = 1,2,\ldots,z, \ t \in (-\infty,\infty), \ i = 1,2,\ldots,k_n, \ j = 1,2,\ldots,l_i$$

Definition 3.16
A multi-state parallel–series system is called regular if

$$l_1 = l_2 = \cdots = l_{k_n} = l_n, \quad l_n \in N$$

The reliability function of the multi-state homogeneous regular parallel–series system is given by

$$\overline{\boldsymbol{R}}_{k_n,l_n}(t,\cdot\,) = [1, \overline{\boldsymbol{R}}_{k_n,l_n}(t,1), \ldots, \overline{\boldsymbol{R}}_{k_n,l_n}(t,z)] \tag{3.30}$$

and

$$\overline{\boldsymbol{R}}_{k_n,l_n}(t,u) = [1 - [F(t,u)]^{l_n}]^{k_n}, \quad t \in (-\infty,\infty), \ \ u = 1,2,\ldots,z \tag{3.31}$$

where k_n is the number of its parallel subsystems linked in series and l_n is the number of components in the parallel subsystems.

Definition 3.17
A multi-state series system is called non-homogeneous if it is composed of a, $1 \le a \le n$, different types of components and the fraction of the ith

type component in the system is equal to q_i, where $q_i > 0$, $\sum_{i=1}^{a} q_i = 1$. Moreover,

$$R^{(i)}(t,u) = 1 - F^{(i)}(t,u), \quad t \in (-\infty, \infty), \; i = 1,2,\ldots,a, \; u = 1,2,\ldots,z \tag{3.32}$$

is the reliability function of the ith type component.

It can be easily proved that the reliability function of the non-homogeneous multi-state series system is given by

$$\overline{\boldsymbol{R}}'_n(t,\cdot) = [1, \overline{\boldsymbol{R}}'_n(t,1), \ldots, \overline{\boldsymbol{R}}'_n(t,z)] \tag{3.33}$$

where

$$\overline{\boldsymbol{R}}'_n(t,u) = \prod_{i=1}^{a} [R^{(i)}(t,u)]^{q_i n}, \quad t \in (-\infty, \infty), \; u = 1,2,\ldots,z \tag{3.34}$$

Definition 3.18

A multi-state parallel system is called non-homogeneous if it is composed of a, $1 \le a \le n$, different types of components and the fraction of the ith type component in the system is equal to q_i, where $q_i > 0$, $\sum_{i=1}^{a} q_i = 1$. Moreover,

$$R^{(i)}(t,u) = 1 - F^{(i)}(t,u), \quad t \in (-\infty, \infty), \; i = 1,2,\ldots,a, \; u = 1,2,\ldots,z \tag{3.35}$$

is the reliability function of the ith type component.

It can be easily proved that the reliability function of the non-homogeneous multi-state parallel system is given by

$$\boldsymbol{R}'_n(t,\cdot) = [1, \boldsymbol{R}'_n(t,1), \ldots, \boldsymbol{R}'_n(t,z)] \tag{3.36}$$

where

$$\boldsymbol{R}'_n(t,u) = 1 - \prod_{i=1}^{a} [F^{(i)}(t,u)]^{q_i n}, \quad t \in (-\infty, \infty), \; u = 1,2,\ldots,z \tag{3.37}$$

Definition 3.19

A multi-state 'm out of n' system is called non-homogeneous if it is composed of a, $1 \le a \le n$, different types of components and the fraction of the ith type component in the system is equal to q_i, where $q_i > 0$, $\sum_{i=1}^{a} q_i = 1$. Moreover,

$$R^{(i)}(t,u) = 1 - F^{(i)}(t,u), \quad t \in (-\infty, \infty), \; i = 1,2,\ldots,a, \; u = 1,2,\ldots,z \tag{3.38}$$

is the reliability function of the ith type component.

It can be easily proved that the reliability function of the non-homogeneous multi-state 'm out of n' system is given either by

$$\boldsymbol{R}'_n(m)(t,\cdot) = \left[1, \boldsymbol{R}'_n(m)(t,1), \ldots, \boldsymbol{R}'_n(m)(t,z)\right] \tag{3.39}$$

where

$$\boldsymbol{R}'_n(m)(t,u) = 1 - \sum_{\substack{0 \le r_i \le q_i n \\ r_1 + r_2 + \cdots + r_a \le m-1}} \prod_{i=1}^{a} \binom{q_i n}{r_i} [R^{(i)}(t,u)]^{r_i} [F^{(i)}(t,u)]^{q_i n - r_i} \tag{3.40}$$

for $t \in (-\infty, \infty)$, $u = 1, 2, \ldots, z$ or by

$$\overline{\boldsymbol{R}}'_n(\overline{m})(t,\cdot) = \left[1, \overline{\boldsymbol{R}}'^{(\overline{m})}_n(t,1), \ldots, \overline{\boldsymbol{R}}'_n(\overline{m})(t,z)\right] \tag{3.41}$$

where

$$\overline{\boldsymbol{R}}^{\overline{m}}_n(t,u) = \sum_{\substack{0 \le r_i \le q_i n \\ r_1 + r_2 + \cdots + r_a \le \overline{m}}} \prod_{i=1}^{a} \binom{q_i n}{r_i} [F^{(i)}(t,u)]^{r_i} [R^{(i)}(t,u)]^{q_i n - r_i}, \quad \overline{m} = n - m \tag{3.42}$$

for $t \in (-\infty, \infty)$, $u = 1, 2, \ldots, z$.

Definition 3.20

A multi-state regular series–parallel system is called non-homogeneous if it is composed of a, $1 \le a \le k_n$, $k_n \in N$, different types of series subsystems and the fraction of the ith type series subsystem is equal to q_i, where $q_i > 0$, $\sum_{i=1}^{a} q_i = 1$. Moreover, the ith type series subsystem consists of e_i, $1 \le e_i \le l_n$, $l_n \in N$, types of components with reliability functions

$$R^{(i,j)}(t,u) = 1 - F^{(i,j)}(t,u), \quad t \in (-\infty, \infty), \ j = 1, 2, \ldots, e_i, \ u = 1, 2, \ldots, z$$

and the fraction of the jth type component in this subsystem is equal to p_{ij}, where $p_{ij} > 0$ and $\sum_{j=1}^{e_i} p_{ij} = 1$.

The reliability function of the multi-state non-homogeneous regular series–parallel system is given by

$$\boldsymbol{R}'_{k_n,l_n}(t,\cdot) = [1, \boldsymbol{R}'_{k_n,l_n}(t,1), \ldots, \boldsymbol{R}'_{k_n,l_n}(t,z)] \tag{3.43}$$

where

$$\boldsymbol{R}'_{k_n,l_n}(t,u) = 1 - \prod_{i=1}^{a} [1 - [R^{(i)}(t,u)]^{l_n}]^{q_i k_n}, \quad t \in (-\infty, \infty), \ u = 1, 2, \ldots, z \tag{3.44}$$

and

$$R^{(i)}(t,u) = \prod_{j=1}^{e_i} [R^{(i,j)}(t,u)]^{p_{ij}}, \quad i = 1,2,\ldots,a \tag{3.45}$$

Definition 3.21

A multi-state regular parallel–series system is called non-homogeneous if it is composed of a, $1 \le a \le k_n$, $k_n \in N$, different types of parallel subsystems and the fraction of the ith type parallel subsystem is equal to q_i, where $q_i > 0, \sum_{i=1}^{a} q_i = 1$. Moreover, the ith type parallel subsystem consists of e_i, $1 \le e_i \le l_n$, $l_n \in N$, types of components with reliability functions

$$R^{(i,j)}(t,u) = 1 - F^{(i,j)}(t,u), \quad t \in (-\infty, \infty), \; j = 1,2,\ldots,e_i, \; u = 1,2,\ldots,z$$

and the fraction of the jth type component in this subsystem is equal to p_{ij}, where $p_{ij} > 0$ and $\sum_{j=1}^{e_i} p_{ij} = 1$.

The reliability function of the multi-state non-homogeneous regular parallel–series system is given by

$$\overline{\boldsymbol{R}}'_{k_n,l_n}(t,\cdot) = [1, \overline{\boldsymbol{R}}'_{k_n,l_n}(t,1),\ldots,\overline{\boldsymbol{R}}'_{k_n,l_n}(t,z)] \tag{3.46}$$

where

$$\overline{\boldsymbol{R}}'_{k_n,l_n}(t,u) = \prod_{i=1}^{a} [1 - [F^{(i)}(t,u)]^{l_n}]^{q_i k_n}, \quad t \in (-\infty,\infty), \; u = 1,2,\ldots,z \tag{3.47}$$

and

$$F^{(i)}(t,u) = \prod_{j=1}^{e_i} [F^{(i,j)}(t,u)]^{p_{ij}}, \quad i = 1,2,\ldots,a \tag{3.48}$$

In the asymptotic approach to multi-state system reliability analysis we are interested in the limit distributions of a standardized random variable

$$(T(u) - b_n(u))/a_n(u), \quad u = 1,2,\ldots,z$$

where $T(u)$ is the lifetime of the system in the state subset $\{u, u+1,\ldots,z\}$ and

$$a_n(u) > 0, b_n(u) \in (-\infty,\infty), \quad u = 1,2,\ldots,z$$

are some suitably chosen numbers called normalizing constants. And, since

$$P((T(u) - b_n(u))/a_n(u) > t) = P(T(u) > a_n(u)t + b_n(u))$$
$$= \boldsymbol{R}_n(a_n(u)t + b_n(u), u), \quad u = 1,2,\ldots,z$$

where

$$\boldsymbol{R}_n(t,\cdot) = [\boldsymbol{R}_n(t,0), \boldsymbol{R}_n(t,1), \ldots, \boldsymbol{R}_n(t,z)], \quad t \in (-\infty, \infty)$$

is the multi-state reliability function of the system composed of n components, then we assume the following definition.

Definition 3.22
A vector

$$\Re(t,\cdot) = [1, \Re(t,1), \ldots, \Re(t,z)], \quad t \in (-\infty, \infty)$$

is called the limit multi-state reliability function of the system with reliability function $\boldsymbol{R}_n(t,\cdot)$ if there exist normalizing constants $a_n(u) > 0$, $b_n(u) \in (-\infty, \infty)$ such that

$$\lim_{n \to \infty} \boldsymbol{R}_n(a_n(u)t + b_n(u), u) = \Re(t,u) \quad \text{for } t \in C_{\Re(u)}, \quad u = 1, 2, \ldots, z$$

where $C_{\Re(u)}$ is the set of continuity points of $\Re(t,u)$.

Knowing the system limit reliability function allows us, for sufficiently large n, to apply the following approximate formula:

$$\boldsymbol{R}_n(t,\cdot) \cong \Re\left(\frac{t - b_n(u)}{a_n(u)}, \cdot\right)$$

i.e.

$$[1, \boldsymbol{R}_n(t,1), \ldots, \boldsymbol{R}_n(t,z)] \cong \left[1, \Re\left(\frac{t - b_n(1)}{a_n(1)}, 1\right), \ldots, \Re\left(\frac{t - b_n(z)}{a_n(z)}, z\right)\right], \quad t \in (-\infty, \infty) \tag{3.49}$$

4 Reliability of Large Two-State Systems

4.1 Reliability Evaluation of Two-State Series Systems

The investigations of limit reliability functions of homogeneous two-state series systems are based on the following auxiliary theorem.

Lemma 4.1
If

i. $\overline{\mathfrak{R}}(t) = \exp[-\overline{V}(t)]$ is a non-degenerate reliability function,
ii. $\overline{\boldsymbol{R}}_n(t)$ is the reliability function of a homogeneous two-state series system defined by (2.1),
iii. $a_n > 0, \quad b_n \in (-\infty, \infty)$,

then

$$\lim_{n \to \infty} \overline{\boldsymbol{R}}_n(a_n t + b_n) = \overline{\mathfrak{R}}(t) \quad \text{for } t \in C_{\overline{\mathfrak{R}}} \tag{4.1}$$

if and only if

$$\lim_{n \to \infty} nF(a_n t + b_n) = \overline{V}(t) \quad \text{for } t \in C_{\overline{V}} \tag{4.2}$$

Lemma 4.1 is an essential tool in finding limit reliability functions of two-state series systems. Its various proofs may be found in Refs. [7,41,93]. It is also the basis for fixing the class of all possible limit reliability functions of these systems. This class is determined by the following theorem proved in Refs. [7,41,93].

Theorem 4.1
The only non-degenerate limit reliability functions of the homogeneous two-state series system are

$$\overline{\mathfrak{R}}_1(t) = \exp[-(-t)^{-\alpha}] \quad \text{for } t < 0, \quad \overline{\mathfrak{R}}_1(t) = 0 \quad \text{for } t \geq 0, \alpha > 0, \tag{4.3}$$

$$\overline{\mathfrak{R}}_2(t) = 1 \quad \text{for } t < 0, \quad \overline{\mathfrak{R}}_2(t) = \exp[-t^{\alpha}] \quad \text{for } t \geq 0, \alpha > 0, \tag{4.4}$$

$$\overline{\mathfrak{R}}_3(t) = \exp[-\exp[t]] \quad \text{for } t \in (-\infty, \infty). \tag{4.5}$$

Reliability of Large Systems. DOI: http://dx.doi.org/10.1016/B978-0-08-099949-4.00004-0

The next auxiliary theorem is an extension of Lemma 4.1 to non-homogeneous two-state series systems.

Lemma 4.2

If

i. $\overline{\mathfrak{R}}'(t) = \exp[-\overline{V}'(t)]$ is a non-degenerate reliability function,

ii. $\overline{\boldsymbol{R}}'_n(t)$ is the reliability function of a non-homogeneous two-state series system defined by (2.8),

iii. $a_n > 0, \quad b_n \in (-\infty, \infty)$,

then

$$\lim_{n\to\infty} \overline{\boldsymbol{R}}'_n(a_n t + b_n) = \overline{\mathfrak{R}}'(t) \quad \text{for } t \in C_{\overline{\mathfrak{R}}'}$$

if and only if

$$\lim_{n\to\infty} n \sum_{i=1}^{a} q_i F^{(i)}(a_n t + b_n) = \overline{V}'(t) \quad \text{for } t \in C_{\overline{V}'}.$$

The proof of Lemma 4.2 is given in Refs. [65,93]. This lemma is a particular case of Lemma 1 proved in Ref. [69]. In Ref. [69] Lemma 2 is also proved. From the latest lemma, as a particular case, it is possible to derive the next auxiliary theorem that is a more convenient tool than Lemma 4.2 for finding limit reliability functions of non-homogeneous series systems and the starting point for fixing limit reliability functions for these systems.

Lemma 4.3

If

i. $\overline{\mathfrak{R}}'(t) = \exp[-\overline{V}'(t)]$ is a non-degenerate reliability function,

ii. $\overline{\boldsymbol{R}}'_n(t)$ is the reliability function of a non-homogeneous two-state series system defined by (2.8),

iii. $a_n > 0, b_n \in (-\infty, \infty)$,

iv. $F(t)$ is one of the distribution functions $F^{(1)}(t), F^{(2)}(t), \ldots, F^{(a)}(t)$ defined by (2.7) such that

v. $\exists N \forall n > N \; F(a_{nt} + b_n) = 0$ for $t < t_0$ and $F(a_{nt} + b_n) \neq 0$ for $t \geq t_0$, where $t_0 \in (-\infty, \infty)$,

vi.

$$\lim_{n\to\infty} \frac{F^{(i)}(a_n t + b_n)}{F(a_n t + b_n)} \leq 1 \quad \text{for } t \geq t_0, i = 1, 2, \ldots, a$$

and moreover there exists a non-decreasing function

vii.

$$\overline{d}(t) = \begin{cases} 0 & \text{for } t < t_0 \\ \lim\limits_{n\to\infty} \sum\limits_{i=1}^{a} q_i \overline{d}_i(a_n t + b_n) & \text{for } t \geq t_0, \end{cases} \tag{4.6}$$

where

viii.
$$\overline{d}_i(a_nt+b_n)=\frac{F^{(i)}(a_nt+b_n)}{F(a_nt+b_n)}$$

then

$$\lim_{n\to\infty}\boldsymbol{\overline{R}}'_n(a_nt+b_n)=\overline{\mathfrak{R}}'(t)\quad\text{for } t\in C_{\overline{\mathfrak{R}}'}\tag{4.7}$$

if and only if

$$\lim_{n\to\infty}nF(a_nt+b_n)\overline{d}(t)=\overline{V}'(t)\quad\text{for } t\in C_{\overline{V}'}.\tag{4.8}$$

On the basis of Theorem 4.1 and Lemma 4.3 in Refs. [65,93], the class of limit reliability functions for non-homogeneous two-state series systems has been fixed. The members of this class are specified in the following theorem [93].

Theorem 4.2
The only non-degenerate limit reliability functions of the non-homogeneous two-state series system, under the assumptions of Lemma 4.3, are

$$\overline{\mathfrak{R}}'_1(t)=\exp[-\overline{d}(t)(-t)^{-\alpha}]\quad\text{for } t<0,\quad\overline{\mathfrak{R}}'_1(t)=0\quad\text{for } t\geq 0,\alpha>0\tag{4.9}$$

$$\overline{\mathfrak{R}}'_2(t)=1\quad\text{for } t<0,\overline{\mathfrak{R}}'_2(t)=\exp[-\overline{d}(t)t^{\alpha}]\quad\text{for } t\geq 0,\alpha>0\tag{4.10}$$

$$\overline{\mathfrak{R}}'_3(t)=\exp[-\overline{d}(t)\exp[t]]\quad\text{for } t\in(-\infty,\infty)\tag{4.11}$$

where $\overline{d}(t)$ is a non-decreasing function dependent on the reliability functions of particular system components and their fractions in the system are defined by (4.6).

The above theorem is a particular case of Theorem 2 proved in Ref. [69].

Corollary 4.1
If the ith type components of the non-homogeneous two-state series system have Weibull reliability functions

$$R^{(i)}(t)=1\quad\text{for } t<0,R^{(i)}(t)=\exp[-\beta_it^{\alpha_i}]\quad\text{for } t\geq 0,\alpha_i>0,\beta_i>0,i=1,2,\ldots,a,\tag{4.12}$$

and

$$a_n=1/(\beta n)^{1/\alpha},b_n=0\tag{4.13}$$

where

$$\alpha=\min_{1\leq i\leq a}\{\alpha_i\},\quad\beta=\max\{\beta_i:\alpha_i=\alpha\}\tag{4.14}$$

then

$$\overline{\mathfrak{R}}'_2(t) = 1 \quad \text{for } t < 0, \quad \overline{\mathfrak{R}}'_2(t) = 1 - \exp[-\overline{d}(t)t^{\alpha}] \quad \text{for } t \geq 0,$$

where

$$\overline{d}(t) = \sum_{(i:\alpha_i = \alpha)} q_i\beta_i/\beta, \tag{4.15}$$

is its limit reliability function.

Motivation: Since, from (4.13), we have

$$a_n t + b_n = (\beta l_n)^{-1/\alpha} t \to 0^- \quad \text{for } t < 0$$

and

$$a_n t + b_n = (\beta l_n)^{-1/\alpha} t \to 0^+ \quad \text{for } t \geq 0 \text{ as } n \to \infty,$$

then, according to (4.12), for each $i = 1,2,...,a$, we get

$$R^{(i)}(a_n t + b_n) = 1 \quad \text{for } t < 0$$

and

$$R^{(i)}(a_n t + b_n) = \exp[-\beta_i(a_n t)^{\alpha_i}] \quad \text{for } t \geq 0$$

Assuming

$$R(t) = 1 \quad \text{for } t < 0 \text{ and } R(t) = \exp[-\beta t^{\alpha}] \quad \text{for } t \geq 0 \tag{4.16}$$

where α and β are defined by (4.14), for all $i = 1,2,\ldots,a$ and $t \geq t_0 = 0$, we obtain

$$\lim_{n\to\infty} \frac{F^{(i)}(a_n t + b_n)}{F(a_n t + b_n)} = \lim_{n\to\infty} \frac{1 - \exp[-\beta_i(a_n t)^{\alpha_i}]}{1 - \exp[-\beta(a_n t)^{\alpha}]}$$

$$= \lim_{n\to\infty} \frac{\beta_i}{\beta}(a_n t)^{\alpha_i - \alpha} \leq 1$$

The above means that condition (vii) of Lemma 4.3 holds with $t_0 = 0$. And, due to condition (viii) of Lemma 4.3

$$\lim_{n\to\infty} d_i(a_n t + b_n) = \beta_i/\beta$$

for i such that $\alpha_i = \alpha$ and

$$\lim_{n\to\infty} d_i(a_n t + b_n) = 0$$

otherwise.

Thus, from (4.6), $\overline{d}(t)$ is given by (4.15) for $t \geq 0$ and equal to 0 for $t < 0$. Moreover, according to (4.16), we get

$$\overline{V}'(t) = \lim_{n\to\infty} nF(a_n t + b_n)\overline{d}(t) = 0 \quad \text{for } t < 0$$

and

$$\begin{aligned}
\overline{V}'(t) &= \lim_{n\to\infty} nF(a_n t + b_n)\overline{d}(t) \\
&= \lim_{n\to\infty} n[1 - \exp[-\beta(a_n t)^\alpha]]\,\overline{d}(t) \\
&= \lim_{n\to\infty} n\left[\beta(a_n t)^\alpha - o\left(\frac{1}{n}\right)\right]\overline{d}(t) \\
&= t^\alpha \overline{d}(t) \quad \text{for } t \geq 0
\end{aligned}$$

which from Lemma 4.3 completes the proof.

Example 4.1 (A Gas Piping System)

Let us consider a gas piping line composed of $n = 100$ pipe segments of four types linked in series. In the system there are the following:

40 segments with reliability functions

$$R^{(1)}(t) = \exp[-0.025t] \quad \text{for } t \geq 0$$

20 segments with reliability functions

$$R^{(2)}(t) = \exp[-0.020t] \quad \text{for } t \geq 0,$$

10 segments with reliability functions

$$R^{(3)}(t) = \exp[-0.0015t^2] \quad \text{for } t \geq 0,$$

and 30 segments with reliability functions

$$R^{(4)}(t) = \exp[-0.001t^2] \quad \text{for } t \geq 0.$$

According to Definition 2.13 the gas piping is a non-homogeneous series system with parameters

$$n = 100, a = 4,$$
$$q_1 = 0.4, q_2 = 0.2, q_3 = 0.1, q_4 = 0.3$$

and from (2.8) its exact reliability function is given by

$$\begin{aligned}\overline{\boldsymbol{R}}'_{100}(t) &= \prod_{i=1}^{4}[R^{(i)}(t)]^{q_i 100}\\ &= \exp[-t - 0.4t - 0.015t^2 - 0.03t^2] \quad \text{for } t \geq 0\end{aligned}$$

Since

$$\alpha_1 = 1,\ \beta_1 = 0.025,\ \alpha_2 = 1,\ \beta_2 = 0.02,\ \alpha_3 = 2,\ \beta_3 = 0.0015,\ \alpha_4 = 2,\ \beta_4 = 0.001$$

then

$$\alpha = \min\{1, 1, 2, 2\} = 1, \quad \beta = \max\{0.025, 0.02\} = 0.025$$

Assuming normalizing constants

$$a_n = 1/\beta n = 0.4, \qquad b_n = 0$$

and according to (4.15), after determining

$$d(t) = \sum_{(i:\alpha_i = \alpha)} q_i \frac{\beta_i}{\beta} = 0.56$$

from Corollary 4.1 it follows that the gas piping system limit reliability function is

$$\overline{\mathfrak{R}}'_2(t) = \exp[-0.56t] \quad \text{for } t \geq 0$$

Hence, according to (1.1), the exact reliability of the considered system may be approximated by the formula

$$\begin{aligned}\overline{\boldsymbol{R}}'_{100}(t) &\cong \exp[-0.56(t/0.4)]\\ &= \exp[-1.4t] \text{ for } t > 0\end{aligned} \tag{4.17}$$

The mean values of particular system component lifetimes in years are as follows:

$$E(T_1) = 1/0.025 = 40,\ E(T_2) = 1/0.020 = 50,$$
$$E(T_3) = \Gamma(3/2)(0.0015)^{-1/2} \cong 23,\ E(T_4) = \Gamma(3/2)(0.001)^{-1/2} \cong 28.$$

The approximate mean value of the gas piping lifetime and its standard deviation calculated on the base of the formula (4.17) are

$$E(T) \cong 1/1.4 \cong 0.71 \text{ years}, \quad \sigma(T) \cong 1/1.4 \cong 0.71 \text{ years}$$

The behaviour of the exact and approximate reliability functions of the gas piping is illustrated in Table 4.1 and Figure 4.1. Moreover, in Table 4.1, the differences between the values of these functions are also given. These differences testify that the approximation of the system's exact reliability function by its limit reliability function is good enough.

Table 4.1 The Values and Differences Between the Exact and Limit Reliability Functions of the Gas Piping System

t	$\overline{R}'_{100}(t)$	$\overline{\mathfrak{R}}'_2((t-b_n)/a_n)$	$\Delta = \overline{R}'_{100} - \overline{\mathfrak{R}}'_2$
0.00	1.000	1.000	0.000
0.10	0.869	0.869	0.000
0.20	0.754	0.756	− 0.001
0.30	0.654	0.657	− 0.003
0.40	0.567	0.571	− 0.004
0.50	0.491	0.497	− 0.006
0.60	0.425	0.432	− 0.007
0.70	0.367	0.375	− 0.008
0.80	0.317	0.326	− 0.009
0.90	0.274	0.284	− 0.010
1.00	0.236	0.247	− 0.011
1.20	0.175	0.186	− **0.012**
1.40	0.129	0.141	− **0.012**
1.60	0.095	0.106	− **0.012**
1.80	0.070	0.080	− 0.011

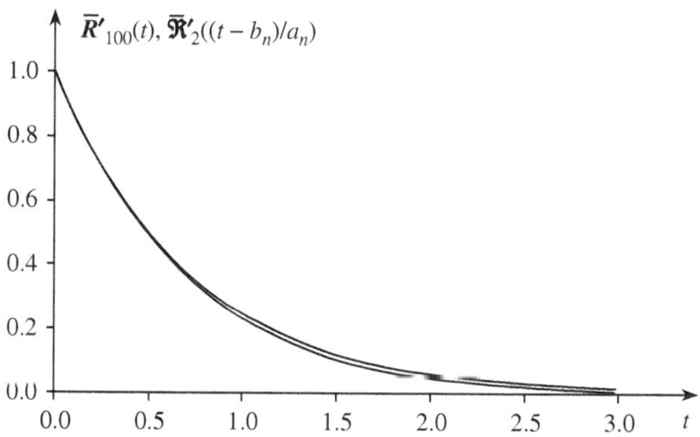

Figure 4.1 The graphs of the exact and limit reliability functions of the gas piping system.

4.2 Reliability Evaluation of Two-State Parallel Systems

The class of limit reliability functions for homogeneous two-state parallel systems may be determined on the basis of the following auxiliary theorem proved, for instance, in Refs. [7,41,93].

Lemma 4.4

If $\overline{\mathfrak{R}}(t)$is the limit reliability function of a homogeneous two-state series system with reliability functions of particular components $\overline{R}(t)$, then

$$\mathfrak{R}(t) = 1 - \overline{\mathfrak{R}}(-t) \quad \text{for } t \in C_{\overline{\mathfrak{R}}}$$

is the limit reliability function of a homogeneous two-state parallel system with reliability functions of particular components

$$R(t) = 1 - \overline{R}(-t) \quad \text{for } t \in C_{\overline{R}}$$

At the same time, if (a_n, b_n) is a pair of normalizing constants in the first case, then $(a_n, -b_n)$ is such a pair in the second case.

Applying the above lemma it is possible to prove an equivalent of Lemma 4.1 that allows us to justify facts on limit reliability functions for homogeneous parallel systems. Its form is as follows [7,65,93].

Lemma 4.5

If

i. $\mathfrak{R}(t) = 1 - \exp[-V(t)]$ is a non-degenerate reliability function,
ii. $\boldsymbol{R}_n(t)$ is the reliability function of a homogeneous two-state parallel system defined by (2.2),
iii. $a_n > 0, b_n \in (-\infty, \infty)$,

then

$$\lim_{n \to \infty} \boldsymbol{R}_n(a_n t + b_n) = \mathfrak{R}(t) \quad \text{for } t \in C_{\mathfrak{R}} \tag{4.18}$$

if and only if

$$\lim_{n \to \infty} nR(a_n t + b_n) = V(t) \quad \text{for } t \in C_V \tag{4.19}$$

By applying Lemma 4.5 and proceeding in an analogous way to the case of homogeneous series systems it is possible to fix the class of limit reliability functions for homogeneous two-state parallel systems. However, it is easier to obtain this result using Lemma 4.4 and Theorem 4.1. Their application immediately results in the following issue.

Theorem 4.3
The only non-degenerate limit reliability functions of the homogeneous parallel system are

$$\mathfrak{R}_1(t) = 1 \quad \text{for } t \leq 0, \quad \mathfrak{R}_1(t) = 1 - \exp[-t^{-\alpha}] \quad \text{for } t > 0, \alpha > 0 \tag{4.20}$$

$$\mathfrak{R}_2(t) = 1 - \exp[-(-t)^{\alpha}] \quad \text{for } t < 0, \quad \mathfrak{R}_2(t) = 0 \quad \text{for } t \geq 0, \alpha > 0 \tag{4.21}$$

$$\mathfrak{R}_3(t) = 1 - \exp[-\exp[-t]] \quad \text{for } t \in (-\infty, \infty) \tag{4.22}$$

Corollary 4.2
If components of the homogeneous two-state parallel system have Weibull reliability functions

$$R(t) = 1 \quad \text{for } t < 0, \quad R(t) = \exp[-\beta t^{\alpha}] \quad \text{for } t \geq 0, \ \alpha > 0, \ \beta > 0$$

and

$$a_n = b_n/(\alpha \log n), \quad b_n = (\log n/\beta)^{1/\alpha}$$

then

$$\mathfrak{R}_3(t) = 1 - \exp[-\exp[-t]], \quad t \in (-\infty, \infty)$$

is its limit reliability function.

Motivation: Since for sufficiently large n and all $t \in (-\infty, \infty)$ we have

$$a_n t + b_n = b_n(t/(\alpha \log n) + 1) > 0$$

then

$$R(a_n t + b_n) = \exp[-\beta(a_n t + b_n)^{\alpha}] \quad \text{for } t \in (-\infty, \infty)$$

Hence

$$\begin{aligned} n\, R(a_n t + b_n) &= n \exp[-\beta(a_n t + b_n)^{\alpha}] \\ &= n \exp[-\beta(b_n)^{\alpha}(t/(\alpha \log n) + 1)^{\alpha}] \\ &= n \exp[-\log n(t/(\alpha \log n) + 1)^{\alpha}] \end{aligned}$$

Further, applying the equality

$$(t/(\alpha \log n) + 1)^{\alpha} = 1 + t/\log n + o(1/\log n) \quad \text{for } t \in (-\infty, \infty)$$

we obtain

$$V(t) = \lim_{n\to\infty} nR(a_n t + b_n)$$
$$= \lim_{n\to\infty} n\exp[-\log n - t - o(1)]$$
$$= \lim_{n\to\infty} \exp[-t - o(1)] = \exp[-t] \quad \text{for } t \in (\infty, \infty)$$

which from Lemma 4.5 completes the proof.

Example 4.2 (An Energetic Cable)

Let us consider an energetic cable composed of 36 wires of the type *AlSi* used in overhead energetic nets and assume that it is able to conduct the current if at least one of its wires is not failed. Under this assumption we may consider the cable as a homogeneous parallel system composed of $n = 36$ basic components. The cross-section of the considered cable is presented in Figure 4.2.

Further, assuming that the cable wires have Weibull reliability functions with parameters

$$\alpha = 2, \quad \beta = (7.07)^{-6}$$

by (2.2), the cable's exact reliability function takes the form

$$\boldsymbol{R}_{36}(t) = 1 \quad \text{for } t < 0, \quad \boldsymbol{R}_{36}(t) = 1 - [1 - \exp[-(7.07)^{-6} t^2]^{36} \quad \text{for } t \geq 0$$

Thus, according to Corollary 4.2, assuming

$$a_n = (7.07)^3 / \left(2\sqrt{\log 36}\right), \quad b_n = (7.07)^3 \sqrt{\log 36}$$

and applying (1.1), we arrive at the approximate formula for the cable reliability function of the form

$$\boldsymbol{R}_{36}(t) \cong \mathfrak{R}_3((t - b_n)/a_n) = 1 - \exp[-\exp[-0.01071t + 7.167]] \quad \text{for } t \in (-\infty, \infty)$$

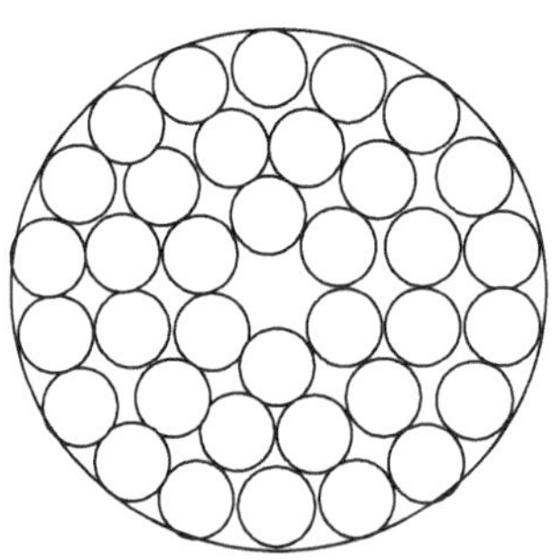

Figure 4.2 The cross-section of the energetic cable.

The expected value of the cable lifetime T and its standard deviation, in months, calculated on the basis of the above approximate result and according to the formulae [15]

$$E[T] = Ca_n + b_n, \quad \sigma = \pi a_n/\sqrt{6}$$

where $C \cong 0.5772$ is Euler's constant, respectively are

$$E[T] \cong 723, \quad \sigma \cong 120$$

The values of the exact and approximate reliability functions of the cable are presented in Table 4.2 and Figure 4.3. Moreover, Table 4.2 also gives the differences

Table 4.2 The Values of the Exact and Approximate Reliability Functions of the Energetic Cable

t	$R_{36}(t)$	$\Re_3\left(\frac{t-b_n}{a_n}\right)$	$\Delta = R_{36} - \Re_3$
0	1.000	1.000	0.000
400	1.000	1.000	0.000
500	0.995	0.988	− 0.003
550	0.965	0.972	− 0.007
600	0.874	0.877	− 0.003
650	0.712	0.707	0.005
700	0.513	0.513	0.000
750	0.330	0.344	− 0.014
800	0.193	0.218	− 0.025
900	0.053	0.081	− **0.028**
1000	0.012	0.029	− 0.017
1100	0.002	0.010	− 0.008
1200	0.000	0.003	− 0.003

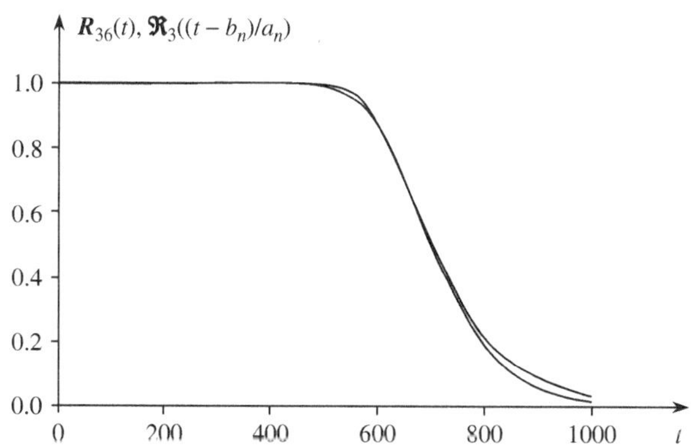

Figure 4.3 The graphs of the exact and approximate reliability functions of the energetic cable.

between those values. The differences are not large, which means that the mistakes in replacing the exact cable reliability function by its approximate form are practically not significant.

The next lemma is a slight modification of Lemma 4.5 proved in Refs. [65,93]. It is also a particular case of Lemma 2, which is proved in Ref. [72].

Lemma 4.6

If $\overline{\mathfrak{R}}'(t)$is the limit reliability function of a non-homogeneous two-state series system with reliability functions of particular components

$$\overline{R}^{(i)}(t), \quad i = 1, 2, \ldots, a$$

then

$$\mathfrak{R}'(t) = 1 - \overline{\mathfrak{R}}'(-t) \quad \text{for } t \in C_{\overline{\mathfrak{R}}'}$$

is the limit reliability function of a non-homogeneous two-state parallel system with reliability functions of particular components

$$R^{(i)}(t) = 1 - \overline{R}^{(i)}(-t) \quad \text{for } t \in C_{\overline{R}^{(i)}}, \quad i = 1, 2, \ldots, a$$

At the same time, if (a_n, b_n) is a pair of normalizing constants in the first case, then $(a_n, -b_n)$ is such a pair in the second case.

Applying the above lemma and Theorem 4.2, it is possible to arrive at the next result [65,72,93].

Lemma 4.7

If

i. $\mathfrak{R}'(t) = 1 - \exp[-V'(t)]$ is a non-degenerate reliability function,
ii. $\boldsymbol{R}'_n(t)$ is the reliability function of a non-homogeneous two-state parallel system defined by (2.10),
iii. $a_n > 0$, $b_n \in (-\infty, \infty)$,

then

$$\lim_{n \to \infty} \boldsymbol{R}'_n(a_n t + b_n) = \mathfrak{R}'(t) \quad \text{for } t \in C_{\mathfrak{R}'}$$

if and only if

$$\lim_{n \to \infty} n \sum_{i=1}^{a} q_i R^{(i)}(a_n t + b_n) = V'(t) \quad \text{for } t \in C_{V'}$$

The next lemma motivated in Refs. [65,93] that is useful in practical applications is a particular case of Lemma 3 proved in Ref. [72].

Lemma 4.8
If

i. $\mathfrak{R}'(t) = 1 - \exp[-V'(t)]$ is a non-degenerate reliability function,
ii. $\boldsymbol{R}'_n(t)$ is the reliability function of a non-homogeneous two-state parallel system defined by (2.10),
iii. $a_n > 0,\ b_n \in (-\infty, \infty)$,
iv. $R(t)$ is one of the reliability functions $R^{(1)}(t), R^{(2)}(t), \ldots, R^{(a)}(t)$ defined by (2.9) such that
v. $\exists N \forall n > N R(a_n t + b_n) \neq 0$ for $t < t_0$ and $R(a_n t + b_n) = 0$ for $t \geq t_0$, where $t_0 \in (-\infty, \infty >$,
vi.

$$\lim_{n\to\infty} \frac{R^{(i)}(a_n t + b_n)}{R(a_n t + b_n)} \leq 1 \quad \text{for } t < t_0, \quad i = 1, 2, \ldots, a$$

and moreover there exists a non-increasing function

vii.

$$d(t) = \begin{cases} \lim\limits_{n\to\infty} \sum\limits_{i=1}^{a} q_i d_i(a_n t + b_n) & \text{for } t < t_0 \\ 0 & \text{for } t \geq t_0 \end{cases} \tag{4.23}$$

where

viii.

$$d_i(a_n t + b_n) = \frac{R^{(i)}(a_n t + b_n)}{R(a_n t + b_n)}$$

then

$$\lim_{n\to\infty} \boldsymbol{R}'_n(a_n t + b_n) = \mathfrak{R}'(t) \quad \text{for } t \in C_{\mathfrak{R}'} \tag{4.24}$$

if and only if

$$\lim_{n\to\infty} nR(a_n t + b_n) d(t) = V'(t) \quad \text{for } t \in C_{V'} \tag{4.25}$$

Starting from this lemma it is possible to fix the class of possible limit reliability for non-homogeneous two-state parallel systems [65,72,93].

Theorem 4.4
The only non-degenerate limit reliability functions of the non-homogeneous two-state parallel system, under the assumptions of Lemma 4.8, are

$$\mathfrak{R}'_1(t) = 1 \quad \text{for } t \leq 0, \quad \mathfrak{R}'_1(t) = 1 - \exp[-d(t) t^{-\alpha}] \quad \text{for } t > 0, \alpha > 0 \tag{4.26}$$

$$\mathfrak{R}'_2(t) = 1 - \exp[-d(t)(-t)^{\alpha}] \quad \text{for } t < 0, \quad \mathfrak{R}'_2(t) = 0 \quad \text{for } t \geq 0, \alpha > 0 \tag{4.27}$$

$$\mathfrak{R}'_3(t) = 1 - \exp[-d(t)\exp[-t]] \quad \text{for } t \in (-\infty, \infty) \tag{4.28}$$

where $d(t)$ is a non-increasing function dependent on the reliability functions of particular system components and their fractions in the system defined by (4.23).

Theorem 4.4 is a particular case of Theorem 1 proved in Ref. [72].

4.3 Reliability Evaluation of Two-State '*m* out of *n*' Systems

The class of limit reliability function for homogeneous two-state 'm out of n' systems may be established by applying the three following auxiliary theorems proved in Refs. [151,93].

Lemma 4.9

If

i. $m = \text{constant}(m/n \to 0 \text{ as } n \to \infty)$

ii. $\mathfrak{R}^{(0)}(t) = 1 - \sum_{i=0}^{m-1} \frac{[V(t)]^i}{i!} \exp[-V(t)]$ is a non-degenerate reliability function,

iii. $\boldsymbol{R}_n^{(m)}(t)$ is the reliability function of a homogeneous two-state 'm out of n' system defined by (2.3),

iv. $a_n > 0$, $b_n \in (-\infty, \infty)$,

then

$$\lim_{n\to\infty} \boldsymbol{R}_n^{(m)}(a_n t + b_n) = \mathfrak{R}^{(0)}(t) \quad \text{for } t \in C_{\mathfrak{R}^{(0)}} \tag{4.29}$$

if and only if

$$\lim_{n\to\infty} nR(a_n t + b_n) = V(t) \quad \text{for } t \in C_V \tag{4.30}$$

Lemma 4.10

If

i. $m/n \to \mu$, $0 < \mu < 1$, as $n \to \infty$,

ii. $\mathfrak{R}^{(\mu)}(t) = 1 - (1/\sqrt{2\pi}) \int_{-\infty}^{-v(t)} e^{-\frac{x^2}{2}} dx$ is a non-degenerate reliability function,

iii. $\boldsymbol{R}_n^{(m)}(t)$ is the reliability function of a homogeneous two-state 'm out of n' system defined by (2.3),

iv. $a_n > 0$, $b_n \in (-\infty, \infty)$,

then

$$\lim_{n\to\infty} \boldsymbol{R}_n^{(m)}(a_n t + b_n) = \mathfrak{R}^{(\mu)}(t) \quad \text{for } t \in C_{\mathfrak{R}^{(\mu)}} \tag{4.31}$$

if and only if

$$\lim_{n\to\infty} \frac{(n+1)R(a_n t + b_n) - m}{\sqrt{(m(n-m+1)/n+1)}} = v(t) \quad \text{for } t \in C_V. \tag{4.32}$$

Lemma 4.11
If

i. $n - m = \overline{m} =$ constant ($m/n \to 1$ as $n \to \infty$),

ii. $\overline{\Re}^{(1)}(t) = \sum_{i=0}^{\overline{m}} \frac{[\overline{V}(t)]^i}{i!} \exp[-\overline{V}(t)]$ is a non-degenerate reliability function,

iii. $\overline{\boldsymbol{R}}_n^{(\overline{m})}(t)$ is the reliability function of a homogeneous two-state 'm out of n' system defined by (2.4),

iv. $a_n > 0$, $b_n \in (-\infty, \infty)$,

then

$$\lim_{n\to\infty} \overline{\boldsymbol{R}}_n^{(\overline{m})}(a_n t + b_n) = \overline{\Re}^{(1)}(t) \quad \text{for } t \in C_{\overline{\Re}^{(1)}}, \tag{4.33}$$

if and only if

$$\lim_{n\to\infty} nF(a_n t + b_n) = \overline{V}(t) \quad \text{for } t \in C_{\overline{V}}. \tag{4.34}$$

The applications of Lemmas 4.9 – 4.11 allow us to establish the class of possible limit reliability functions for homogeneous two-state 'm out of n' systems pointed out in the following theorem [93,151].

Theorem 4.5
The only non-degenerate limit reliability functions of the homogeneous two-state 'm out of n' system are as follows:

Case 1. $m =$ constant ($m/n \to 0$ as $n \to \infty$).

$$\Re_1^{(0)}(t) = 1 \text{ for } t \le 0, \quad \Re_1^{(0)}(t) = 1 - \sum_{i=0}^{m-1} \frac{t^{-i\alpha}}{i!} \exp[-t^{-\alpha}] \text{ for } t > 0, \alpha > 0, \tag{4.35}$$

$$\Re_2^{(0)}(t) = 1 - \sum_{i=0}^{m-1} \frac{(-t)^{i\alpha}}{i!} \exp[-(-t)^{\alpha}] \text{ for } t < 0, \Re_2^{(0)}(t) = 0 \text{ for } t \ge 0, \alpha > 0, \tag{4.36}$$

$$\Re_2^{(0)}(t) = 1 - \sum_{i=0}^{m-1} \frac{\exp[-it]}{i!} \exp[-\exp[-t]] \text{ for } t \in (-\infty, \infty). \tag{4.37}$$

Case 2. $m/n = \mu + o(1/\sqrt{n})$, $0 < \mu < 1$, ($m/n \to \mu$ as $n \to \infty$).

$$\Re_4^{(\mu)}(t) = 1 \text{ for } t < 0, \Re_4^{(\mu)}(t) = 1 - \frac{1}{\sqrt{2\pi}} \int_{-\infty}^{ct^{\alpha}} e^{-\frac{x^2}{2}} dx \text{ for } t \ge 0, c > 0, \alpha > 0 \tag{4.38}$$

$$\mathfrak{R}_5^{(\mu)}(t) = 1 - \frac{1}{\sqrt{2\pi}} \int_{-\infty}^{-c|t|^{\alpha}} e^{-\frac{x^2}{2}} dx \ \text{ for } t < 0, \mathfrak{R}_5^{(\mu)}(t) = 0 \ \text{ for } t \geq 0, c > 0, \alpha > 0 \tag{4.39}$$

$$\mathfrak{R}_6^{(\mu)}(t) = 1 - \frac{1}{\sqrt{2\pi}} \int_{-\infty}^{-c_1|t|^{\alpha}} e^{-\frac{x^2}{2}} dx \ \text{ for } t < 0, c_1 > 0, \alpha > 0 \tag{4.40}$$

$$\mathfrak{R}_6^{(\mu)}(t) = \frac{1}{2} - \frac{1}{\sqrt{2\pi}} \int_0^{c_2 t^{\alpha}} e^{-\frac{x^2}{2}} dx \ \text{ for } t \geq 0, c_2 > 0, \alpha > 0 \tag{4.41}$$

$$\mathfrak{R}_7^{(\mu)}(t) = 1 \ \text{ for } t < -1, \ \ \mathfrak{R}_7^{(\mu)}(t) = \frac{1}{2} \ \text{ for } -1 \leq t < 0, \ \ \mathfrak{R}_7^{(\mu)}(t) = 0 \ \text{ for } t \geq 0 \tag{4.42}$$

Case 3. $n - m = \overline{m} =$ constant ($m/n \to 1$ as $n \to \infty$).

$$\overline{\mathfrak{R}}_8^{(1)}(t) = \sum_{i=0}^{\overline{m}} \frac{(-t)^{-i\alpha}}{i!} \exp[-(-t)^{-\alpha}] \ \text{ for } t < 0, \overline{\mathfrak{R}}_8^{(1)}(t) = 0 \ \text{ for } t \geq 0, \alpha > 0 \tag{4.43}$$

$$\overline{\mathfrak{R}}_9^{(1)}(t) = 1 \ \text{ for } t < 0, \ \ \mathfrak{R}_9^{(1)}(t) = \sum_{i=0\overline{m}} \frac{t^{i\alpha}}{i!} \exp[-t^{\alpha}] \ \text{ for } t \geq 0, \alpha > 0 \tag{4.44}$$

$$\overline{\mathfrak{R}}_9^{(1)}(t) = \sum_{i=0}^{\overline{m}} \frac{\exp[it]}{i!} \exp[-\exp[t]] \ \text{ for } t \in (-\infty, \infty). \tag{4.45}$$

Corollary 4.3

If components of the homogeneous two-state 'm out of n' system have exponential reliability functions

$$R(t) = 1 \quad \text{for } t < 0, \quad R(t) = \exp[-\lambda t] \quad \text{for } t \geq 0, \lambda > 0,$$

m tends to infinity in such a way that

$$m/n \to \mu, 0 < \mu < 1, \text{ as } n \to \infty,$$

and

$$a_n = \frac{1}{\lambda}\sqrt{\frac{n-m+1}{(n+1)m}}, \quad b_n = \frac{1}{\lambda}\log\frac{n+1}{m},$$

then

$$\Re_6^{(\mu)}(t) = 1 - \frac{1}{\sqrt{2\pi}} \int_{-\infty}^{t} e^{-\frac{x^2}{2}}\, dx, t \in (-\infty, \infty)$$

is its limit reliability function.

Motivation: Since for sufficiently large n and all $t \in (-\infty, \infty)$ we have

$$a_n t + b_n = \frac{t}{\lambda}\sqrt{\frac{n-m+1}{(n+1)m}} + \frac{1}{\lambda}\log\frac{n+1}{m} > 0,$$

then for sufficiently large n

$$\begin{aligned} R(a_n t + b_n) &= \exp[-\lambda(a_n t + b_n)] \\ &= \exp\left[-t\sqrt{\frac{n-m+1}{(n+1)m}} - \log\frac{n+1}{m}\right] \\ &= \left[1 - t\sqrt{\frac{n-m+1}{(n+1)m}} + o\left(\frac{1}{\sqrt{n}}\right)\right]\frac{m}{n+1} \quad \text{for } t \in (-\infty, \infty) \end{aligned}$$

Hence

$$\begin{aligned} v(t) &= \lim_{n\to\infty} \frac{(n+1)R(a_n t + b_n) - m}{\sqrt{m(n-m+1)/n+1}} \\ &= \lim_{n\to\infty}\left[-t + o\left(\frac{1}{\sqrt{n}}\right)\sqrt{\frac{(n+1)m}{n-m+1}}\right] \\ &= -t \quad \text{for } t \in (-\infty, \infty), \end{aligned}$$

which from Lemma 4.10 completes the proof.

Example 4.3 (A Lighting System)

We consider a lighting system composed of $n = 35$ identical lighting points that is not failed if at least $m = 16$ of the points are not failed. Assuming that the lighting points have exponential reliability functions with the failure rate $\lambda = 1$/year, from (2.3), the exact system reliability function is given by

$$\boldsymbol{R}_{35}^{(16)}(t) = 1 \text{ for } t < 0, \quad \boldsymbol{R}_{35}^{(16)}(t) = 1 - \sum_{i=0}^{15}\binom{35}{i}\exp[-it][1 - \exp[-t]]^{35-i} \text{ for } t \geq 0,$$

Table 4.3 The Values of the Exact and Approximate Reliability Function of the Lighting System

t	$\Re_6^{(\mu)}\left(\frac{t-b_n}{a_n}\right)$	$R_{35}^{(16)}(t)$	$\Delta = R_{35}^{(16)} - \Re_6^{(\mu)}$
0.0	0.99999	1.00000	0.00001
0.1	0.99994	1.00000	0.00006
0.2	0.99948	1.00000	0.00052
0.3	0.99695	0.99980	0.00285
0.4	0.98629	0.99507	0.00878
0.5	0.95241	0.96253	0.01012
0.6	0.87118	0.86178	− 0.00940
0.7	0.72419	0.68205	− 0.04214
0.8	0.52339	0.46713	**− 0.05626**
0.9	0.31633	0.27703	− 0.03929
1.0	0.15513	0.14394	− 0.01119
1.1	0.06041	0.06653	0.00612
1.2	0.01840	0.02777	0.00937
1.3	0.00433	0.01062	0.00628
1.4	0.00078	0.00376	0.00298

According to Corollary 4.3, assuming

$$a_n = \frac{\sqrt{5}}{12} \cong 0.1863, \quad b_n = \log\frac{9}{4} \cong 0.8109,$$

after applying (1.1), we get the approximate formula for the reliability function of the lighting system in the form

$$R_{35}^{(16)}(t) \cong \Re_6^{(\mu)}((t-b_n)/a_n)$$

$$= 1 - \frac{1}{\sqrt{2\pi}}\int_{-\infty}^{5.37t-4.35} e^{-\frac{x^2}{2}}\,dx \quad \text{for } t\in(-\infty,\infty).$$

The mean value of the lighting system lifetime T and its standard deviation, in years, calculated from the above formula are

$$E[T] \cong 0.811, \quad \sigma = 0.186$$

The values of the exact and approximate reliability functions of the lighting system are presented in Table 4.3 and their graphs in Figure 4.4. The differences between the exact and approximate reliability functions of the system given in Table 4.3 testify good accuracy of the approximation.

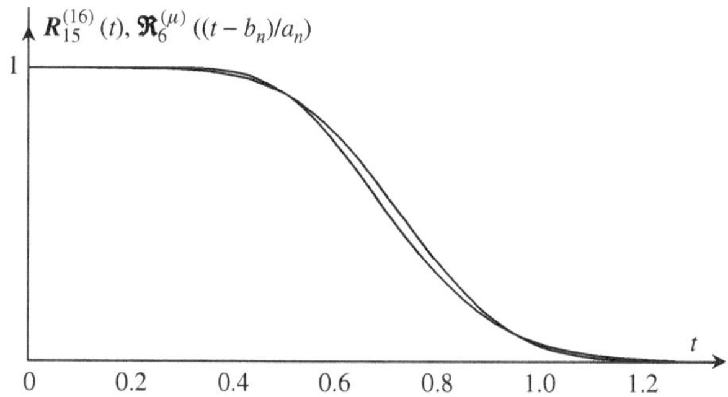

Figure 4.4 The graphs of the exact and approximate reliability function of the lighting system.

Unfortunately, there are no generalizations of Lemmas 4.9 − 4.11 for the non-homogeneous two-state 'm out of n' systems. Each particular case of a non-homogeneous two-state 'm out of n' system has to be considered separately and a suitable auxiliary theorem and corollary have to be formulated and proved and then applied to the reliability evaluation of a real system.

4.4 Reliability Evaluation of Two-State Series–Parallel Systems

Prior to the formulation of the overall results for the classes of limit reliability functions for two-state regular series–parallel systems, we introduce some assumptions for all cases of the considered systems shapes. These assumptions distinguish all possible relationships between the number of their series subsystems k_n and the number of components l_n in these subsystems. In the assumptions for two-state regular parallel–series systems, considered in the next section, k_n is the number of parallel subsystems and l_n is the number of components in these subsystems.

Assumption 4.1
Here are considered the relationships between k_n and l_n of the form

$$k_n = n, \ \ l_n = c(\log n)^{\rho(n)}, \ \ n \in (0, \infty), \ \ c > 0, \tag{4.46}$$

with the following cases distinguished:

Case 1. $k_n = n$, $\left| l_n - c \log n \right| \gg s$, $s > 0$, $c > 0$.

1^0 $l_n \ll c \log n$

$$|\rho(\tau_\nu) - \rho(n)| \underset{\approx}{\ll} \frac{\delta \cdot \log \nu}{\log n \cdot [\log(\log n)]} \quad \text{for all natural } \nu > 1,$$

where $0 < \delta \neq 1$ and $\tau_\nu / n = \nu^{1/1-\rho(n)}$,

2^0 $l_n \approx c \log n$ and $\left|l_n - c \log n\right| \gg \log(\log n)$,

$$|\rho(\tau_\nu - \rho(n))| \underset{\approx}{\ll} \frac{\delta \cdot \log \nu}{\log n \cdot [\log(\log n)]} \quad \text{for all natural } \nu > 1,$$

where $0 < \delta \neq 1$ and $\tau_\nu / n = \nu^{1/1-\rho(n)}$,

3^0 $s \ll \left|l_n - c \log n\right| \underset{\approx}{\ll} C \log(\log n), \quad s > 0, C > 0,$

$$|\rho(\tau_\nu - \rho(n)| \approx \frac{\delta \cdot \log \nu}{\log n \cdot [\log(\log n)]} \quad \text{for all natural } \nu > 1$$

where $\delta > 0$ and $\tau_\nu / n = \nu^{1/(1-\rho(n))\cdot\log(\log n)}$,

4^0 $l_n \gg c \log n$ and $\rho(n) \ll (\log n)^\lambda$ for all $\lambda > 0$,

$$|\rho(\tau_\nu) - \rho(n)| \underset{\approx}{\ll} \frac{\delta \cdot \log \nu}{\log n \cdot [\log(\log n)]} \quad \text{for all natural } \nu > 1,$$

where $0 < \delta \neq 1$ and $\tau_\nu / n = \nu^{1/1-\rho(n)}$,

5^0 $\rho(n) \underset{\approx}{\gg} (\log n)^\lambda$, $\lambda > 0$,

$$|\rho(\tau_\nu) - \rho(n)| \underset{\approx}{\ll} \lim_{n \to \infty} \quad \text{for all natural } \nu > 1,$$

where $\delta > 0$ and $\tau_\nu / n = \nu^{1/(1-\rho(n))A(t)}$,

$$A(n) \approx \prod_{i=1}^{\nu} f_i(\rho(n)),$$

where $f_i(n)$ for $i = 1,2,\ldots,\nu$ is the ith superposition of a function $\log n$, and ν is such that

$$f_{\nu+1}(\rho(n)) \ll \log(\log n).$$

Case 2. $k_n = n$, $l_n - c \log n \approx s$, $s \in (-\infty, \infty)$, $c > 0$.
Case 3. $k_n \to k$, $k > 0$, $l_n \to \infty$.

The proofs of the theorems on limit reliability functions for homogeneous regular series–parallel systems and methods of finding such functions for individual systems are based on the following essential lemma.

Lemma 4.12
If

i. $k_n \to \infty$,
ii. $\Re(t) = 1 - \exp[-V(t)]$ is a non-degenerate reliability function,
iii. $\boldsymbol{R}_{k_n,l_n}(t)$ is the reliability function of a homogeneous regular two-state series–parallel system defined by (2.5),
iv. $a_n > 0$, $b_n \in (-\infty, \infty)$,

then

$$\lim_{n\to\infty} \boldsymbol{R}_{k_n,l_n}(a_n t + b_n) = \Re(t) \quad \text{for } t \in C_{\Re}, \tag{4.47}$$

if and only if

$$\lim_{n\to\infty} k_n[R(a_n t + b_n)]^{l_n} = V(t) \quad \text{for } t \in C_V. \tag{4.48}$$

The proof of Lemma 4.12 is given in Refs. [53,56,71].

The justification of the next auxiliary theorem that follows from Lemma 4.12 may be found in Refs. [65,93].

Lemma 4.13
If

i. $k_n \to k$, $k > 0$, $l_n \to \infty$,
ii. $\Re(t)$ is a non-degenerate reliability function,
iii. $\boldsymbol{R}_{k_n,l_n}(t)$ is the reliability function of a homogeneous regular two-state series–parallel system defined by (2.5),
iv. $a_n > 0$, $b_n \in (-\infty, \infty)$,

then

$$\lim_{n\to\infty} \boldsymbol{R}_{k_n,l_n}(a_n t + b_n) = \Re(t) \quad \text{for } t \in C_{\Re}, \tag{4.49}$$

if and only if

$$\lim_{n\to\infty} [R(a_n t + b_n)]^{l_n} = \Re_0(t) \quad \text{for } t \in C_{\Re_0}, \tag{4.50}$$

where $\Re_0(t)$ is a non-degenerate reliability function and moreover

$$\Re(t) = 1 - [1 - \Re_0(t)]^k \quad \text{for } t \in (-\infty, \infty). \tag{4.51}$$

The results achieved in Refs. [62–64,68] and based on Lemma 4.12 and 4.13 may be formulated in the form of the following theorem [63,66,72].

Theorem 4.6
The only non-degenerate limit reliability functions of the homogeneous regular two-state series–parallel system are as follows:

Case 1. $k_n = n$, $|l_n - c \log n| \gg s$, $s > 0$, $c > 0$ (under Assumption 4.1).

$$\mathfrak{R}_1(t) = 1 \quad \text{for } t \le 0, \quad \mathfrak{R}_1(t) = 1 - \exp[-t^{-\alpha}] \quad \text{for } t > 0, \alpha > 0, \tag{4.52}$$

$$\mathfrak{R}_2(t) = 1 - \exp[-(-t)^{\alpha}] \quad \text{for } t < 0, \quad \mathfrak{R}_2(t) = 0 \quad \text{for } t \ge 0, \alpha > 0, \tag{4.53}$$

$$\mathfrak{R}_3(t) = 1 - \exp[-\exp[-t]] \quad \text{for } t \in (-\infty, \infty), \tag{4.54}$$

Case 2. $k_n = n$, $l_n - c \log n \approx s$, $s \in (-\infty, \infty)$, $c > 0$.

$$\mathfrak{R}_4(t) = 1 \quad \text{for } t < 0, \quad \mathfrak{R}(t) = 1 - \exp[-\exp[-t^{\alpha} - s/c]] \quad \text{for } t \ge 0, \alpha > 0, \tag{4.55}$$

$$\mathfrak{R}_5(t) = 1 - \exp[-\exp[(-t)^{\alpha} - s/c]] \quad \text{for } t < 0, \quad \mathfrak{R}_5(t) = 0 \quad \text{for } t \ge 0, \alpha > 0, \tag{4.56}$$

$$\mathfrak{R}_6(t) = 1 - \exp[-\exp[\beta(-t)^{\alpha} - s/c]] \quad \text{for } t < 0,$$

$$\mathfrak{R}_6(t) = 1 - \exp[-\exp[-t^{\alpha} - s/c]] \quad \text{for } t \ge 0, \alpha > 0, \beta > 0, \tag{4.57}$$

$$\mathfrak{R}_7(t) = 1 \quad \text{for } t < t_1, \quad \mathfrak{R}_7(t) = 1 - \exp[-\exp[-s/c]] \quad \text{for } t_1 \le t < t_2,$$

$$\mathfrak{R}_7(t) = 0 \quad \text{for } t \ge t_2,\ t_1 < t_2, \tag{4.58}$$

Case 3. $k_n \to k$, $k > 0$, $l_n \to \infty$.

$$\mathfrak{R}_8(t) = 1 - [1 - \exp[-(-t)^{-\alpha}]]^k \quad \text{for } t < 0, \quad \mathfrak{R}_8(t) = 0 \quad \text{for } t \ge 0, \alpha > 0, \tag{4.59}$$

$$\mathfrak{R}_9(t) = 1 \quad \text{for } t < 0, \quad \mathfrak{R}_9(t) = 1 - [1 - \exp[-t^{\alpha}]]^k \quad \text{for } t \ge 0, \alpha > 0, \tag{4.60}$$

$$\mathfrak{R}_{10}(t) = 1 - [1 - \exp[-\exp t]]^k \quad \text{for } t \in (-\infty, \infty). \tag{4.61}$$

Corollary 4.4
If components of the homogeneous regular two-state series–parallel system have Weibull reliability functions

$$R(t) = 1 \quad \text{for } t < 0,\ R(t) = \exp[-\beta t^{\alpha}] \quad \text{for } t \ge 0,\ \alpha > 0,\ \beta > 0, \tag{4.62}$$

and

$$k_n \to k, \quad l_n > 0, \tag{4.63}$$

$$a_n = 1/(\beta l_n)^{1/\alpha}, \quad b_n = 0, \tag{4.64}$$

then

$$\mathfrak{R}_9(t) = 1 \quad \text{for } t < 0, \quad \mathfrak{R}_9(t) = 1 - [1 - \exp[-t^\alpha]]^k \quad \text{for } t \geq 0,$$

is its limit reliability function.

Motivation: Since, according to (4.63) and (4.64), we have

$$a_n t + b_n = (\beta l_n)^{-1/\alpha}\, t < 0 \quad \text{for } t < 0$$

and

$$a_n t + b_n = (\beta l_n)^{-1/\alpha}\, t \geq 0 \quad \text{for } t \geq 0$$

then from (4.62) the equalities

$$R(a_n t + b_n) = 1 \quad \text{for } t < 0$$

and

$$R(a_n t + b_n) = \exp[-\beta(a_n t + b_n)^\alpha] = \exp[-t^\alpha / l_n] \quad \text{for } t \geq 0$$

are satisfied.

Further, we have

$$\lim_{n \to \infty} [R(a_n t + b_n)]^{l_n} = 1 \quad \text{for } t < 0$$

and

$$\lim_{n \to \infty} [R(a_n t + b_n)]^{l_n} = \exp[-t^\alpha] \quad \text{for } t \geq 0.$$

Thus, from Lemma 4.13, $\mathfrak{R}_9(t)$ is the limit reliability function of the system.

Example 4.4 ***(a gas distribution system)***
The gas distribution system consists of $k_n = 2$ piping lines, each of them composed of $l_n = 1000$ identical pipe segments having Weibull reliability functions

$$R(t) = \exp[-0.0002 t^3] \quad \text{for } t \geq 0.$$

The system is a homogeneous regular series–parallel system and according to (2.5) its exact reliability function is given by

$$\boldsymbol{R}_{2,1000}(t) = 1 - [1-\exp[-0.2t^3]]^2 \quad \text{for } t \geq 0.$$

Assuming, according to (4.64), the normalizing constants

$$a_n = (0.0002 \cdot 1000)^{-1/3} = 1.71, \quad b_n = 0,$$

from Corollary 4.4, we conclude that the limit reliability function of the system is given by the formula

$$\mathfrak{R}_9(t) = 1 - [1-\exp[-t^3]]^2 \quad \text{for } t \geq 0.$$

Thus, according to (1.1), the approximate formula (it is easy to check that it is exact in this case) takes the form

$$\boldsymbol{R}_{2,1000}(t) \cong \mathfrak{R}_9((t-b_n)/a_n) = 1 - [1-\exp[-0.2t^3]]^2 \quad \text{for } t \geq 0.$$

The expected values of the lifetimes of pipe segments are [93]

$$E(T_i) = \Gamma(4/3)(0.0002)^{-1/3} \cong 15.3 \text{ years},$$

while the mean value of the system lifetime is [93]

$$E(T) = 2\Gamma(4/3)(0.2)^{-1/3} - \Gamma(4/3)(0.04)^{-1/3} \cong 2.4 \text{ years}.$$

The behaviour of the reliability function of the gas distribution system is illustrated in Table 4.4 and Figure 4.5.

The proofs of the facts concerned with limit reliability functions of non-homogeneous two-state series–parallel systems are based on the following auxiliary theorems formulated and proved in Refs. [65,69,72,93].

Lemma 4.14
If

i. $k_n \to \infty$,
ii. $\mathfrak{R}'(t) = 1 - \exp[-V'(t)]$ is a non-degenerate reliability function,
iii. $\boldsymbol{R}'_{k_n,l_n}(t)$ is the reliability function of a non-homogeneous regular two-state series–parallel system defined by (2.14) and (2.15),
iv. $a_n > 0$, $b_n \in (-\infty, \infty)$,

then

$$\lim_{n\to\infty} \boldsymbol{R}'_{k_n,l_n}(a_n t + b_n) = \mathfrak{R}'(t) \quad \text{for } t \in C_{\mathfrak{R}'}$$

Table 4.4 The Behaviour of the Exact and Approximate Reliability Function of the Gas Distribution System

t	$R_{2,1000}(t) = \Re_{9}((t - b_n)/a_n)$
0.0	1.0000
0.2	1.0000
0.4	0.9998
0.6	0.9982
0.8	0.9905
1.0	0.9671
1.2	0.9146
1.4	0.8216
1.6	0.6873
1.8	0.5259
2.0	0.3630
2.2	0.2236
2.4	0.1220

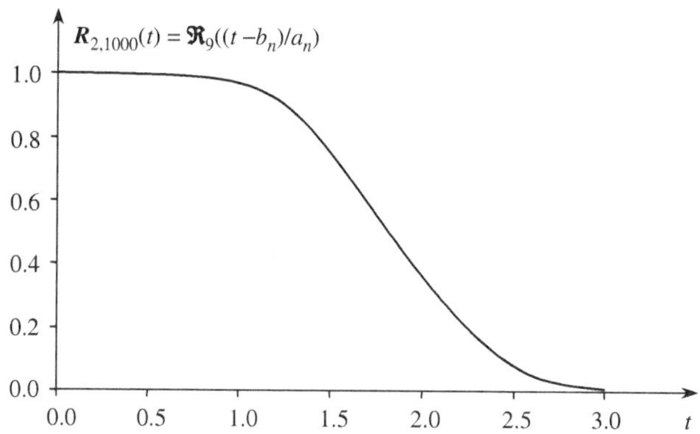

Figure 4.5 The graphs of the reliability function of the gas distribution system.

if and only if

$$\lim_{n\to\infty} k_n \sum_{i=1}^{a} q_i[R^{(i)}(a_n t + b_n)]^{l_n} = V(t) \quad \text{for } t \in C_{V'}.$$

Lemma 4.15
If

i. $k_n \to \infty$,
ii. $\Re'(t) = 1 - \exp[-V'(t)]$ is a non-degenerate reliability function,

iii. $\boldsymbol{R}'_{k_n,l_n}(t)$ is the reliability function of a non-homogeneous regular two-state series–parallel system defined by (2.14) and (2.15),

iv. $a_n > 0$, $b_n \in (-\infty, \infty)$,

v. $R(t)$ is one of the reliability functions $R^{(1)}(t), R^{(2)}(t), \ldots, R^{(a)}(t)$ defined by (2.15) such that

vi. $\exists N \forall n > N R(a_n t + b_n) \neq 0$ for $t < t_0$ and $R(a_n t + b_n) = 0$ for $t \geq t_0$, where $t_0 \in (-\infty, \infty)$,

$$\lim_{n\to\infty} \frac{R^{(i)}(a_n t + b_n)}{R(a_n t + b_n)} \leq 1 \quad \text{for } t < t_0, i = 1, 2, \ldots, a$$

and moreover there exists a non-increasing function

vii.

$$d(t) = \begin{cases} \lim\limits_{n\to\infty} \sum\limits_{i=1}^{a} q_i d_i(a_n t + b_n) & \text{for } t < t_0 \\ 0 & \text{for } t \geq t_0, \end{cases} \tag{4.65}$$

where

$$d_i(a_n t + b_n) = \left[\frac{R^{(i)}(a_n t + b_n)}{R(a_n t + b_n)}\right]^{l_n},$$

then

$$\lim_{n\to\infty} \boldsymbol{R}'_{k_n,l_n}(a_n t + b_n) = \mathfrak{R}'(t) \quad \text{for } t \in C_{\mathfrak{R}'} \tag{4.66}$$

if and only if

$$\lim_{n\to\infty} k_n [R(a_n t + b_n)]^{l_n} d(t) = V'(t) \quad \text{for } t \in C_{V'}. \tag{4.67}$$

Lemma 4.16

If

i. $k_n \to k$, $k > 0$, $l_n \to \infty$,

ii. $\mathfrak{R}'(t)$ is a non-degenerate reliability function,

iii. $\boldsymbol{R}'_{k_n,l_n}(t)$ is the reliability function of a non-homogeneous regular two-state series–parallel system defined by (2.14) and (2.15),

iv. $a_n > 0$, $b_n \in (-\infty, \infty)$,

v. $R(t)$ is one of reliability functions $R^{(1)}(t), R^{(2)}(t), \ldots, R^{(a)}(t)$ defined by (2.15) such that

vi. $\exists N \forall n > N R(a_n t + b_n) \neq 0$ for $t < t_0$ and $R(a_n t + b_n) = 0$ for $t \geq t_0$, where $t_0 \in (-\infty, \infty >$,

vii.

$$\lim_{n\to\infty} \frac{R^{(i)}(a_n t + b_n)}{R(a_n t + b_n)} \leq 1 \quad \text{for } t < t_0, i = 1, 2, \ldots, a$$

and moreover there exist non-increasing functions

viii.

$$d_i(t) = \begin{cases} \lim_{n\to\infty} d_i(a_n t + b_n) & \text{for } t < t_0 \\ 0 & \text{for } t \geq t_0, \end{cases} \tag{4.68}$$

where

$$d_i(a_n t + b_n) = \left[\frac{R^{(i)}(a_n t + b_n)}{R(a_n t + b_n)}\right]^{l_n},$$

then

$$\lim_{n\to\infty} \boldsymbol{R}'_{k_n,l_n}(a_n t + b_n) = \mathfrak{R}'(t) \quad \text{for } t \in C_{\mathfrak{R}'} \tag{4.69}$$

if and only if

$$\lim_{n\to\infty} [R(a_n t + b_n)]^{l_n} = \mathfrak{R}_0(t) \quad \text{for } t \in C_{\mathfrak{R}_0} \tag{4.70}$$

where $\mathfrak{R}_0(t)$ is a non-degenerate reliability function and moreover

$$\mathfrak{R}'(t) = 1 - \prod_{i=1}^{a} [1 - d_i(t)\, \mathfrak{R}_0(t)]^{q_i k}, t \in (-\infty, \infty). \tag{4.71}$$

Theorem 4.6, Lemma 4.15 and 4.16 determine the class of limit reliability functions for non-homogeneous regular series–parallel systems whose members are pointed out in the following theorem [65,66,69,93].

Theorem 4.7
The only non-degenerate limit reliability functions of the non-homogeneous regular two-state series–parallel system are:

Case 1. $k_n = n$, $|l_n - c \log n| \gg s$, $s > 0$, $c > 0$ (under Assumption 4.1 and the assumptions of Lemma 4.15).

$$\mathfrak{R}'_1(t) = 1 \quad \text{for } t \leq 0, \quad \mathfrak{R}'_1(t) = 1 - \exp[-d(t) t^{-\alpha}] \quad \text{for } t > 0, \alpha > 0, \tag{4.72}$$

$$\mathfrak{R}'_2(t) = 1 - \exp[-d(t)(-t)^{\alpha}] \quad \text{for } t < 0, \quad \mathfrak{R}'_2(t) = 0 \quad \text{for } t \geq 0, \alpha > 0, \tag{4.73}$$

$$\mathfrak{R}'_3(t) = 1 - \exp[-d(t) \exp[-t]] \quad \text{for } t \in (-\infty, \infty), \tag{4.74}$$

Case 2. $k_n = n$, $l_n - c \log n \approx s$, $s \in (-\infty, \infty)$, $c > 0$ (under the assumptions of Lemma 4.15).

$$\mathfrak{R}'_4(t) = 1 \text{ for } t < 0, \quad \mathfrak{R}'_4(t) = 1 - \exp[-d(t) \exp[-t^{\alpha} - s/c]] \text{ for } t \geq 0, \alpha > 0, \tag{4.75}$$

$$\mathfrak{R}'_5(t) = 1 - \exp[-d(t)\exp[(-t)^{\alpha} - s/c]] \quad \text{for } t < 0,$$

$$\mathfrak{R}'_5(t) = 0 \quad \text{for } t \geq 0, \alpha > 0, \tag{4.76}$$

$$\mathfrak{R}'_6(t) = 1 - \exp[-d(t)\exp[\beta(-t)^{\alpha} - s/c]] \quad \text{for } t < 0,$$

$$\mathfrak{R}'_6(t) = 1 - \exp[-d(t)\exp[-t^{\alpha} - s/c]] \quad \text{for } t \geq 0, \alpha > 0, \beta > 0, \tag{4.77}$$

$$\mathfrak{R}'_7(t) = 1 \quad \text{for } t < t_1, \quad \mathfrak{R}'_7(t) = 1 - \exp[-d(t)\exp[-s/c]] \quad \text{for } t_1 \leq t \leq t_2,$$

$$\mathfrak{R}'_7(t) = 0 \quad \text{for } t \geq t_2, t_1 < t_2, \tag{4.78}$$

Case 3. $k_n \to k$, $k > 0$, $l_n \to \infty$ (under the assumptions of Lemma 4.16).

$$\mathfrak{R}'_8(t) = 1 - \prod_{i=1}^{a} [1 - d_i(t)\exp[-(-t)^{-\alpha}]]^{q_i k} \quad \text{for } t < 0,$$

$$\mathfrak{R}'_8(t) = 0 \quad \text{for } t \geq 0, \alpha > 0, \tag{4.79}$$

$$\mathfrak{R}'_9(t) = 1 \quad \text{for } t < 0, \quad \mathfrak{R}'_9(t) = 1 - \prod_{i=1}^{a} [1 - d_i(t)\exp[-t^{\alpha}]]^{q_i k} \quad \text{for } t \geq 0, \alpha > 0 \tag{4.80}$$

$$\mathfrak{R}'_{10}(t) = 1 - \prod_{i=1}^{a} [1 - d_i(t)\exp[-\exp t]]^{q_i k} \quad \text{for } t \in (-\infty, \infty) \tag{4.81}$$

where $d(t)$ and $d_i(t)$ are non-increasing functions dependent on the reliability functions of the system's particular components and their fractions in the system defined by (4.65) and (4.68), respectively.

Corollary 4.5

If components of the non-homogeneous regular two-state series–parallel system have Weibull reliability functions

$$R^{(i,j)}(t) = 1 \quad \text{for } t < 0, R^{(i,j)}(t) = \exp[-\beta_{ij} t^{\alpha_{ij}}]$$
$$\text{for } t \geq 0, \alpha_{ij} > 0, \beta_{ij} > 0, i = 1, 2, \ldots, a, j = 1, 2, \ldots, e_i \tag{4.82}$$

and

$$k_n \to k, l_n \to \infty \tag{4.83}$$

$$a_n = 1/(\beta l_n)^{-1/\alpha}, \quad b_n = 0 \tag{4.84}$$

where

$$\alpha_i = \min_{1 \le j \le e_i} \{\alpha_{ij}\}, \quad \beta_i = \sum_{(j:\alpha_{ij} = a_i)} p_{ij}\beta_{ij}, \tag{4.85}$$

$$\alpha = \max_{1 \le i \le a} \{\alpha_i\}, \quad \beta = \min\{\beta_i : \alpha_i = \alpha\}, \tag{4.86}$$

then

$$\mathfrak{R}'_9(t) = 1 \quad \text{for } t < 0, \quad \mathfrak{R}'_9(t) = 1 - \prod_{(i:\alpha_i = \alpha)} \left[1 - \exp\left[-\frac{\beta_i}{\beta} t^\alpha\right]\right]^{q_i k} \quad \text{for } t \ge 0$$

is its limit reliability function.

Motivation: Since, according to (4.83) and (4.84), we have

$$a_n t + b_n = (\beta l_n)^{-1/\alpha} t \to 0^- \quad \text{for } t < 0$$

and

$$a_n t + b_n = (\beta l_n)^{-1/\alpha} t \to 0^+ \quad \text{for } t \ge 0 \text{ as } n \to \infty,$$

then for every $i = 1,2,\ldots,a$, we get

$$R^{(i)}(a_n t + b_n) = 1 \quad \text{for } t < 0$$

and applying (2.15), (4.82) and (4.84), we obtain

$$\begin{aligned} R^{(i)}(a_n t + b_n) &= \exp\left[-\sum_{j=1}^{e_i} p_{ij}\beta_{ij}(a_n t)^{\alpha_{ij}}\right] \\ &= \exp\left[-(a_n t)^{\alpha_i} \sum_{j=1}^{e_i} p_{ij}\beta_{ij}(a_n t)^{\alpha_{ij} - \alpha_i}\right] \\ &= \exp[-\beta_i (a_n t)^{\alpha_i} + o(1)] \quad \text{for } t \ge 0. \end{aligned}$$

Letting

$$R(t) = 1 \quad \text{for } t < 0 \text{ and } R(t) = \exp[-\beta t^\alpha] \quad \text{for } t \ge 0,$$

where α and β are defined by (4.86), for all $i = 1,2,\ldots,a$, we have

$$\lim_{n \to \infty} \frac{R^{(i)}(a_n t + b_n)}{R(a_n t + b_n)} = 1 \quad \text{for } t < 0$$

and

$$\lim_{n\to\infty}\frac{R^{(i)}(a_nt+b_n)}{R(a_nt+b_n)}=\lim_{n\to\infty}\frac{\exp[-\beta_i(a_nt+b_n)^{\alpha_i}]}{\exp[-\beta(a_nt+b_n)^{\alpha}]}$$

$$=\lim_{n\to\infty}\exp\left[-\beta(a_nt)^{\alpha}\left[\frac{\beta_i}{\beta}(a_nt)^{\alpha_i-\alpha}-1\right]\right]\le 1 \quad \text{for } t\ge 0,$$

The above means that condition (vii) of Lemma 4.16 holds with $t_0=\infty$. Further, according to (4.68), we get

$$d_i(t)=\begin{cases}1 & \text{for } t<0\\ \exp\left[-\left(\frac{\beta_i}{\beta}-1\right)t^{\alpha}\right] & \text{for } t\ge 0\end{cases}$$

for i such that $\alpha_{i=}\alpha$ and

$$d_i(t)=\begin{cases}1 & \text{for } t<0\\ 0 & \text{for } t\ge 0\end{cases}$$

otherwise.

Moreover, we have

$$\lim_{n\to\infty}[R(a_nt+b_n)]^{l_n}=1 \quad \text{for } t<0$$

and

$$\lim_{n\to\infty}[R(a_nt+b_n)]^{l_n}=\lim_{n\to\infty}\exp[-l_n\beta(a_nt)^{\alpha}]$$

$$=\exp[-t^{\alpha}] \quad \text{for } t\ge 0,$$

which from Lemma 4.16 completes the proof.

Example 4.5 (*a water supply system*)

The water supply system consists of $k_n=3$ lines of segment pipes. Each line is composed of $l_n=100$ segments. The scheme of the considered system is shown in Figure 4.6.

In two of the lines there are 40 segment pipes with a reliability function

$$R^{(1,1)}(t)=\exp[-0.05t] \quad \text{for } t\ge 0,$$

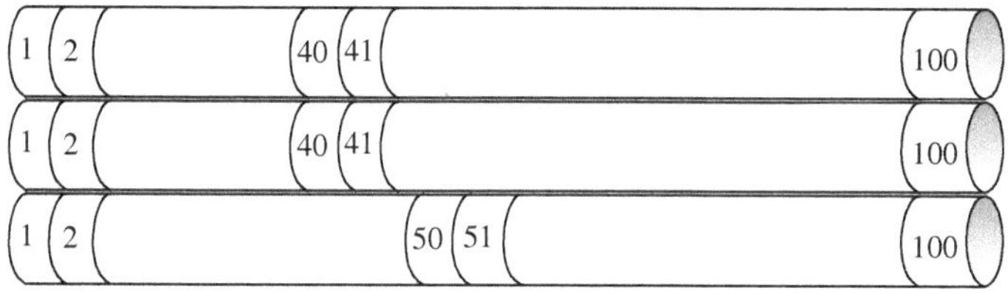

Figure 4.6 The model of a non-homogeneous regular series–parallel water supply system.

and 60 segment pipes with a reliability function

$$R^{(1,2)}(t) = \exp[-0.0015t^2] \quad \text{for } t \geq 0,$$

In the third line there are 50 segment pipes with a reliability function

$$R^{(2,1)}(t) = \exp[-0.0007t^3] \quad \text{for } t \geq 0,$$

and 50 segment pipes with a reliability function

$$R^{(2,2)}(t) = \exp[-0.2\sqrt{t}] \quad \text{for } t \geq 0,$$

Thus, according to Definition 2.16, the water supply system is a non-homogeneous regular series–parallel system with the following parameters:

$$k_n = k = 3, l_n = 100, a = 2, q_1 = 2/3, q_2 = 1/3,$$

Considering (2.14), we have

$$\begin{aligned}\boldsymbol{R}'_{3,100}(\mathrm{t}) &= 1 - \prod_{i=1}^{2}[1-(R^{(i)}(t))^{100}]^{q_i 3} \\ &= 1 - [1-(R^{(1)}(t))^{100}]^2[1-(R^{(2)}(t))^{100}],\end{aligned}$$

where after considering the substitutions

$$\begin{aligned}&e_1 = 1, \;\; p_{11} = 0.4, \;\; p_{12} = 0.6, \\ &\alpha_{11} = 1, \;\; \beta_{11} = 0.05, \;\; \alpha_{12} = 2, \;\; \beta_{12} = 0.0015, \\ &e_2 = 2, \;\; p_{21} = 0.5, \;\; p_{22} = 0.5, \\ &\alpha_{21} = 3, \;\; \beta_{21} = 0.0007, \;\; \alpha_{22} = 0.5, \;\; \beta_{22} = 0.2,\end{aligned}$$

and the formula (2.15)

$$\begin{aligned}R^{(1)}(t) &= \prod_{j=1}^{e_1}(R^{(1,j)}(t))^{p_{1j}} \\ &= (R^{(1,1)}(t))^{0.4}(R^{(1,2)}(t))^{0.6} = \exp[-0.02t - 0.0009t^2]\end{aligned}$$

and

$$R^{(2)}(t) = \prod_{j=1}^{e_2} (R^{(2,j)}(t))^{p_{2j}}$$
$$= (R^{(2,1)}(t))^{0.5} (R^{(2,2)}(t))^{0.5} = \exp\left[-0.00035t^3 - 0.1\sqrt{t}\right].$$

From the above it follows that the exact reliability function of the system is given by

$$\boldsymbol{R}'_{3,100}(t) = 1 - [1 - \exp[-2t - 0.09t^2]]^2 \cdot \left[1 - \exp\left[-0.035t^3 - 10\sqrt{t}\right]\right] \quad \text{for } t \geq 0,$$

Further, according to (4.84)–(4.86), we have

$$\alpha_1 = \min\{\alpha_{11}, \alpha_{12}\} = \min\{1, 2] = 1,$$
$$\beta_1 = p_{11}\beta_{11} = 0.4 \cdot 0.05 = 0.02,$$
$$\alpha_2 = \min\{\alpha_{21}, \alpha_{22}\} = \min\{3, 0.5\} = 0.5,$$
$$\beta_2 = p_{22}\beta_{22} = 0.5 \cdot 0.5 = 0.1,$$
$$\alpha = \max\{\alpha_1, \alpha_2\} = \max\{1, 0.5\} = 1,$$
$$\beta = \min\{\beta_1\} = \min\{0.02\} = 0.02,$$
$$a_n = (0.02 \cdot 100)^{-1} = 0.5, b_n = 0,$$

and from Corollary 4.5 the limit reliability function of the system is given by

$$\mathfrak{R}'_9(t) = 1 - [1 - \exp[-t]]^2 \quad \text{for } t \geq 0.$$

Hence, after considering (1.1), the reliability function of the system is approximately given by

$$\boldsymbol{R}'_{3,100}(t) \cong \mathfrak{R}'_9((t - b_n)/a_n) = 1 - [1 - \exp[-2t]]^2 \quad \text{for } t \geq 0. \tag{4.87}$$

The reliability data of the system components have come from experts. According to their opinions the mean lifetimes of the pipe segments, depending on their types, vary in a range from 10 up to 50 years and are as follows [93]:

$$E(T_{11}) = 1/0.05 = 20, \;\; E(T_{12}) = \Gamma(3/2)(0.0015)^{-1/2} \cong 23,$$
$$E(T_{21}) = \Gamma(4/3)(0.0007)^{-1/3} \cong 10, \;\; E(T_{22}) = \Gamma(3)(0.2)^{-2} \cong 50.$$

The water supply system lifetime and its standard deviation calculated on the basis of the approximate formula (4.87) are

$$E(T) \cong 0.75 \text{ years}, \;\; \sigma(T) \cong 0.56 \text{ years}.$$

The values of the exact and approximate reliability functions of the system and the differences between them are presented in Table 4.5 and Figure 4.7.

Table 4.5 The Behaviour of the Exact and Approximate Reliability Functions of the Water Supply System

t	$R'_{3,100}(t)$	$\mathfrak{R}'_9\left(\frac{t-b_n}{a_n}\right)$	$\Delta = R'_{3,100} - \mathfrak{R}'_9$
0.0	1.0000	1.0000	0.0000
0.2	0.8910	0.8913	− 0.0003
0.4	0.6902	0.6968	− 0.0066
0.6	0.4984	0.5117	− 0.0133
0.8	0.3449	0.3630	− 0.0181
1.0	0.2321	0.2524	**− 0.0202**
1.2	0.1530	0.1732	**− 0.0202**
1.4	0.0994	0.1179	− 0.0186
1.6	0.0637	0.0799	− 0.0162
1.8	0.0404	0.0539	− 0.0135
2.0	0.0254	0.0363	− 0.0109
2.2	0.0158	0.0244	− 0.0086
2.4	0.0098	0.0164	− 0.0066

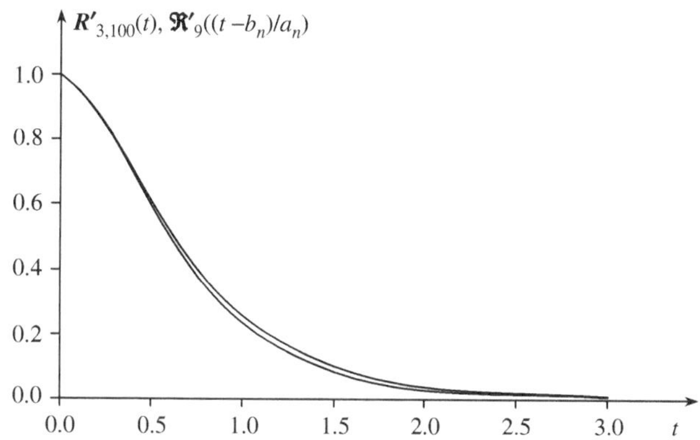

Figure 4.7 The graphs of the exact and approximate reliability functions of the water supply system.

4.5 Reliability Evaluation of Two-State Parallel–Series Systems

The class of limit reliability functions for homogeneous regular two-state parallel–series systems is successively fixed in Refs. [62 – 65,68] on the basis of the following lemmas.

Lemma 4.17

If $\mathfrak{R}(t)$ is the limit reliability function of a homogeneous regular two-state series–parallel system composed of components with a reliability function $R(t)$, then

$$\overline{\mathfrak{R}}(t) = 1 - \mathfrak{R}(-t) \quad \text{for } t \in C_{\mathfrak{R}}$$

is the limit reliability function of a homogeneous regular two-state parallel–series system composed of components with a reliability function

$$\overline{R}(t) = 1 - R(-t) \quad \text{for } t \in C_R.$$

At the same time, if (a_n, b_n) is the pair of normalizing constants in the first case, then $(a_n, -b_n)$ is such a pair in the second case.

Lemma 4.18

If

i. $k_n \to \infty$,
ii. $\overline{\mathfrak{R}}(t) = \exp[-\overline{V}(t)]$ is a non-degenerate reliability function,
iii. $\overline{\boldsymbol{R}}_{k_n l_n}(t)$ is the reliability function of a homogeneous regular two-state parallel–series system defined by (2.6),
iv. $a_n > 0$, $b_n \in (-\infty, \infty)$,

then

$$\lim_{n \to \infty} \overline{\boldsymbol{R}}_{k_n l_n}(a_n t + b_n) = \overline{\mathfrak{R}}(t) \quad \text{for } t \in C_{\overline{\mathfrak{R}}} \tag{4.88}$$

if and only if

$$\lim_{n \to \infty} k_n [F(a_n t + b_n)]^{l_n} = \overline{V}(t) \quad \text{for } t \in C_{\overline{V}}. \tag{4.89}$$

Lemma 4.19

If

i. $k_n \to k$, $k > 0$, $l_n \to \infty$,
ii. $\overline{\mathfrak{R}}(t)$ is a non-degenerate reliability function,
iii. $\overline{\boldsymbol{R}}_{k_n l_n}(t)$ is the reliability function of a homogeneous regular two-state parallel–series system defined by (2.6),
iv. $a_n > 0$, $b_n \in (-\infty, \infty)$,

then

$$\lim_{n \to \infty} \overline{\boldsymbol{R}}_{k_n l_n}(a_n t + b_n) = \overline{\mathfrak{R}}(t) \quad \text{for } t \in C_{\overline{\mathfrak{R}}} \tag{4.90}$$

if and only if

$$\lim_{n\to\infty}[F(a_nt+b_n)]^{l_n}=\mathfrak{I}_0(t) \quad \text{for } t\in C_{\mathfrak{I}_0} \tag{4.91}$$

where $\mathfrak{I}_0$(t) is a non-degenerate distribution function and moreover

$$\overline{\mathfrak{R}}(t)=[1-\mathfrak{I}_0(t)]^k \quad \text{for } t\in(-\infty,\infty). \tag{4.92}$$

By applying Lemma 4.18 and 4.19 and proceeding in the same way as in the case of homogeneous regular series–parallel systems, it is possible to fix the class of limit reliability functions for homogeneous regular parallel–series systems. This class is presented in Refs. [72,74] as the successive results given in Refs. [71,73,77,79]. However, it is much easier to obtain this result by applying Lemma 4.17 and Theorem 4.6. Their direct application immediately results in the following theorem [63,65,93].

Theorem 4.8

The only non-degenerate limit reliability functions of the homogeneous regular two-state parallel–series system are as follows:

Case 1. $k_n=n$, $|l_n-c\log n|\gg s$, $s>0$, $c>0$ (under Assumption 4.1).

$$\overline{\mathfrak{R}}_1(t)=\exp[-(-t)^{-\alpha}] \quad \text{for } t<0, \quad \overline{\mathfrak{R}}_1(t)=0, \quad \text{for } t\geq 0, \alpha>0 \tag{4.93}$$

$$\overline{\mathfrak{R}}_2(t)=1 \quad \text{for } t<0, \quad \overline{\mathfrak{R}}_2(t)=\exp[-t^{\alpha}], \quad \text{for } t\geq 0, \alpha>0 \tag{4.94}$$

$$\overline{\mathfrak{R}}_3(t)=\exp[-\exp[t]] \quad \text{for } t\in(-\infty,\infty) \tag{4.95}$$

Case 2. $k_n=n$, $l_n-c\log n\approx s$, $s\in(-\infty,\infty)$, $c>0$;

$$\overline{\mathfrak{R}}_4(t)=\exp[-\exp[-(-t)^{\alpha}-s/c]] \quad \text{for } t<0, \quad \overline{\mathfrak{R}}_4(t)=0 \quad \text{for } t\geq 0, \alpha>0 \tag{4.96}$$

$$\overline{\mathfrak{R}}_5(t)=1 \quad \text{for } t<0, \quad \overline{\mathfrak{R}}_5(t)=\exp[-\exp[t^{\alpha}-s/c]] \quad \text{for } t\geq 0, \alpha>0 \tag{4.97}$$

$$\overline{\mathfrak{R}}_6(t)=\exp[-\exp[-(-t)^{\alpha}-s/c]] \quad \text{for } t<0$$

$$\overline{\mathfrak{R}}_6(t)=\exp[-\exp[\beta t^{\alpha}-s/c]] \quad \text{for } t\geq 0, \alpha>0, \beta>0 \tag{4.98}$$

$$\overline{\mathfrak{R}}_7(t)=1 \quad \text{for } t<t_1, \quad \overline{\mathfrak{R}}_7(t)=\exp[-\exp[-s/c]] \quad \text{for } t_1\leq t\leq t_2$$

$$\overline{\mathfrak{R}}_7(t)=0 \quad \text{for } t\geq t_2, t_1<t_2 \tag{4.99}$$

Case 3. $k_n \to k, k > 0, l_n \to \infty$.

$$\overline{\mathfrak{R}}_8(t) = 1 \quad \text{for } t \leq 0, \quad \overline{\mathfrak{R}}_8(t) = [1 - \exp[-t^{-\alpha}]]^k \quad \text{for } t > 0, \alpha > 0 \tag{4.100}$$

$$\overline{\mathfrak{R}}_9(t) = [1 - \exp[-(-t)^{\alpha}]]^k \quad \text{for } t < 0, \quad \overline{\mathfrak{R}}_9(t) = 0 \quad \text{for } t \geq 0, \alpha > 0 \tag{4.101}$$

$$\overline{\mathfrak{R}}_{10}(t) = [1 - \exp[-\exp[-t]]]^k \quad \text{for } t(-\infty, \infty) \tag{4.102}$$

Corollary 4.6
If components of the homogeneous regular two-state parallel–series system have Weibull reliability functions

$$R(t) = 1 \quad \text{for } t < 0, R(t) = \exp[-\beta_t^{\alpha}] \quad \text{for } t \geq 0, \alpha > 0, \beta > 0$$

and

$$k_n = n, l_n - c \log n \gg s, c > 0, s > 0$$

$$a_n = b_n/(\alpha\beta(b_n)^{\alpha} \log n), b_n = [(1/\beta)\log(l_n/\log n)]^{1/\alpha}$$

then

$$\overline{\mathfrak{R}}_3(t) = \exp[-\exp[t]], \quad t \in (-\infty, \infty]$$

is its limit reliability function.

Motivation: Since for sufficiently large n and all $t \in (-\infty, \infty)$, we have

$$a_n t + b_n > 0$$

and

$$a_n/b_n \to 0 \text{ as } n \to \infty$$

then

$$\begin{aligned} F(a_n t + b_n) &= 1 - \exp[-\beta(a_n t + b_n)^{\alpha}] \\ &= 1 - \exp[-\beta(b_n)^{\alpha}(1 + (a_n/b_n)t)^{\alpha}] \\ &= 1 - \exp[[-\beta(b_n)^{\alpha}(1 + \alpha(a_n/b_n t) + o(a_n/b_n)]. \end{aligned}$$

Moreover

$$\alpha\beta(b_n)^{\alpha} a_n/b_n = 1/\log n \to \text{as } n \to \infty,$$

and therefore

$$F(a_n t + b_n) = 1 - \exp[-\log(l_n/\log n) - t/\log n + o(1/\log n)]$$
$$= 1 - (\log n)/l_n(1 - t/\log n + o(1/\log n)$$
$$= 1 - (\log n)/l_n + t/l_n - o(1/l_n) \quad \text{for } t \in (-\infty, \infty)$$

and

$$\overline{V}(t) = \lim_{n\to\infty} k_n[F(a_n t + b_n)]^{l_n}$$
$$= \lim_{n\to\infty} n[1-(\log n)/l_n + t/l_n - o(1/l_n)]^{l_n}$$
$$- \lim_{n\to\infty} \exp[t - l_n o(1/l_n)] = \exp[t] \quad \text{for } t \in (-\infty, \infty)$$

which from Lemma 4.18 completes the proof.

Example 4.6 (A Model Parallel–Series System)
If the shape of a homogeneous regular parallel–series system is such that

$$k_n = 30, \quad l_n = 60$$

and its components have Weibull reliability functions with parameters

$$\beta = 1/100, \quad \alpha = 1$$

then, according to (2.6), its exact reliability function is given by

$$\overline{\boldsymbol{R}}_{30,60}(t) = 1 \quad \text{for } t < 0, \quad \overline{\boldsymbol{R}}_{30,60}(t) = [1-[1-\exp[-0.01t]]^{60}]^{30} \quad \text{for } t \geq 0.$$

Applying Corollary 4.6 with the normalizing constants

$$R_{ij}(t,) = 1 \quad \text{for } t < 0, \quad R_{ij}(t,) = \exp[-\lambda t], \quad \text{for } t \geq 0, \lambda > 0$$

from (1.1), we get the following approximate expression for the reliability function of the system

$$\overline{\boldsymbol{R}}_{30,60}(t) \cong \overline{\mathfrak{R}}_3((t - b_n)/a_n) = \exp[-\exp[0.034t - 9.76]] \quad \text{for } t \in (-\infty, \infty). \tag{4.103}$$

The mean values of the system component lifetimes T_i are here

$$E[T_i] = 1/\beta = 100h,$$

while the expected value and the standard deviation of the system lifetime calculated from (4.103) are [15,93] are

$$E[T] \cong -0.5772a_n + b_n \cong 270h, \quad \sigma(T) = \pi a_n/\sqrt{6} = 38h.$$

The behaviour of the exact and approximate reliability functions of the considered system is illustrated in Table 4.6 and Figure 4.8.

The generalizations of Lemmas 4.18 − 4.19 are the next lemmas proved in Refs. [65,69,70,72], giving the way of finding limit reliability functions for non-homogeneous regular parallel−series systems.

Table 4.6 The Values of the Exact and Approximate Reliability Functions of the Model Parallel−Series System

t	$\overline{R}_{30,60}(t)$	$\overline{\mathfrak{R}}_3((t-b_n)/a_n)$	$\Delta = \overline{R}_{30,60} - \overline{\mathfrak{R}}_3$
0	1.0000	1.0000	0.0000
100	1.0000	0.9983	0.0017
150	1.0000	0.9906	0.0084
200	0.9961	0.9495	0.0456
220	0.9742	0.9028	0.0714
240	0.9049	0.8172	0.0877
260	0.7453	0.6713	0.0742
280	0.4947	0.3973	0.0974
300	0.2382	0.2117	0.0265
320	0.0760	0.0467	0.0293
340	0.0151	0.0024	0.0127
360	0.0018	0.0000	0.0018

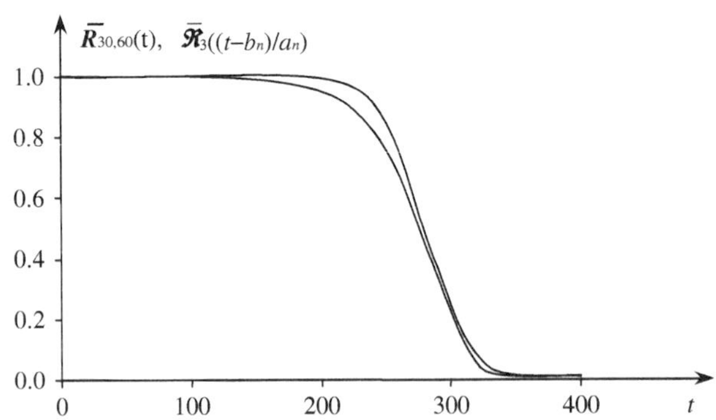

Figure 4.8 The graphs of the exact and approximate reliability functions of the model parallel−series system.

Lemma 4.20

If $\overline{\Re}'(t)$ is the limit reliability function of a non-homogeneous regular two-state series–parallel system composed of components with reliability functions

$$R^{(i,j)}(t), \quad i=1,2,\ldots,a,\ j=1,2,\ldots,e_i$$

then

$$\overline{\Re}'(t)=1-\Re'(-t) \quad \text{for } t \in C_{\Re'}$$

is the limit reliability function of a non-homogeneous regular two-state parallel–series system composed of components with reliability functions

$$\overline{R}^{(i,j)}(t)=1-R^{(i,j)}(-t) \quad \text{for } t \in C_{R^{(i,j)}}, \quad i=1,2,\ldots,a,\ j=1,2,\ldots,e_i$$

At the same time, if (a_n, b_n) is the pair of normalizing constants in the first case, then $(a_n, -b_n)$ is such a pair in the second case.

Lemma 4.21

If

- **i.** $\overline{\Re}'(t)=\exp[-\overline{V}'(t)]$ is a non-degenerate reliability function,
- **ii.** $\overline{\boldsymbol{R}}'_{k_n l_n}(t)$ is the reliability function of a non-homogeneous regular two-state parallel–series system defined by (2.16) and (2.17),
- **iii.** $a_n>0,\ b_n \in (-\infty,\infty)$,

then

$$\lim_{n\to\infty} \overline{\boldsymbol{R}}'_{k_n l_n}(a_n t+b_n)=\overline{\Re}'(t) \quad \text{for } t \in C_{\overline{\Re}'}$$

if and only if

$$\lim_{n\to\infty} k_n \sum_{i=1}^{a} q_i[F^{(i)}(a_n t+b_n)]^{l_n}=\overline{V}'(t) \quad \text{for } t \in C_{\overline{V}'}.$$

Lemma 4.22

If

- **i.** $k_n \to \infty$,
- **ii.** $\overline{\Re}'(t)=\exp[-\overline{V}'(t)]$ is a non-degenerate reliability function,
- **iii.** $\overline{\boldsymbol{R}}'_{k_n l_n}(t)$ is the reliability function of a non-homogeneous regular two-state parallel–series system defined by (2.16) and (2.17),
- **iv.** $a_n>0,\ b_n \in (-\infty,\infty)$,
- **v.** $F(t)$ is one of the distribution functions $F^{(1)}(t), F^{(2)}(t),\ldots, F^{(a)}(t)$ defined by (2.17) such that

vi. $\exists N \forall n > N \; F(a_n t + b_n) = 0$ for $t < t_0$ and $F(a_n t + b_n) \neq 0$ for $t \geq t_0$, where $t_0 \in < -\infty, \infty)$,

vii.

$$\lim_{n\to\infty} \frac{F^{(i)}(a_n t + b_n)}{F(a_n t + b_n)} \leq 1 \quad \text{for } t \geq t_0,\ i = 1, 2, \ldots, a$$

and moreover there exists a non-decreasing function

viii.

$$\bar{d}(t) = \begin{cases} 0 & \text{for } t < t_o \\ \lim\limits_{n\to\infty} \sum\limits_{i=1}^{a} q_i \bar{d}_i(a_n t + b_n) & \text{for } t \geq t_o \end{cases} \tag{4.104}$$

where

$$\bar{d}_i(a_n t + b_n) = \left[\frac{F^{(i)}(a_n t + b_n)}{F(a_n t + b_n)}\right]^{l_n}$$

then

$$\lim_{n\to\infty} \boldsymbol{R}'_{k_n l_n}(a_n t + b_n) = \overline{\mathfrak{R}}'(t) \quad \text{for } t \in C_{\mathfrak{R}'} \tag{4.105}$$

if and only if

$$\lim_{n\to\infty} k_n [F(a_n t + b_n)]^{l_n} \bar{d}(t) = \overline{V}'(t) \quad \text{for } t \in C_{\overline{V}'}. \tag{4.106}$$

Lemma 4.23

If

i. $k_n \to k$, $k > 0$, $l_n \to \infty$,

ii. $\overline{\mathfrak{R}}'(t)$ is a non-degenerate reliability function,

iii. $\overline{\boldsymbol{R}}_{k_n l_n}(t)$ is the reliability function of a non-homogeneous regular two-state parallel–series system defined by (2.16) and (2.17),

iv. $a_n > 0$, $b_n \in (-\infty, \infty)$,

v. $F(t)$ is one of the distribution functions $F^{(1)}(t), F^{(2)}(t), \ldots, F^{(a)}(t)$ defined by (2.17) such that

vi. $\exists N \forall n > N \; F(a_n t + b_n) = 0$ for $t < t_0$ and $F(a_n t + b_n) \neq 0$ for $t \geq t_0$, where $t_0 \in (-\infty, \infty)$,

vii.

$$\lim_{n\to\infty} \frac{F^{(i)}(a_n t + b_n)}{F(a_n t + b_n)} \leq 1 \quad \text{for } t > t_0,\ i = 1, 2, \ldots, a$$

and moreover there exist non-decreasing functions

viii.

$$\bar{d}_i(t) = \begin{cases} 0 & \text{for } t < t_0 \\ \lim\limits_{n\to\infty} \bar{d}_i(a_n t + b_n) & \text{for } t \geq t_0 \end{cases} \tag{4.107}$$

where

$$\bar{d}_i(a_nt + b_n) = \left[\frac{F^{(i)}(a_nt+b_n)}{F(a_nt+b_n)}\right]^{l_n}$$

then

$$\lim_{n\to\infty} \bar{\boldsymbol{R}}'_{k_nl_n}(a_nt + b_n) = \bar{\mathfrak{R}}'(t) \quad \text{for } t \in C_{\bar{\mathfrak{R}}'} \tag{4.108}$$

if and only if

$$\lim_{n\to\infty} [F(a_nt+b_n)]^{l_n} = \mathfrak{I}_0(t) \quad \text{for } t \in C_{\mathfrak{I}_0} \tag{4.109}$$

where $\mathfrak{I}_0(t)$ is a non-degenerate distribution function and moreover

$$\bar{\mathfrak{R}}'(t) = \prod_{i=1}^{a} [1 - \bar{d}_i(t)\mathfrak{I}_0(t)]^{q_ik}, \quad t \in (-\infty, \infty) \tag{4.110}$$

Theorem 4.8, Lemmas 4.22 and 4.23 determine the class of limit reliability functions for non-homogeneous regular two-state parallel–series system pointed out in the following theorem [65,69,93].

Theorem 4.9
The only non-degenerate limit reliability functions of the non-homogeneous regular two-state parallel–series system are as follows:

Case 1. $k_n = n$, $|l_n - c \log n| \gg s$, $s > 0$, $c > 0$ (under Assumption 4.1 and the assumptions of Lemma 4.22).

$$\bar{\mathfrak{R}}'_1(t) = \exp[-\bar{d}(t)(-t)^{-\alpha}] \quad \text{for } t < 0, \quad \bar{\mathfrak{R}}'_1(t) = 0 \quad \text{for } t \geq 0,\ \alpha > 0 \tag{4.111}$$

$$\bar{\mathfrak{R}}'_2(t) = 1 \quad \text{for } t < 0, \quad \bar{\mathfrak{R}}'_2(t) = \exp[-\bar{d}(t)t^{\alpha}] \quad \text{for } t \geq 0,\ \alpha > 0 \tag{4.112}$$

$$\bar{\mathfrak{R}}'_3(t) = \exp[-\bar{d}(t)\exp[t]] \quad \text{for } t \in (-\infty, \infty) \tag{4.113}$$

Case 2. $k_n = n$, $l_n - c \log n \approx s$, $s \in (-\infty, \infty)$, $c > 0$ (under the assumptions of Lemma 4.22).

$$\bar{\mathfrak{R}}'_4(t) = \exp[-\bar{d}(t)\exp[-(-t)^{\alpha} - s/c]] \quad \text{for } t < 0$$

$$\bar{\mathfrak{R}}'_4(t) = 0 \quad \text{for } t \geq 0,\ \alpha > 0 \tag{4.114}$$

$$\bar{\mathfrak{R}}'_5(t) = 1 \quad \text{for } t < 0, \quad \bar{\mathfrak{R}}'_5(t) = \exp[-\bar{d}(t)\exp[t^{\alpha} - s/c]] \quad \text{for } t \geq 0,\ \alpha > 0 \tag{4.115}$$

$$\overline{\mathfrak{R}}'_6(t) = \exp[-\overline{d}(t)\exp[-(-t)^{\alpha} - s/c]] \quad \text{for } t<0$$

$$\overline{\mathfrak{R}}'_6(t) = \exp[-\overline{d}(t)\exp[\beta t^{\alpha} - s/c]] \quad \text{for } t\geq 0,\ \alpha>0,\ \beta>0 \tag{4.116}$$

$$\overline{\mathfrak{R}}'_7(t) = 1 \quad \text{for } t<t_1, \quad \overline{\mathfrak{R}}'_7(t) = \exp[-\overline{d}(t)\exp[-s/c]] \quad \text{for } t_1 \leq t \leq t_2$$

$$\overline{\mathfrak{R}}'_7(t) = 0 \quad \text{for } t\geq t_2,\ t_1<t_2 \tag{4.117}$$

Case 3. $k_n \to k,\ k>0,\ l_n \to \infty$ (under the assumptions of Lemma 4.23).

$$\overline{\mathfrak{R}}'_8(t) = 1 \quad \text{for } t\leq 0, \quad \overline{\mathfrak{R}}'_8(t) = \prod_{i=1}^{a}[1-\overline{d}_i(t)\exp[-t^{-\alpha}]]^{q_i k} \quad \text{for } t>0,\ \alpha>0 \tag{4.118}$$

$$\overline{\mathfrak{R}}'_9(t) = \prod_{i=1}^{a}[1-\overline{d}_i(t)\exp[-(-t)^{\alpha}]]^{q_i k} \quad \text{for } t<0$$

$$\overline{\mathfrak{R}}'_9(t) = 0 \quad \text{for } t\geq 0,\ \alpha>0 \tag{4.119}$$

$$\overline{\mathfrak{R}}'_{10}(t) = \prod_{i=1}^{a}[1-\overline{d}_i(t)\exp[-\exp(-t)]]^{q_i k} \quad \text{for } t\in(-\infty, \infty) \tag{4.120}$$

where $\overline{d}(t)$ and $\overline{d}_i(t)$ are non-decreasing functions dependent on the reliability functions of particular system components and their fractions in the system defined by (4.104) and (4.107), respectively.

4.6 Reliability Evaluation of Other Two-State Systems

The procedure of the reliability evaluation of more complicated large systems whose structures are compositions of large two-state systems considered in this chapter's previous sections depends on applying and modifying suitably selected lemmas and corollaries presented in Sections 4.1–4.5. This procedure is illustrated in this section to the exact and approximate reliability evaluation of the two-state non-homogeneous system composed of a large series–parallel system and a large series- 'm out of k' system linked in series and the two-state homogeneous system composed of three large 'm out of l' systems linked in series.

Example 4.7
We consider an exemplary system S that consists of two subsystems S_1, S_2 with the reliability structure presented in Figure 4.9.

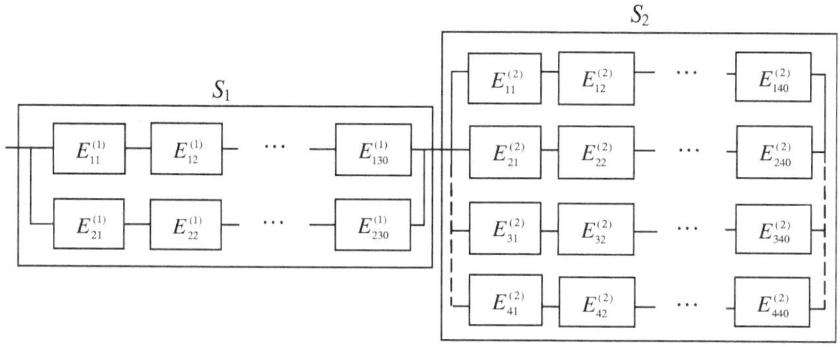

Figure 4.9 The scheme of the exemplary system S reliability structure.

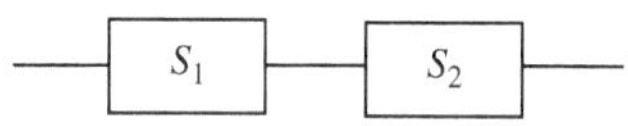

Figure 4.10 The general scheme of the exemplary system S reliability structure.

The subsystem S_1 is composed of two series subsystems, each of them composed of 30 components, denoted respectively by

$$E_{ij}^{(1)}, \quad i = 1, 2, \quad j = 1, 2, \ldots, 30.$$

The subsystem S_2 is composed of four series subsystems, each of them composed of 40 components, denoted respectively by

$$E_{ij}^{(2)}, \quad i = 1, 2, 3, 4, \quad j = 1, 2, \ldots, 40.$$

The subsystems S_1 and S_2 are forming a general series reliability structure of the system S presented in Figure 4.10.

Further, we assume that the system is a series system composed of the subsystem S_1 and S_2, with the scheme showed in Figure 4.10, while the subsystem S_1 is a series–parallel system with the scheme given in Figure 4.9 and the subsystem S_2 illustrated in Figure 4.9 is a series-'2 out of 4' system.

From the above, the subsystems S_υ, $\upsilon = 1, 2$, are composed of two-state components $E_{ij}^{(\upsilon)}$, $\upsilon = 1,2$, having the reliability functions that by the assumption are exponential of the forms

$$R_{ij}^{(\upsilon)}(t) = \exp[-\lambda_{ij}^{(\upsilon)} t].$$

Thus, the considered exemplary system is a two-state series system composed of the two-state subsystems S_1 and S_2. The subsystem S_1 is a series–parallel system and the subsystem S_2 is a series-'2 out of 4' system.

The subsystem S_1 consists of $k=2$ identical series subsystems, each composed of 30 components, i.e. $l_1 = l_2 = 30$, with the exponential reliability functions. In both series subsystems of the subsystem S_1 there are respectively

- the components $E_{ij}^{(1)}$, $i = 1, 2$, $j = 1, 2, \ldots, 20$, with the exponential reliability functions of the form

$$R_{ij}^{(1)}(t) = \exp[-0.0008t], \quad i = 1,2, \ j = 1,2,\ldots,20;$$

- the components $E_{ij}^{(1)}$, $i = 1, 2$, $j = 21, 22, \ldots, 30$, with the exponential reliability functions of the form

$$R_{ij}^{(1)}(t) = \exp[-0.0011t], \quad i = 1,2, \ j = 21,22,\ldots,30.$$

The subsystem S_1 is a series–parallel system and according to the formulae appearing after Definition 2.7, its exact reliability function is given by

$$\begin{aligned}\boldsymbol{R}^{(1)}(t) = \boldsymbol{R}_{2;30,30}(t) &= 1 - \prod_{i=1}^{2}\left[1 - \prod_{j=1}^{30} R_{ij}^{(1)}(t)\right] = 1 - \prod_{i=1}^{2}\left[1 - \exp\left[1 - \sum_{j=1}^{30}\lambda_{ij}^{(1)} t\right]\right]\\ &= 1 - [1-\exp[-[0.0008\cdot 20 + 0.0011\cdot 10]t]]^2 = 1 - [1-\exp[-0.027t]]^2\\ &= 2\exp[-0.027t] - \exp[-0.054t].\end{aligned} \quad (4.121)$$

To find the subsystem S_1 limit reliability function, we modify Corollary 4.5 into its following particular case for a non-homogeneous exponential series–parallel system.

Corollary 4.7
If components of the non-homogeneous regular two-state series–parallel system have exponential reliability functions

$$R^{(i,j)}(t) = 1 \quad \text{for } t < 0, \ R^{(i,j)}(t) = \exp[-\lambda_{ij} t] \quad \text{for } t \geq 0, \ \lambda_{ij} > 0,$$
$$i = 1,2,\ldots,a, \ j = 1,2,\ldots,e_i$$

and

$$k_n \to k, \ l_n \to \infty,$$

$$a_n = \frac{1}{\lambda l_n}, b_n = 0, \quad u = 1,2,\ldots,z \quad (4.122)$$

where

$$\lambda_i = \sum_{j=1}^{e_i} p_{ij}\lambda_{ij}, \quad i = 1, 2, \ldots, a \tag{4.123}$$

$$\lambda = \min_{1 \le i \le a} \{\lambda_i\} \tag{4.124}$$

then

$$\mathfrak{R}_9'(t) = 1 \quad \text{for } t < 0, \quad \mathfrak{R}_9'(t) = 1 - \prod_{i=1}^{a}\left[1 - \exp\left[-\frac{\lambda_i}{\lambda}t\right]\right]^{q_i k} \quad \text{for } t \ge 0,$$

is its limit reliability function.

The subsystem S_1 is a non-homogeneous, regular two-state series–parallel system in which, according to Definition 2.16, we have

$$k_n = k = 2, \; l_n = 30, \; a = 1, \; q_1 = 1, \; e_1 = 2, \; p_{11} = 2/3, \; p_{12} = 1/3.$$

Further, applying Corollary 4.7 and considering (4.122)–(4.124), we have

$$\lambda_i = \frac{2}{3}0.0008 + \frac{1}{3}0.0011 = 0.0009 \quad \text{for } i = 1, 2$$

$$\lambda = \min\{0.0009\} = 0.0009$$

$$a_n = 1/(0.0009 \cdot 30) = 1/0.027, b_n = 0 \tag{4.125}$$

and we conclude that the system limit reliability function is

$$\mathfrak{R}_9'(t) = 1 - [1 - \exp[-t]]^2 \quad \text{for } t \ge 0 \tag{4.126}$$

Thus, considering (4.125) and (4.126), from (1.1), the approximate formula for the subsystem S_1 reliability function takes the form

$$\begin{aligned} \boldsymbol{R}^{(1)}(t) &\cong \mathfrak{R}_9'((t - b_n)/a_n) \\ &= 1 - [1 - \exp[-0.027t]]^2 \\ &= 2\exp[-0.027t] - \exp[-0.054t] \quad \text{for } t \ge 0 \end{aligned} \tag{4.127}$$

It is easy to see that the above result is identical with the subsystem S_1 exact reliability given by (4.121).

The subsystem S_2 consists of $k = 4$ identical two-state series subsystems, each composed of 40 components, i.e. $l_1 = l_2 = l_3 = l_4 = 40$, with the exponential reliability functions. In all series subsystems of the subsystem S_2 there are respectively

- the components $E_{ij}^{(2)}$, $i = 1, 2, 3, 4, j = 1, 2, \ldots, 20$, with the exponential reliability functions of the form

$$R_{ij}^{(2)}(t) = \exp[-0.0014t], \quad i = 1,2,3,4, \ j = 1,2,\ldots,20;$$

- the components $E_{ij}^{(2)}$, $i = 1, 2, 3, 4, j = 1, 2, \ldots, 20$, with the exponential reliability functions of the form

$$R_{ij}^{(2)}(t) = \exp[-0.0018t], \quad i = 1,2,3,4, \ j = 21,22,\ldots,40.$$

The subsystem S_2 is a two-state series-'2 out of 4' system and according to the formulae appearing after Definition 2.5, its exact reliability function is given by

$$\boldsymbol{R}^{(2)}(t) = \boldsymbol{R}^{2}_{4;40,40,40,40}(t) = 1 - \sum_{\substack{r_1,r_2,r_3,r_4=0 \\ r_1+r_2+r_3+r_4 \le 1}}^{1} \prod_{i=1}^{4} \left[\prod_{j=1}^{40} R_{ij}^{(2)}(t) \right]^{r_i} \left[1 - \prod_{j=1}^{40} R_{ij}^{(2)}(t) \right]^{1-r_i}$$

$$= 1 - \sum_{\substack{r_1,r_2,r_3,r_4=0 \\ r_1+r_2+r_3+r_4 \le 1}}^{1} \prod_{i=1}^{4} \exp\left[-r_i \sum_{j=1}^{40} \lambda_{ij}^{(2)} t \right] \left[1 - \exp\left[-\sum_{j=1}^{40} \lambda_{ij}^{(2)} t \right] \right]^{1-r_i}$$

$$= 1 - \sum_{\substack{r_1,r_2,r_3,r_4=0 \\ r_1+r_2+r_3+r_4 \le 1}}^{1} \prod_{i=1}^{4} \exp[-r_i[0.0014 \cdot 20 + 0.0018 \cdot 20]t]$$

$$[1 - \exp[-[0.0014 \cdot 20 + 1.0018 \cdot 20]t]]^{1-r_i}$$

$$= 1 - \sum_{\substack{r_1,r_2,r_3,r_4=0 \\ r_1+r_2+r_3+r_4 \le 1}}^{1} \prod_{i=1}^{4} \exp[-r_i 0.064t][1 - \exp[-0.064t]]^{1-r_i}$$

$$= 1 - \sum_{i=0}^{1} \binom{4}{i} \exp[-i \cdot 0.064t][1 - \exp[-0.064t]]^{4-i}$$

$$= 1 - \exp[-0 \cdot 0.064t][1 - \exp[-0.0064t]^4$$

$$- 4\exp[-1 \cdot 0.064t][1 - \exp[-0.0064t]^3$$

$$= 6\exp[-0.128t] - 8\exp[-0.192t] + 3\exp[-0.256t] \tag{4.128}$$

To find the subsystem S_2 limit reliability function, we modify Corollary 4.1 into its following particular case for non-homogeneous exponential series system.

Corollary 4.8
If components of the non-homogeneous, regular two-state series system have exponential reliability functions

$$R^{(i)}(t) = 1 \quad \text{for } t < 0, \; R^{(i)}(t) = \exp[-\lambda_i t] \quad \text{for } t \geq 0, \; \lambda_i > 0, \; i = 1, 2, \ldots, a$$

and

$$a_n = \frac{1}{\lambda n}, \; b_n = 0 \tag{4.129}$$

where

$$\lambda = \max_{1 \leq i \leq a} \{\lambda_i\} \tag{4.130}$$

then

$$\overline{\mathfrak{R}}_2'(t) = 1 \quad \text{for } t < 0, \quad \overline{\mathfrak{R}}_2'(t) = \exp\left[-\sum_{i=1}^{a} q_i \frac{\lambda_i}{\lambda} t\right] \quad \text{for } t \geq 0 \tag{4.131}$$

is its limit reliability function.

The series systems of the subsystem S_2 are identical non-homogeneous, regular two-state series systems in which, according to Definition 2.13, we have

$$n = 40, \; a = 2, \; q_1 = 0.5, \; q_2 = 0.5$$

Further, applying Corollary 4.8 and considering (4.129) and (4.130), we have

$$\lambda = \max\{0.0014, 0.0018\} = 0.0018$$

$$a_n = 1/(0.0018 \cdot 40), \; b_n = 0 \tag{4.132}$$

and considering (4.131), we conclude that their limit reliability function is

$$\overline{\mathfrak{R}}_2'(t) = \exp\left[-\left(0.5\frac{0.0014}{0.0018} + 0.5\right)t\right] \quad \text{for } t \geq 0 \tag{4.133}$$

Thus, considering (4.132), from (1.1), the approximate formulae for the series system reliability functions take the form

$$\begin{aligned} \overline{R}_{40}'(t) &\cong \overline{\mathfrak{R}}_2'((t - b_n)/a_n) \\ &= \exp[-0.064t] \quad \text{for } t \geq 0. \end{aligned} \tag{4.134}$$

Further, according to (2.3) and (2.4), the approximate formula for the subsystem S_2 reliability function takes the form

$$\boldsymbol{R}^{(2)}(t) = 1 - \sum_{i=0}^{1} \binom{4}{i} \exp[-i \cdot 0.064t][1 - \exp[-0.064t]]^{4-i}$$
$$= 6\exp[-0.128t] - 8\exp[-0.192t] + 3\exp[-0.256t] \quad \text{for } t \geq 0, \tag{4.135}$$

which is identical with the subsystem S_2 exact reliability given by (4.128).

Taking into account that the systems S_1 and S_2 are two-state subsystems creating a non-homogeneous series system S, after applying the formulae appearing after Definition 2.1 and the results (4.127) and (4.135), the considered system reliability function is given by

$$\boldsymbol{R}(t) = \overline{\boldsymbol{R}}_2(t,) = \boldsymbol{R}^{(1)}(t)\boldsymbol{R}^{(2)}(t)$$
$$= 12\exp[-0.155t] - 6\exp[-0.182t] - 16\exp[-0.219t] \tag{4.136}$$
$$+ 8\exp[-0.264t] + 6\exp[-0.283t] - 3\exp[-0.310t] \quad \text{for } t \geq 0.$$

The exact and approximate mean values of the system lifetime are the same and after considering (4.136) amount

$$M = \int_0^\infty \boldsymbol{R}(t)\mathrm{d}t = 12/0.155 - 6/0.182 - 16/0.219 + 8/0.246 + 6/0.283 - 3/0.310$$
$$\cong 15.44$$

Example 4.8

We consider a two-state homogeneous '24 out of 36'-series system composed of independent components E_{ij}, $i = 1,2,3$, $j = 1,2,\ldots,$ 36. The system is composed of three '24 out of 36' subsystems linked in series and composed of components E_{ij}, $i = 1,2,3$, $j = 1,2,\ldots,$ 36, with the reliability structure presented in Figure 4.11.

Consequently, we define the reliability functions of the system components E_{ij}, $i = 1, 2, 3, j = 1, 2, \ldots, 36$, in the form

$$R_{ij}(t,) = \exp[-\lambda t[] = \exp[-3t], \quad i = 1,2,3, \; j = 1,2,\ldots,36 \tag{4.137}$$

Since the shape parameters of the considered '24 out of 36'-series system are

$$k = 3, \; l = 36, \; m = 24$$

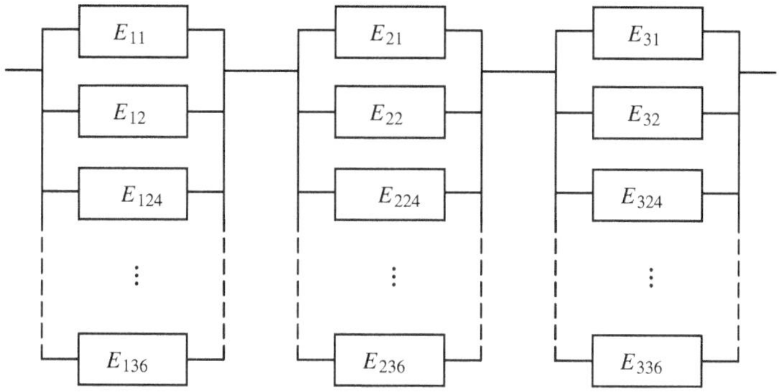

Figure 4.11 The scheme of the '24 out of 36'-series system.

then considering (2.3) and (2.1), we may conclude that its exact reliability function is given by

$$\boldsymbol{R}(t) = \left[1 - \sum_{i=0}^{23} \binom{36}{i} \exp[-3it][1-\exp[-3t]]^{36-i}\right]^3, \qquad (4.138)$$

To find the system limit reliability function, we put together Lemma 4.10 and the formulae (2.1) and get the following result for 'm out of l'-series exponential system.

Corollary 4.9

If components of the homogeneous, regular two-state 'm out of l'-series system have exponential reliability functions

$$R_{ij}(t,) = 1 \quad \text{for } t < 0, \quad R_{ij}(t) = \exp[-\lambda t], \quad \text{for } t \geq 0, \ \lambda > 0,$$
$$i = 1, 2, \ldots, k, \quad j = 1, 2, \ldots, l \qquad (4.139)$$

$$k = \text{constant}, \ l \to \infty \text{ as } n \to \infty,$$

$$m/l \to \mu, \ 0 < \mu < 1, \text{ as } n \to \infty, \qquad (4.140)$$

$$a_n = \frac{1}{\lambda \mu \sqrt{l+1}} \sqrt{\mu(1-\mu)}, \ b_n = -\frac{1}{\lambda} \log \mu, \qquad (4.141)$$

then

$$\mathfrak{R}^{(\mu)}(t) = \left[1 - \frac{1}{\sqrt{2\pi}} \int_{-\infty}^{t} e^{-\frac{x^2}{2}} dx\right]^k \quad \text{for } t \in (-\infty, \infty) \qquad (4.142)$$

is the limit multi-state reliability function of this system.

Motivation: Since for sufficiently large n

$$a_n t + b_n = \frac{1}{\lambda\mu\sqrt{l+1}}\sqrt{\mu(1-\mu)}t - \frac{1}{\lambda}\log\mu > 0$$

for all $t \in (-\infty, \infty)$, $u = 1, 2, \ldots, z$, then for sufficiently large n

$$\begin{aligned} R(a_n t + b_n) &= \exp[-\lambda(a_n t + b_n)] \\ &= \left[1 - a_n t + o\left(\frac{1}{\sqrt{l+1}}\right)\right]\exp[-\lambda b_n)] \\ &= \left[1 - \lambda a_n t + o\left(\frac{1}{\sqrt{l+1}}\right)\right]\exp[-\lambda b_n)] \\ &= \mu - \frac{1}{\sqrt{l+1}}\sqrt{\mu(1-\mu)}t + o\left(\frac{1}{\sqrt{l+1}}\right) \quad \text{for all } t \in (-\infty, \infty) \end{aligned}$$

and considering that result in (4.32), we have

$$\begin{aligned} &\lim_{n\to\infty}\frac{(l+1)R(a_n t + b_n) - m}{\sqrt{\dfrac{m(l-m+1)}{l+1}}} \\ &= \lim_{n\to\infty}\frac{(l+1)\left[\mu - \dfrac{1}{\sqrt{l+1}}\sqrt{\mu(1-\mu)}t + o\left(\dfrac{1}{\sqrt{l+1}}\right)\right] - m}{\sqrt{\dfrac{m(l-m+1)}{l+1}}} \\ &= \lim_{n\to\infty}\frac{\mu - \dfrac{1}{\sqrt{l+1}}\sqrt{\mu(1-\mu)}t + o\left(\dfrac{1}{\sqrt{l+1}}\right) - \dfrac{m}{l+1}}{\dfrac{1}{\sqrt{l+1}}\sqrt{\dfrac{m(l+1-m)}{(l+1)(l+1)}}} \\ &= \lim_{n\to\infty}\frac{\mu - \dfrac{1}{\sqrt{l+1}}\sqrt{\mu(1-\mu)}t + o\left(\dfrac{1}{\sqrt{l+1}}\right) - \dfrac{m}{l+1}}{\dfrac{1}{\sqrt{l+1}}\sqrt{\dfrac{m}{(l+1)}\left(1 - \dfrac{m}{l+1}\right)}} = -t \end{aligned}$$

for $t \in (-\infty, \infty)$, $u = 1, 2, \ldots, z$.

The above result, by Lemma 4.10, means that $\mathfrak{R}^{(\mu)}(t)$ given by (4.142) is the considered system limit reliability function, which completes the proof.

Further, applying Corollary 4.9, since the shape parameters of the considered '24 out of 36'-series system are

$$k = 3,\ l = 36,\ m = 24$$

and by (4.140) and (4.141)

$$\mu = \frac{24}{36} = \frac{2}{3}$$

$$a_n = \frac{1}{3\frac{2}{3}\sqrt{35}}\sqrt{\frac{2}{3}\left(1-\frac{2}{3}\right)} = \frac{1}{6\sqrt{17.5}},\ b_n = -\frac{1}{3}\log\frac{2}{3},$$

and considering (1.1), the approximate formula for the system reliability function takes the form

$$\boldsymbol{R}(t,\cdot) \cong \mathfrak{R}^{(2/3)}(t-b_n)/a_n)$$

$$= \left[1 - \frac{1}{\sqrt{2\pi}} \int_{-\infty}^{6\sqrt{17.5}t+2\sqrt{17.5}\log(2/3)} e^{-x^2/2}\,dx\right]^3 \tag{4.143}$$

It is obvious that using the approximate formula (4.143) is much easier than using the exact formula (4.138) for the considered system reliability function. The accuracy of this approximation is illustrated in Table 4.7, where the values and differences between the system exact reliability function $\boldsymbol{R}(t)$ given by (4.138) and the system approximate reliability function $\mathfrak{R}^{(2/3)}(t - b_n)/a_n))$ given by (4.143) are presented and in the results (4.144) and (4.145) appearing after this table and concerned with the system lifetime mean values, calculated with application of the system exact and approximate reliability functions.

The exact and approximate mean values of the system lifetime, after applying the trapezium methods of integration and data given in Tables 4.7, are

$$M = \int_0^\infty \boldsymbol{R}(t)dt \cong \frac{0.01}{2}[R(0.00) + 2[R(0.01) + R(0.02) + \cdots + R(0.19)] + R(0.20)]$$

$$= \frac{0.01}{2}[1.0000 + 2[1.0000 + 1.000 + \cdots + 0.0031] + 0.0000] \cong 0.1134 \tag{4.144}$$

Table 4.7 The Values and Differences Between the Exemplary System S Exact and Approximate Reliability Functions

t	$R(t)$	$\Re^{(2/3)}(t-b_n)/a_n))$	$\Delta = R(t) - \Re^{(2/3)}(t-b_n)/a_n))$
0.00	1.000	0.9991	0.0009
0.01	1.000	0.9976	0.0024
0.02	1.000	0.9958	0.0042
0.03	1.000	0.9892	0.0108
0.04	0.9997	0.9750	0.0247
0.05	0.9978	0.9522	0.0456
0.06	0.9895	0.9144	0.0751
0.07	0.9653	0.8560	0.1093
0.08	0.9123	0.7691	0.1432
0.09	0.8206	0.6603	0.1603
0.10	0.6909	0.5327	0.1582
0.11	0.5374	0.3982	0.1392
0.12	0.3831	0.2721	0.1110
0.13	0.2492	0.1679	0.0813
0.14	0.1479	0.0925	0.0554
0.15	0.0802	0.0450	0.0352
0.16	0.0399	0.0192	0.0207
0.17	0.0183	0.0071	0.0112
0.18	0.0078	0.0022	0.0056
0.19	0.0031	0.0006	0.0025
0.20	0.0011	0.0000	0.0011

$$M = \int_0^{\infty} \Re^{(2/3)}(t)\mathrm{d}t \cong \frac{0.01}{2}\left[\Re^{(2/3)}(0.00) + 2\left[\Re^{(2/3)}(0.01) + \Re^{(2/3)}(0.02) + \cdots + \Re^{(2/3)}(0.19)\right] + \Re^{(2/3)}(0.20)\right]$$
$$= \frac{0.01}{2}[0.9991 + 2[0.9976 + 0.9958 + \cdots + 0.0006] + 0.0011] \cong 0.1015 \quad (4.145)$$

5 Reliability of Large Multi-State Systems

5.1 Reliability Evaluation of Multi-State Series Systems

In proving facts on limit reliability functions of homogeneous multi-state series systems we apply the following obvious extension of Lemma 4.1 [81].

Lemma 5.1
If

i. $\overline{\mathfrak{R}}(t,u) = \exp[-\overline{V}(t,u)],\ \ u = 1,2,\ldots,z,$ is a non-degenerate reliability function,
ii. $\overline{\boldsymbol{R}}_n(t,\cdot) = [1, \overline{\boldsymbol{R}}_n(t,1),\ldots,\overline{\boldsymbol{R}}_n(t,z)],\ \ t \in (-\infty, \infty)$, is the reliability function of a homogeneous multi-state series system defined by Eqs. (3.20) and (3.21),
iii. $a_n(u) > 0,\ b_n(u) \in (-\infty, \infty),\ u = 1,2,\ldots,z$

then

$$\overline{\mathfrak{R}}(t,\cdot) = [1, \overline{\mathfrak{R}}(t,1),\ldots,\overline{\mathfrak{R}}(t,z)],\ \ t \in (-\infty, \infty)$$

is the multi-state limit reliability function of this system, i.e.

$$\lim_{n\to\infty} \overline{\boldsymbol{R}}_n(a_n(u)t + b_n(u), u) = \overline{\mathfrak{R}}(t,u) \quad \text{for } t \in C_{\overline{\mathfrak{R}}(u)},\ \ u = 1,2,\ldots,z \tag{5.1}$$

if and only if

$$\lim_{n\to\infty} nF(a_n(u)t + b_n(u), u) = \overline{V}(t,u) \quad \text{for } t \in C_{\overline{V}(u)},\ \ u = 1,2,\ldots,z \tag{5.2}$$

Motivation: For each fixed u, $u = 1, 2, \ldots, z$, assumptions (i)–(iii) of Lemma 5.1 are identical to assumptions (i)–(iii) of Lemma 4.1, condition (5.1) is identical to condition (4.1) and moreover condition (5.2) is identical to condition (4.2). Since, from Lemma 4.1, conditions (4.1) and (4.2) are equivalent, then conditions (5.1) and (5.2) are also equivalent.

Lemma 5.1 and Theorem 4.1 allow us to fix the class of all limit reliability functions for homogeneous multi-state series systems. Their application results in the following theorem [81].

Reliability of Large Systems. DOI: http://dx.doi.org/10.1016/B978-0-08-099949-4.00005-2

Theorem 5.1

The class of limit non-degenerate reliability functions of the homogeneous multi-state series system is composed of 3^z reliability functions of the form

$$\overline{\mathfrak{R}}(t,\cdot)=[1,\overline{\mathfrak{R}}(t,1),\ldots,\overline{\mathfrak{R}}(t,z)],\quad t\in(-\infty,\infty) \tag{5.3}$$

where

$$\overline{\mathfrak{R}}(t,u)\in\{\overline{\mathfrak{R}}_1(t),\overline{\mathfrak{R}}_2(t),\overline{\mathfrak{R}}_3(t)\},\quad u=1,2,\ldots,z \tag{5.4}$$

and $\overline{\mathfrak{R}}_i(t)$, $i = 1, 2, 3$, are defined by Eqs. (4.3) – (4.5).

Motivation: For each fixed u, $u = 1, 2, \ldots, z$, the coordinate $\overline{\mathfrak{R}}(t,u)$ of the vector $\overline{\mathfrak{R}}(t,\cdot)$ defined by Eq. (5.3) from Theorem 4.1 that is the consequence of Lemma 4.1, may be one of the three types of reliability functions defined by Eqs. (4.3) – (4.5). Thus the number of different multi-state limit reliability functions of the considered system is equal to the number of z-term variations of the 3-component set (5.4), i.e. 3^z, and they are of the form (5.3).

Corollary 5.1

If the homogeneous multi-state series system is composed of components having Weibull reliability functions

$$R(t,\cdot)=[1,R(t,1),\ldots,R(t,z)],\quad t\in(-\infty,\infty)$$

where

$$R(t,u)=1 \quad \text{for } t<0,\ R(t,u)=\exp[-\beta(u)t^{\alpha(u)}] \quad \text{for } t\geq 0,\ \alpha(u)>0,$$
$$\beta(u)>0,\ u=1,2,\ldots,z$$

and

$$a_n(u)=(n\beta(u))^{-1/\alpha(u)},\ b_n(u)=0,\quad u=1,2,\ldots,z$$

then

$$\overline{\mathfrak{R}}_2(t,\cdot)=[1,\overline{\mathfrak{R}}_2(t,1),\ldots,\overline{\mathfrak{R}}_2(t,z)],\quad t\in(-\infty,\infty)$$

where

$$\overline{\mathfrak{R}}_2(t,u)=1 \quad \text{for } t<0,\ \overline{\mathfrak{R}}_2(t,u)=\exp[-t^{\alpha(u)}] \quad \text{for } t\geq 0,\ u=1,2,\ldots,z$$

is its limit reliability function.

Motivation: Since for each fixed u we have

$$a_n(u)t + b_n(u) < 0 \quad \text{for } t < 0$$

and

$$a_n(u)t + b_n(u) \geq 0 \quad \text{for } t \geq 0$$

then

$$F(a_n(u)t + b_n(u), u) = 0 \quad \text{for } t < 0$$

and

$$F(a_n(u)t + b_n(u), u) = 1 - \exp[-\beta(u)(a_n(u)t + b_n(u))^{\alpha(u)}]$$
$$= 1 - \exp[-t^{\alpha(u)}/n] = t^{\alpha(u)}/n - o(1/n) \quad \text{for } t \geq 0$$

Hence

$$\overline{V}(t,u) = \lim_{n\to\infty} nF(a_n(u)t + b_n(u), u) = 0 \quad \text{for } t < 0$$

and

$$\overline{V}(t,u) = \lim_{n\to\infty} nF(a_n(u)t + b_n(u), u) = t^{\alpha(u)} \quad \text{for } t \geq 0$$

which from Lemma 5.1 completes the proof.

Example 5.1 (A Piping Transportation System)
The piping system is composed of $n = 1000$ identical pipe segments with reliability functions

$$R(t,u) = \exp[-0.0002ut^3] \quad \text{for } t \geq 0, \quad u = 1, 2, 3, 4$$

Since it is a homogeneous five-state system, then according to Eqs. (3.20) and (3.21) its multi-state reliability function is given by

$$\overline{\boldsymbol{R}}_{1000}(t,\cdot) = [1, \exp[-0.2t^3], \exp[-0.4t^3], \exp[-0.6t^3], \exp[-0.8t^3]] \quad \text{for } t \geq 0$$

Assuming normalizing constants

$$a_n(u) = (0.0002 \cdot 1000u)^{1/3}, \quad b_n(u) = 0, \quad u = 1, 2, 3, 4$$

on the basis of Corollary 5.1, we conclude that the system limit reliability function is

$$\overline{\Re}_2(t,\cdot)=[1,\exp[-t^3],\ \exp[-t^3],\ \exp[-t^3],\ \exp[-t^3]]\quad \text{for } t\geq 0$$

Hence, considering Eq. (3.49), we arrive at the following approximate formula (it is exact in this case)

$$\begin{aligned}\overline{\boldsymbol{R}}_{1000}(t,\cdot)&\cong\overline{\Re}_2((t-b_n(u))/a_n(u),\cdot)\\&=[1,\exp[-0.2t^3],\ \exp[-0.4t^3],\ \exp[-0.6t^3],\ \exp[-0.8t^3]]\\&\quad\text{for } t\geq 0\end{aligned}$$

The mean values of the sojourn times $T_i(u)$ in the state subsets in years, according to Eq. (3.4), are

$$M_i(u)=E[T_i(u)]=\Gamma(4/3)(0.0002u)^{-1/3},\quad u=1,2,3,4$$

i.e.

$$M_i(1)=15.3,\ \ M_i(2)=12.1,\ \ M_i(3)=10.6,\ \ M_i(4)=9.6$$

and according to Eq. (3.8), their mean sojourn lifetimes in the particular states are

$$\overline{M}_i(1)=3.2,\ \ \overline{M}_i(2)=1.5,\ \ \overline{M}_i(3)=1.0,\ \ \overline{M}_i(4)=9.6$$

The expected values of the system sojourn times $T(u)$ in the state subsets, according to Eq. (3.13), are

$$M(u)=E[T(u)]=\Gamma(4/3)(0.2u)^{-1/3},\quad u=1,2,3,4$$

i.e.

$$M(1)\cong 1.53,\ \ M(2)\cong 1.21,\ \ M(3)\cong 1.06,\ \ M(4)\cong 0.96$$

Thus, from Eq. (3.17), the expected values of the system sojourn times in the particular states are

$$\overline{M}(1)\cong 0.32,\ \ \overline{M}(2)\cong 0.15,\ \ \overline{M}(3)\cong 0.10,\ \ \overline{M}(4)\cong 0.96$$

If the critical state is $r=2$, then from Eq. (3.18), the system risk function is given by

$$r(t)\cong 1-\exp[-0.4t^3]\quad \text{for } t\geq 0$$

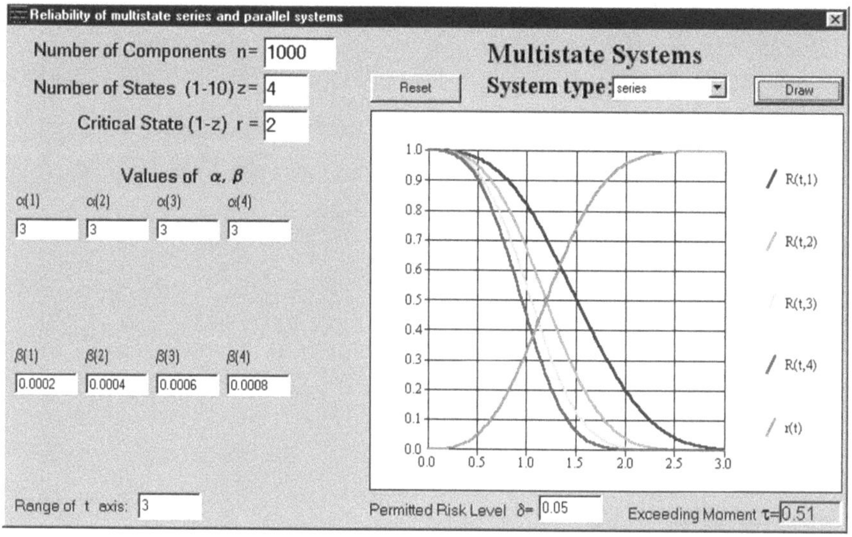

Figure 5.1 The graphs of the piping system reliability function and risk function.

The moment when the risk exceeds a permitted level $\delta = 0.05$, calculated according to Eq. (3.19), is

$$\tau = \boldsymbol{r}^{-1}(\delta) \cong (-\log(1-\delta)/0.4)^{1/3} = 0.5 \text{ years}$$

The graphs of the piping system limit multi-state reliability function and its risk function are plotted using a computer program [71] and presented in Figure 5.1.

Corollary 5.2

If the homogeneous multi-state series system is composed of components having Erlang's reliability function of order 2 given by

$$R(t,\cdot) = [1, R(t,1), \ldots, R(t,z)], \quad t \in (-\infty, \infty)$$

where

$$R(t,u) = 1 \quad \text{for } t < 0, \quad R(t,u) = [1 + \lambda(u)]\exp[-\lambda(u)t] \quad \text{for } t \geq 0, \quad \lambda > 0,$$
$$u = 1, 2, \ldots, z$$

and

$$a_n(u) = \frac{\sqrt{2}}{\lambda(u)\sqrt{n}}, \quad b_n(u) - 0, \quad u - 1, 2, \ldots, z$$

then

$$\overline{\mathfrak{R}}_2(t,\cdot) = [1, \overline{\mathfrak{R}}_2(t,1), \ldots, \overline{\mathfrak{R}}_2(t,z)], \quad t \in (-\infty, \infty)$$

where

$$\overline{\mathfrak{R}}_2(t,u) = 1 \quad \text{for } t < 0, \ \overline{\mathfrak{R}}_2(t,u) = \exp[-t^2] \quad \text{for } t \geq 0, \ u = 1, 2, \ldots, z$$

is its limit reliability function.

Motivation: Since for each fixed u, we have

$$a_n(u)t + b_n(u) < 0 \quad \text{for } t < 0$$

and

$$a_n(u)t + b_n(u) \geq 0 \quad \text{for } t \geq 0$$

then

$$F(a_n(u)t + b_n(u), u) = 0 \quad \text{for } t < 0$$

and

$$\begin{aligned} F(a_n(u)t + b_n(u), u) &= 1 - [1 - \lambda(u)a_n(u)t]\exp[-\lambda(u)(a_n(u)t)] \\ &= 1 - \left[1 + \sqrt{2}t/\sqrt{n}\right]\exp\left[-\sqrt{2}t/\sqrt{n}\right] \\ &= 1 - \left[1 + \sqrt{2}t/\sqrt{n}\right]\left[1 - \sqrt{2}t/\sqrt{n} + t^2/n - o(1/n)\right] \\ &= t^2/n + o(1/n) \quad \text{for } t \geq 0 \end{aligned}$$

Hence

$$\overline{V}(t,u) = \lim_{n\to\infty} nF(a_n(u)t + b_n(u), u) = 0 \quad \text{for } t < 0$$

and

$$\overline{V}(t,u) = \lim_{n\to\infty} nF(a_n(u)t + b_n(u), u) = \lim_{n\to\infty} n[t^2/n - o(1/n)] = t2 \quad \text{for } t \geq 0$$

which from Lemma 5.1 completes the proof.

Example 5.2 (A Model Telecommunication Network)
The telecommunication network operating for telephone subscribers is composed of $n = 2000$ subscriber terminals, subscriber cables and one head linking subscriber cables with distributing cables. We analyse the reliability of the network part that

consists of subscriber cables only. Thus the considered system is composed of $n = 2000$ double cables that consist of one basic cable and one cable in a cold reserve. The cables are five-state ($z = 4$) components of the system having exponential reliability functions with the following transition rates between the state subsets:

$$\lambda(u) = \frac{1}{60 - 10u}, \quad u = 1, 2, 3, 4$$

Under this assumption, since the sojourn time of a double cable in the state subsets is the sum of two lifetimes having exponential distributions, then it has Erlang's distribution of order 2, that is its reliability function is given by

$$R(t, u) = 1 \quad \text{for } t < 0, \quad R(t, u) = \left[1 + \frac{1}{60 - 10u} t\right] \exp\left[-\frac{1}{60 - 10u} t\right]$$
$$\text{for } t \geq 0, \quad u = 1, 2, 3, 4$$

Thus the considered part of the telecommunication network is a homogeneous five-state series system and according to Corollary 5.2, assuming the normalizing constants

$$a_n(u) = 0.6 - 0.1u, \quad b_n(u) = 0 \quad \text{for } u = 1, 2, 3, 4$$

we conclude that its limit reliability function is

$$\overline{\mathfrak{R}}_2(t, \cdot) = [1, \exp[-t^2], \; \exp[-t^2], \; \exp[-t^2], \; \exp[-t^2]] \quad \text{for } t \geq 0$$

Hence, from Eq. (3.49), we get the following approximate formula:

$$\overline{\boldsymbol{R}}_{20{,}000}(t, \cdot) \cong \overline{\mathfrak{R}}_2((t - b_n(u))/a_n(u), \cdot)$$
$$= [1, \; \exp[-0.781t^2], \; \exp[-1.389t^2], \; \exp[-3.125t^2],$$
$$\exp[-12.5t^2]] \quad \text{for } t \geq 0$$

The mean lifetimes $T_i(u)$ of the system components in the state subsets in years, according to Eq. (3.4), are

$$M_i(u) = E[T_i(u)] = 60 - 10u, \quad u = 1, 2, 3, 4$$

i.e.

$$M_i(1) = 50, \; M_i(2) = 40, \; M_i(3) = 30, \; M_i(4) = 20$$

and from Eq. (3.8), the system component mean lifetimes in particular states are

$$\overline{M}_i(1) = 10, \; \overline{M}_i(2) = 10, \; \overline{M}_i(3) = 10, \; \overline{M}_i(4) = 20$$

The expected values of the network sojourn times $T(u)$ in the state subsets, according to Eq. (3.13), are

$$M(u) = E[T(u)] = \Gamma(3/2)(0.6 - 0.1u), \quad u = 1, 2, 3, 4$$

i.e.

$$M(1) \cong 0.44, \ M(2) \cong 0.35, \ M(3) \cong 0.27, \ M(4) \cong 0.18$$

Thus, from Eq. (3.17), the lifetime expected values of the network in particular states are

$$\overline{M}(1) \cong 0.09, \ \overline{M}(2) \cong 0.08, \ \overline{M}(3) \cong 0.09, \ \overline{M}(4) \cong 0.18$$

If the critical state is $r = 2$, then from Eq. (3.18), the network risk function is given by

$$\boldsymbol{r}(t) \cong 1 - \exp[-6.25t^2] \quad \text{for } t \geq 0$$

The moment when the network risk exceeds an admissible level $\delta = 0.05$, from Eq. (3.19), is

$$\tau = \boldsymbol{r}^{-1}(\delta) \cong [-\log(1-\delta)/6.25)^{1/2} = 0.09 \text{ years}$$

Corollary 5.3
If components of the homogeneous multi-state series system have reliability functions

$$R(t, \cdot) = [1, R(t, 1), \ldots, R(t, z)], \quad t \in (-\infty, \infty)$$

where

$$R(t, u) = 1 \quad \text{for } t < 0, \ R(t, u) = r_1 \exp[-\lambda_1(u)t] + r_2 \exp[-\lambda_2(u)t] \quad \text{for } t \geq 0,$$
$$u = 1, 2, \ldots, z, \ \lambda_1(u) > 0, \ \lambda_2(u) > 0, \ 0 \leq r_1 \leq 1, \ r_2 = 1 - r_1$$

and

$$a_n(u) = \frac{1}{[r_1 \lambda_1(u) + r_2 \lambda_2(u)]n}, \quad b_n(u) = 0, \quad u = 1, 2, \ldots, z$$

then

$$\overline{\mathfrak{R}}_2(t, \cdot) = [1, \overline{\mathfrak{R}}_2(t, 1), \ldots, \overline{\mathfrak{R}}_2(t, z)], \quad t \in (-\infty, \infty)$$

where

$$\overline{\Re}_2(t,u) = 1 \quad \text{for } t < 0, \quad \overline{\Re}_2(t,\cdot) = \exp[-t] \quad \text{for } t \geq 0, \quad u = 1, 2, \ldots, z$$

is its limit reliability function.

Motivation: Since for each fixed u, we have

$$a_n(u)t + b_n(u) < 0 \quad \text{for } t < 0$$

and

$$a_n(u)t + b_n(u) \geq 0 \quad \text{for } t \geq 0$$

then

$$F(a_n(u)t + b_n(u), u) = 0 \quad \text{for } t < 0$$

and

$$\begin{aligned} F(a_n(u)t + b_n(u), u) &= 1 - [r_1 \exp[-\lambda_1(u)(a_n(u)t)] + r_2 \exp[-\lambda_2(u)(a_n(u)t)]] \\ &= [r_1\lambda_1(u) + r_2\lambda_2(u)][a_n(u)t - o(a_n(u))] \\ &= t/n - o(1/n) \quad \text{for } t \geq 0 \end{aligned}$$

Hence

$$\overline{V}(t,u) = \lim_{n\to\infty} nF(a_n(u)t + b_n(u), u) = 0 \quad \text{for } t < 0$$

and

$$\overline{V}(t,u) = \lim_{n\to\infty} nF(a_n(u)t + b_n(u), u) = \lim_{n\to\infty} n[t/n - o(1/n)] = t \quad \text{for } t \geq 0$$

which from Lemma 5.1 completes the proof.

Example 5.3 (A Bus Transportation System)

The city transportation system is composed of n, $n \geq 1$, buses necessary to perform its communication tasks. We assume that the bus lifetimes are independent random variables and that the system is operating in successive cycles (days) $c = 1, 2, \ldots$ In each of the cycles the following three operating phases of all components are distinguished:

f_1: components waiting for inclusion in the operation process, lasting from the moment t_0 up to the moment t_1,
f_2: components' activation for the operation process, lasting from t_1 up to t_2,
f_3: components operating, lasting from t_2 up to $t_3 = t_0$.

Each of the system components during the waiting phase may be damaged because of the circumstances at the stoppage place. We assume that the probability that at the end moment t_1 of the first phase the ith component is not failed is equal to $p_i^{(1)}$, where $0 \le p_i^{(1)} \le 1$, $i = 1, 2, \ldots, n$. Since component lifetimes are independent then the system availability at the end moment t_1 of phase f_1 is given by

$$p^{(1)} = \prod_{i=1}^{n} p_i^{(1)} \tag{5.5}$$

In the activation phase f_2, system components are prepared for the operation process by the service. They are checked and small flaws are removed. Sometimes the flaws cannot be removed and particular components are not prepared to fulfil their tasks. We assume that the probability that at the end moment t_2 of the first phase the ith component is not failed is equal to $p_i^{(2)}$, where $0 \le p_i^{(2)} \le 1$, $i = 1, 2, \ldots, n$. Since component lifetimes are independent then the system availability at the end moment t_2 of the phase f_2 is given by

$$p^{(2)} = \prod_{i=1}^{n} p_i^{(2)} \tag{5.6}$$

Thus, finally, the system S availability after two phases is given by

$$p^{(1,2)} = p^{(1)} p^{(2)} \tag{5.7}$$

where $p^{(1)}$ and $p^{(2)}$ are defined respectively by Eqs. (5.5) and (5.6).

In the operating phase f_3, during the time $t_4 = t_3 - t_2$, each of the system components is performing one of two tasks:

z_1: a first task (working at normal communication conditions),
z_2: a second task (working at a communication peak) with probabilities respectively equal to r_1 and r_2, where $0 \le r_1 \le 1$, $r_2 = 1 - r_1$.

Let

$$R^{(1)}(t, u) = 1 \quad \text{for } t < 0, \quad R^{(1)}(t, u) = \exp\left[-\frac{1}{15 - 5u} t \right] \quad \text{for } t \ge 0, \quad u = 1, 2$$

be the reliability function of the ith component during performance of task z_1 and

$$R^{(2)}(t, u) = 1 \quad \text{for } t < 0, \quad R^{(2)}(t, u) = \exp\left[-\frac{1}{10 - 2u} t \right] \quad \text{for } t \ge 0, \quad u = 1, 2$$

be the reliability function of the ith component during performance of task z_2. Then, according to the formula for total probability,

$$R^{(1,2)}(t) = 1 \quad \text{for } t < 0$$

$$R^{(1,2)}(t,u) = r_1 \exp\left[-\frac{1}{15-5u}t\right] + r_2 \exp\left[-\frac{1}{10-2u}t\right] \quad \text{for } t \geq 0, \quad u = 1,2$$

is the reliability function of the ith component performing two tasks.

Thus the considered transportation system is a homogeneous three-state series system and according to Corollary 5.3, assuming the normalizing constants

$$a_n(u) = 1/[[r_1/(15-5u) + r_2/(10-2u)]n], \quad b_n(u) = 0 \quad \text{for } u = 1,2$$

we conclude that its limit reliability function is

$$\overline{\mathfrak{R}}_2(t,\cdot) = [1, \exp[-t], \exp[-t]] \quad \text{for } t \geq 0$$

Hence, from Eq. (3.49), we get the following approximate formula:

$$\begin{aligned}\overline{\boldsymbol{R}}_n(t,\cdot) &\cong \overline{\mathfrak{R}}_2((t-b_n(u))/a_n(u),\cdot)\\ &= [1, \exp[-(r_1/10 + r_2/8)nt], \exp[-(r_1/5 + r_2/6)nt]] \quad \text{for } t \geq 0\end{aligned}$$

The mean values of the system lifetimes $T(u)$ in the state subsets, according to Eq. (3.13), are

$$M(u) = E[T(u)] = 1/[[r_1/(15-5u) + r_2/(10-2u)]n], \quad u = 1,2$$

If we assume that

$$n = 30, \quad r_1 = 0.8, \quad r_2 = 0.2$$

then

$$\overline{\boldsymbol{R}}_{30}(t,\cdot) \cong [1, \exp[-3.15t], \exp[-5.80t]] \quad \text{for } t \geq 0 \tag{5.8}$$

and

$$M(1) \cong 0.32, \quad M(2) \cong 0.17$$

Thus, considering Eq. (3.17), the expected values of the sojourn times in the particular states are

$$\overline{M}(1) \cong 0.15, \quad \overline{M}(2) \cong 0.17$$

If a critical state is $r = 1$, then according to Eq. (3.18), the system risk function is given by

$$\boldsymbol{r}(t) \cong 1 - \exp[-3.15t] \quad \text{for } t \geq 0$$

The moment when the system risk exceeds a permitted level $\delta = 0.05$, according to Eq. (3.19), is

$$\tau = \boldsymbol{r}^{-1}(\delta) \cong -\log(1-\delta)/3.15 = 0.016 \text{ years} \cong 6 \text{ days}$$

At the end moment of the system activation phase, which is simultaneously the starting moment of the system operating phase t_2, the system is able to perform its tasks with the probability$p^{(1,2)}$ defined by Eq. (5.7). Therefore, after applying the formula (5.8), we conclude that the system reliability in c cycles, $c = 1, 2, \ldots,$ is given by the following formula:

$$G(c,\cdot) \cong [1, p^{(1,2)} \exp[-3.15ct_4], p^{(1,2)} \exp[-5.80ct_4]]$$

where $t_4 = t_3 - t_2$ is the time duration of the system operating phase f_3. Further, assuming for instance

$$p^{(1,2)} = p^{(1)}p^{(2)} = 0.99 \times 0.99 = 0.98, \quad t_4 = 18 \text{ h} = 0.002055 \text{ years}$$

for the number of cycles

$$c = 7 \text{ days} = 1 \text{ week}$$

we get

$$G(7,\cdot) \cong [1, 0.966, 0.902]$$

This result means that during 7 days the considered transportation system will be able to perform its tasks in a state not worse than the first state with probability 0.966, whereas it will be able to perform its tasks in the second state with probability 0.902.

In finding the class of limit reliability functions for non-homogeneous multi-state series systems, we use an obvious extension of Lemma 4.3 formulated as follows [81].

Lemma 5.2

If

i. $\overline{\mathfrak{R}}'(t,u) = \exp[-\overline{V}'(t,u)], \quad u = 1, 2, \ldots, z,$ is a non-degenerate reliability function,

ii. $\overline{\boldsymbol{R}}'_n(t,\cdot) = [1, \overline{\boldsymbol{R}}'_n(t,1), \ldots, \overline{\boldsymbol{R}}'_n(t,z)], \quad t \in (-\infty, \infty),$ is the reliability function of a non-homogeneous multi-state series system defined by Eqs. (3.33) and (3.34),

iii. $a_n(u) > 0$, $b_n(u) \in (-\infty, \infty)$, $u = 1,2,\ldots,z,$

iv. $F(t,u)$ for each fixed u is one of distribution functions $F^{(1)}(t,u)$, $F^{(2)}(t,u)$, $\ldots$, $F^{(a)}(t,u)$ defined by Eq. (3.32) such that

v. $\exists N(u) \, \forall n > N(u) F(a_n(u)t + b_n(u), u) = 0 \quad \text{for } t < t_0(u), \; F(a_n(u)u)t + b_n(u), u \neq 0$ for $t \geq t_0(u)$, where $t_0(u) \in < -\infty, \infty$

vi. $\lim\limits_{n\to\infty} \dfrac{F^{(i)}(a_n(u)t + b_n(u), u)}{F(a_n(u)t + b_n(u), u)} \leq 1 \quad \text{for } t \geq t_0(u), \; i = 1, 2, \ldots, a, \; u = 1, 2, \ldots, z$

and moreover there exist non-decreasing functions.

vii.

$$\overline{d}(t,u) = \begin{cases} 0 & \text{for } t < t_0(u) \\ \lim\limits_{n\to\infty} \sum\limits_{i=1}^{a} q_i \overline{d}_i(a_n(u)t + b_n(u), u) & \text{for } t \geq t_0(u) \end{cases} \tag{5.9}$$

where

viii.

$$\overline{d}_i(a_n(u)t + b_n(u), u) = \frac{F^{(i)}(a_n(u)t + b_n(u), u)}{F(a_n(u)t + b_n(u), u)} \tag{5.10}$$

then

$$\overline{\mathfrak{R}}'(t,\cdot) = [1, \overline{\mathfrak{R}}'(t,1), \ldots, \overline{\mathfrak{R}}'(t,z)], \quad t \in (-\infty, \infty)$$

is the multi-state limit reliability function of this system, i.e.

$$\lim_{n\to\infty} \overline{\boldsymbol{R}}'_n(a_n(u)t + b_n(u), u) = \overline{\mathfrak{R}}'(t,u) \quad \text{for } t \in C_{\overline{\mathfrak{R}}'(u)}, \ u = 1, 2, \ldots, z \tag{5.11}$$

if and only if

$$\lim_{n\to\infty} nF(a_n(u)t + b_n(u), u)\overline{d}(t,u) = \overline{V}'(t,u) \ \text{ for } t \in C_{\overline{V}'(u)}, \ u = 1, 2, \ldots, z \tag{5.12}$$

Motivation: For each fixed u, $u = 1, 2, \ldots, z$, assumptions (i)–(viii) of Lemma 5.2 are identical to assumptions (i)–(viii) of Lemma 4.3, condition (5.11) is identical to condition (4.7) and condition (5.12) is identical to condition (4.8). Moreover, since from Lemma 4.3, conditions (4.7) and (4.8) are equivalent, then Lemma 5.2 is valid.

Lemma 5.2 and Theorem 4.2 establish the class of limit reliability functions for non-homogeneous multi-state series systems pointed out in the form of the next theorem [81].

Theorem 5.2

The class of limit non-degenerate reliability functions of the non-homogeneous multi-state series system, under the assumptions of Lemma 5.2, is composed of 3^z reliability functions of the form

$$\overline{\mathfrak{R}}'(t,\cdot) = [1, \overline{\mathfrak{R}}'(t,1), \ldots, \overline{\mathfrak{R}}'(t,z)], \quad t \in (-\infty, \infty) \tag{5.13}$$

where

$$\overline{\mathfrak{R}}'(t,u) \in \{\overline{\mathfrak{R}}'_1(t), \overline{\mathfrak{R}}'_2(t), \overline{\mathfrak{R}}'_3(t)\}, \quad u = 1, 2, \ldots, z \tag{5.14}$$

and $\overline{\mathfrak{R}}'_i(t)$, $i = 1, 2, 3$ are given by Eqs. (4.9) – (4.11) with $\overline{d}(t) = \overline{d}(t,u)$, $u = 1, 2, \ldots, z$, where $\overline{d}(t,u)$ are defined by Eq. (5.9).

Motivation: For each fixed u, $u = 1, 2, \ldots, z$, the coordinate $\overline{\mathfrak{R}}'(t,u)$ of the vector $\overline{\mathfrak{R}}'(t,\cdot)$ defined by Eq. (5.13), from Theorem 4.2 that is the consequence of Lemma 4.3, may be one of the three types of reliability functions given by Eqs. (4.9) – (4.11) with $\overline{d}(t) = \overline{d}(t,u)$, where $\overline{d}(t,u)$ are defined by Eq. (5.9). Therefore, the number of different multi-state limit reliability functions of the considered system is equal to the number of z-term variations of the 3-component set (5.14), i.e. 3^z, and they are of the form (5.13).

Corollary 5.4

If the non-homogeneous multi-state series system is composed of components having Weibull reliability functions

$$R^{(i)}(t,\cdot) = [1, R^{(i)}(t,1), \ldots, R^{(i)}(t,z)], \quad t \in (-\infty, \infty)$$

where

$$R^{(i)}(t,u) = 1 \quad \text{for } t < 0, \ R^{(i)}(t,u) = \exp[-\beta_i(u) t^{\alpha_i(u)}] \quad \text{for } t \geq 0, \ \alpha_i(u) > 0,$$
$$\beta_i(u) > 0, \ i = 1,2,\ldots,a, \ u = 1,2,\ldots,z$$

and

$$a_n(u) = (\beta(u)n)^{-1/\alpha(u)}, \ b_n(u) = 0 \quad \text{for } u = 1,2,\ldots,z$$

where

$$\alpha(u) = \min_{1 \leq i \leq a} \{\alpha_i(u)\}, \beta(u) = \max_{i:\alpha_i(u) = \alpha(u)} \{\beta_i(u)\} \quad \text{for } u = 1,2,\ldots,z$$

then

$$\overline{\mathfrak{R}}'_2(t,\cdot) = [1, \overline{\mathfrak{R}}'_2(t,1), \ldots, \overline{\mathfrak{R}}'_2(t,z)]$$

where

$$\overline{\mathfrak{R}}'_2(t,u) = 1 \quad \text{for } t < 0, \ \overline{\mathfrak{R}}'_2(t,u) = \exp[-\overline{d}(t,u) t^{\alpha(u)}] \quad \text{for } t \geq 0, \ u = 1,2,\ldots,z$$

and

$$\overline{d}(t,u) = \sum_{i:\alpha_i(u) = \alpha(u)} q_i \beta_i(u)/\beta(u)$$

is its limit reliability function.

Motivation: Since for all fixed u, we have

$$a_n(u)t + b_n(u) = a_n(u)t < 0 \quad \text{for } t < 0 \text{ and } a_n(u)t + b_n(u) = a_n(u)t \geq 0 \quad \text{for } t \geq 0$$

then

$$F^{(i)}(a_n(u)t + b_n(u), u) = 0 \quad \text{for } t < 0$$

and

$$F^{(i)}(a_n(u)t + b_n(u), u) = 1 - \exp[-\beta_i(u)(a_n(u)t + b_n(u))t^{\alpha_i(u)}] \quad \text{for } t \geq 0$$

Assuming

$$F(t, u) = 0 \quad \text{for } t < 0 \text{ and } F(t, u) = 1 - \exp[-\beta(u)t^{\alpha(u)}] \quad \text{for } t \geq 0$$

for all $i = 1, 2, \ldots, a$ and $t \geq t_0(u) = 0$, we have

$$\lim_{n\to\infty} \frac{F^{(i)}(a_n(u)t + b_n(u), u)}{F(a_n(u)t + b_n(u), u)} = \lim_{n\to\infty} \frac{1 - \exp[-\beta_i(u)(a_n(u)t)^{\alpha_i(u)}]}{1 - \exp[-\beta(u)(a_n(u)t)^{\alpha(u)}]}$$
$$= \lim_{n\to\infty} \frac{\beta_i(u)}{\beta(u)}(a_n(u)t)^{\alpha_i(u) - \alpha(u)} \leq 1$$

which means that condition (vi) of Lemma 5.2 holds. Moreover, from Eq. (5.9), we have

$$\overline{d}(t, u) = \begin{cases} 0 & \text{for } t < 0 \\ \sum\limits_{i:\alpha_i(u) = \alpha(u)} q_i\beta_i(u)/\beta(u) & \text{for } t \geq 0 \end{cases}$$

Therefore

$$\overline{V}'(t, u) = \lim_{n\to\infty} nF(a_n(u)t + b_n(u), u)\overline{d}(t, u) = 0 \quad \text{for } t < 0, \quad u = 1, 2, \ldots, z$$

and

$$\begin{aligned} \overline{V'}(t, u) &= \lim_{n\to\infty} nF(a_n(u)t + b_n(u), u)\overline{d}(t, u) \\ &= \lim_{n\to\infty} n[1 - \exp[-\beta(u)(a_n(u)t)^{\alpha(u)}]]\overline{d}(t, u) \\ &= \lim_{n\to\infty} n\beta(u)(a_n(u)t)^{\alpha(u)}\overline{d}(t, u) = \overline{d}(t, u)t^{\alpha(u)} \quad \text{for } t \geq 0, \quad u = 1, 2, \ldots, z \end{aligned}$$

which from Lemma 5.2 completes the proof.

Example 5.4 (A Piping Transportation System)
The pipeline system is composed of $n = 80$ five-state pipe segments, i.e. $z = 4$. There are four types of pipe segments in the system, as follows:

20 segments with exponential reliability functions

$$R^{(1)}(t,1) = \exp[-0.01t], \ \ R^{(1)}(t,2) = \exp[-0.0120t],$$
$$R^{(1)}(t,3) = \exp[-0.015t], \ \ R^{(1)}(t,4) = \exp[-0.025t]$$

20 segments with exponential reliability functions

$$R^{(2)}(t,1) = \exp[-0.018t], \ \ R^{(2)}(t,2) = \exp[-0.019t],$$
$$R^{(2)}(t,3) = \exp[-0.020t], \ \ R^{(2)}(t,4) = \exp[-0.023t]$$

10 segments with Weibull reliability functions

$$R^{(3)}(t,1) = \exp[-0.0005t^2], \ \ R^{(3)}(t,2) = \exp[-0.0006t^2],$$
$$R^{(3)}(t,3) = \exp[-0.0010t^2], \ \ R^{(3)}(t,4) = \exp[-0.0015t^2]$$

and 30 segments with Weibull reliability functions

$$R^{(4)}(t,1) = \exp[-0.00061t^3], \ \ R^{(4)}(t,2) = \exp[-0.00062t^3],$$
$$R^{(4)}(t,3) = \exp[-0.00064t^3], \ \ R^{(4)}(t,4) = \exp[-0.00070t^3].$$

According to Definition 3.17, the considered system is a non-homogeneous multi-state series system with parameters

$$n = 80, \ \ a = 4, \ \ q_1 = 2/8, \ \ q_2 = 2/8, \ \ q_3 = 1/8, \ \ q_4 = 3/8$$

Thus, from Eqs. (3.33) and (3.34), we have

$$\overline{\boldsymbol{R}}'_{80}(t,\cdot) = [1, \overline{\boldsymbol{R}}'_{80}(t,1), \overline{\boldsymbol{R}}'_{80}(t,2), \overline{\boldsymbol{R}}'_{80}(t,3), \overline{\boldsymbol{R}}'_{80}(t,4)]$$

where

$$\overline{\boldsymbol{R}}'_{80}(t,1) = \exp[-0.2t - 0.34t - 0.005t^2 - 0.00183t^3],$$

$$\overline{\boldsymbol{R}}'_{80}(t,2) = \exp[-0.24t - 0.38t - 0.006t^2 - 0.00186t^3],$$

$$\overline{\boldsymbol{R}}'_{80}(t,3) = \exp[-0.3t - 0.4t - 0.01t^2 - 0.00192t^3],$$

$$\overline{\boldsymbol{R}}'_{80}(t,4) = \exp[-0.5t - 0.46t - 0.015t^2 - 0.0021t^3] \quad \text{for } t \geq 0$$

Further, from Corollary 5.4, we find

$$\alpha(u) = \min\{1,1,2,3\} = 1 \quad \text{for } u = 1,2,3,4$$

$$\beta(1) = \max\{0.01, 0.018\} = 0.018, \quad \beta(2) = \max\{0.012, 0.019\} = 0.019,$$

$$\beta(3) = \max\{0.015, 0.02\} = 0.020, \quad \beta(4) = \max\{0.025, 0.023\} = 0.025$$

$$a_n(1) = 1/(0.018.80) = 0.694, \quad a_n(2) = 1/(0.019.80) = 0.658,$$

$$a_n(3) = 1/(0.020.80) = 0.625, \quad a_n(4) = 1/(0.025.80) = 0.500, \quad b_n(u) = 0$$

$$\text{for } u = 1, 2, 3, 4$$

$$d(t, 1) = (2/8)(0.018/0.018) + (2/8)(0.010/0.018) = 0.389,$$

$$d(t, 2) = (2/8)(0.019/0.019) + (2/8)(0.012/0.019) = 0.408,$$

$$d(t, 3) = (2/8)(0.020/0.020) + (2/8)(0.015/0.020) = 0.438,$$

$$d(t, u) = (2/8)(0.025/0.025) + (2/8)(0.023/0.025) = 0.480$$

and we conclude that the system limit reliability function takes the form

$$\overline{\mathfrak{R}}'_2(t, \cdot) = [\exp[-0.389t], \exp[-0.408t], \exp[-0.438t], \exp[-0.48t]] \quad \text{for } t \geq 0$$

Hence, according to Eq. (3.49), the approximate formula for $t \geq 0$ is given by

$$\begin{aligned}\overline{\boldsymbol{R}}'_{80}(t, \cdot) &\cong \overline{\mathfrak{R}}'_2((t - b_n(u))/a_n(u), \cdot) \\ &= [\exp[-0.56t], \exp[-0.62t], \exp[-0.7t], \exp[-0.96t]]\end{aligned}$$

The mean values of the system sojourn times in the state subsets in years (for instance, for segments of the second type), after applying Eq. (3.4), are

$$M_2(1) = 1/0.018 = 55.56, \quad M_2(2) = 1/0.019 = 52.63,$$

$$M_2(3) = 1/0.020 = 50.00, \quad M_2(4) = 1/0.023 = 43.48$$

and their mean times in the particular states, from Eq. (3.8), are

$$\overline{M}_2(1) = 2.93, \quad \overline{M}_2(2) = 2.63, \quad \overline{M}_2(3) = 6.52, \quad \overline{M}_2(4) = 43.48$$

The approximate expected values of the pipeline system sojourn times in the state subsets calculated from the approximate reliability function according to Eq. (3.13) are

$$M(1) = 1/0.56 = 1.79, \quad M(2) = 1/0.62 = 1.61,$$

$$M(3) = 1/0.7 = 1.43, \quad M(4) = 1/0.96 = 1.04$$

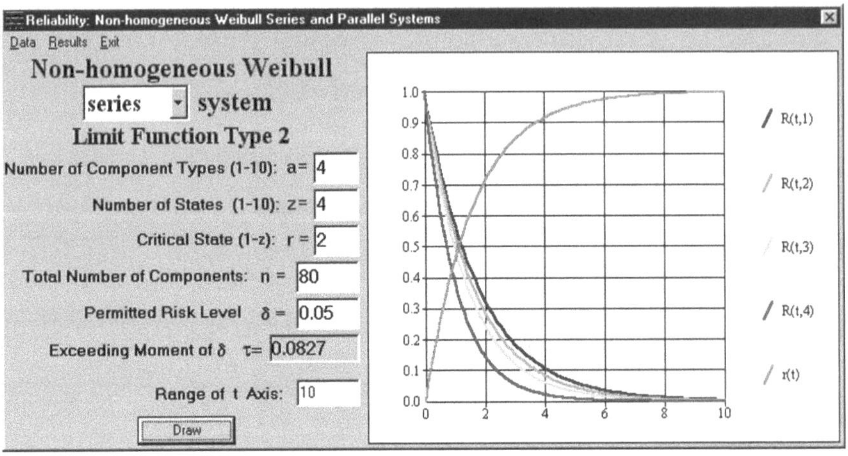

Figure 5.2 The graphs of multi-state reliability function and risk function of the piping system.

and from Eq. (3.17), the system lifetimes in the particular states are

$$\overline{M}(1) = 0.18,\ \overline{M}(2) = 0.18,\ \overline{M}(3) = 0.39,\ \overline{M}(4) = 1.04$$

If the critical state is $r = 2$, then the system risk function, according to Eq. (3.18), is given by

$$\boldsymbol{r}(t) = 1 - \exp[-0.62t] \quad \text{for } t \geq 0$$

The moment when the system risk function exceeds a permitted level $\delta = 0.05$, from Eq. (3.19), is

$$\tau = \boldsymbol{r}^{-1}(\delta) = -(1/0.62)\log(1-\delta) = 0.0827 \text{ years}$$

The graphs of the piping reliability function and its risk function have been drawn by a computer program [93] and are presented in Figure 5.2.

5.2 Reliability Evaluation of Multi-State Parallel Systems

In proving facts on limit reliability functions of homogeneous parallel multi-state systems, the following extension of Lemma 4.5 is useful [81].

Lemma 5.3
If

i. $\Re(t,u) = 1 - \exp[-V(t,u)]$, $u = 1, 2, \ldots, z$, is a non-degenerate reliability function,
ii. $\boldsymbol{R}_n(t,\cdot) = [1, \boldsymbol{R}_n(t,1), \ldots, \boldsymbol{R}_n(t,z)]$, $t \in (-\infty, \infty)$, is the reliability function of a homogeneous multi-state parallel system defined by Eqs. (3.22) and (3.23),
iii. $a_n(u) > 0$, $b_n(u) \in (-\infty, \infty)$, $u = 1, 2, \ldots, z$,

then

$$\mathfrak{R}(t,\cdot) = [1, \mathfrak{R}(t,1), \ldots, \mathfrak{R}(t,z)], \quad t \in (-\infty, \infty)$$

is the multi-state limit reliability function of this system, i.e.

$$\lim_{n\to\infty} \boldsymbol{R}_n(a_n(u)t + b_n(u), u) = \mathfrak{R}(t,u) \quad \text{for } t \in C_{\mathfrak{R}(u)}, \ u = 1, 2, \ldots, z \tag{5.15}$$

if and only if

$$\lim_{n\to\infty} nR(a_n(u)t + b_n(u), u) = V(t,u) \quad \text{for } t \in C_{V(u)}, \ u = 1, 2, \ldots, z \tag{5.16}$$

Motivation: For each fixed u, $u = 1, 2, \ldots, z$, assumptions (i)–(iii) of Lemma 5.3 are identical to assumptions (i)–(iii) of Lemma 4.5, condition (5.15) is identical to condition (4.18) and condition (5.16) is identical to condition (4.19). Since, from Lemma 4.5, conditions (4.18) and (4.19) are equivalent, then conditions (5.15) and (5.16) are also equivalent.

Lemma 5.3 and Theorem 4.3 determine the class of all non-degenerate limit reliability functions for homogeneous multi-state parallel systems, namely, their application results in the following theorem [74].

Theorem 5.3
The class of limit non-degenerate reliability functions of the homogeneous multi-state parallel system is composed of 3^z reliability functions of the form

$$\mathfrak{R}(t,\cdot) = [1, \mathfrak{R}(t,1), \ldots, \mathfrak{R}(t,z)], \quad t \in (-\infty, \infty) \tag{5.17}$$

where

$$\mathfrak{R}(t,u) \in \{\mathfrak{R}_1(t), \mathfrak{R}_2(t), \mathfrak{R}_3(t)\}, \quad u = 1, 2, \ldots, z \tag{5.18}$$

and $\mathfrak{R}_i(t)$, $i = 1, 2, 3$, are defined by Eqs. (4.20) – (4.22).

Motivation: For each fixed u, $u = 1, 2, \ldots, z$, the coordinate $\mathfrak{R}(t,u)$ of the vector $\mathfrak{R}(t,\cdot)$ defined by Eq. (5.17), from Theorem 4.3 that is the consequence of Lemma 4.5, may be one of the three types of reliability functions given by Eqs. (4.20) – (4.22). It means that the number of different limit multi-state reliability functions of the considered system is equal to the number of z-term variations of the 3-component set (5.18), i.e. 3^z, and they are of the form (5.17).

Corollary 5.5
If components of the homogeneous multi-state parallel system have Weibull reliability functions

$$R(t,\cdot) = [1, R(t,1), \ldots, R(t,z)], \quad t \in (-\infty, \infty)$$

where

$$R(t,u)=1 \quad \text{for } t<0, \quad R(t,u)=\exp[-\beta(u)t^{\alpha(u)}] \quad \text{for } t\geq 0, \quad \alpha(u)>0,$$
$$\beta(u)>0, \quad u=1,2,\ldots,z$$

and

$$a_n(u)=b_n(u)/(\alpha(u)\log n), \quad b_n(u)=(\log n/\beta(u))^{1/\alpha(u)}, \qquad u=1,2,\ldots,z$$

then

$$\mathfrak{R}_3(t,\cdot)=[1,\mathfrak{R}_3(t,1),\ldots,\mathfrak{R}_3(t,z)], \quad t\in(-\infty,\infty)$$

where

$$\mathfrak{R}_3(t,u)=1-\exp[-\exp[-t]] \quad \text{for } t\in(-\infty,\infty), \quad u=1,2,\ldots,z$$

is its limit reliability function.

Motivation: Since for each fixed u, sufficiently large n and all $t\in(-\infty,\infty)$, we have

$$a_n(u)t+b_n(u)=b_n(u)(t/(\alpha(u)\log n)+1)>0$$

then

$$R(a_n(u)t)+b_n(u),u)=\exp[-\beta(u)(a_n(u)t+b_n(u))^{\alpha(u)}] \quad \text{for } t\in(-\infty,\infty)$$

Hence

$$\begin{aligned} nR(a_n(u)t+b_n(u),u) &= n\exp[-\beta(u)(a_n(u)t)+b_n(u))^{\alpha(u)}] \\ &= n\exp[-\beta(u)(b_n(u))^{\alpha(u)}(t/(\alpha(u)\log n)+1)^{\alpha(u)}] \\ &= n\exp[-\log n[t/(\alpha(u)\log n)+1]^{\alpha(u)}] \end{aligned}$$

Further, applying the equality

$$[t/(\alpha(u)\log n)+1]^{\alpha(u)}=1+t/\log n+\mathrm{o}(1/\log n) \quad \text{for } t\in(-\infty,\infty)$$

we get

$$\begin{aligned} V(t,u) &= \lim_{n\to\infty} nR(a_n(u)t)+b_n(u),u \\ &= \lim_{n\to\infty} n\exp[-\log n-t-o(1)] \\ &= \lim_{n\to\infty} \exp[-t-o(1)]=\exp[-t] \quad \text{for } t\in(-\infty,\infty) \end{aligned}$$

which from Lemma 5.3 completes the proof.

Example 5.5 (An Energetic Cable)
Let us consider a model energetic network stretched between two poles and composed of three energetic cables, six insulators and two bearers and analyse the reliability of a single cable. The cable consists of 36 identical wires. Assuming that the cable is able to conduct the current if at least one of its wires is not failed we conclude that it is a homogeneous parallel system composed of $n = 36$ basic components. Further, assuming that the wires are four-state components, i.e. $z = 3$, having Weibull reliability functions with parameters

$$\alpha(u) = 2, \quad \beta(u) = (7.07)^{2u-8}, \quad u = 1, 2, 3$$

after applying Corollary 5.5 with normalizing constants

$$a_n(u) = (7.07)^{4-u} / \left(2\sqrt{\log 36}\right), \quad b_n(u) = (7.07)^{4-u}\sqrt{\log 36}, \quad u = 1, 2, 3$$

and considering Eq. (3.49), we obtain, for $t \in (-\infty, \infty)$, the following approximate formula for the reliability function of the considered energetic cable

$$\begin{aligned}[\boldsymbol{R}_{36}(t,0), \boldsymbol{R}_{36}(t,1), \boldsymbol{R}_{36}(t,2), \boldsymbol{R}_{36}(t,3)] \cong [1, 1 - \exp[-\exp[-0.01071t + 7.167]],\\ 1 - \exp[-\exp[-0.07572t + 7.167]],\\ 1 - \exp[-\exp[-0.53543t + 7.167]]]\end{aligned}$$

The mean values of the cable lifetimes $T(u)$ in the state subsets in months and their standard deviations calculated according to the formulae [15,93]

$$M(u) = E[T(u)] = Ca_n(u) + b_n(u), \quad \sigma(u) = \pi a_n/\sqrt{6}, \quad u = 1, 2, 3$$

where $C \cong 0.5772$ is Euler's constant, are

$$M(1) \cong 7.23, \quad M(2) \cong 102, \quad M(3) \cong 14.5, \quad \sigma(1) \cong 120, \quad \sigma(2) \cong 17, \quad \sigma(3) \cong 2.4$$

while the mean values of the cable lifetimes in the particular states, from Eq. (3.17), are

$$\overline{M}(1) \cong 621, \quad \overline{M}(2) \cong 8.75, \quad \overline{M}(3) \cong 14.5$$

If the critical state is $r = 2$, then from Eq. (3.18), the cable risk function is given by

$$\boldsymbol{r}(t) \cong \exp[-\exp[-0.07572t + 7.167]]$$

The moment when the cable risk function exceeds a permitted level $\delta = 0.05$, according to Eq. (3.19), is

$$\tau = \boldsymbol{r}^{-1}(\delta) \cong [7.167 - \log[-\log \delta]]/0.07572 = 80 \text{ months}$$

The values of the cable risk function are given in Table 5.1. The graphs of the cable multi-state reliability function and its risk function plotted by the computer program [93] are given in Figure 5.3.

In fixing the class of limit reliability functions for non-homogeneous multi-state parallel systems, we apply the following extension of Lemma 4.8.

Table 5.1 The Values of the Energetic Cable Risk Function

t	$r(t)$
0	0.000
50	0.000
60	0.000
70	0.017
80	0.048
90	0.276
100	0.514
110	0.731
120	0.856
130	0.934
140	0.968
150	0.985
160	0.993

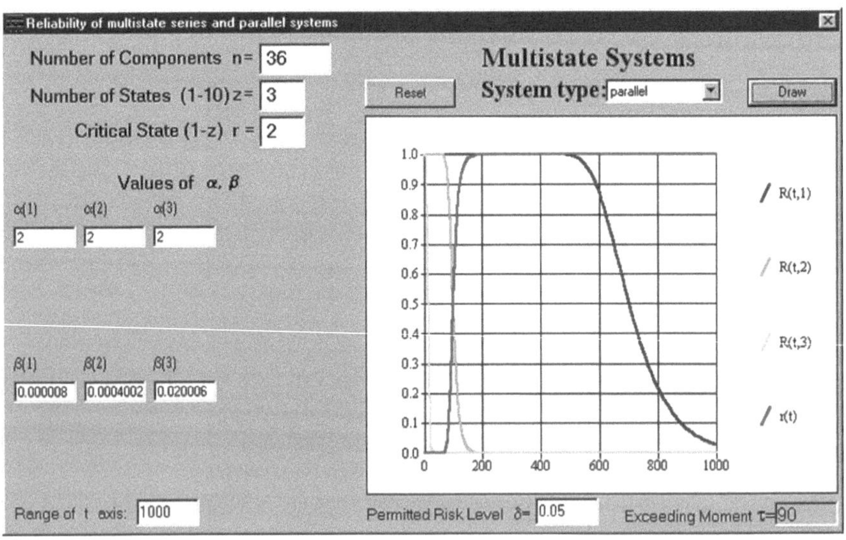

Figure 5.3 The graphs of the energetic cable reliability function and risk function.

Lemma 5.4
If

i. $\overline{\Re}(t,u) = 1 - \exp[-V'(t,u)]$, $u = 1, 2, \ldots, z$, is a non-degenerate reliability function,
ii. $\boldsymbol{R}'_n(t,\cdot) = [1, \boldsymbol{R}'_n(t,1), \ldots, \boldsymbol{R}'_n(t,z)]$, $t \in (-\infty, \infty)$, is the reliability function of a non-homogeneous multi-state parallel system defined by Eqs. (3.36) and (3.37),
iii. $a_n(u) > 0$, $b_n(u) \in (-\infty, \infty)$, $u = 1, 2, \ldots, z$,
iv. $R(t,u)$ for each fixed u is one of the reliability functions $R^{(1)}(t,u)$, $R^{(2)}(t,u)$, …, $R^{(a)}(t,u)$ defined by Eq. (3.35) such that
v. $\exists N(u)'n > N(u)$, $R(a_n(u)t + b_n(u), u) \neq 0$ for $t < t_0(u)$, $R(a_n(u)t + b_n(u), u) = 0$ for $t \geq t_0(u)$, where $t_0(u) \in (-\infty, \infty >$
vi.
$$\lim_{n\to\infty} \frac{R^{(i)}(a_n(u)t + b_n(u), u)}{R(a_n(u)t + b_n(u), u)} \leq 1 \quad \text{for } t < t_0(u), \; i = 1, 2, \ldots, a, \; u = 1, 2, \ldots, z$$
and moreover there exist non-increasing functions
vii.
$$d(t,u) = \begin{cases} \lim\limits_{n\to\infty} \sum\limits_{i=1}^{a} q_i d_i(a_n(u)t + b_n(u), u) & \text{for } t < t_0(u) \\ 0 & \text{for } t \geq t_0(u) \end{cases} \tag{5.19}$$
where
viii.
$$d_i(a_n(u)t + b_n(u), u) = \frac{R^{(i)}(a_n(u)t + b_n(u), u)}{R(a_n(u)t + b_n(u), u)}$$

then

$$\Re'(t,\cdot) = [1, \Re'(t,1), \ldots, \Re'(t,z)], \quad t \in (-\infty, \infty)$$

is the multi-state limit reliability function of this system, i.e.

$$\lim_{n\to\infty} \boldsymbol{R}'_n(a_n(u)t + b_n(u), u) = \Re'(t,u) \quad \text{for } t \in C_{\Re'(u)}, \; u = 1, 2, \ldots, z \tag{5.20}$$

if and only if

$$\lim_{n\to\infty} nR(a_n(u)t + b_n(u), u)d(t,u) = V'(t,u) \quad \text{for } t \in C_{V'(u)}, \; u = 1, 2, \ldots, z \tag{5.21}$$

Motivation: For each fixed u, $u = 1, 2, \ldots, z$, assumptions (i)–(viii) of Lemma 5.4 are identical to assumptions (i)–(viii) of Lemma 4.8, condition (5.20) is identical to condition (4.24) and condition (5.21) is identical to condition (4.25). And, since from Lemma 4.8, conditions (4.24) and (4.25) are equivalent, then Lemma 5.4 is valid.

Lemma 5.4 and Theorem 4.4 allow us to fix the class of all possible non-degenerate limit reliability functions of the non-homogeneous multi-state parallel system listed in the following theorem [81].

Theorem 5.4
The class of limit non-degenerate reliability functions of the non-homogeneous multi-state parallel system, under the assumptions of Lemma 5.4, is composed of 3^z reliability functions of the form

$$\mathfrak{R}'(t,\cdot) = [1, \mathfrak{R}'(t,1), \ldots, \mathfrak{R}'(t,z)], \quad t \in (-\infty, \infty) \tag{5.22}$$

where

$$\mathfrak{R}'(t,u) \in \{\mathfrak{R}'_1(t), \mathfrak{R}'_2(t), \mathfrak{R}'_3(t)\}, \quad u = 1, 2, \ldots, z \tag{5.23}$$

and $\mathfrak{R}'_i(t)$, $i = 1, 2, 3$, are given by Eqs. (4.26)–(4.28) with $d(t) = d(t,u)$, $u = 1, 2, \ldots, z$, where $d(t,u)$ are defined by Eq. (5.19).

Motivation: For each fixed u, $u = 1, 2, \ldots, z$, coordinate $\mathfrak{R}'(t,u)$ of the vector $\mathfrak{R}'(t,\cdot)$ defined by Eq. (5.22), from Theorem 4.4 that is a consequence of Lemma 4.8, may be one of the three types of reliability functions given by Eqs. (4.26)–(4.28) with $d(t) = d(t,u)$, where $d(t,u)$ are defined by Eq. (5.19). Thus the number of different multi-state limit reliability functions of the considered system is equal to the number of z-term variations of the 3-element set (5.23), i.e. 3^z, and they are of the form (5.22).

Corollary 5.6
If components of the non-homogeneous multi-state parallel system have Weibull reliability functions

$$R^{(i)}(t,\cdot) = [1, R^{(i)}(t,1), \ldots, R^{(i)}(t,z)], \quad t \in (-\infty, \infty)$$

where

$$R^{(i)}(t,u) = 1 \quad \text{for } t < 0, \quad R^{(i)}(t,u) = \exp[-\beta_i(u) t^{\alpha_i(u)}]$$
$$\text{for } t \geq 0, \quad \alpha_i(u) > 0, \quad \beta_i(u) > 0, \quad i = 1, 2, \ldots, a, \quad u = 1, 2, \ldots, z$$

and

$$a_n(u) = b_n(u)/(\alpha(u)\log n), \quad b_n(u) = (\log n/\beta(u))^{1/\alpha(u)}, \quad u = 1, 2, \ldots, z$$

where

$$\alpha(u) = \min_{1 \leq i \leq a} \{\alpha_i(u)\}, \quad \beta(u) = \max_{(i:\alpha_i(u) = \alpha(u))} \{\beta_i(u)\} \quad \text{for } u = 1, 2, \ldots, z$$

then

$$\mathfrak{R}'_3(t,\cdot) = [1, \mathfrak{R}'_3(t,1), \ldots, \mathfrak{R}'_3(t,z)], \quad t \in (-\infty, \infty)$$

where

$$\mathfrak{R}_3'(t,u) = 1 - \exp[-d(t,u)\exp[-t]] \quad \text{for } t \in (-\infty, \infty), \ \ u = 1, 2, \ldots, z$$

and

$$d(t,u) = \sum_{i:\alpha_i(u) = \alpha(u), \ \beta_i(u) = \beta(u)} q_i \quad \text{for } t \in (-\infty, \infty), \ \ u = 1, 2, \ldots, z$$

is its limit reliability function.

Motivation: Since for each fixed u, we have

$$a_n(u)t + b_n(u) = b_n(u)\left(1 + \frac{a_n(u)}{b_n(u)}t\right) \to \infty \quad \text{as } n \to \infty$$

then defining

$$R(t,u) = 1 \ \text{ for } t < 0 \text{ and } R(t,u) = 1 - \exp[-\beta(u)t^{\alpha(u)}] \ \text{ for } t \geq 0, \ u = 1, 2, \ldots, z$$

for all $i = 1, 2, \ldots, a$ and $t \in (-\infty, \infty)$, we have

$$\lim_{n\to\infty} \frac{R^{(i)}(a_n(u)t + b_n(u), u)}{R(a_n(u)t + b_n(u), u)} = \lim_{n\to\infty} \frac{\exp[-\beta_i(u)(a_n(u)t + b_n(u))^{\alpha_i(u)}]}{\exp[-\beta(u)(a_n(u)t + b_n(u))^{\alpha(u)}]}$$

$$= \lim_{n\to\infty} \exp\left[-\beta(u)(a_n(u)t + b_n(u))^{\alpha(u)}\left[\frac{\beta_i(u)}{\beta(u)}(a_n(u)t + b_n(u))^{\alpha_i(u)-\alpha(u)} - 1\right]\right] \leq 1$$

which means that condition (vi) of Lemma 5.4 holds with $t_0(u) = \infty$. Moreover, according to Eq. (5.19), we have

$$d(t,u) = \sum_{i:\alpha_i(u) = \alpha(u), \ \beta_i(u) = \beta(u)} q_i$$

Therefore

$$\begin{aligned}
V'(t,u) &= \lim_{n\to\infty} nR(a_n(u)t + b_n(u), u)d(t,u) \\
&= \lim_{n\to\infty} n \exp[-\beta(u)(a_n(u)t + b_n(u))^{\alpha(u)}]]d(t,u) \\
&= \lim_{n\to\infty} \exp\left[-\log n\left(1 + \frac{t}{\alpha(u)\log n}\right)^{\alpha(u)} + \log n\right] d(t,u) \\
&= d(t,u)\exp[-t] \quad \text{for } t \in (-\infty, \infty), \ \ u = 1, 2, \ldots, z
\end{aligned}$$

which from Lemma 5.4 completes the proof.

Example 5.6 (A Three-Stratum Rope, Durability)

Let us consider the steel rope of type *M*-80-200-10 described in the Polish Norm [145]. It is a three-stratum rope composed of 36 strands: 18 outer strands, 12 inner strands and 6 more inner strands. All strands consist of seven still wires. The rope cross section is presented in Figure 5.4.

Considering the strands as basic components we conclude that the rope is a parallel system composed of $n = 36$ components (strands). Due to Trade Norm [170] concerned with the evaluation of wear level, the following reliability states of the strands are distinguished:

- state 3: a strand is new, without any defects,
- state 2: the number of broken wires in the strand is >0% and <25% of all its wires, or corrosion of wires is >0% and <25%, abrasion is up to 25% and strain is up to 50%,
- state 1: the number of broken wires in the strand is ≥25% and <50% of all its wires, or corrosion of wires is ≥25% and <50%, abrasion is up to 50% and strain is up to 50%,
- state 0: otherwise (a strand is failed).

Since outer strands are more subject to damage than inner strands, then two types of strands are distinguished. Namely, it is assumed that all strands have Weibull reliability functions with two different parameters:

$$\alpha_1(u) = 2, \quad \beta_1(u) = 0.10u, \quad u = 1, 2, 3$$

for outer strands and

$$\alpha_2(u) = 1.5, \quad \beta_2(u) = 0.25u, \quad u = 1, 2, 3$$

for inner strands, which means that the reliability functions of the components of the first and second types respectively are given by

$$R^{(1)}(t, u) = 1 \quad \text{for } t < 0, \quad R^{(1)}(t, u) = \exp[-0.10ut^2] \quad \text{for } t \geq 0, \quad u = 1, 2, 3$$

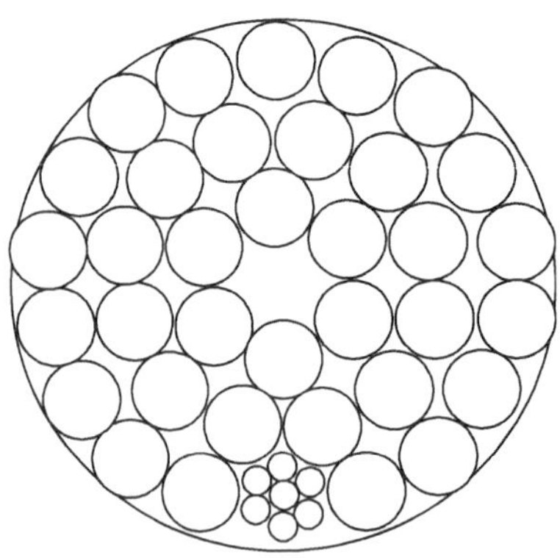

Figure 5.4 The steel rope *M*-80-200-10 cross section.

and

$$R^{(2)}(t,u) = 1 \quad \text{for } t < 0, \;\; R^{(2)}(t,u) = \exp[-0.25ut^{3/2}] \quad \text{for } t \geq 0, \;\; u = 1,2,3$$

Thus, since the considered system is non-homogeneous, according to Definition 3.18, we conclude that the rope is a non-homogeneous four-state, i.e. $z = 3$, parallel system with parameters $n = 36, a = 2, q_1 = 1/2, q_2 = 1/2$

Then, according to Eqs. (3.36) and (3.37), its reliability function is given by

$$\boldsymbol{R}'_{36}(t,\cdot) = [1, \boldsymbol{R}'_{36}(t,1), \boldsymbol{R}'_{36}(t,2), \boldsymbol{R}'_{36}(t,3)]$$

where

$$\boldsymbol{R}'_{36}(t,u) = 1 \quad \text{for } t < 0,$$

$$\boldsymbol{R}'_{36}(t,u) = 1 - \prod_{i=1}^{2}(F^{(i)}(t,u))^{q_i n}$$
$$= [1 - \exp[-0.10ut^2]]^{18}[1 - \exp[-0.25ut^{3/2}]]^{18} \quad \text{for } t \geq 0, \;\; u = 1,2,3$$

Since

$$\alpha(u) = \min\{2, 1.5\} = 1.5$$

$$\beta(u) = \min\{0.25u\} = 0.25u, \quad u = 1,2,3$$

then applying Corollary 5.6 with normalizing constants

$$a_n(u) = b_n(u)/(\alpha(u)\log n) = 2/3(0.25u)^{2/3}(\log 36)^{1/3}$$

$$b_n(u) = \left[\frac{1}{\beta(u)}\log n\right]^{1/\alpha(u)} = (\log 36/0.25u)^{2/3}, \quad u = 1,2,3$$

and considering that according to Eq. (5.19)

$$d(t,u) = 1/2 \quad \text{for } u = 1,2,3$$

we conclude that the rope limit reliability function is

$$\mathfrak{R}'_3(t,\cdot) = [1, 1 - \exp[-0.5\exp[-t]], 1 - \exp[-0.5\exp[-t]],$$
$$1 - \exp[-0.5\exp[-t]]]$$

for $t \in (-\infty, \infty)$.

Hence, using Eq. (3.49), since

$$a_n(1) = 1.098, \quad b_n(1) = 5.901$$

$$a_n(2) = 0.692, \quad b_n(2) = 3.717$$

$$a_n(3) = 0.528, \quad b_n(3) = 2.837$$

then the approximate formula for the rope reliability function takes the form

$$\begin{aligned} \boldsymbol{R}'_{36}(t,\cdot) &\cong \mathfrak{R}'_3\left(\frac{t-b_n(u)}{a_n(u)},\cdot\right) \\ &= [1, 1-\exp[-0.5\exp[-0.911t+5.372]], \\ &\quad 1-\exp[-0.5\exp[-1.445t+5.372]], \\ &\quad 1-\exp[-0.5\exp[-1.894t+5.372]]], \quad t\in(-\infty,\infty) \end{aligned}$$

The approximate mean values of the rope lifetimes $T(u)$ in the state subsets and their standard deviations in years, from Eq. (3.13), calculated according to the formulae [15,93]

$$M(u) = E[T(u)] = Ca_n(u) + b_n(u) - \log 2/a_n(u)$$

$$\sigma(u) = \pi a_n(u)/\sqrt{6}, \quad u = 1, 2, 3$$

where $C \cong 0.5772$ is Euler's constant, are

$$M(1) \cong 5.90, \quad M(2) \cong 3.11, \quad M(3) \cong 1.83$$

$$\sigma(1) \cong 1.41, \quad \sigma(2) \cong 0.89, \quad \sigma(3) \cong 0.68$$

whereas the rope expected lifetimes in the particular states, from Eq. (3.17), are

$$\overline{M}(1) \cong 2.79, \quad \overline{M}(2) \cong 1.28, \quad \overline{M}(3) \cong 1.83$$

If the critical rope reliability state is $r = 2$, then according to Eq. (3.18) its risk function is given by the following approximate formula:

$$\boldsymbol{r}(t) \cong \exp[-0.5\exp[-1.445t+5.371]], \quad t\in(-\infty,\infty)$$

The moment when the rope risk function exceeds an admissible level $\delta = 0.05$, from Eq. (3.19), is

$$\tau = (5.372 - \log(-2\log\delta))/1.445 = 2.48 \text{ years}$$

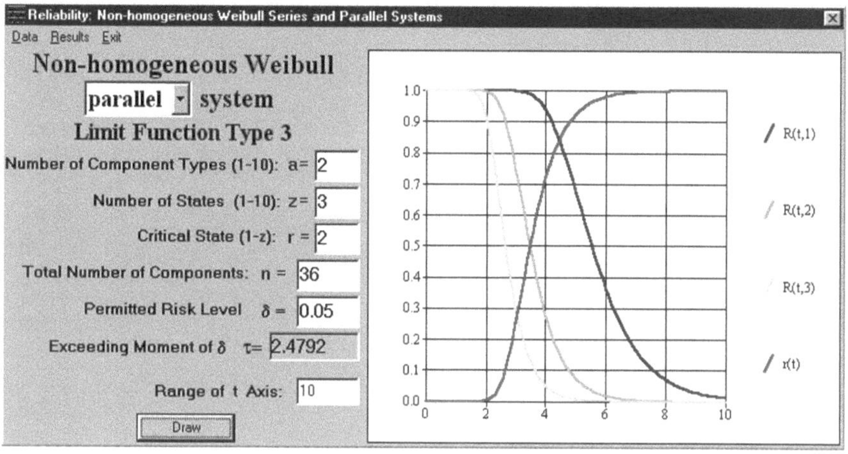

Figure 5.5 The graphs of the rope multi-state reliability function and risk function.

The graphs of the rope multi-state reliability function and its risk function plotted by the computer program [93] are given in Figure 5.5.

5.3 Reliability Evaluation of Multi-State '*m* out of *n*' Systems

In proving facts on limit reliability functions of the homogeneous multi-state 'm out of n' systems, the following two extensions of Lemmas 4.9 – 4.11 are used.

Lemma 5.5
If

i. $m = \text{constant}$ ($m/n \to 0$ as $n \to \infty$),
ii. $\mathfrak{R}^{(0)}(t,u) = 1 - \sum_{i=0}^{m-1}([V(t,u)]^i/i!)\exp[-V(t,u)]$, $u = 1, 2, \ldots, z$, is a non-degenerate reliability function,
iii. $\boldsymbol{R}_n^{(m)}(t,\cdot) = [1, \boldsymbol{R}_n^{(m)}(t,1), \ldots, \boldsymbol{R}_n^{(m)}(t,z)]$, $t \in (-\infty, \infty)$, is the reliability function of a homogeneous multi-state 'm out of n' system defined by Eqs. (3.24) and (3.25),
iv. $a_n(u) > 0$, $b_n(u) \in (-\infty, \infty)$, $u = 1, 2, \ldots, z$

then

$$\mathfrak{R}^{(0)}(t,\cdot) = [1, \mathfrak{R}^{(0)}(t,1), \ldots, \mathfrak{R}^{(0)}(t,z)], \quad t \in (-\infty, \infty)$$

is the limit multi-state reliability function of this system, i.e.

$$\lim_{n\to\infty} \boldsymbol{R}_n^{(m)}(a_n(u)t + b_n(u), u) = \mathfrak{R}^{(0)}(t,u) \quad \text{for } t \in C_{\mathfrak{R}^{(0)}(u)}, \quad u = 1, 2, \ldots, z \qquad (5.24)$$

if and only if

$$\lim_{n\to\infty} nR(a_n(u)t + b_n(u), u) = V(t,u) \quad \text{for } t \in C_{V(u)}, \quad u = 1,2,\ldots,z \tag{5.25}$$

Motivation: For each fixed u, $u = 1, 2, \ldots, z$, assumptions (i)–(iv) of Lemma 5.5 are identical to assumptions (i)–(iv) of Lemma 4.9, condition (5.24) is identical to condition (4.29) and condition (5.25) is identical to condition (4.30). And since, from Lemma 4.9, conditions (4.29) and (4.30) are equivalent, then conditions (5.24) and (5.25) are also equivalent.

Lemma 5.6
If

- **i.** $m/n \to \mu$, $0 < \mu < 1$, as $n \to \infty$
- **ii.** $\mathfrak{R}^{(\mu)}(t) = 1 - (1/\sqrt{2\pi})\int_{-\infty}^{-v(t,u)} e^{-x^2/2}\,dx$, $u = 1,2,\ldots,z$ is a non-degenerate reliability function,
- **iii.** $\boldsymbol{R}_n^{(m)}(t,\cdot) = [1, \boldsymbol{R}_n^{(m)}(t,1), \ldots, \boldsymbol{R}_n^{(m)}(t,z)]$, $t \in (-\infty, \infty)$ is the reliability function of a homogeneous multi-state 'm out of n' system defined by Eqs. (3.24) and (3.25),
- **iv.** $a_n(u) > 0$, $b_n(u) \in (-\infty, \infty)$, $u = 1,2,\ldots,z$

then

$$\mathfrak{R}^{(\mu)}(t,\cdot) = [1, \mathfrak{R}^{(\mu)}(t,1), \ldots, \mathfrak{R}^{(\mu)}(t,z)], \quad t \in (-\infty, \infty)$$

is the limit multi-state reliability function of this system, i.e.

$$\lim_{n\to\infty} \boldsymbol{R}_n^{(m)}(a_n(u)t + b_n(u), u) = \mathfrak{R}^{(\mu)}(t,u) \quad \text{for } t \in C_{\mathfrak{R}^{(\mu)}(u)}, \quad u = 1,2,\ldots,z \tag{5.26}$$

if and only if

$$\lim_{n\to\infty} \frac{(n+1)R(a_n(u)t + b_n(u), u) - m}{\sqrt{m(n-m+1)/(n+1)}} = v(t,u) \quad \text{for } t \in C_{v(u)} \tag{5.27}$$

Motivation: For each fixed u, $u = 1, 2, \ldots, z$, assumptions (i)–(iv) of Lemma 5.6 are identical to assumptions (i)–(iv) of Lemma 4.10, condition (5.26) is identical to condition (4.31) and condition (5.27) is identical to condition (4.32). Moreover, since, from Lemma 4.10, conditions (4.31) and (4.32) are equivalent, then conditions (5.26) and (5.27) are also equivalent.

Lemma 5.7
If

- **i.** $n - m = \overline{m} = \text{constant}$ $(m/n \to 1$ as $n \to \infty)$
- **ii.** $\overline{\mathfrak{R}}^{(1)}(t,u) = \sum_{i=0\overline{m}}([\overline{V}(t)]^i/i!)\exp[-\overline{V}(t)]$, $u = 1,2,\ldots,z$ is a non-degenerate reliability function,

iii. $\overline{\boldsymbol{R}}_n^{(m)}(t,\cdot)=[1,\overline{\boldsymbol{R}}_n^{(m)}(t,1),\ldots,\overline{\boldsymbol{R}}_n^{(m)}(t,z)]$, $t\in(-\infty,\infty)$, is the reliability function of a homogeneous multi-state 'm out of n' system defined by Eqs. (3.26) and (3.27),
iv. $a_n(u)>0$, $b_n(u)\in(-\infty,\infty)$, $u=1,2,\ldots,z$

then

$$\overline{\mathfrak{R}}^{(1)}(t,\cdot)=[1,\overline{\mathfrak{R}}^{(1)}(t,1),\ldots,\overline{\mathfrak{R}}^{(1)}(t,z)],\quad t\in(-\infty,\infty)$$

is the limit multi-state reliability function of this system, i.e.

$$\lim_{n\to\infty}\overline{\boldsymbol{R}}_n^{(m)}(a_n(u)t+b_n(u),u)=\overline{\mathfrak{R}}^{(1)}(t,u)\quad \text{for } t\in C_{\overline{\mathfrak{R}}^{(1)}(u)},\quad u=1,2,\ldots,z \tag{5.28}$$

if and only if

$$\lim_{n\to\infty} nF(a_n(u)t+b_n(u),u)=\overline{V}(t,u)\quad \text{for } t\in C_{\overline{V}(u)},\quad u=1,2,\ldots,z \tag{5.29}$$

Motivation: For each fixed u, $u=1, 2, \ldots, z$, assumptions (i) – (iv) of Lemma 5.7 are identical to assumptions (i) – (iv) of Lemma 4.11, condition (5.28) is identical to condition (4.33) and condition (5.29) is identical to condition (4.34). And since, from Lemma 4.9, conditions (4.33) and (4.34) are equivalent, then conditions (5.28) and (5.29) are also equivalent.

Lemmas 5.5 – 5.7 and Theorem 4.5 allow us to establish the class of possible limit reliability functions of the homogeneous multi-state 'm out of n' systems pointed out in the next theorem.

Theorem 5.5

The class of limit non-degenerate reliability functions of the homogeneous multi-state 'm out of n' system is composed of $3^z+4^z+3^z$ reliability functions of the following form:

Case 1. $m=\text{constant}$ $(m/n\to 0$ as $n\to\infty)$.

$$\mathfrak{R}^{(0)}(t,\cdot)=[1,\mathfrak{R}^{(0)}(t,1),\ldots,\mathfrak{R}^{(0)}(t,z)],\quad t\in(-\infty,\infty) \tag{5.30}$$

where

$$\mathfrak{R}^{(0)}(t,u)\in\{\mathfrak{R}_1^{(0)}(t),\mathfrak{R}_2^{(0)}(t),\mathfrak{R}_3^{(0)}(t)\},\quad u=1,2,3,\ldots,z \tag{5.31}$$

and $\mathfrak{R}_i^{(0)}(t)$, $i=1, 2, 3$, are defined by Eqs. (4.35) – (4.37).

Case 2. $m/n=\mu+o(1/\sqrt{n})$, $0<\mu<1$ $(m/n\to\mu$ as $n\to\infty)$

$$\mathfrak{R}^{(\mu)}(t,\cdot)=[1,\mathfrak{R}^{(\mu)}(t,1),\ldots,\mathfrak{R}^{(\mu)}(t,z)],\quad t\in(-\infty,\infty) \tag{5.32}$$

where

$$\Re^{(\mu)}(t,u) \in \{\Re_4^{(\mu)}(t), \Re_5^{(\mu)}(t), \Re_6^{(\mu)}(t), \Re_7^{(\mu)}(t)\}, \quad u = 1, 2, 3, \ldots, z \tag{5.33}$$

and $\Re_i^{(\mu)}(t)$, $i = 4, 5, 6, 7$, are defined by Eqs. (4.38) – (4.42).

Case 3. $n - m = \overline{m} =$ constant $(m/n \to 1$ as $n \to \infty)$.

$$\overline{\Re}^{(1)}(t, \cdot) = [1, \overline{\Re}^{(1)}(t, 1), \ldots, \overline{\Re}^{(1)}(t, z)], \quad t \in (-\infty, \infty) \tag{5.34}$$

where

$$\overline{\Re}^{(1)}(t,u) \in \{\overline{\Re}_8^{(1)}(t), \overline{\Re}_9^{(1)}(t), \overline{\Re}_{10}^{(1)}(t)\}, \quad u = 1, 2, \ldots, z \tag{5.35}$$

and $\overline{\Re}_i^{(1)}(t)$, $i = 8, 9, 10$, are defined by Eqs. (4.43) – (4.45).

Motivation: For each fixed u, $u = 1, 2, \ldots, z$, coordinate $\Re^{(0)}(t,u)$ of the vector $\Re^{(0)}(t, \cdot)$ defined by Eq. (5.30), coordinate $\Re^{(\mu)}(t,u)$ of the vector $\Re^{(\mu)}(t, \cdot)$ defined by Eq. (5.32) and coordinate $\overline{\Re}^{(1)}(t,u)$ of the vector $\Re^{(1)}(t, \cdot)$ defined by Eq. (5.34), from Theorem 4.5 that is the consequence of Lemmas 4.9 – 4.11, may be one of the three types of reliability functions defined by Eqs. (4.35) – (4.37), or one of the four types of reliability functions defined by Eqs. (4.38) – (4.42), or one of the three types of reliability functions defined by Eqs. (4.43) – (4.45) respectively. Thus the number of different limit multi-state reliability functions of the considered system is equal to the sum of the number of z-term variations of the 3-component set (5.31), the number of z-term variations of the 4-component set (5.33) and the number of z-term variations of the 3-component set (5.35). It means that this number is equal to $3^z + 4^z + 3^z$ and they are of the forms (5.30), (5.32) and (5.34) respectively.

Corollary 5.7

If components of the homogeneous multi-state 'm out of n' system have exponential reliability functions

$$R(t, \cdot) = [1, R(t, 1), \ldots, R(t, z)], \quad t \in (-\infty, \infty)$$

where

$$R(t,u) = 1 \text{ for } t < 0, \; R(t,u) = \exp[-\lambda(u)t] \text{ for } t \geq 0, \; \lambda(u) > 0, \; u = 1, 2, \ldots, z$$

and

$$m = \text{constant}$$

$$a_n(u)\frac{1}{\lambda(u)}, \; b_n(u)\frac{1}{\lambda(u)}\log n, \; u = 1, 2, \ldots, z, \frac{n!}{r!(n-r)!}$$

then

$$\mathfrak{R}_3^{(0)}(t,\cdot) = [1, \mathfrak{R}_3^{(0)}(t,1), \ldots, \mathfrak{R}_3^{(0)}(t,z)], \quad t \in (-\infty, \infty)$$

where

$$\mathfrak{R}_3^{(0)}(t,u) = 1 - \sum_{i=0}^{m-1} \frac{\exp[-it]}{i!} \exp[-\exp[t]] \quad \text{for } t \in (-\infty, \infty), \quad u = 1, 2, \ldots, z$$

is its limit reliability function.

Motivation: Since for each fixed u, sufficiently large n and all $t \in (-\infty, \infty)$, we have

$$a_n(u)t + b_n(u) = \frac{t + \log n}{\lambda(u)} > 0$$

then for sufficiently large n

$$R(a_n(u)t + b_n(u), u) = \exp[-\lambda(u)(a_n(u)t + b_n(u))] \quad \text{for } t \in (-\infty, \infty)$$

Hence

$$\begin{aligned} V(t,u) &= \lim_{n\to\infty} nR(a_n(u)t + b_n(u), u) \\ &= \lim_{n\to\infty} n \exp[-\lambda(u)(a_n(u)t + b_n(u))] \\ &= \lim_{n\to\infty} n \exp[-t - \log n] \\ &= \exp[-t] \quad \text{for } t \in (-\infty, \infty) \end{aligned}$$

which from Lemma 5.5 completes the proof.

Example 5.7 (A Steel Rope, Durability)
Let us consider a steel rope composed of $n = 36$ four-state, i.e. $z = 3$, identical wires having exponential reliability functions with transition rates between the state subsets

$$\lambda(u) = 0.2u/\text{year}, \quad u = 1, 2, 3$$

Assuming that the rope is in the state subset $\{u, u+1, \ldots, z\}$ if at least $m = 10$ of its wires are in this state subset, according to Definition 3.9, we conclude the rope is a homogeneous four-state '10 out of 36' system. Thus, according to Eqs. (3.24) and (3.25), its reliability function is given by

$$\boldsymbol{R}_{36}^{(10)}(t,\cdot) = [1, \boldsymbol{R}_{36}^{(10)}(t,1), \boldsymbol{R}_{36}^{(10)}(t,2), \boldsymbol{R}_{36}^{(10)}(t,3)]$$

where

$$\boldsymbol{R}_{36}^{(10)}(t,u)=1 \quad \text{for } t<0$$

$$\boldsymbol{R}_{3}^{(10)}(t,u)=1-\sum_{i=0}^{9}\binom{36}{i}\exp[-i0.2ut][1-\exp[-0.2ut]]^{36-i}, \quad u=1,2,3$$

Applying Corollary 5.7 with normalizing constants

$$a_n(u)=\frac{5}{u}, \;\; b_n(u)=\frac{5}{u}\log 36, \quad u=1,2,3$$

we conclude that the rope limit reliability function is given by

$$\mathfrak{R}_3^{(0)}(t,\cdot)=[1,\mathfrak{R}_3^{(0)}(t,1),\mathfrak{R}_3^{(0)}(t,2),\mathfrak{R}_3^{(0)}(t,3)(t,3)], \quad t\in(-\infty,\infty)$$

where

$$\mathfrak{R}_3^{(0)}(t,u)=1-\sum_{i=0}^{9}\frac{\exp[-it]}{i!}\exp[-\exp[-t]]$$

Hence, considering Eq. (3.49), since

$$a_n(1)=5.00, \;\; b_n(1)=17.92,$$

$$a_n(2)=2.5, \;\; b_n(2)=8.96,$$

$$a_n(3)=1.67, \;\; b_n(3)=5.97$$

then the approximate formula for the rope reliability function takes the form

$$\begin{aligned}\boldsymbol{R}_{36}^{(10)}(t,\cdot)&\cong\mathfrak{R}_3^{(0)}\left(\frac{t-b_n(u)}{a_n(u)},\cdot\right)\\&=\left[1,1-\sum_{i=0}^{9}\frac{\exp[-i(0.2t-3.58)]}{i!}\exp[-\exp[-0.2t+3.58]]\right.\\&\quad 1-\sum_{i=0}^{9}\frac{\exp[-i(0.4t-3.58)]}{i!}\exp[-\exp[-0.4t+3.58]],\\&\quad\left.1-\sum_{i=0}^{9}\frac{\exp[-i(0.6t-3.58)]}{i!}\exp[-\exp[-0.6t+3.58]]\right],\\&\quad t\in(-\infty,\infty)\end{aligned}$$

The approximate mean values of the rope lifetimes $T(u)$ in the state subsets and their standard deviations in years, by Eq. (3.13), are

$$M(1) \cong 6.66, \quad M(2) \cong 3.33, \quad M(3) \cong 2.22$$

$$\sigma(1) \cong 1.62, \quad \sigma(2) \cong 0.81, \quad \sigma(3) \cong 0.54$$

whereas, from Eq. (3.17), the approximate mean values of the rope sojourn times in the particular reliability states are

$$\overline{M}(1) \cong 3.33, \quad \overline{M}(2) \cong 1.11, \quad \overline{M}(3) \cong 2.22$$

If the critical state is $r = 2$, then from Eq. (3.18) the rope risk function is approximately given by

$$\boldsymbol{r}(t) \cong \sum_{i=0}^{9} \frac{\exp[-i(0.4t - 3.58)]}{i!} \exp[-\exp[-0.4t + 3.58]], \quad t \in (-\infty, \infty)$$

The moment when the risk exceeds an admissible level $\delta = 0.05$, after applying Eq. (3.19), is $\tau \cong 2.0738$ years

The behaviour of the rope system reliability function and its risk function are given in Table 5.2 and Figure 5.6.

Table 5.2 The Values of the Still Rope Multi-State Reliability Function and Risk Function

t	$\mathfrak{R}_3^{(0)}(t - b_n(1)/a_n(1), 1)$	$\mathfrak{R}_3^{(0)}(t - b_n(2)/a_n(2), 2)$	$\mathfrak{R}_3^{(0)}(t - b_n(3)/a_n(3), 3)$	$r(t)$
0.2	1.000000	0.999999	0.999998	0.000000
0.6	0.999998	0.999977	0.999795	0.000023
1.0	0.999990	0.999609	0.994249	0.000391
1.4	0.999950	0.996412	0.945898	0.003588
1.8	0.999795	0.980140	0.776749	0.019860
2.2	0.999283	0.927919	0.493317	0.072081
2.6	0.997831	0.815200	0.231067	0.184800
3.0	0.994249	0.642210	0.080576	0.357790
3.4	0.986488	0.444152	0.021678	0.555848
3.8	0.971569	0.267825	0.004693	0.732175
4.2	0.945898	0.141305	0.000851	0.858695
4.6	0.906025	0.065843	0.000134	0.934157
5.0	0.849686	0.027422	0.000019	0.972578
5.4	0.776749	0.010338	0.000002	0.989662
5.8	0.689654	0.003571	0.000000	0.996429
6.2	0.593139	0.001143	0.000000	0.998857
6.6	0.493317	0.000342	0.000000	0.999658
7.0	0.396455	0.000097	0.000000	0.999903
7.4	0.307841	0.000026	0.000000	0.999974
7.8	0.231067	0.000007	0.000000	0.999993

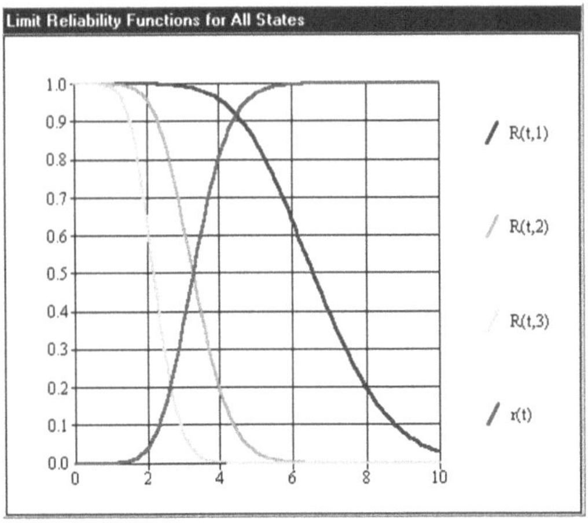

Figure 5.6 The graphs of the still rope multi-state reliability function and risk function.

Unfortunately, similarly to the case of two-state systems, there are no extensions of Lemmas 4.5 − 4.7 to non-homogeneous multi-state 'm out of n' systems.

5.4 Reliability Evaluation of Multi-State Series–Parallel Systems

In proving facts on limit reliability functions for homogeneous regular multi-state series–parallel systems, the following extensions of Lemmas 4.12 − 4.13 are used [81].

Lemma 5.8
If

i. $k_n \to \infty$,
ii. $\mathfrak{R}(t,u) = 1 - \exp[-V(t,u)], \ u = 1,2,\ldots,z$, is a non-degenerate reliability function,
iii. $\boldsymbol{R}_{k_n,l_n}(t,\cdot) = [1, \boldsymbol{R}_{k_n,l_n}(t,1),\ldots,\boldsymbol{R}_{k_n,l_n}(t,z)], \ t \in (-\infty,\infty)$, is the reliability function of a homogeneous regular multi-state series–parallel system defined by Eqs. (3.28) and (3.29),
iv. $a_n(u) > 0, \ b_n(u) \in (-\infty,\infty), \quad u = 1,2,\ldots,z$

then

$$\mathfrak{R}(t,\cdot) = [1, \mathfrak{R}(t,1),\ldots,\mathfrak{R}(t,z)], \quad t \in (-\infty,\infty)$$

is the limit multi-state reliability function of this system, i.e.

$$\lim_{n\to\infty} \boldsymbol{R}_{k_n,l_n}(a_n(u)t + b_n(u),u) = \mathfrak{R}(t,u) \quad \text{for } t \in C_{\mathfrak{R}(u)}, \ u = 1,2,\ldots,z \qquad (5.36)$$

if and only if

$$\lim_{n\to\infty} k_n[R(a_n(u)t + b_n(u),u)]^{l_n} = V(t,u) \quad \text{for } t \in C_{V(u)}, \ u = 1,2,\ldots,z \qquad (5.37)$$

Motivation: For each fixed u, $u = 1, 2, \ldots, z$, assumptions (i)–(iv) of Lemma 5.8 are identical to assumptions (i)–(iv) of Lemma 4.12, condition (5.36) is identical to condition (4.47) and condition (5.37) is identical to condition (4.48). Moreover, since from Lemma 4.12, conditions (4.47) and (4.48) are equivalent, then conditions (5.36) and (5.37) are also equivalent.

Lemma 5.9

If

i. $k_n \to k$, $k > 0$, $l_n \to \infty$,
ii. $\Re(t, u)$, $u = 1, 2, \ldots, z$, is a non-degenerate reliability function,
iii. $\boldsymbol{R}_{k_n,l_n}(t, \cdot) = [1, \boldsymbol{R}_{k_n,l_n}(t, 1), \ldots, \boldsymbol{R}_{k_n,l_n}(t, z)]$, $t \in (-\infty, \infty)$, is the reliability function of a homogeneous regular multi-state series–parallel system defined by Eqs. (3.28) and (3.29),
iv. $a_n(u) > 0$, $b_n(u) \in (-\infty, \infty)$, $u = 1, 2, \ldots, z$

then

$$\Re(t, \cdot) = [1, \Re(t, 1), \ldots, \Re(t, z)], \quad t \in (-\infty, \infty)$$

is its multi-state limit reliability function, i.e.

$$\lim_{n \to \infty} \boldsymbol{R}_{k_n,l_n}(a_n(u)t + b_n(u), u) = \Re(t, u) \quad \text{for } t \in C_{\Re(u)}, \ u = 1, 2, \ldots, z \tag{5.38}$$

if and only if

$$\lim_{n \to \infty} [R(a_n(u)t + b_n(u), u)]^{l_n} = \Re_0(t, u) \quad \text{for } t \in C_{\Re(u)}, \ u = 1, 2, \ldots, z \tag{5.39}$$

where $\Re_0(t,u)$ is a non-degenerate reliability function and moreover

$$\Re(t, u) = 1 - [1 - \Re_0(t, u)]^k \quad \text{for } t \in (-\infty, \infty), \ u = 1, 2, \ldots, z \tag{5.40}$$

Motivation: For each fixed u, $u = 1, 2, \ldots, z$, assumptions (i)–(iv) of Lemma 5.9 are identical to assumptions (i)–(iv) of Lemma 4.13. Moreover, condition (5.38) is equivalent to condition (4.49) and condition (5.39) is equivalent to condition (4.50). And moreover, since from Lemma 4.13, conditions (4.49) and (4.50) are equivalent, then conditions (5.38) and (5.39) are also equivalent and moreover, considering Eq. (4.51), the equality (5.40) holds.

Lemmas 5.8 and 5.9 and Theorem 4.6 are the basis for formulating the next theorem [81].

Theorem 5.6

The class of limit non-degenerate reliability functions of the homogeneous regular multi-state series–parallel system is composed of $3^z + 4^z + 3^z$ reliability functions of the form

$$\Re(t, \cdot) = [1, \Re(t, 1), \ldots, \Re(t, z)], \quad t \in (-\infty, \infty) \tag{5.41}$$

where

Case 1. $k_n = n$, $|l_n - c\log n| \gg s$, $s > 0$, $c > 0$ (under Assumption 4.1).

$$\mathfrak{R}(t,u) \in \{\mathfrak{R}_1(t), \mathfrak{R}_2(t), \mathfrak{R}_3(t)\}, \quad u = 1, 2, \ldots, z \tag{5.42}$$

and $\mathfrak{R}_i(t)$, $i = 1, 2, 3$, are defined by Eqs. (4.52)–(4.54).

Case 2. $k_n = n$, $l_n - c\log n \sim s$, $s \in (-\infty, \infty)$, $c > 0$.

$$\mathfrak{R}(t,u) \in \{\mathfrak{R}_4(t), \mathfrak{R}_5(t), \mathfrak{R}_6(t), \mathfrak{R}_7(t)\}, \quad u = 1, 2, \ldots, z \tag{5.43}$$

and $\mathfrak{R}_i(t)$, $i = 4, 5, 6, 7$, are defined by Eqs. (4.55)–(4.58).

Case 3. $k_n \to k$, $k > 0$, $l_n \to \infty$.

$$\mathfrak{R}(t,u) \in \{\mathfrak{R}_8(t), \mathfrak{R}_9(t), \mathfrak{R}_{10}(t)\}, \quad u = 1, 2, \ldots, z \tag{5.44}$$

and $\mathfrak{R}_i(t)$, $i = 8, 9, 10$, are defined by Eqs. (4.59)–(4.61).

Motivation: For each fixed u, $u = 1, 2, \ldots, z$, coordinate $\mathfrak{R}(t,u)$ of the vector $\mathfrak{R}(t,\cdot)$ defined by Eq. (5.41), from Theorem 4.6 that is the consequence of Lemmas 4.12 and 4.13, can be a reliability function that is one of three types defined by Eqs. (4.52)–(4.54), or one of four types defined by Eqs. (4.55)–(4.58), or one of three types defined by Eqs. (4.59)–(4.61). Thus the number of different multi-state reliability functions of the considered system is equal to the sum of the number of z-term variations of the 3-component sets defined by Eqs. (5.42) and (5.44) and the 4-component set defined by Eq. (5.43), i.e. $3^z + 4^z + 3^z$, and they are of the form (5.41).

Corollary 5.8

If components of the homogeneous regular multi-state series–parallel system have Weibull reliability functions

$$R(t,\cdot) = [1, R(t,1), \ldots, R(t,z)], \quad t \in (-\infty, \infty)$$

where

$$R(t,u) = 1 \quad \text{for } t < 0, \quad R(t,u) = \exp[-\beta(u)t^{\alpha(u)}] \quad \text{for } t \geq 0, \quad \alpha(u) > 0,$$
$$\beta(u) > 0, \quad u = 1, 2, \ldots, z$$

and

$$k_n = n, \quad l_n > 0$$

$$a_n(u) = b_n(u)/(\alpha(u)\log n), \quad b_n(u) = (\log n/(\beta(u)l_n))^{1/\alpha(u)}, \quad u = 1, 2, \ldots, z$$

then

$$\mathfrak{R}_3(t, \cdot) = [1, \mathfrak{R}_3(t, 1), \ldots, \mathfrak{R}_3(t, z)], \quad t \in (-\infty, \infty)$$

where

$$\mathfrak{R}_3(t, u) = 1 - \exp[-\exp[-t] \quad \text{for } t \in (-\infty, \infty), \quad u = 1, 2, \ldots, z$$

is its limit reliability function.

Motivation: Since for each fixed u, sufficiently large n and all $t \in (-\infty, \infty)$, we have

$$a_n(u)t + b_n(u) = b_n(u)(t/(\alpha(u)\log n) + 1) > 0$$

then for $t \in (-\infty, \infty)$ we get

$$\begin{aligned} R(a_n(u)t + b_n(u), u) &= \exp[-\beta(u)(a_n(u)t) + b_n(u))^{\alpha(u)}] \\ &= \exp[-(\log n)/l_n \cdot (t/(\alpha(u)\log n) + 1)^{\alpha(u)}] \\ &= \exp[-(\log n)/l_n - t/l_n - o(1/ln)] \end{aligned}$$

Further, for all $t \in (-\infty, \infty)$, we have

$$\begin{aligned} V(t, u) &= \lim_{n \to \infty} k_n [R(a_n(u)t + b_n(u), u)]^{l_n} \\ &= \lim_{n \to \infty} n\exp[-t - \log n + l_n o(1/l_n)] = \exp[-t] \end{aligned}$$

which from Lemma 5.8 completes the proof.

Example 5.8 (A Model Series–Parallel System)
If the homogeneous regular multi-state series–parallel system is such that

$$k_n = 30, \quad l_n = 10, \quad z = 5$$

and its components have Weibull reliability functions with parameters

$$\beta(u) = 10^{-5}, \quad \alpha(u) = (11 + u)/4, \quad u = 1, 2, \ldots, 5$$

then, according to Corollary 5.8, assuming normalizing constants

$$a_n(u) = 4b_n(u)/((11 + u) \cdot \log 30), \quad b_n(u) = (10^4 \cdot \log 30)^{4/(11+u)}, \quad u = 1, 2, \ldots, 5$$

considering Eq. (3.49), we get the following approximate formula:

$$[1, \boldsymbol{R}_{30,10}(t,1), \boldsymbol{R}_{30,10}(t,2), \boldsymbol{R}_{30,10}(t,3), \boldsymbol{R}_{30,10}(t,4), \boldsymbol{R}_{30,10}(t,5)]$$
$$\cong [1, 1-\exp[-\exp[-0.0315t+10.204]], 1-\exp[-\exp[-0.446t+11.054]],$$
$$1-\exp[-\exp[-0.604t+11.904]], 1-\exp[-\exp[-0.789t+12.755]],$$
$$1-\exp[-\exp[-1.002t+13.605]]] \quad \text{for } t \in (-\infty, \infty)$$

According to Eq. (3.4), the expected values of the system components' sojourn times $T_{ij}(u)$ in the state subsets are given by the formula [93]:

$$M_{ij}(u) = E[T_{ij}(u)] = (\beta(u))^{-1/\alpha(u)}\Gamma((\alpha(u)+1)/\alpha(u))$$
$$= 10^{20/(11+u)}\Gamma((15+u)/(11+u)), \quad u = 1, 2, \ldots, 5$$

Hence, in particular, we have

$$M_{ij}(1) \cong 41.46, \ M_{ij}(2) \cong 31.01, \ M_{ij}(3) \cong 24.19, \ M_{ij}(4) \cong 19.40, \ M_{ij}(5) \cong 16.12$$

and according to Eq. (3.8) the expected values of the system components' lifetimes in particular states are

$$\overline{M}_{ij}(1) \cong 10.45, \ \overline{M}_{ij}(2) \cong 6.82, \ \overline{M}_{ij}(3) \cong 4.79, \ \overline{M}_{ij}(4) \cong 3.28, \ \overline{M}_{ij}(5) \cong 16.12$$

The mean values of the system sojourn times $T(u)$ in the state subsets after applying the formula (3.13) are given by [15,93]

$$M(u) = E[T(u)] \cong 0.5772a_n(u) + b_n(u), \quad u = 1, 2, \ldots, 5$$

i.e.

$$M(1) \cong 34.23, \ M(2) \cong 26.09, \ M(3) \cong 20.67, \ M(4) \cong 16.89, \ M(5) \cong 14.16$$

Hence, from Eq. (3.17), the mean values of the system lifetimes in particular states are

$$\overline{M}(1) \cong 8.14, \ \overline{M}(2) \cong 5.42, \ \overline{M}(3) \cong 3.78, \ \overline{M}(4) \cong 2.73, \ \overline{M}(5) \cong 14.16$$

If the critical state is $r = 2$, then from Eq. (3.18) the system risk function is

$$\boldsymbol{r}(t) \cong \exp[-\exp[-0.446t + 11.054]]$$

The moment when the system risk exceeds an admissible level $\delta = 0.05$, from Eq. (3.19), is

$$\tau = \boldsymbol{r}^{-1}(\delta) \cong [11.054 - \log[-\log \delta]]/0.446) = 22.32$$

Corollary 5.9

If components of the homogeneous regular multi-state series–parallel system have Weibull reliability functions

$$R(t,\cdot) = [1, R(t,1), \ldots, R(t,z)], \quad t \in (-\infty, \infty)$$

where

$$R(t,u) = 1 \quad \text{for } t < 0, \quad R(t,u) = \exp[-\beta(u)t^{\alpha(u)}] \quad \text{for } t \geq 0, \quad \alpha(u) > 0, \\ \beta(u) > 0, \quad u = 1, 2, \ldots, z$$

and

$$k_n \to k, \quad l_n \to \infty$$

$$a_n(u) = (\beta(u)l_n)^{-1/\alpha(u)}, \quad b_n(u) = 0 \tag{5.45}$$

then

$$\mathfrak{R}_9(t,\cdot) = [1, \mathfrak{R}_9(t,1), \ldots, \mathfrak{R}_9(t,z)]$$

where

$$\mathfrak{R}_9(t,u) = 1 \quad \text{for } t < 0, \quad \mathfrak{R}_9(t,u) = 1 - [1 - \exp[-t^{\alpha(u)}]]^3 \quad \text{for } t \geq 0$$

is its limit reliability function.

Motivation: Corollary 5.9 is a particular case of more general Corollary 5.10, which will be proved in a later part of this chapter. Therefore, we omit its proof.

Example 5.9 (A Pipeline System)

Let us consider the pipeline system composed of $k_n = 3$ lines of pipe segments linked in parallel, each of them composed of $l_n = 100$ five-state identical segments linked in series. Considering pipe segments as basic components of the pipeline system, according to Definitions 3.12 – 3.13, we conclude that it is a homogeneous regular five-state series–parallel system. Therefore, from Eqs. (3.28) and (3.29), the pipeline system reliability function is given by

$$\boldsymbol{R}_{3,100}(t,\cdot) = [1, \boldsymbol{R}_{3,1000}(t,1), \boldsymbol{R}_{3,100}(t,2), \boldsymbol{R}_{3,100}(t,3), \boldsymbol{R}_{3,100}(t,4)] \tag{5.46}$$

where

$$\boldsymbol{R}_{3,100}(t,u) = 1 - [1 - [R(t,u)]^{100}]^3, \quad t \in (-\infty, \infty), \quad u = 1, 2, 3, 4 \tag{5.47}$$

Taking into account pipe segment reliability data given in their technical certificates and expert opinions, we assume that they have Weibull reliability functions

$$R(t,u)=1 \quad \text{for } t<0, \quad R(t,u)=\exp[-\beta(u)t^{\alpha(u)}] \quad \text{for } t\geq 0, \quad u=1,2,3,4$$

with the following parameters:

$$\alpha(1)=3, \quad \beta(1)=0.00001, \quad \alpha(2)=2.5, \quad \beta(2)=0.0001,$$
$$\alpha(3)=2, \quad \beta(3)=0.0016, \quad \alpha(4)=1, \quad \beta(4)=0.05$$

Hence and from Eqs. (5.46) and (5.47) it follows that the pipeline system exact reliability function is given by

$$\boldsymbol{R}_{3,1000}(t,\cdot)=[1,1-[1-\exp[-0.001t^3]]^3, 1-[1-\exp[-0.01t^{5/2}]]^3,$$
$$1-[1-\exp[-0.16t^2]]^3, 1-[1-\exp[-5t]]^3] \quad \text{for } t\geq 0$$

From Eq. (3.4), the mean values $M_{ij}(u)$, $u = 1, 2, 3, 4$, of the pipe segments in the state subsets in years are

$$M_{ij}(1)=\Gamma(4/3)(0.00001)^{-1/3}\cong 41.45, \quad M_{ij}(2)=\Gamma(7/5)(0.0001)^{-2/5}\cong 35.32,$$
$$M_{ij}(3)=\Gamma(3/2)(0.0016)^{-1/2}\cong 22.16, \quad M_{ij}(4)=\Gamma(2)(0.5)^{-2}\cong 20.00$$

while from Eq. (3.8), the mean values $\overline{M}_{ij}(u)$, $u = 1, 2, 3, 4$, of the pipe segments in particular states are

$$\overline{M}_{ij}(1)\cong 6.13, \quad \overline{M}_{ij}(2)\cong 13.16, \quad \overline{M}_{ij}(3)\cong 2.16, \quad \overline{M}_{ij}(4)\cong 20.00$$

Assuming, according to Eq. (5.45), normalizing constants

$$a_n(u)=(\beta(u)l_n)^{-1/\alpha(u)}, \quad b_n(u)=0, \quad u=1,2,3,4$$

and applying Corollary 5.9, we conclude that the limit reliability function of the pipeline system is

$$\mathfrak{R}_9(t,\cdot)=[1,\mathfrak{R}_9(t,1),\mathfrak{R}_9(t,2),\mathfrak{R}_9(t,3),\mathfrak{R}_{9,}(t,4)], \quad t\in(-\infty,\infty)$$

where

$$\mathfrak{R}_9(t,1)=1-[1-\exp[-t^3]]^3, \quad \mathfrak{R}_9(t,2)=1-[1-\exp[-t^{5/2}]]^3,$$
$$\mathfrak{R}_9(t,3)=1-[1-\exp[-t^2]]^3, \quad \mathfrak{R}_9(t,4)=1-[1-\exp[-t]]^3 \quad \text{for } t\geq 0$$

Since, in particular from Eq. (5.45), we have

$$a_n(1)=10, \quad a_n(2)=6.31, \quad a_n(3)=2.5, \quad a_n(4)=0.2, \quad b_n(u)=0, \quad u=1,2,3,4$$

then applying the approximate formula (3.49) for $t \geq 0$, we get

$$\begin{aligned} \boldsymbol{R}_{3,100}(t,\cdot) &\cong \mathfrak{R}_9((t-b_n(u))/a_n(u,\cdot) \\ &= [1, 1-[1-\exp[-0.001t^3]]^3, [1-\exp[-0.01t^{5/2}]]^3, \\ &\quad 1-[1-\exp[-0.16t^2]]^3, 1-[1-\exp[-5t]]^3] \end{aligned} \tag{5.48}$$

The expected values $M(u)$, $u=1, 2, 3, 4$, of the system sojourn times in the state subsets in years, calculated on the basis of the approximate formula (5.48), according to Eq. (3.13), are

$$M(1) = \Gamma(4/3)[3(0.001)^{-1/3} - 3(0.002)^{-1/3} + (0.003)^{-1/3}] \cong 11.72$$

$$M(3) = \Gamma(3/2)[3(0.16)^{-1/2} - 3(0.32)^{-1/2} + (0.48)^{-1/2}] \cong 3.23$$

$$M(2) = \Gamma(7/5)[3(0.01)^{-2/5} - 3(0.02)^{-2/5} + (0.03)^{-2/5}] \cong 7.67$$

$$M(4) = \Gamma(2)[3(5)^{-1} - 3(10)^{-1} + (15)^{-1}] \cong 0.37$$

Hence, from Eq. (3.17), the system mean lifetimes $\overline{M}(u)$ in particular states are

$$\overline{M}(1) \cong 4.05, \quad \overline{M}(2) \cong 4.44, \quad \overline{M}(3) \cong 2.86, \quad \overline{M}(4) \cong 0.37$$

If the critical state is $r=2$, then the system risk function, according to Eq. (3.18), is given by

$$\boldsymbol{r}(t) = [1-\exp[-0.01t^{5/2}]]^3$$

The moment when the system risk exceeds an admissible level $\delta = 0.05$, from Eq. (3.19), is

$$\tau = \boldsymbol{r}^{-1}(\delta) \cong \left[-100\log\left(1-\sqrt[3]{\delta}\right)\right]^{2/5} \cong 4.62$$

The behaviour of the exact and approximate multi-state system reliability function coordinate $u=2$ and the risk function are presented in Tables 5.3 and 5.4 and Figures 5.7 and 5.8.

The following extensions of Lemmas 4.15–4.16 are necessary tools in determination of limit reliability functions of non-homogeneous regular multi-state series–parallel systems [81].

Lemma 5.10

If

i. $k_n \to \infty$,

ii. $\mathfrak{R}'(t,u) = 1 - \exp[-V'(t,u)]$, $u = 1, 2, \ldots, z$ is a non-degenerate reliability function,

Table 5.3 The Values of the Component $u = 2$ of the Exact and Approximate Piping System Reliability Function

t	$\boldsymbol{R_{3,100}(t,2) = \Re_9((t-b_n(2))/a_n(2))}$
0.0	1.000
1.5	1.000
3.0	0.997
4.5	0.957
6.0	0.799
7.5	0.515
9.0	0.242
10.5	0.082
12.0	0.020
13.5	0.004
15.0	0.000

Table 5.4 The Values of the Piping System Risk Function

t	$r(t)$
0.0	0.000
1.5	0.000
3.0	0.003
4.5	0.043
6.0	0.201
7.5	0.485
9.0	0.758
10.5	0.918
12.0	0.980
13.5	0.996
15.0	1.000

iii. $\boldsymbol{R}'_{k_n,l_n}(t,\cdot) = [1, \boldsymbol{R}'_{k_n,l_n}(t,1), \ldots, \boldsymbol{R}'_{k_n,l_n}(t,z)]$, $t \in (-\infty, \infty)$ is the reliability function of a non-homogeneous regular multi-state series–parallel system defined by Eqs. (3.43)–(3.45),

iv. $a_n(u) > 0$, $b_n(u) \in (-\infty, \infty)$, $u = 1, 2, \ldots, z$

v. $R(t,u)$ for each fixed u is one of reliability functions $R^{(1)}(t,u)$, $R^{(2)}(t,u)$, …, $R^{(a)}(t,u)$ defined by Eq. (3.45) such that

vi. $\exists N(u)\, \forall\, n > N(u) R(a_n(u)t + b_n(u), u) \neq 0$ for $t < t_0(u)$, $R(a_n(u)u)t + b_n(u), u) = 0$ for $t \geq t_0(u)$, where $t_0(u) \in (-\infty, \infty >$

vii. $$\lim_{n\to\infty} \frac{R^{(i)}(a_n(u)t + b_n(u), u)}{R(a_n(u)t + b_n(u), u)} \leq 1 \quad \text{for } t \geq t_0(u),\ i = 1, 2, \ldots, a,\ u = 1, 2, \ldots, z$$
and moreover there exist non-increasing functions

viii. $$d(t,u) = \begin{cases} \lim\limits_{n\to\infty} \sum\limits_{i=1}^{a} q_i d_i(a_n(u)t + b_n(u), u) & \text{for } t < t_0(u) \\ 0 & \text{for } t \geq t_0(u) \end{cases} \tag{5.49}$$

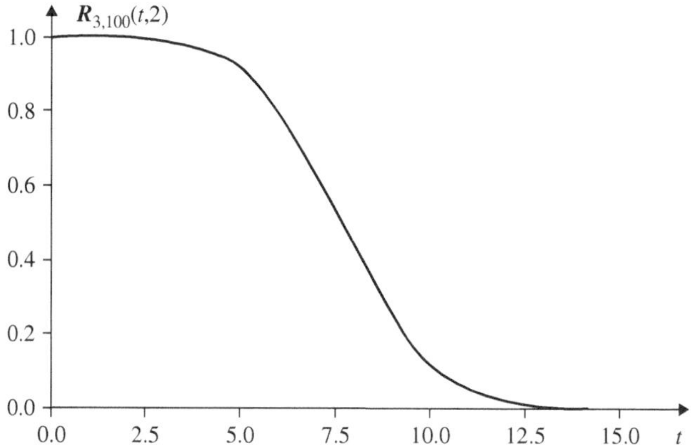

Figure 5.7 The graph of the component $u = 2$ of the exact and approximate piping system reliability function.

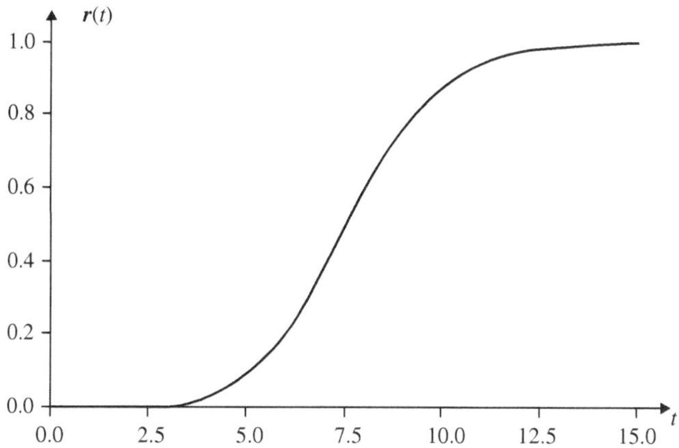

Figure 5.8 The graph of the piping system risk function.

ix. where

$$d_i(a_n(u)t + b_n(u), u) = \left[\frac{R^{(i)}(a_n(u)t+b_n(u), u)}{R(a_n(u)t+b_n(u), u)}\right]^{l_n}$$

then

$$\overline{\mathfrak{R}}'(t,\cdot) = [1, \overline{\mathfrak{R}}'(t,1), \ldots, \overline{\mathfrak{R}}'(t,z)], t \in (-\infty, \infty)$$

is its multi-state limit reliability function, i.e.

$$\lim_{n\to\infty} \overline{\boldsymbol{R}}'_{k_n,l_n}(a_n(u)t + b_n(u), u) = \mathfrak{R}'(t,u) \quad \text{for } t \in C_{\mathfrak{R}'(u)}, \; u = 1, 2, \ldots, z \qquad (5.50)$$

if and only if

$$\lim_{n\to\infty} k_n[R(a_n(u)t+b_n(u),u)]^{l_n} d(t,u) = V'(t,u) \quad \text{for } t\in C_{V'(u)},\ u=1,2,\ldots,z \qquad (5.51)$$

Motivation: Since for each fixed u, $u=1, 2, \ldots, z$, assumptions (i)−(viii) of Lemma 5.10 are identical to assumptions (i)−(viii) of Lemma 4.15, condition (5.50) is identical to condition (4.66) and condition (5.51) is identical to condition (4.67), which from Lemma 4.15 are equivalent, then Lemma 5.10 is valid.

Lemma 5.11

If

i. $k_n \to k$, $k>0$, $l_n \to \infty$,

ii. $\mathfrak{R}'(t,u)$, $u=1,2,\ldots,z$, is a non-degenerate reliability function,

iii. $\boldsymbol{R}'_{k_n,l_n}(t,\cdot) = [1, \boldsymbol{R}'_{k_n,l_n}(t,1),\ldots,\boldsymbol{R}'_{k_n,l_n}(t,z)]$, $t\in(-\infty,\infty)$, is the reliability function of a non-homogeneous regular multi-state series−parallel system defined by Eqs. (3.43)−(3.45),

iv $a_n(u)>0$, $b_n(u)\in(-\infty,\infty)$, $u=1,2,\ldots,z$

v. $R(t,u)$ for each fixed u is one of reliability functions $R^{(1)}(t,u)$, $R^{(2)}(t,u)$, …, $R^{(a)}(t,u)$ defined by Eq. (3.45) such that

vi. $\exists N(u)\,\forall\, n>N(u) R(a_n(u)t+b_n(u),u)\neq 0$ for $t<t_0(u)$, $R(a_n(u)t+b_n(u),u)=0$ for $t\geq t_0(u)$, where $t_0(u)\in(-\infty,\infty>$

vii.

$$\lim_{n\to\infty}\frac{R^{(i)}(a_n(u)t+b_n(u),u)}{R(a_n(u)t+b_n(u),u)}\leq 1 \quad \text{for } t<t_0(u),\ i=1,2,\ldots,a,\ u=1,2,\ldots,z$$

and moreover there exist non-increasing functions

viii.

$$d_i(t,u)=\begin{cases}\lim\limits_{n\to\infty} d_i(a_n(u)t+b_n(u),u) & \text{for } t<t_0(u)\\ 0 & \text{for } t\geq t_0(u)\end{cases} \qquad (5.52)$$

where

ix.

$$d_i(a_n(u)t+b_n(u),u)=\left[\frac{R^{(i)}(a_n(u)t+b_n(u),u)}{R(a_n(u)t+b_n(u),u)}\right]^{l_n} \qquad (5.53)$$

then

$$\mathfrak{R}'(t,\cdot)=[1,\mathfrak{R}'(t,1),\ldots,\mathfrak{R}'(t,z)], \quad t\in(-\infty,\infty)$$

is its multi-state limit reliability function, i.e.

$$\lim_{n\to\infty}\boldsymbol{R}'_{k_n,l_n}(a_n(u)t+b_n(u),u)=\mathfrak{R}'(t,u) \quad \text{for } t\in C_{\mathfrak{R}'(u)},\ u=1,2,\ldots,z \qquad (5.54)$$

if and only if

$$\lim_{n\to\infty}[R(a_n(u)t+b_n(u),u)]^{l_n}=\mathfrak{R}_0(t,u) \quad \text{for } t\in C_{\mathfrak{R}_0(u)},\ u=1,2,\ldots,z \qquad (5.55)$$

where $\Re_0(t,u)$ is a non-degenerate reliability function and moreover

$$\Re'(t,u) = 1 - \prod_{i=1}^{a}[1 - d_i(t,u)\Re_0(t,u)]^{q_i k}, \quad t \in (-\infty, \infty), \; u = 1,2,\ldots,z \tag{5.56}$$

Motivation: For each fixed u, $u = 1, 2, \ldots, z$, assumptions (i)–(viii) of Lemma 5.11 and assumptions (i)–(viii) of Lemma 4.16 are equivalent. Condition (5.54) is identical to condition (4.69) and condition (5.55) is identical to condition (4.70). Moreover, from Lemma 4.16, conditions (4.69) and (4.70) are equivalent, which means that conditions (5.54) and (5.55) are also equivalent and from Eq. (4.71) the equality (5.56) holds.

Lemmas 5.10 and 5.11 and Theorem 4.7 establish the class of limit reliability functions for non-homogeneous regular multi-state series–parallel systems given in the following theorem [81].

Theorem 5.7

The class of limit non-degenerate reliability functions of the non-homogeneous regular multi-state series–parallel system is composed of $3^z + 4^z + 3^z$ reliability functions of the form

$$\Re'(t, \cdot) = [1, \Re'(t,1), \ldots, \Re'(t,z)], \quad t \in (-\infty, \infty) \tag{5.57}$$

where

Case 1. $k_n = n$, $|l_n - c \log n| \gg s$, $s > 0$, $c > 0$ (under Assumption 4.1 and the assumptions of Lemma 5.10).

$$\Re'(t,u) \in \{\Re'_1(t), \Re'_2(t), \Re'_3(t)\}, \quad u = 1,2,\ldots,z \tag{5.58}$$

and $\Re'_i(t)$, $i = 1, 2, 3$, are defined by Eqs. (4.72)–(4.74) with $d(t) = d(t,u)$, $u = 1, 2, \ldots, z$, where $d(t,u)$ are defined by Eq. (5.49).

Case 2. $k_n = n$, $l_n - c \log n \sim s$, $s \in (-\infty, \infty)$, $c > 0$ (under the assumptions of Lemma 5.10).

$$\Re'(t,u) \in \{\Re'_4(t), \Re'_5(t), \Re'_6(t), \Re'_7(t)\}, \quad u = 1,2,\ldots,z \tag{5.59}$$

and $\Re'_i(t)$, $i = 4, 5, 6, 7$, are defined by Eqs. (4.75)–(4.78) with $d(t) = d(t,u)$, $u = 1, 2, \ldots, z$, where $d(t,u)$ are defined by Eq. (5.49).

Case 3. $k_n \to k$, $k > 0$, $l_n \to \infty$ (under the assumptions of Lemma 5.11).

$$\Re'(t,u) \in \{\Re'_8(t), \Re'_9(t), \Re'_{10}(t)\}, \quad u = 1,2,\ldots,z \tag{5.60}$$

and $\mathfrak{R}_i'(t)$, $i = 8, 9, 10$, are defined by Eqs. (4.79)–(4.81) with $d_i(t) = d_i(t,u)$, $u = 1, 2, \ldots, z$, where $d_i(t,u)$ are defined by Eq. (5.52).

Motivation: For each fixed u, $u = 1, 2, \ldots, z$, coordinate $\mathfrak{R}'(t,u)$ of the vector $\mathfrak{R}'(t,\cdot)$ defined by Eq. (5.57), according to Theorem 4.7, that is a consequence of Lemmas 4.15 and 4.16, can be one of the three types of reliability functions given by Eqs. (4.72)–(4.74), or one of the four types of reliability function given by Eqs. (4.75)–(4.78) with $d(t) = d(t,u)$, where $d(t,u)$ are defined by Eq. (5.49), or one of the three types of reliability functions given by Eqs. (4.79)–(4.81) with $d_i(t) = d_i(t,u)$, where $d_i(t,u)$ are defined by Eq. (5.52). Thus the number of different limit reliability functions of the considered system is equal to the sum of the number of z-term variations of the 3-element sets defined by Eqs. (5.58) and (5.60) and the 4-element set defined by Eq. (5.59), i.e. $3^z + 4^z + 3^z$, and they are of the form (5.57).

Corollary 5.10

If components of the non-homogeneous regular multi-state series–parallel system have Weibull reliability functions

$$R^{(i,j)}(t,\cdot) = [1, R^{(i,j)}(t,1), \ldots, R^{(i,j)}(t,z)], \quad t \in (-\infty, \infty)$$

where

$$R^{(i,j)}(t,u) = 1 \quad \text{for } t < 0, R^{(i,j)}(t,u) = \exp[-\beta_{ij}(u)t^{\alpha_{ij}(u)}] \quad \text{for } t \geq 0, \quad \alpha_{ij}(u) > 0,$$
$$\beta_{ij}(u) > 0, \quad u = 1,2,\ldots,z, \quad i = 1,2,\ldots,a, \quad j = 1,2,\ldots,e_i$$

and

$$k_n \to k, \quad l_n \to \infty$$

$$a_n(u) = (\beta(u)l_n)^{-1/\alpha(u)}, \quad b_n(u) = 0 \tag{5.61}$$

where

$$\alpha_i(u) = \min_{1 \leq j \leq e_i} \{\alpha_{ij}(u)\}, \beta_i(u) = \sum_{j:\alpha_{ij}(u) = \alpha_i(u)} p_{ij}\beta_{ij}(u) \tag{5.62}$$

$$\alpha(u) = \max_{1 \leq i \leq a} \{\alpha_i(u)\}, \quad \beta(u) = \min\{\beta_i(u):\alpha_i(u) = \alpha(u)\} \tag{5.63}$$

then

$$\mathfrak{R}_9'(t,\cdot) = [1, \mathfrak{R}_9'(t,1), \ldots, \mathfrak{R}_9'(t,z)]$$

where

$$\Re_9'(t,u) = 1 \quad \text{for } t < 0$$

$$\Re_9'(t,u) = 1 - \prod_{i:\alpha_i(u)=\alpha(u)} [1 - \exp[-(\beta_i(u)/\beta(u))t^{\alpha(u)}]]^{q_i k} \quad \text{for } t \geq 0$$

is its limit reliability function.

Motivation: Since for each fixed u, according to Eq. (5.61), we have

$$a_n(u)t + b_n(u) \geq 0 \quad \text{for } t \geq 0$$

and

$$a_n(u)t + b_n(u) \to 0^+ \quad \text{for } t \geq 0 \text{ as } n \to \infty$$

then for all $i = 1, 2, \ldots, a$, we get

$$R^{(i)}(a_n(u)t) + b_n(u), u) = 1 \quad \text{for } t < 0$$

and considering Eq. (5.62)

$$R^{(i)}(a_n(u)t + b_n(u), u) = \exp\left[-\sum_{j=1}^{e_i} p_{ij}\beta_{ij}(u)(a_n(u)t)^{\alpha_{ij}(u)}\right]$$

$$= \exp\left[-(a_n(u)t)^{\alpha_i(u)}\sum_{j=1}^{e_i} p_{ij}\beta_{ij}(u)(a_n(u)t)^{\alpha_{ij}(u)-\alpha_i(u)}\right]$$

$$= \exp[-\beta_i(a_n(u)t)^{\alpha_i(u)} + o(1)] \quad \text{for } t \geq 0$$

Defining

$$R(t,u) = 1 \quad \text{for } t < 0, \quad R(t,u) = \exp[-\beta(u)t^{\alpha(u)}] \quad \text{for } t \geq 0, \quad u = 1, 2, \ldots, z$$

for all $i = 1, 2, \ldots, a$, by Eq. (5.63), we get

$$\lim_{n\to\infty} \frac{R^{(i)}(a_n(u)t + b_n(u), u)}{R(a_n(u)t + b_n(u), u)} = 1 \quad \text{for } t < 0$$

and for $t \geq 0$

$$\lim_{n\to\infty} \frac{R^{(i)}(a_n(u)t + b_n(u), u)}{R(a_n(u)t + b_n(u), u)} = \lim_{n\to\infty} \frac{\exp[-\beta_i(u)(a_n(u)t + b_n(u))^{\alpha_i(u)}]}{\exp[-\beta(u)(a_n(u)t + b_n(u))^{\alpha}]}$$

$$= \lim_{n\to\infty} \exp\left[-\beta(u)(a_n(u)t)^{\alpha(u)}\left[\frac{\beta_i(u)}{\beta(u)}(a_n(u)t)^{\alpha_i(u)-\alpha(u)} - 1\right]\right] \leq 1$$

The above means that condition (vii) of Lemma 5.11 is satisfied with $t_0(u) = \infty$ and moreover, from Eq. (5.53) it follows that

$$d_i(t,u) = \begin{cases} 1 & \text{for } t<0 \\ \exp[-(\beta_i(u)/\beta(u) - 1]t^{\alpha(u)} & \text{for } t \geq 0 \end{cases}$$

for i such that

$$\alpha_i(u) = \alpha(u)$$

and

$$d_i(t,u) = \begin{cases} 1 & \text{for } t<0 \\ 0 & \text{for } t \geq 0 \end{cases}$$

otherwise. Further, we have

$$\mathfrak{R}_0(t,u) = \lim_{n\to\infty} (R(a_n t + b_n))^{l_n} = 1 \quad \text{for } t<0$$

and

$$\mathfrak{R}_0(t,u) = \lim_{n\to\infty} [R(a_n(u)t + b_n(u))]^{l_n}$$
$$= \lim_{n\to\infty} \exp[-l_n\beta(u)(a_n(u)t)^{\alpha(u)}] = \exp[-t^{\alpha(u)}] \quad \text{for } t \geq 0$$

which from Lemma 5.11 completes the proof.

Example 5.10 (A Piping System)
The piping system is composed of $k_n = 3$ pipeline lines linked in parallel, each of them composed of $l_n = 100$ five-state pipe segments. In two of the pipelines, there are 40 pipe segments with exponential reliability functions

$$R^{(1,1)}(t,1) = \exp[-0.025t], \quad R^{(1,1)}(t,2) = \exp[-0.026t],$$
$$R^{(1,1)}(t,3) = \exp[-0.028t], \quad R^{(1,1)}(t,4) = \exp[-0.30t] \quad \text{for } t \geq 0$$

and 60 pipe segments with Weibull reliability functions

$$R^{(1,2)}(t,1) = \exp[-0.0015t^2], \quad R^{(1,2)}(t,2) = \exp[-0.0016t^2],$$
$$R^{(1,2)}(t,3) = \exp[-0.002t^2], \quad R^{(1,2)}(t,4) = \exp[-0.0025t^2] \quad \text{for } t \geq 0$$

The third pipeline is composed of 50 pipe segments with Weibull reliability functions

$$R^{(2,1)}(t,1)=\exp[-0.0007t^3],\ R^{(2,1)}(t,2)=\exp[-0.0008t^3],$$
$$R^{(2,1)}(t,3)=\exp[-0.0010t^3],\ R^{(2,1)}(t,4)=\exp[-0.0016t^3]\quad \text{for } t\geq 0$$

and 50 pipe segments with Weibull reliability functions

$$R^{(2,2)}(t,1)=\exp\left[-0.15\sqrt{t}\right],\ R^{(2,2)}(t,2)=\exp\left[-0.16\sqrt{t}\right],$$
$$R^{(2,2)}(t,3)=\exp\left[-0.18\sqrt{t}\right],\ R^{(2,2)}(t,4)=\exp\left[-0.2\sqrt{t}\right]\quad \text{for } t\geq 0$$

Thus the piping system is a non-homogeneous regular multi-state series–parallel system in which, according to Definition 3.20, we have

$$k_n=k=3,\ l_n=100,\ a=2,\ q_1=2/3,\ q_2=1/3$$

Therefore, from Eq. (3.44), we get

$$\boldsymbol{R}'_{3,100}(t,u)=1-\prod_{i=1}^{2}[1-(R^{(i)}(t,u))^{100}]^{q_i 3}$$
$$=1-[1-R^{(1)}(t,u))^{100}]^2[1-(R^{(2)}(t,u))^{100}]$$

where substituting $e_1=2,\ p_{11}=0.4,\ p_{12}=0.6$ according to Eq. (3.45), we get

$$R^{(1)}(t,u)=\prod_{j=1}^{e_1}(R^{(1,j)}(t,u))^{p_{1j}}=(R^{(1,1)}(t,u))^{0.4}(R^{(1,2)}(t,u))^{0.6}$$

and substituting $e_2=2,\ p_{21}=0.4,\ p_{22}=0.5$, we get

$$R^{(2)}(t,u)=\prod_{j=1}^{e_2}(R^{(2,j)}(t,u))^{p_{2j}}=(R^{(2,1)}(t,u))^{0.5}(R^{(2,2)}(t,u))^{0.5}$$

Hence, from Eq. (3.43), the exact reliability function of the piping system takes the form

$$\boldsymbol{R}'_{3,100}(t,\cdot)=[1,1-[1-\exp[-t-0.09t^2]]^2\cdot[1-\exp[-0.035t^3-7.5\sqrt{t}],$$
$$1-[1-\exp[-1.04t-0.096t^2]]^2\cdot[1-\exp[-0.04t^3-8\sqrt{t}],$$
$$1-[1-\exp[-1.12t-0.12t^2]]^2\cdot[1-\exp[-0.05t^3-10\sqrt{t}],$$
$$1-[1-\exp[-1.2t-0.15t^2]]^2\cdot[1-\exp[-0.08t^3-10\sqrt{t}]]\quad \text{for } t\geq 0$$

Further, applying Corollary 5.10 and considering Eqs. (5.61)–(5.63), we have

$$\alpha_1(u)=\min\{1,2\}=1\quad \text{for } u=1,2,3,4$$

$$\beta_1(1) = 0.4 \cdot 0.025 = 0.01, \;\; \beta_1(2) = 0.4 \cdot 0.026 = 0.0104,$$
$$\beta_1(3) = 0.4 \cdot 0.028 = 0.0112, \;\; \beta_1(4) = 0.4 \cdot 0.03 = 0.012$$

$$a_2(u) = \min\{3, 0.5\} = 0.5 \quad \text{for } u = 1, 2, 3, 4$$

$$\beta_2(1) = 0.5 \cdot 0.15 = 0.075, \;\; \beta_2(2) = 0.5 \cdot 0.16 = 0.08,$$
$$\beta_2(3) = 0.5 \cdot 0.18 = 0.09, \;\; \beta_2(1) = 0.5 \cdot 0.2 = 0.1$$

$$\alpha(u) = \max\{1, 0.5\} = 1 \quad \text{for } u = 1, 2, 3, 4$$

$$\beta(1) = \min\{0.01\} = 0.01, \;\; \beta(2) = \min\{0.0104\} = 0.0104,$$
$$\beta(3) = \min\{0.0112\} = 0.0112, \;\; \beta(1) = \min\{0.012\} = 0.012$$

$$a_n(1) = 1/(0.01 \cdot 100) = 1, \;\; a_n(2) = 1/(0.0104 \cdot 100) = 0.962,$$
$$a_n(3) = 1/(0.0112 \cdot 100) = 0.893, \;\; a_n(4) = 1/(0.012 \cdot 100) = 0.833$$

$$b_n(u) = 0 \quad \text{for } u = 1, 2, 3, 4$$

and we conclude that the system limit reliability function is

$$\mathfrak{R}'_9(t, \cdot) = [1, 1 - [1 - \exp[-t]]^2, 1 - [1 - \exp[-t]]^2, 1 - [1 - \exp[-t]]^2$$
$$1 - [1 - \exp[-t]]^2 \quad \text{for } \geq 0$$

Thus, from Eq. (3.49), the approximate formula for the piping system reliability function takes the form

$$\boldsymbol{R}'_{3,100}(t, \cdot) \cong \overline{\mathfrak{R}}_9((t - b_n(u))/a_n(u), \cdot)$$
$$= [1, 1 - [1 - \exp[-t]]^2, 1 - [1 - \exp[-1.04t]]^2, 1 - [1 - \exp[-1.12t]]^2,$$
$$1 - [1 - \exp[-1.2t]]^2 \quad \text{for } t \geq 0.$$

For instance, the expected values of the first type pipe segments' lifetimes in the state subsets, according to Eq. (3.4), are

$$M_{11}(1) = 1/0.025 = 40, \;\; M_{11}(2) = 1/0.026 \cong 38.46,$$
$$M_{11}(3) = 1/0.028 \cong 35.71, \;\; M_{11}(4) = 1/0.030 \cong 33.33$$

and their lifetimes in particular states, according to Eq. (3.8), are

$$\overline{M}_{11}(1) \cong 1.54, \;\; \overline{M}_{11}(2) \cong 2.75, \;\; \overline{M}_{11}(3) \cong 2.38, \;\; \overline{M}_{11}(4) \cong 33.33$$

The approximate mean values of the piping system sojourn times in the state subsets, according to Eq. (3.13), are

$$M(1) \cong 1.5, \;\; M(2) \cong 1.44, \;\; M(3) \cong 1.34, \;\; M(4) \cong 1.25$$

Hence, from Eq. (3.17), the mean values of the piping system lifetime in particular states are

$$\overline{M}(1) \cong 0.06, \quad \overline{M}(2) \cong 0.10, \quad \overline{M}(3) \cong 0.09, \quad \overline{M}(4) \cong 1.25$$

If the critical reliability state of the system is $r = 2$, then according to Eq. (3.18) its risk function is given by

$$\boldsymbol{r}(t) \cong 1 - \exp[-1.04t]]^2$$

Further, from Eq. (3.19), the moment when the system risk exceeds an admissible level $\delta = 0.05$ is

$$\tau = \boldsymbol{r}^{-1}(\delta) = -(1/1.04)\log(1 - \sqrt{\delta}) \cong 0.24 \text{ years}$$

5.5 Reliability Evaluation of Multi-State Parallel–Series Systems

In proving facts on limit reliability functions for homogeneous regular multi-state parallel–series systems, the following slight extensions of Lemmas 4.18 – 4.19 are used [81].

Lemma 5.12
If

- **i.** $k_n \to \infty$,
- **ii.** $\mathfrak{R}(t,u) = \exp[-\overline{V}(t,u)], \quad u = 1,2,\ldots,z$, is a non-degenerate reliability function,
- **iii.** $\overline{\boldsymbol{R}}_{k_n,l_n}(t,\cdot) = [1, \overline{\boldsymbol{R}}_{k_n,l_n}(t,1),\ldots,\overline{\boldsymbol{R}}_{k_n,l_n}(t,z)], \quad t \in (-\infty,\infty)$, is the reliability function of a homogeneous regular multi-state parallel–series system defined by Eqs. (3.30) and (3.31),
- **iv.** $a_n(u) > 0, \quad b_n(u) \in (-\infty,\infty), \quad u = 1,2,\ldots,z$

then

$$\overline{\mathfrak{R}}(t,\cdot) = [1, \overline{\mathfrak{R}}(t,1),\ldots,\overline{\mathfrak{R}}(t,z)], \quad t \in (-\infty,\infty)$$

is its limit multi-state reliability function, i.e.

$$\lim_{n\to\infty} \overline{\boldsymbol{R}}_{k_n,l_n}(a_n(u)t + b_n(u),u) = \overline{\mathfrak{R}}(t,u) \quad \text{for } t \in C_{\overline{\mathfrak{R}}(u)}, \quad u = 1,2,\ldots,z \tag{5.64}$$

if and only if

$$\lim_{n\to\infty} k_n[F(a_n(u)t + b_n(u),u)]^{l_n} = \overline{V}(t,u) \quad \text{for } t \in C_{\overline{V}(u)}, \quad u = 1,2,\ldots,z \tag{5.65}$$

Motivation: For each fixed u, $u = 1, 2, \ldots, z$, assumptions (i)–(iv) of Lemma 5.12 are identical to assumptions (i)–(iv) of Lemma 4.18. Moreover, condition (5.64) is identical to condition (4.88) and condition (5.65) is identical to condition (4.89). Therefore, since from Lemma 4.18, conditions (4.88) and (4.89) are equivalent, then conditions (5.64) and (5.65) are also equivalent.

Lemma 5.13
If

i. $k_n \to k$, $k > 0$, $l_n \to \infty$,
ii. $\overline{\Re}(t,u)$, $u = 1, 2, \ldots, z$ is a non-degenerate reliability function,
iii. $\overline{\boldsymbol{R}}_{k_n,l_n}(t,\cdot) = [1, \overline{\boldsymbol{R}}_{k_n,l_n}(t,1), \ldots, \overline{\boldsymbol{R}}_{k_n,l_n}(t,z)]$, $t \in (-\infty, \infty)$ is the reliability function of a homogeneous regular multi-state parallel–series system defined by Eqs. (3.30) and (3.31),
iv. $a_n(u) > 0$, $b_n(u) \in (-\infty, \infty)$, $u = 1, 2, \ldots, z$

then

$$\overline{\Re}(t,\cdot) = [1, \overline{\Re}(t,1), \ldots, \overline{\Re}(t,z)], \quad t \in (-\infty, \infty)$$

is its limit multi-state reliability function, i.e.

$$\lim_{n\to\infty} \overline{\boldsymbol{R}}_{k_n,l_n}(a_n(u)t + b_n(u), u) = \overline{\Re}(t,u) \quad \text{for } t \in C_{\overline{\Re}(u)}, \quad u = 1, 2, \ldots, z \tag{5.66}$$

if and only if

$$\lim_{n\to\infty} [F(a_n(u)t + b_n(u), u)]^{l_n} = \Im_0(t,u) \quad \text{for } t \in C_{\Im_0(u)}, \quad u = 1, 2, \ldots, z \tag{5.67}$$

where $\Im_0(t,u)$ is a non-degenerate distribution function and moreover

$$\overline{\Re}(t,u) = [1 - \Im_0(t,u)]^k \quad \text{for } t \in (-\infty, \infty), \quad u = 1, 2, \ldots, z \tag{5.68}$$

Motivation: For each fixed u, $u = 1, 2, \ldots, z$, assumptions (i)–(iv) of Lemma 5.13 are identical to assumptions (i)–(iv) of Lemma 4.19. Moreover, condition (5.66) is identical to condition (4.90) and condition (5.67) is identical to condition (4.91). Since from Lemma 4.19, conditions (4.90) and (4.91) are equivalent, then conditions (5.66) and (5.67) are also equivalent and moreover, due to Eq. (4.92), the equality (5.68) holds.

Lemmas 5.12 and 5.13 and Theorem 4.8 yield the next theorem [81].

Theorem 5.8
The class of limit non-degenerate reliability functions of the homogeneous regular multi-state parallel–series system is composed of $3^z + 4^z + 3^z$ reliability functions of the form

$$\overline{\Re}(t,\cdot) = [1, \overline{\Re}(t,1), \ldots, \overline{\Re}(t,z)], \quad t \in (-\infty, \infty) \tag{5.69}$$

where

Case 1. $k_n = n$, $|l_n - c \log n| \gg s$, $s > 0$, $c > 0$ (under Assumption 4.1).

$$\overline{\Re}(t, u) \in \{\overline{\Re}_1(t), \overline{\Re}_2(t), \overline{\Re}_3(t)\}, \quad u = 1, 2, \ldots, z \tag{5.70}$$

and $\overline{\Re}_i(t)$, $i = 1, 2, 3$, are defined by Eqs. (4.93)–(4.95).

Case 2. $k_n = n$, $l_n - c \log n \sim s$, $s \in (-\infty, \infty)$, $c > 0$.

$$\overline{\Re}(t, u) \in \{\overline{\Re}_4(t), \overline{\Re}_5(t), \overline{\Re}_6(t), \overline{\Re}_7(t)\}, \quad u = 1, 2, \ldots, z \tag{5.71}$$

and $\overline{\Re}_i(t)$, $i = 4, 5, 6, 7$, are defined by Eqs. (4.96)–(4.99).

Case 3. $k_n \to k$, $k > 0$, $l_n \to \infty$.

$$\overline{\Re}(t, u) \in \{\overline{\Re}_8(t), \overline{\Re}_9(t), \overline{\Re}_{10}(t)\}, \quad u = 1, 2, \ldots, z \tag{5.72}$$

and $\overline{\Re}_i(t)$, $i = 8, 9, 10$, are defined by Eqs. (4.100)–(4.102).

Motivation: For each fixed u, $u = 1, 2, \ldots, z$, coordinate $\overline{\Re}(t, u)$ of the vector $\overline{\Re}(t, \cdot)$ defined by Eq. (5.69), according to Theorem 4.8 that is the consequence of Lemmas 4.18 and 4.19, can be one of the three types of reliability function given by Eqs. (4.93)–(4.95), or one of the four types of reliability function given by Eqs. (4.96)–(4.99), or one of the three types of reliability function given by Eqs. (4.100)–(4.102). Thus the number of different limit multi-state reliability functions of the considered system is equal to the sum of the numbers of z-term variations of the 3-component sets defined by Eqs. (5.70) and (5.72) and the number of the 4-component set defined by Eq. (5.71). It means that this number is equal to $3^z + 4^z + 3^z$ and they are of the form (5.69).

Corollary 5.11
If components of the homogeneous regular multi-state parallel–series system have Weibull reliability functions

$$R(t, \cdot) = [1, R(t, 1), \ldots, R(t, z)], \quad t \in (-\infty, \infty)$$

where

$$R(t, u) = 1 \quad \text{for } t < 0, \quad R(t, u) = \exp[-\beta(u) t^{\alpha(u)}] \quad \text{for } t \geq 0, \\ \alpha(u) > 0, \quad \beta(u) > 0, \quad u = 1, 2, \ldots, z \tag{5.73}$$

and

$$k_n \to k, \quad k > 0, \quad l_n \to \infty \tag{5.74}$$

$$a_n(u) = b_n(u)/(\alpha(u)\beta(u)(b_n(u))^{\alpha(u)}, \quad b_n(u) = [(\log l_n/\beta(u)]^{1/\alpha(u)}, \\ u = 1, 2, \ldots, z \tag{5.75}$$

then

$$\overline{\mathfrak{R}}_{10}(t, \cdot) = [1, \overline{\mathfrak{R}}_{10}(t, 1), \ldots, \overline{\mathfrak{R}}_{10}(t, z)], \quad t \in (-\infty, \infty)$$

where

$$\overline{\mathfrak{R}}_{10}(t, u) = [1 - \exp[-\exp[-t]]^k \quad \text{for } t \in (-\infty, \infty), \; u = 1, 2, \ldots, z$$

is its limit reliability function.

Motivation: Since for each fixed u, sufficiently large n and all $t \in (-\infty, \infty)$, according to Eqs. (5.74) and (5.75), we have

$$a_n(u)t + b_n(u) > 0$$

and from Eq. (5.73)

$$F(a_n(u)t + b_n(u), u) = 1 - \exp[-\beta(u)(a_n(u)t + b_n(u))^{\alpha(u)}] \quad \text{for } t \in (-\infty, \infty)$$

and further from Eq. (5.67)

$$\begin{aligned} \mathfrak{J}_0(t, u) &= \lim_{n \to \infty} [F(a_n(u)t + b_n(u), u]^{l_n} \\ &= \lim_{n \to \infty} [1 - \exp[-\beta(u)(b_n(u))^{\alpha(u)}((1 + ((a_n(u)/b_n(u))t)^{\alpha(u)}]]^{l_n} \\ &= \lim_{n \to \infty} [1 - \exp[-(\log l_n)(1 + t/(\alpha(u)\log l_n))^{\alpha(u)}]]^{l_n} \\ &= \lim_{n \to \infty} [1 - (1/l_n)\exp[-t + o(1)]]^{l_n} \\ &= \exp[-\exp[-t]] \quad \text{for } t \in (-\infty, \infty) \end{aligned}$$

Thus, after considering Eq. (5.68), from Lemma 5.13, $\overline{\mathfrak{R}}_{10}(t, \cdot)$ is the limit reliability function of the considered system.

Example 5.11 (An Electrical Energy Distribution System)
Let us consider a model energetic network stretched between two poles and composed of three energetic cables, six insulators and two bearers and analyse the reliability of all cables. Each cable consists of 36 identical wires. Assuming that the cable is able to conduct the current if at least one of its wires is not failed we conclude that it is a homogeneous parallel–series system composed of $k_n = 3$ parallel subsystems linked in series, each of them consisting of $l_n = 36$ basic components.

Further, assuming that the wires are four-state components, i.e. $z=3$, having Weibull reliability functions with parameters

$$\alpha(u)=2, \quad \beta(u)=(7.07)^{2u-8}, \quad u=1,2,3$$

according to Corollary 5.11, assuming normalizing constants

$$a_n(u)=(7.07)^{4-u}/\left(2\sqrt{\log 36}\right), \quad b_n(u)=(7.07)^{4-u}/\sqrt{\log 36}, \quad u=1,2,3$$

and applying Eq. (3.49) we obtain the following approximate form of the system multi-state reliability function

$$\begin{aligned}\overline{\boldsymbol{R}}_{3,36}(t,\cdot)&=[1,\overline{\boldsymbol{R}}_{3,36}(t,1),\overline{\boldsymbol{R}}_{3,36}(t,2),\overline{\boldsymbol{R}}_{3,36}(t,3)]\\&\cong[1,[1-\exp[-\exp[-0.01071t+7.167]]]^3,\\&\quad[1-\exp[-\exp[-0.07572t+7.167]]]^3,\\&\quad[1-\exp[-\exp[-0.53543t+7.167]]]^3] \quad \text{for } t\in(-\infty,\infty)\end{aligned}$$

The values of the system sojourn times $T(u)$ in the state subsystems in months, after applying Eq. (3.13), are given by

$$E[T(u)]\cong\int_0^\infty[1-\exp[-\exp[-(7.07)^{u-4}2\sqrt{\log 36}t+2\log 36]]]^3 dt, \quad u=1,2,3$$

and particularly

$$M(1)\cong 650, M(2)\cong 100, M(3)\cong 15$$

Hence, from Eq. (3.17), the system mean lifetimes in particular states are

$$\overline{M}(1)\cong 550, \quad \overline{M}(2)\cong 85, \quad \overline{M}(3)\cong 15$$

If the critical reliability state of the system is $r=2$, then its risk function, according to Eq. (3.18), is given by

$$\boldsymbol{r}(t)\cong 1-[1-\exp[-\exp[-0.07572t+7.167]]]^3$$

The moment when the system risk exceeds an admissible level $\delta=0.05$, calculated due to Eq. (3.19), is

$$\tau=\boldsymbol{r}^{-1}(\delta)\cong[7.167-\log[-\log[1-(1-\delta)^{1/3}]]]/0.07572=76 \text{ months}$$

The extensions of Lemmas 4.22 and 4.23 are essential tools for finding limit reliability functions for non-homogeneous regular multi-state parallel–series systems. They may be formulated as follows [81].

Lemma 5.14

If

i. $k_n \to \infty$,

ii. $\overline{\mathfrak{R}}'(t,u) = \exp[-\overline{V}'(t,u)]$, $u = 1, 2, \ldots, z$, is a non-degenerate reliability function,

iii. $\overline{\boldsymbol{R}}'_{k_n,l_n}(t,\cdot) = [1, \overline{\boldsymbol{R}}'_{k_n,l_n}(t,1), \ldots, \overline{\boldsymbol{R}}'_{k_n,l_n}(t,z)]$, $t \in (-\infty, \infty)$, is the reliability function of a non-homogeneous regular multi-state parallel–series system defined by Eqs. (3.46)–(3.48),

iv. $a_n(u) > 0, b_n(u) \in (-\infty, \infty)$, $u = 1, 2, \ldots, z$

v. $F(t,u)$ for each fixed u is one of the distribution functions $F^{(1)}(t,u)$, $F^{(2)}(t,u)$, ..., $F^{(a)}(t,u)$ defined by Eq. (3.48) such that

vi. $\exists N(u) \, \forall n > N(u) F(a_n(u)t + b_n(u), u) = 0$ for $t < t_0(u)$, $F(a_n(u)t + b_n(u), u) \neq 0$ for $t \geq t_0(u)$, where $t_0(u) \in < -\infty, \infty)$

vii. $\lim\limits_{n\to\infty} \dfrac{F^{(i)}(a_n(u)t + b_n(u), u)}{F(a_n(u)t + b_n(u), u)} \leq 1$ for $t \geq t_0(u)$, $i = 1, 2, \ldots, a$, $u = 1, 2, \ldots, z$

and moreover there exist non-decreasing functions

viii.

$$\overline{d}(t,u) = \begin{cases} 0 & \text{for } t < t_0(u) \\ \lim\limits_{n\to\infty} \sum\limits_{i=1}^{a} q_i \overline{d}_i(a_n(u)t + b_n(u), u) & \text{for } t \geq t_0(u) \end{cases} \tag{5.76}$$

where

$$\overline{d}_i(a_n(u)t + b_n(u), u) = \left[\frac{F^{(i)}(a_n(u)t + b_n(u), u)}{F(a_n(u)t + b_n(u), u)}\right]^{l_n}$$

then

$$\overline{\mathfrak{R}}'(t,\cdot) = [1, \overline{\mathfrak{R}}'(t,1), \ldots, \overline{\mathfrak{R}}'(t,z)], \quad t \in (-\infty, \infty)$$

is its limit multi-state reliability function, i.e.

$$\lim_{n\to\infty} \overline{\boldsymbol{R}}'_{k_n,l_n}(a_n(u)t + b_n(u), u) = \overline{\mathfrak{R}}'(t,u) \quad \text{for } t \in C_{\overline{\mathfrak{R}}'(u)}, \quad u = 1, 2, \ldots, z \tag{5.77}$$

if and only if

$$\lim_{n\to\infty} k_n (F(a_n(u)t + b_n(u), u))^{l_n} \overline{d}(t,u) = \overline{V}'(t,u) \quad \text{for } t \in C_{\overline{V}'(u)}, \; u = 1, 2, \ldots, z \tag{5.78}$$

Motivation: Since for each fixed u, $u = 1, 2, \ldots, z$, assumptions (i)–(viii) of Lemma 5.14 are identical to assumptions (i)–(viii) of Lemma 4.22, condition (5.77) is identical to condition (4.105) and condition (5.78) is identical to condition (4.106), which from Lemma 4.22 are equivalent, then Lemma 5.14 is valid.

Lemma 5.15
If

i. $k_n \to k$, $k > 0$, $l_n \to \infty$,
ii. $\overline{\mathfrak{R}}'(t,\cdot)$, $u = 1, 2, \ldots, z$, is a non-degenerate reliability function,
iii. $\overline{\boldsymbol{R}}'_{k_n,l_n}(t,\cdot) = [1, \overline{\boldsymbol{R}}'_{k_n,l_n}(t,1), \ldots, \overline{\boldsymbol{R}}'_{k_n,l_n}(t,z)]$, $t \in (-\infty, \infty)$, is the reliability function of a non-homogeneous regular multi-state parallel−series system defined by Eqs. (3.46)–(3.48),
iv. $a_n(u) > 0$, $b_n(u) \in (-\infty, \infty)$, $u = 1, 2, \ldots, z$
v. $F(t,u)$ for each fixed u is one of the distribution functions $F^{(1)}(t,u)$, $F^{(2)}(t,u)$, ...,
vi. $F^{(a)}(t,u)$ defined by Eq. (3.48) such that
vii. $\exists N(u)\, \forall\, n > N(u)$ $F(a_n(u)t + b_n(u), u) = 0$ for $t < t_0(u)$, $F(a_n(u)t + b_n(u), u) \neq 0$ for $t \geq t_0(u)$, where $t_0(u) \in < -\infty, \infty)$
viii. $\lim\limits_{n\to\infty} \dfrac{F^{(i)}(a_n(u)t + b_n(u), u)}{F(a_n(u)t + b_n(u), u)} \leq 1$ for $t \geq t_0(u)$, $i = 1, 2, \ldots, a$, $u = 1, 2, \ldots, z$
and moreover there exist non-decreasing functions
ix.

$$\overline{d}_i(t,u) = \begin{cases} 0 & \text{for } t < t_0(u) \\ \lim\limits_{n\to\infty} \overline{d}_i(a_n(u)t + b_n(u), u) & \text{for } t \geq t_0(u) \end{cases} \tag{5.79}$$

where

$$\overline{d}_i(a_n(u)t + b_n(u), u) = \left[\frac{F^{(i)}(a_n(u)t + b_n(u), u)}{F(a_n(u)t + b_n(u), u)}\right]^{l_n}$$

then

$$\overline{\mathfrak{R}}'(t,\cdot) = [1, \overline{\mathfrak{R}}'(t,1), \ldots, \overline{\mathfrak{R}}'(t,z)], \quad t \in (-\infty, \infty)$$

is its limit multi-state reliability function, i.e.

$$\lim_{n\to\infty} \overline{\boldsymbol{R}}'_{k_n,l_n}(a_n(u)t + b_n(u), u) = \overline{\mathfrak{R}}'(t,u) \quad \text{for } t \in C_{\overline{\mathfrak{R}}'(u)}, \quad u = 1, 2, \ldots, z \tag{5.80}$$

if and only if

$$\lim_{n\to\infty} [F(a_n(u)t + b_n(u), u)]^{l_n} = \mathfrak{I}_0(t,u) \quad \text{for } t \in C_{\mathfrak{I}_0(u)}, \quad u = 1, 2, \ldots, z \tag{5.81}$$

where $\mathfrak{I}_0(t,u)$ is a non-degenerate distribution function and moreover

$$\overline{\mathfrak{R}}'(t,u) = \prod_{i=1}^{a} [1 - \overline{d}_i(t,u)\mathfrak{I}_0(t,u)]^{q_i k}, \quad t \in (-\infty, \infty), \quad u = 1, 2, \ldots, z \tag{5.82}$$

Motivation: For each fixed u, $u = 1, 2, \ldots, z$, assumptions (i)–(viii) of Lemma 5.15 are identical to assumptions (i)–(viii) of Lemma 4.23, condition (5.80) is identical to condition (4.108) and condition (5.81) is identical to condition (4.109). Since, from Lemma 4.23, conditions (4.108) and (4.109) are equivalent, then conditions (5.80) and (5.81) are also equivalent and moreover from Eq. (4.110), the equality (5.82) holds.

The application of Lemmas 5.14 and 5.15 and Theorem 4.9 results in the following theorem [74].

Theorem 5.9

The class of limit non-degenerate reliability functions of the non-homogeneous regular multi-state parallel–series system is composed of $3^z + 4^z + 3^z$ reliability functions of the form

$$\overline{\mathfrak{R}}'(t,\cdot) = [1, \overline{\mathfrak{R}}'(t,1), \ldots, \overline{\mathfrak{R}}'(t,z)], \quad t \in (-\infty, \infty) \tag{5.83}$$

where

Case 1. $k_n = n$, $|l_n - c\log n| \gg s$, $s > 0$, $c > 0$ (under Assumption 4.1 and the assumptions of Lemma 5.14).

$$\overline{\mathfrak{R}}'(t,u) \in \{\overline{\mathfrak{R}}'_1(t), \overline{\mathfrak{R}}'_2(t), \overline{\mathfrak{R}}'_3(t)\}, \quad u = 1, 2, \ldots, z \tag{5.84}$$

and $\overline{\mathfrak{R}}'_i(t)$, $i = 1, 2, 3$, are given by Eqs. (4.111)–(4.113) with $\overline{d}_i(t) = \overline{d}_i(t,u)$, $u = 1, 2, \ldots, z$, where $\overline{d}_i(t,u)$ are defined by Eq. (5.76).

Case 2. $k_n = n$, $l_n - c\log n \sim s$, $s \in (-\infty, \infty)$, $c > 0$ (under the assumptions of Lemma 5.14).

$$\overline{\mathfrak{R}}'(t,u) \in \{\overline{\mathfrak{R}}'_4(t), \overline{\mathfrak{R}}'_5(t), \overline{\mathfrak{R}}'_6(t), \overline{\mathfrak{R}}'_7(t)\}, \quad u = 1, 2, \ldots, z \tag{5.85}$$

and $\overline{\mathfrak{R}}'_i(t)$, $i = 4, 5, 6, 7$, are given by Eqs. (4.114)–(4.117) with $\overline{d}(t) = \overline{d}(t,u)$, $u = 1, 2, \ldots z$, where $\overline{d}(t,u)$ are defined by Eq. (5.76),

Case 3. $k_n \to k$, $k > 0$, $l_n \to \infty$ (under the assumptions of Lemma 5.15).

$$\overline{\mathfrak{R}}'(t,u) \in \{\overline{\mathfrak{R}}'_8(t), \overline{\mathfrak{R}}'_9(t), \overline{\mathfrak{R}}'_{10}(t)\}, \quad u = 1, 2, \ldots, z \tag{5.86}$$

and $\overline{\mathfrak{R}}'_i(t)\overline{\mathfrak{R}}'_i(t)$, $i = 8, 9, 10$, are given by Eqs. (4.118)–(4.120) with $\overline{d}_i(t) = \overline{d}_i(t,u)$, $u = 1, 2, \ldots, z$, where $\overline{d}_i(t,u)$ are defined by Eq. (5.79).

Motivation: For each fixed u, $u = 1, 2, \ldots, z$, coordinate $\overline{\mathfrak{R}}'(t,u)$ of the vector $\overline{\mathfrak{R}}'(t,\cdot)$ defined by Eq. (5.83), from Theorem 4.9 that is the consequence of Lemmas 4.22 and 4.23, may be one of the three types of reliability function defined by Eqs. (4.111)–(4.113), or one of the four types of reliability function defined by Eqs. (4.114)–(4.117) with $\overline{d}(t) = \overline{d}(t,u)$, where $\overline{d}(t,u)$ are given by Eq. (5.76), or one of the three types of reliability function given by Eqs. (4.118)–(4.120) with $\overline{d}_i(t) = \overline{d}_i(t,u)$, where $\overline{d}_i(t,u)$ are defined by Eq. (5.79). Thus the number of different limit multi-state reliability functions of the system is equal to the sum of the number of z-term variations of the 3-element sets given by Eqs. (5.84) and (5.86) and the number of the 4-element set given by Eq. (5.85). It means this number is equal to $3^z + 4^z + 3^z$ and they are of the form (5.83).

5.6 Reliability Evaluation of Other Multi-State Systems

The procedure of the reliability evaluation of more complicated large systems whose structures are compositions of large multi-state systems depends on applying and modifying suitably selected lemmas and corollaries presented in Sections 5.1–5.5. This procedure is illustrated in this section to the exact and approximate reliability evaluation of the multi-state non-homogeneous system composed of a large series–parallel system and a large series–'m out of k' system linked in series and the multi-state homogeneous system composed of three 'm out of l' systems linked in series.

Example 5.12
We consider an exemplary system S that consists of two subsystems S_1, S_2 with the reliability structure presented in Figure 5.9.

The subsystem S_1 is composed of two series subsystems, each of which is composed of 30 components, denoted respectively by

$$E_{ij}^{(1)}, \quad i = 1, 2, \ j = 1, 2, \ldots, 30$$

The subsystemS_2 is composed of four series subsystems, each of which is composed of 40 components, denoted respectively by

$$E_{ij}^{(2)}, \quad i = 1, 2, 3, 4, \ j = 1, 2, \ldots, 40$$

The subsystems S_1 and S_2 are forming a general series reliability structure of the system S presented in Figure 5.10.

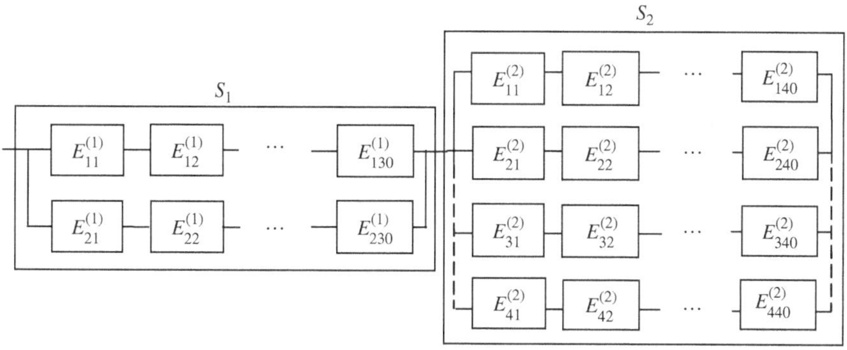

Figure 5.9 The scheme of the exemplary system S reliability structure.

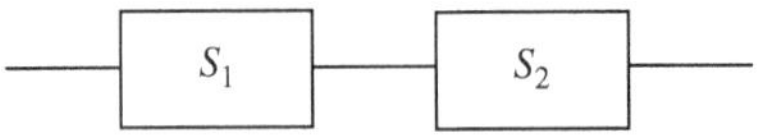

Figure 5.10 The general scheme of the exemplary system S reliability structure.

Further, we assume that the system is a series system composed of the subsystem S_1 and S_2, with the scheme shown in Figure 5.10, while the subsystem S_1 is a series–parallel system with the scheme given in Figure 5.9 and the subsystem S_2 illustrated in Figure 5.9 is a series–'2 out of 4' system. Moreover, we arbitrarily distinguish the following four reliability states ($z = 3$) of the system and its components:

- a reliability state 3: the system operation is fully effective,
- a reliability state 2: the system operation is less effective because of ageing,
- a reliability state 1: the system operation is less effective because of ageing and more dangerous for the environment,
- a reliability state 0: the system is destroyed.

To ensure ageing of the system and its components, we assume that there are possible the transitions between the components reliability states only from better to worse ones. Moreover, we assume that the system and its components' critical reliability state is $r = 2$.

From the above, the subsystems S_υ, $\upsilon = 1, 2$, are composed of four-state, i.e. $z = 3$, components $E_{ij}^{(\upsilon)}$, $\upsilon = 1, 2$, having the four-state reliability functions

$$R_{ij}^{(\upsilon)}(t, \cdot) = [1, R_{ij}^{(\upsilon)}(t, 1), R_{ij}^{(\upsilon)}(t, 2), R_{ij}^{(\upsilon)}(t, 3)]$$

with the coordinates that by the assumption are exponential of the forms

$$R_{ij}^{(\upsilon)}(t,1) = \exp[-\lambda_{ij}^{(\upsilon)}(1)t],\ R_{ij}^{(\upsilon)}(t,2) = \exp[-\lambda_{ij}^{(\upsilon)}(2)t],\ R_{ij}^{(\upsilon)}(t,3) = \exp[-\lambda_{ij}^{(\upsilon)}(3)t]$$

Thus, the considered exemplary system is a four-state series system composed of the four-state subsystems S_1 and S_2. The subsystem S_1 is a four-state series–parallel system and the subsystem S_2 is a four-state series–'2 out of 4' system.

The subsystem S_1 consists of $k = 2$ identical four-state series subsystems, each composed of 30 components, i.e. $l_1 = l_2 = 30$, with the exponential reliability functions. In both series subsystems of the subsystem S_1 there are respectively

- the components $E_{ij}^{(1)}$, $i = 1, 2, j = 1, 2, \ldots, 20$, with the exponential reliability function coordinates

$$R_{ij}^{(1)}(t, 1) = \exp[-0.0008t],\ R_{ij}^{(1)}(t, 2) = \exp[-0.0009t],\ R_{ij}^{(1)}(t, 3) = \exp[-0.0010t],$$
$$i = 1, 2,\ j = 1, 2, \ldots, 20$$

- the components $E_{ij}^{(1)}$, $i = 1, 2$, $j = 21, 22, \ldots, 30$, with the exponential reliability function coordinates

$$R_{ij}^{(1)}(t, 1) = \exp[-0.0011t],\ R_{ij}^{(1)}(t, 2) = \exp[-0.0012t],\ R_{ij}^{(1)}(t, 3) = \exp[-0.0013t],$$
$$i = 1, 2,\ j = 21, 22, \ldots, 30$$

The subsystem S_1 is a four-state series–parallel system and according to the formulae appearing after Definition 3.11 its exact reliability function is given by the vector

$$\boldsymbol{R}^{(1)}(t,\cdot) = [1, \boldsymbol{R}^{(1)}(t,1), \boldsymbol{R}^{(1)}(t,2), \boldsymbol{R}^{(1)}(t,3)], \quad t \geq 0 \tag{5.87}$$

where

$$\begin{aligned}\boldsymbol{R}^{(1)}(t,1) &= \boldsymbol{R}_{2;30,30}(t,1) = 1 - \prod_{i=1}^{2}\left[1 - \prod_{j=1}^{30} R_{ij}^{(1)}(t,1)\right] = 1 - \prod_{i=1}^{2}\left[1 - \exp\left[-\sum_{j=1}^{30}\lambda_{ij}^{(1)}(1)t\right]\right]\\ &= 1-[1-\exp[-[0.0008\cdot 20+0.0011\cdot 10]t]]^2 = 1-[1-\exp[-0.027t]]^2\\ &= 2\exp[-0.027t]-\exp[-0.054t]\end{aligned} \tag{5.88}$$

$$\begin{aligned}\boldsymbol{R}^{(1)}(t,2) &= \boldsymbol{R}_{2;30,30}(t,2) = 1 - \prod_{i=1}^{2}\left[1 - \prod_{j=1}^{30} R_{ij}^{(1)}(t,2)\right] = 1 - \prod_{i=1}^{2}\left[1 - \exp\left[-\sum_{j=1}^{30}\lambda_{ij}^{(1)}(2)t\right]\right]\\ &= 1-[1-\exp[-[0.0009\cdot 20+0.0012\cdot 10]t]]^2 = 1-[1-\exp[-0.030t]]^2\\ &= 2\exp[-0.030t]-\exp[-0.060t]\end{aligned} \tag{5.89}$$

$$\begin{aligned}\boldsymbol{R}^{(1)}(t,3) &= \boldsymbol{R}_{2;30,30}(t,3) = 1 - \prod_{i=1}^{2}\left[1 - \prod_{j=1}^{30} R_{ij}^{(1)}(t,3)\right] = 1 - \prod_{i=1}^{2}\left[1 - \exp\left[-\sum_{j=1}^{30}\lambda_{ij}^{(1)}(3)t\right]\right]\\ &= 1-[1-\exp[-[0.0010\cdot 20+0.0013\cdot 10]t]]^2 = 1-[1-\exp[-0.033t]]^2\\ &= 2\exp[-0.033t]-\exp[-0.066t]\end{aligned} \tag{5.90}$$

To find the subsystem S_1 limit reliability function, we modify Corollary 5.10 into its following particular case for non-homogeneous exponential series–parallel system.

Corollary 5.12

If components of the non-homogeneous regular multi-state series–parallel system have exponential reliability functions

$$R^{(i,j)}(t,\cdot) = [1, R^{(i,j)}(t,1), \ldots, R^{(i,j)}(t,z)], \quad t \in (-\infty, \infty)$$

where

$$R^{(i,j)}(t,u) = 1 \quad \text{for } t<0,\ R^{(i,j)}(t,u) = \exp[-\lambda_{ij}(u)t] \quad \text{for } t \geq 0,\ \lambda_{ij}(u) > 0,$$
$$u = 1,2,\ldots,z, \quad i = 1,2,\ldots,a, \quad j = 1,2,\ldots,e_i$$

and

$$k_n \to k, \;\; l_n \to \infty$$

$$a_n(u) = \frac{1}{\lambda(u) l_n}, \;\; b_n(u) = 0, \quad u = 1, 2, \ldots, z \tag{5.91}$$

where

$$\lambda_i(u) = \sum_{j=1}^{e_i} p_{ij} \lambda_{ij}(u), \quad u = 1, 2, \ldots, z, \;\; i = 1, 2, \ldots, a \tag{5.92}$$

$$\lambda(u) = \min_{1 \le i \le a} \{\lambda_i(u)\}, \quad u = 1, 2, \ldots, z \tag{5.93}$$

then

$$\mathfrak{R}'_9(t, \cdot) = [1, \mathfrak{R}'_9(t, 1), \ldots, \overline{\mathfrak{R}'}_9(t, z)]$$

where

$$\mathfrak{R}'_9(t, u) = 1 \quad \text{for } t < 0$$

$$\mathfrak{R}'_9(t, u) = 1 - \prod_{i=1}^{a} \left[1 - \exp\left[-\frac{\lambda_i(u)}{\lambda(u)} t \right] \right]^{q_i k} \quad \text{for } t \ge 0$$

is its limit reliability function.

The subsystem S_1 is a non-homogeneous regular multi-state series–parallel system in which, according to Definition 3.20, we have

$$k_n = k = 2, \;\; l_n = 30, \;\; a = 1, \;\; q_1 = 1, \;\; e_1 = 2, \;\; p_{11} = 2/3, \;\; p_{12} = 1/3$$

Further, applying Corollary 5.12 and considering Eqs. (5.91)–(5.93), we have

$$\lambda_i(1) = \frac{2}{3} 0.0008 + \frac{1}{3} 0.0011 = 0.0009$$

$$\lambda_i(2) = \frac{2}{3} 0.0009 + \frac{1}{3} 0.0012 = 0.0010$$

$$\lambda_i(3) = \frac{2}{3} 0.0010 + \frac{1}{3} 0.0013 = 0.0011 \quad \text{for } i = 1, 2$$

$$\lambda(1) = \min\{0.0009\} = 0.0009$$

$$\lambda(2) = \min\{0.0010\} = 0.0010$$

$$\lambda(3) = \min\{0.0011\} = 0.0011$$

$$a_n(1) = 1/(0.0009 \cdot 30) = 1/0.027, \quad a_n(2) = 1/(0.0010 \cdot 30) = 0.030 \tag{5.94}$$

$$a_n(3) = 1/(0.0011 \cdot 30) = 0.033, \quad b_n(u) = 0 \quad \text{for } u = 1, 2, 3 \tag{5.95}$$

and we conclude that the system limit reliability function is

$$\begin{aligned}\mathfrak{R}'_9(t, \cdot) = [1, 1 - [1 - \exp[-t]]^2, 1 - [1 - \exp[-t]]^2, 1 - [1 - \exp[-t]]^2 \\ 1 - [1 - \exp[-t]]^2 \quad \text{for } t \geq 0\end{aligned} \tag{5.96}$$

Thus, considering Eqs. (5.94)–(5.96), from Eq. (3.49), the approximate formula for the subsystem S_1 reliability function takes the form

$$\begin{aligned}\boldsymbol{R}^{(1)}(t, \cdot) &\cong \mathfrak{R}'_9((t - b_n(u))/a_n(u), \cdot) \\ &= [1, 1 - [1 - \exp[-0.027t]]^2, 1 - [1 - \exp[-030t]]^2, \\ &\quad 1 - [1 - \exp[-0.033t]]^2 \\ &= [1, 2\exp[-0.027t] - \exp[-0.054t], 2\exp[-0.030t] - \exp[-0.060t], \\ &\quad 2\exp[-0.033tt] - \exp[-0.066t]] \quad \text{for } t \geq 0\end{aligned} \tag{5.97}$$

It is easy to see that the above result is identical to the subsystem S_1 exact reliability given by Eqs. (5.87)–(5.90).

The subsystem S_2 consists of $k = 4$ identical four-state series subsystems, each composed of 40 components, i.e. $l_1 = l_2 = l_3 = l_4 = 40$, with the exponential reliability functions. In all series subsystems of the subsystem S_2 there are respectively

- the components $E_{ij}^{(2)}$, $i = 1, 2, 3, 4, j = 1, 2, \ldots, 20$, with the exponential reliability function coordinates

$$R_{ij}^{(2)}(t, 1) = \exp[-0.0014t], \quad R_{ij}^{(2)}(t, 2) = \exp[-0.0015t],$$
$$R_{ij}^{(2)}(t, 3) = \exp[-0.0016t], \quad i = 1, 2, 3, 4, \quad j = 1, 2, \ldots, 20$$

- the components $E_{ij}^{(2)}$, $i = 1, 2, 3, 4, j = 1, 2, \ldots, 20$, with the exponential reliability function coordinates

$$R_{ij}^{(2)}(t, 1) = \exp[-0.0018t], \quad R_{ij}^{(2)}(t, 2) = \exp[-0.0020t],$$
$$R_{ij}^{(2)}(t, 3) = \exp[-0.0022t], \quad i = 1, 2, 3, 4, \quad j = 21, 22, \ldots, 40$$

The subsystem S_2 is a four-state series–'2 out of 4' system and according to the formulae appearing after Definition 3.8, its exact reliability function is given by the vector

$$\boldsymbol{R}^{(2)}(t, \cdot) = [1, \boldsymbol{R}^{(2)}(t, 1), \boldsymbol{R}^{(2)}(t, 2), \boldsymbol{R}^{(2)}(t, 3)], \quad t \geq 0 \tag{5.98}$$

where

$$\boldsymbol{R}^{(2)}(t,1) = \boldsymbol{R}^{2}_{4;40,40,40,40}(t,1)$$

$$= 1 - \sum_{\substack{r_1,r_2,r_3,r_4=0 \\ r_1+r_2+r_3+r_4 \le 1}}^{1} \prod_{i=1}^{4} \left[\prod_{j=1}^{40} R_{ij}^{(2)}(t,1)\right]^{r_i} \left[1-\prod_{j=1}^{40} R_{ij}^{(2)}(t,1)\right]^{1-r_i}$$

$$= 1 - \sum_{\substack{r_1,r_2,r_3,r_4=0 \\ r_1+r_2+r_3+r_4 \le 1}}^{1} \prod_{i=1}^{4} \exp\left[-r_i \sum_{j=1}^{40} \lambda_{ij}^{(2)}(1)t\right] \left[1-\exp\left[-\sum_{j=1}^{40} \lambda_{ij}^{(2)}(1)t\right]\right]^{1-r_i}$$

$$= 1 - \sum_{\substack{r_1,r_2,r_3,r_4=0 \\ r_1+r_2+r_3+r_4 \le 1}}^{1} \prod_{i=1}^{4} \exp[-r_i[0.0014\cdot 20 + 0.0018\cdot 20]t]$$

$$[1-\exp[-[0.0014\cdot 20+1.0018\cdot 20]t]]^{1-r_i}$$

$$= 1 - \sum_{\substack{r_1,r_2,r_3,r_4=0 \\ r_1+r_2+r_3+r_4 \le 1}}^{1} \prod_{i=1}^{4} \exp[-r_i 0.064t][1-\exp[-0.064t]]^{1-r_i}$$

$$= 1 - \sum_{i=0}^{1} \binom{4}{i} \exp[-i\cdot 0.064t][1-\exp[-0.064t]]^{4-i}$$

$$= 1 - \exp[-0\cdot 0.064t][1 - \exp[-0.0064t]^{4}$$

$$-4\exp[-1\cdot 0.064t][1 - \exp[-0.0064t]^{3}$$

$$= 6\exp[-0.128t] - 8\exp[-0.192t] + 3\exp[-0.256t] \qquad (5.99)$$

$$
\begin{aligned}
\boldsymbol{R}^{(2)}(t,1) &= \boldsymbol{R}^{2}_{4;40,40,40,40}(t,2) \\
&= 1 - \sum_{\substack{r_1,r_2,r_3,r_4=0 \\ r_1+r_2+r_3+r_4\le 1}}^{1} \prod_{i=1}^{4}\left[\prod_{j=1}^{40} R_{ij}^{(2)}(t,2)\right]^{r_i}\left[1-\prod_{j=1}^{40} R_{ij}^{(2)}(t,2)\right]^{1-r_i} \\
&= 1 - \sum_{\substack{r_1,r_2,r_3,r_4=0 \\ r_1+r_2+r_3+r_4\le 1}}^{1} \prod_{i=1}^{4}\exp\left[-r_i\sum_{j=1}^{40}\lambda_{ij}^{(2)}(2)t\right]\left[1-\exp\left[-\sum_{j=1}^{40}\lambda_{ij}^{(2)}(2)t\right]\right]^{1-r_i} \\
&= 1 - \sum_{\substack{r_1,r_2,r_3,r_4=0 \\ r_1+r_2+r_3+r_4\le 1}}^{1} \prod_{i=1}^{4}\exp[-r_i[0.0015\cdot 20+0.0020\cdot 20]t] \\
&\quad [1-\exp[-[0.0015\cdot 20+1.0020\cdot 20]t]]^{1-r_i} \\
&= 1 - \sum_{\substack{r_1,r_2,r_3,r_4=0 \\ r_1+r_2+r_3+r_4\le 1}}^{1} \prod_{i=1}^{4}\exp[-r_i 0.070t][1-\exp[-0.070t]]^{1-r_i} \\
&= 1 - \sum_{i=0}^{1}\binom{4}{i}\exp[-i\cdot 0.070t][1-\exp[-0.070t]]^{4-i} \\
&= 1-\exp[-0\cdot 0.070t][1-\exp[-0.0070t]^{4} \\
&\quad -4\exp[-1\cdot 0.070t][1-\exp[-0.0070t]^{3} \\
&= 6\exp[-0.140t]-8\exp[-0.210t]+3\exp[-0.280t]
\end{aligned}
\tag{5.100}
$$

$$\boldsymbol{R}^{(2)}(t,1) = \boldsymbol{R}^{2}_{4;40,40,40,40}(t,3)$$

$$= 1 - \sum_{\substack{r_1,r_2,r_3,r_4=0 \\ r_1+r_2+r_3+r_4 \le 1}}^{1} \prod_{i=1}^{4} \left[\prod_{j=1}^{40} R_{ij}^{(2)}(t,3)\right]^{r_i} \left[1 - \prod_{j=1}^{40} R_{ij}^{(2)}(t,3)\right]^{1-r_i}$$

$$= 1 - \sum_{\substack{r_1,r_2,r_3,r_4=0 \\ r_1+r_2+r_3+r_4 \le 1}}^{1} \prod_{i=1}^{4} \exp\left[-r_i \sum_{j=1}^{40} \lambda_{ij}^{(2)}(3)t\right] \left[1 - \exp\left[-\sum_{j=1}^{40} \lambda_{ij}^{(2)}(3)t\right]\right]^{1-r_i}$$

$$= 1 - \sum_{\substack{r_1,r_2,r_3,r_4=0 \\ r_1+r_2+r_3+r_4 \le 1}}^{1} \prod_{i=1}^{4} \exp[-r_i[0.0016 \cdot 20 + 0.0022 \cdot 20]t]$$

$$[1 - \exp[-[0.0016 \cdot 20 + 1.0022 \cdot 20]t]]^{1-r_i}$$

$$= 1 - \sum_{\substack{r_1,r_2,r_3,r_4=0 \\ r_1+r_2+r_3+r_4 \le 1}}^{1} \prod_{i=1}^{4} \exp[-r_i 0.076t][1 - \exp[-0.076t]]^{1-r_i}$$

$$1 - \sum_{i=0}^{1} \binom{4}{i} \exp[-i \cdot 0.076t][1 - \exp[-0.076t]]^{4-i}$$

$$= 1 - \exp[-0 \cdot 0.076t][1 - \exp[-0.0076t]^{4}$$

$$- 4\exp[-1 \cdot 0.076t][1 - \exp[-0.0076t]^{3}$$

$$= 6\exp[-0.152t] - 8\exp[-0.228t] + 3\exp - 0.304t] \tag{5.101}$$

To find the subsystem S_2 limit reliability function, we modify Corollary 5.4 into its following particular case for a non-homogeneous exponential series system.

Corollary 5.13

If components of the non-homogeneous, regular multi-state series system have exponential reliability functions

$$R^{(i)}(t,\cdot) = [1, R^{(i)}(t,1), \ldots, R^{(i)}(t,z)], \quad t \in (-\infty, \infty)$$

where

$$R^{(i)}(t,u) = 1 \quad \text{for } t < 0, \quad R^{(i)}(t,u) = \exp[-\lambda_i(u)t]$$
$$\text{for } t \geq 0, \quad \lambda_i(u) > 0, \quad u = 1, 2, \ldots z, \quad i = 1, 2, \ldots, a$$

and

$$a_n(u) = \frac{1}{\lambda(u)n}, \quad b_n(u) = 0, \quad u = 1, 2, \ldots, z \tag{5.102}$$

where

$$\lambda(u) = \max_{1 \leq i \leq a} \{\lambda_i(u)\} \quad u = 1, 2, \ldots, z \tag{5.103}$$

then

$$\overline{\mathfrak{R}'}_2(t,\cdot) = [1, \overline{\mathfrak{R}'}_2(t,1), \ldots, \mathfrak{R}'_2(t,z)]$$

where

$$\overline{\mathfrak{R}'}_2(t,u) = 1 \quad \text{for } t < 0$$

$$\overline{\mathfrak{R}'}_2(t,u) = \exp\left[-\sum_{i=1}^{a} q_i \frac{\lambda_i(u)}{\lambda(u)} t\right] \quad \text{for } t \geq 0, \quad u = 1, 2, \ldots, z \tag{5.104}$$

is its limit reliability function.

The series systems of the subsystem S_2 are identical, non-homogeneous, regular multi-state series systems in which, according to Definition 3.17, we have

$$n = 40, \quad a = 2, \quad q_1 = 0.5, \quad q_2 = 0.5$$

Further, applying Corollary 5.13 and considering Eqs. (5.102) and (5.103), we have

$$\lambda(1) = \max\{0.0014, \ 0.0018\} = 0.0018$$

$$\lambda(2) = \max\{0.0015, \ \ 0.0020\} = 0.0020$$

$$\lambda(3) = \max\{0.0016, \ \ 0.0022\} = 0.0022$$

$$a_n(1) = 1/(0.0018 \cdot 40), \ a_n(2) = 1/(0.0020 \cdot 40), \ a_n(3) = 1/(0.0022 \cdot 40) \quad (5.105)$$

$$b_n(u) = 0 \ \text{ for } u = 1,2,3,4 \quad (5.106)$$

and considering Eq. (5.104), we conclude that their limit reliability function is

$$\overline{\mathfrak{R}}'_2(t,\cdot) = \left[1, \exp\left[-\left(0.5\frac{0.0014}{0.0018} + 0.5\right)t\right], \exp\left[-\left(0.5\frac{0.0015}{0.0020} + 0.5\right)t\right]\right.$$
$$\left.\exp\left[-(0.5\frac{0.0016}{0.0022} + 0.5)t\right]\right] \quad \text{for } t \geq 0$$
$$(5.107)$$

Thus, considering Eqs. (5.105)–(5.107), from Eq. (3.49), the approximate formulae for the series system reliability functions take the form

$$\overline{\mathfrak{R}}'_{40}(t,\cdot) \cong \overline{\mathfrak{R}}'_2((t - b_n(u))/a_n(u),\cdot)$$
$$= [1, \exp[-0.064t],], \exp[-0.070t], \exp[-0.076t]] \quad \text{for } t \geq 0$$
$$(5.108)$$

Further, the approximate formula for the subsystem S_2 reliability function takes the form

$$\boldsymbol{R}^{(2)}(t,\cdot) = [1, \boldsymbol{R}^{(2)}(t,1), \boldsymbol{R}^{(2)}(t,2), \boldsymbol{R}^{(2)}(t,3)], \quad t \geq 0 \quad (5.109)$$

where according to Eqs. (3.24) and (3.25)

$$\boldsymbol{R}^{(2)}(t,1) = 1 - \sum_{i=0}^{1}\binom{4}{i}\exp[-i \cdot 0.064t][1-\exp[-0.064t]]^{4-i}$$
$$= 6\exp[-0.128t] - 8\exp[-0.192t] + 3\exp[-0.256t] \quad (5.110)$$

$$\boldsymbol{R}^{(2)}(t,2) = 1 - \sum_{i=0}^{1}\binom{4}{i}\exp[-i \cdot 0.070t][1-\exp[-0.070t]]^{4-i}$$
$$= 6\exp[-0.140t] - 8\exp[-0.210t] + 3\exp[-0.280t] \quad (5.111)$$

$$\boldsymbol{R}^{(2)}(t,3) = 1 - \sum_{i=0}^{1} \binom{4}{i} \exp[-i \cdot 0.076t][1-\exp[-0.076t]]^{4-i}$$
$$= 6\exp[-0.152t] - 8\exp[-0.228t] + 3\exp[-0.304t] \quad (5.112)$$

which is identical to the subsystem S_2 exact reliability given by Eqs. (5.98)–(5.101).

Taking into account that the systems S_1 and S_2 are four-state subsystems creating a non-homogeneous series system S, after applying the formulae appearing after Definition 3.4, the considered system reliability function is given by

$$\boldsymbol{R}(t,\cdot) = [1, \boldsymbol{R}(t,1), \boldsymbol{R}(t,2), \boldsymbol{R}(t,3)], \quad t \geq 0 \quad (5.113)$$

where according to Eqs. (5.97) and (5.109)–(5.112) we have

$$\boldsymbol{R}(t,1) = \overline{\boldsymbol{R}}_2(t,1) = \boldsymbol{R}^{(1)}(t,1)\boldsymbol{R}^{(2)}(t,1)$$
$$= 12\exp[-0.155t] - 6\exp[-0.182t] - 16\exp[-0.219t] \quad (5.114)$$
$$+ 8\exp[-0.246t] + 6\exp[-0.283t] - 3\exp[-0.310t]$$

$$\boldsymbol{R}(t,2) = \overline{\boldsymbol{R}}_2(t,2) = \boldsymbol{R}^{(1)}(t,2)\boldsymbol{R}^{(2)}(t,2)$$
$$= 12\exp[-0.170t] - 6\exp[-0.200t] - 16\exp[-0.240t] \quad (5.115)$$
$$+ 8\exp[-0.270t] + 6\exp[-0.310t] - 3\exp[-0.340t]$$

$$\boldsymbol{R}(t,3) = \overline{\boldsymbol{R}}_2(t,3) = \boldsymbol{R}^{(1)}(t,3)\boldsymbol{R}^{(2)}(t,3)$$
$$= 12\exp[-0.185t] - 6\exp[-0.218t] - 16\exp[-0.261t] \quad (5.116)$$
$$+ 8\exp[-0.294t] + 6\exp[-0.337t] - 3\exp[-0.370t]$$

The coordinates of the considered four-state system reliability function are illustrated in Figure 5.11.

The mean values of the system sojourn times $T(u)$ in the reliability state subsets after applying the formulae (3.13) and (5.114)–(5.116) are

$$M(1) = \int_0^\infty \boldsymbol{R}(t,1)\mathrm{d}t = 12/0.155 - 6/0.182 - 16/0.219 + 8/0.246$$
$$+ 6/0.283 - 3/0.310 \quad (5.117)$$
$$\cong 15.44$$

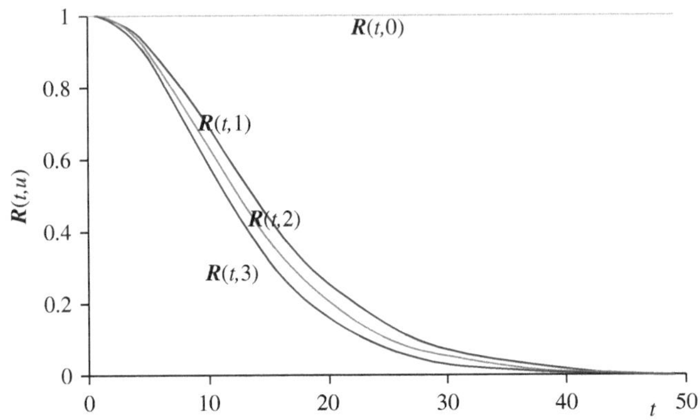

Figure 5.11 The graph of the exemplary system reliability function $\boldsymbol{R}(t,\ \cdot)$ coordinates.

$$\begin{aligned} M(2) = \int_0^\infty \boldsymbol{R}(t,2)\mathrm{d}t &= 12/0.170 - 6/0.200 - 16/0.240 + 8/0.270 \\ &+ 6/0.310 - 3/0.340 \\ &\cong 14.08 \end{aligned} \tag{5.118}$$

$$\begin{aligned} M(3) = \int_0^\infty \boldsymbol{R}(t,3)\mathrm{d}t &= 12/0.185 - 6/0.218 - 16/0.261 + 8/0.294 \\ &+ 6/0.337 - 3/0.370 \\ &\cong 12.95 \end{aligned} \tag{5.119}$$

Hence, applying Eq. (3.17), the mean values of the system lifetimes in the particular reliability states are

$$\begin{aligned} &\overline{M}(1) \cong 15.44 - 14.08 = 1.34, \quad \overline{M}(2) \cong 14.08 - 12.95 = 1.13, \\ &\overline{M}(3) \cong 12.95 \end{aligned} \tag{5.120}$$

Since the critical reliability state is $r = 2$, then the system risk function, according to Eq. (3.18), is given by

$$\begin{aligned} \boldsymbol{r}(t) &= 1 - \boldsymbol{R}(t,2) \\ &= 1 - [12\exp[-0.170t] - 6\exp[-0.200t] - 16\exp[-0.240t] \\ &\quad + 8\exp[-0.270t] + 6\exp[-0.310t] - 3\exp[-0.340t]] \quad \text{for } t \geq 0 \end{aligned} \tag{5.121}$$

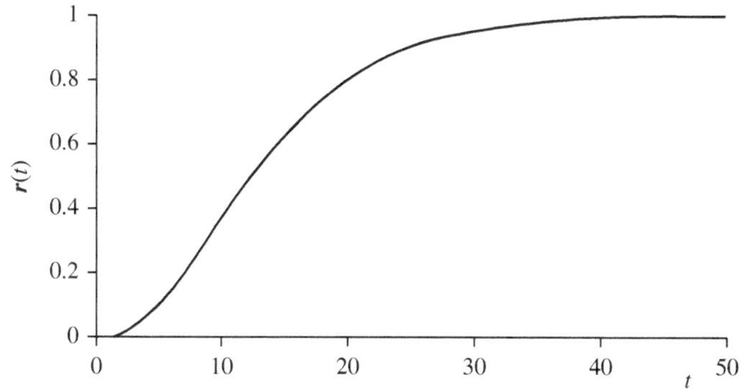

Figure 5.12 The graph of the risk function $\boldsymbol{r}(t)$ of the exemplary system S.

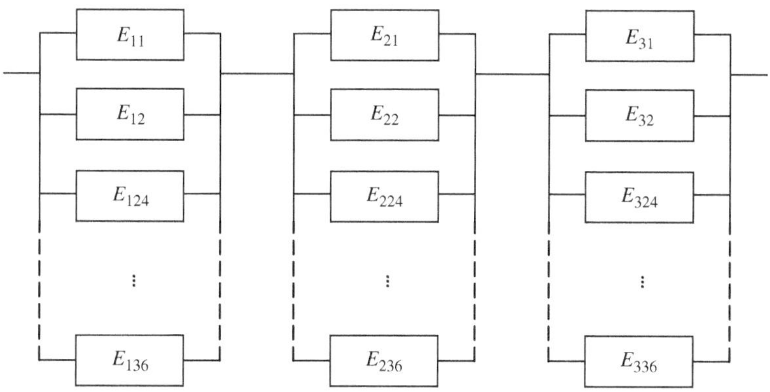

Figure 5.13 The scheme of the '24 out of 36'−series system.

Hence, the moment when the system risk function exceeds a permitted level, for instance $\delta = 0.002$, by Eq. (3.19), is

$$\tau = \boldsymbol{r}^{-1}(\delta) \cong 0.99 \tag{5.122}$$

The exemplary four-state system risk function is illustrated in Figure 5.12.

Example 5.13

We consider a '24 out of 36'−series system composed of independent components E_{ij}, $i = 1, 2, 3, j = 1, 2, \ldots, 36$. The system is composed of three '24 out of 36' subsystems linked in series and composed of components E_{ij}, $i = 1, 2, 3, j = 1, 2, \ldots, 36$, with the reliability structure presented in Figure 5.13.

We arbitrarily distinguish four safety states of the system components 0, 1, 2, 3, i.e. $z = 3$, and we fix that the critical reliability state is $r = 2$. Consequently, we define the four-state reliability functions of the system components $E_{ij}i = 1,2,3$, $j = 1,2,\ldots,36$, in the form of the vector

$$R_{ij}(t,\cdot) = [1, R_{ij}(t,1), R_{ij}(t,2), R_{ij}(t,3)], \quad i = 1,2,3, \ j = 1,2,\ldots 36 \tag{5.123}$$

with the exponential coordinates

$$R_{ij}(t,u) = \exp[-\lambda(u)t[\,] = \exp[-3ut], \quad u = 1,2,3, \ i = 1,2,3, \ j = 1,2,\ldots,36 \tag{5.124}$$

Since the shape parameters of the considered '24 out of 36'−series system are

$$k = 3, l = 36, m = 24$$

then considering Eqs. (3.24)−(3.25) and (3.20)−(3.21), we may conclude that its exact reliability function is given by the vector

$$\boldsymbol{R}(t,\cdot) = [1, \boldsymbol{R}(t,1), \boldsymbol{R}(t,2), \boldsymbol{R}(t,3)] \tag{5.125}$$

where

$$\boldsymbol{R}(t,u) = \left[1 - \sum_{i=0}^{m-1} \binom{l}{i} \exp[-i\lambda(u)t][1-\exp[-\lambda(u)t]]^{l-i}\right]^k \quad \text{for } t \geq 0, \quad u = 1,\ldots,z$$

and particularly

$$\boldsymbol{R}(t,1) = \left[1 - \sum_{i=0}^{23} \binom{36}{i} \exp[-3it][1-\exp[-3t]]^{36-i}\right]^3 \tag{5.126}$$

$$\boldsymbol{R}(t,2) = \left[1 - \sum_{i=0}^{23} \binom{36}{i} \exp[-6it][1-\exp[-6t]]^{36-i}\right]^3 \tag{5.127}$$

$$\boldsymbol{R}(t,3) = \left[1 - \sum_{i=0}^{23} \binom{36}{i} \exp[-9it][1-\exp[-9t]]^{36-i}\right]^3 \tag{5.128}$$

To find the system limit reliability function, we put together Lemma 5.6 and the formulae (3.20) and (3.21) and get the following result for 'm out of l'−series exponential system.

Corollary 5.14
If components of the homogeneous, regular multi-state 'm out of l'−series system have exponential reliability functions

$$R_{ij}(t,\cdot) = [1, R_{ij}(t,1), R_{ij}(t,2), \ldots, R_{ij}(t,z)], \quad i = 1,2,\ldots,k, \ j = 1,2,\ldots,l$$

with the exponential coordinates

$$R_{ij}(t,u)=1 \quad \text{for } t<0,\;\; R_{ij}(t,u)=\exp[-\lambda(u)t] \quad \text{for } t\geq 0,\;\; \lambda(u)>0,$$
$$u=1,2,\ldots z,\;\; i=1,2,\ldots,k,\;\; j=1,2,\ldots,l$$
$$k=\text{constant},\;\; l\to\infty \;\text{ as }\; n\to\infty,$$

$$m/l\to\mu,\quad 0<\mu<1,\text{ as } n\to\infty \tag{5.129}$$

$$a_n(u)=\frac{1}{\lambda(u)\mu\sqrt{l+1}}\sqrt{\mu(1-\mu)},\; b_n(u)=-\frac{1}{\lambda(u)}\log\mu,\;\; u=1,2,\ldots,z \tag{5.130}$$

then

$$\mathfrak{R}^{(\mu)}(t,\cdot)=[1,\mathfrak{R}^{(\mu)}(t,1),\ldots,\mathfrak{R}^{(\mu)}(t,z)],\quad t\in(-\infty,\infty) \tag{5.131}$$

where

$$\mathfrak{R}^{(\mu)}(t,\cdot)=\left[1-\frac{1}{\sqrt{2\pi}}\int_{-\infty}^{t}e^{-x^2/2}\,\mathrm{d}x\right]^k \quad \text{for } t\in(-\infty,\infty),\; u=1,2,\ldots,z \tag{5.132}$$

is the limit multi-state reliability function of this system.

Motivation: Since for sufficiently large n

$$a_n(u)t+b_n(u)=\frac{1}{\lambda(u)\mu\sqrt{l+1}}\sqrt{\mu(1-\mu)}t-\frac{1}{\lambda(u)}\log\mu>0$$

for all $t\in(-\infty,\infty)$, $u=1,2,\ldots,z$, then for sufficiently large n

$$\begin{aligned}R(a_n(u)t+b_n(u))&=\exp[-\lambda(u)(a_n(u)t+b_n(u))]\\&=\left[1-a_n(u)t+o\left(\frac{1}{\sqrt{l+1}}\right)\right]\exp[-\lambda(u)b_n(u)]\\&=\left[1-\lambda(u)a_n(u)t+o\left(\frac{1}{\sqrt{l+1}}\right)\right]\exp[-\lambda(u)b_n(u)]\\&=\mu-\frac{1}{\sqrt{l+1}}\sqrt{\mu(1-\mu)}t+o\left(\frac{1}{\sqrt{l+1}}\right)\quad\text{for all } t\in(-\infty,\infty)\end{aligned}$$

and considering that result in Eq. (5.27), we have

$$\lim_{n\to\infty}\frac{(l+1)R(a_n(u)t+b_n(u),u)-m}{\sqrt{m(l-m+1)/(l+1)}}$$

$$=\lim_{n\to\infty}\frac{(l+1)\left[\mu-(1/\sqrt{l+1})\sqrt{\mu(1-\mu)}t+o\left((1/\sqrt{l+1})\right)\right]-m}{\sqrt{m(l-m+1)/(l+1)}}$$

$$=\lim_{n\to\infty}\frac{\mu-\frac{1}{\sqrt{l+1}}\sqrt{\mu(1-\mu)}t+o((1)/\sqrt{l+1})-(m/(l+1))}{(1/\sqrt{l+1})\sqrt{(m(l+1-m))/((l+1)(l+1))}}$$

$$=\lim_{n\to\infty}\frac{\mu-(1/\sqrt{l+1})\sqrt{\mu(1-\mu)}t+o((1/\sqrt{l+1}))-(m/(l+1)}{(1/\sqrt{l+1})\sqrt{(m/(l+1))(1-(m/l+1))}}=-t$$

for $t\in(-\infty,\infty)$, $u=1,2,\ldots,z$.

The above result by Lemma 5.6 means that $\mathfrak{R}^{(\mu)}(t,\cdot)$ given by Eqs. (5.131) and (5.132) is the considered system limit reliability function, which completes the proof.

Further applying Corollary 5.14, since the shape parameters of the considered '24 out of 36'−series system are

$$k=3,\ \ l=36,\ \ m=24$$

the number of reliability states is $z=3$, and by Eqs. (5.129) and (5.130)

$$\mu=\frac{24}{36}=\frac{2}{3}$$

$$a_n(1)=\frac{1}{3\frac{2}{3}\sqrt{35}}\sqrt{\frac{2}{3}(1-\frac{2}{3})}=\frac{1}{6\sqrt{17.5}},\ \ b_n(1)=-\frac{1}{3}\log\frac{2}{3}$$

$$a_n(2)=\frac{1}{6\frac{2}{3}\sqrt{35}}\sqrt{\frac{2}{3}(1-\frac{2}{3})}=\frac{1}{12\sqrt{17.5}},\ \ b_n(2)=-\frac{1}{6}\log\frac{2}{3}$$

$$a_n(3)=\frac{1}{9\frac{2}{3}\sqrt{35}}\sqrt{\frac{2}{3}(1-\frac{2}{3})}=\frac{1}{18\sqrt{17.5}},\ \ b_n(1)=-\frac{1}{9}\log\frac{2}{3}$$

and considering Eq. (3.49), the approximate formula for the system reliability function takes the form

$$\begin{aligned}\boldsymbol{R}(t,\cdot)&\cong\mathfrak{R}^{(2/3)}(t-b_n(u))/a_n(u),\cdot))\\&=[\mathfrak{R}(t,1),\mathfrak{R}(t,2),\mathfrak{R}(t,3)],\quad t\in(-\infty,\infty)\end{aligned}\tag{5.133}$$

where

$$\mathfrak{R}(t,1) = \mathfrak{R}^{(2/3)}(t - b_n(1))/a_n(1),1)\left[1 - \frac{1}{\sqrt{2\pi}}\int\limits_{-\infty}^{6\sqrt{17.5}t+2\sqrt{17.5}\log(2/3)} e^{-x^2/2}\,\mathrm{d}x\right]^3 \tag{5.134}$$

$$\mathfrak{R}(t,2) = \mathfrak{R}^{(2/3)}(t - b_n(1))/a_n(2),2) = \left[1 - \frac{1}{\sqrt{2\pi}}\int\limits_{-\infty}^{12\sqrt{17.5}t+2\sqrt{17.5}\log(2/3)} e^{-x^2/2}\,\mathrm{d}x\right]^3 \tag{5.135}$$

$$\mathfrak{R}(t,3) = \mathfrak{R}^{(2/3)}(t - b_n(3))/a_n(3),3) = \left[1 - \frac{1}{\sqrt{2\pi}}\int\limits_{-\infty}^{18\sqrt{17.5}t+2\sqrt{17.5}\log(2/3)} e^{-x^2/2}\,\mathrm{d}x\right]^3 \tag{5.136}$$

It is obvious that using the approximate formulae (5.133)–(5.136) is much easier than using the exact formulae (5.125)–(5.128) for the considered system reliability function. The accuracy of the approximation of the exact reliability function by its approximate form is illustrated in Tables 5.5–5.7 and in the results (5.137)–(5.142) and (5.143)–(5.148) appearing after them and concerned with the system lifetimes mean values and the system risk function, calculated with application of the system exact and approximate reliability functions respectively.

In Table 5.5, the values and differences between the coordinate $\boldsymbol{R}(t,1)$ of the system exact reliability function given by Eq. (5.126) and the coordinate $\mathfrak{R}(t,1)$ of the system approximate reliability function given by Eq. (5.134) are presented.

In Table 5.6, the values and differences between the coordinate $\boldsymbol{R}(t,2)$ of the system exact reliability function given by Eq. (5.127) and the coordinate $\mathfrak{R}(t,2)$ of the system approximate reliability function given by Eq. (5.135) and the values of the approximate system risk function $\boldsymbol{r}(t)$ given by Eq. (5.141) are presented.

In Table 5.7, the values and differences between the coordinate $\boldsymbol{R}(t,3)$ of the system exact reliability function given by Eq. (5.128) and the coordinate $\mathfrak{R}(t,3)$ of the system approximate reliability function given by Eq. (5.136) are presented.

The approximate mean values of the system lifetimes $T(u)$ in the reliability state subsets after applying the formulae (3.13) and (5.126)–(5.128), the

Table 5.5 The Values and Differences Between the First Coordinates of the '24 Out of 36'–Series System Exact and Approximate Reliability Functions

t	$\boldsymbol{R}(t,1)$	$\boldsymbol{\Re}(t,1)$	$\boldsymbol{R}(t,1)-\boldsymbol{\Re}(t,1)$
0.00	1.000	0.9991	0.0009
0.01	1.000	0.9976	0.0024
0.02	1.000	0.9958	0.0042
0.03	1.000	0.9892	0.0108
0.04	0.9997	0.9750	0.0247
0.05	0.9978	0.9522	0.0456
0.06	0.9895	0.9144	0.0751
0.07	0.9653	0.8560	0.1093
0.08	0.9123	0.7691	0.1432
0.09	0.8206	0.6603	0.1603
0.10	0.6909	0.5327	0.1582
0.11	0.5374	0.3982	0.1392
0.12	0.3831	0.2721	0.1110
0.13	0.2492	0.1679	0.0813
0.14	0.1479	0.0925	0.0554
0.15	0.0802	0.0450	0.0352
0.16	0.0399	0.0192	0.0207
0.17	0.0183	0.0071	0.0112
0.18	0.0078	0.0022	0.0056
0.19	0.0031	0.0006	0.0025
0.20	0.0011	0.0000	0.0011

trapezium methods of integration and suitable data given respectively in Tables 5.5–5.7, are

$$M(1)=\int_0^\infty \boldsymbol{R}(t,1)\mathrm{d}t \cong \frac{0.01}{2}[\boldsymbol{R}(0.00,1)+2[\boldsymbol{R}(0.01,1)+\boldsymbol{R}(0.02,1)+\cdots$$
$$+\boldsymbol{R}(0.19,1)]+\boldsymbol{R}(0.20,1)] \tag{5.137}$$
$$=\frac{0.01}{2}[1.0000+2[1.0000+1.0000+\cdots+0.0031]+0.0011]$$
$$\cong 0.1134$$

$$M(2)=\int_0^\infty \boldsymbol{R}(t,2)\mathrm{d}t \cong \frac{0.005}{2}[\boldsymbol{R}(0.000,2)+2[\boldsymbol{R}(0.005,2)+\boldsymbol{R}(0.010,2)+\cdots$$
$$+\boldsymbol{R}(0.095,2)]+\boldsymbol{R}(0.100,2)] \tag{5.138}$$
$$=\frac{0.005}{2}[1.0000+2[1.0000+1.0000+\cdots+0.0031]+0.0011]$$
$$\cong 0.0567$$

Table 5.6 The Values and Differences Between the Second Coordinates of the System Exact and Approximate Reliability Functions and the Values of the '24 Out of 36'–Series System Approximate Risk Function

t	$\boldsymbol{R}(t,2)$	$\mathfrak{R}(t,2)$	$\boldsymbol{R}(t,2)-\mathfrak{R}(t,2)$	$r(t)$
0.000	1.000	0.9991	0.0009	0.0009
0.005	1.000	0.9976	0.0024	0.0024
0.010	1.000	0.9958	0.0042	0.0042
0.015	1.000	0.9892	0.0108	0.0108
0.020	0.9997	0.9750	0.0247	0.0250
0.025	0.9978	0.9522	0.0456	0.0478
0.030	0.9895	0.9144	0.0751	0.0856
0.035	0.9653	0.8560	0.1093	0.1440
0.040	0.9123	0.7691	0.1432	0.2309
0.045	0.8206	0.6603	0.1603	0.3397
0.050	0.6909	0.5327	0.1582	0.4673
0.055	0.5374	0.3982	0.1392	0.6018
0.060	0.3831	0.2721	0.1110	0.7279
0.065	0.2492	0.1679	0.0813	0.8321
0.070	0.1479	0.0925	0.0554	0.9075
0.075	0.0802	0.0450	0.0352	0.9550
0.080	0.0399	0.0192	0.0207	0.9808
0.085	0.0183	0.0071	0.0112	0.9929
0.090	0.0078	0.0022	0.0056	0.9978
0.095	0.0031	0.0006	0.0025	0.9994
0.100	0.0011	0.0000	0.0011	1.0000

$$
\begin{aligned}
M(3) = \int_0^\infty \boldsymbol{R}(t,3)\mathrm{d}t &\cong \frac{0.003}{2}[\boldsymbol{R}(0.00,3) + 2[\boldsymbol{R}(0.003,3) + \boldsymbol{R}(0.006,3) + \cdots \\
&\quad + \boldsymbol{R}(0.66,3)] + \boldsymbol{R}(0.069,3)] \\
&= \frac{0.003}{2}[1.0000 + 2[1.0000 + 1.0000 + \cdots + 0.0078] + 0.0034] \\
&\cong 0.0408
\end{aligned}
\tag{5.139}
$$

Hence, applying Eq. (3.17), the mean values of the system lifetimes in the particular reliability states are

$$
\begin{aligned}
&\overline{M}(1) \cong 0.1134 - 0.0567 = 0.0568, \\
&\overline{M}(2) \cong 0.0567 - 0.0408 = 0.0159, \quad \overline{M}(3) \cong 0.0408
\end{aligned}
\tag{5.140}
$$

Table 5.7 The Values and Differences Between the Third Coordinates of the '24 Out of 36'–Series System Exact and Approximate Reliability Functions

t	$\boldsymbol{R}(t,3)$	$\mathfrak{R}(t,3)$	$\boldsymbol{R}(t,3)-\mathfrak{R}(t,3)$
0.000	1.0000	0.9991	0.0009
0.003	1.0000	0.9976	0.0024
0.006	1.0000	0.9952	0.0048
0.009	1.0000	0.9898	0.0102
0.012	1.0000	0.9809	0.0191
0.015	0.9999	0.9647	0.0352
0.018	0.9992	0.9392	0.0600
0.021	0.9957	0.8984	0.0973
0.024	0.9845	0.8383	0.1462
0.027	0.9574	0.7613	0.1961
0.030	0.9049	0.6603	0.2446
0.033	0.8206	0.5485	0.2721
0.036	0.7052	0.4248	0.2804
0.039	0.5690	0.3106	0.2584
0.042	0.4281	0.2064	0.2217
0.045	0.2993	0.1250	0.1743
0.048	0.1942	0.0704	0.1238
0.051	0.1170	0.0348	0.0822
0.054	0.0657	0.0159	0.0498
0.060	0.0344	0.0022	0.0322
0.063	0.0168	0.0007	0.0161
0.066	0.0078	0.0002	0.0076
0.069	0.0034	0.0000	0.0034

Since the critical reliability state is $r=2$, then the approximate system risk function, according to Eq. (3.18), is given by

$$r(t)=1-\boldsymbol{R}(t,2)$$
$$=1-\left[1-\sum_{i=0}^{23}\binom{36}{i}\exp[-6it][1-\exp[-6t]]^{36-i}\right]^3 \quad \text{for } t\geq 0 \qquad (5.141)$$

Hence, the moment when the system risk function exceeds a permitted level, for instance $\delta=0.025$, by Eq. (3.19), is

$$\tau=\boldsymbol{r}^{-1}(\delta)\cong 0.033 \qquad (5.142)$$

The approximate mean values of the system lifetimes $T(u)$ in the reliability state subsets after applying the formulae (3.13) and (5.134)–(5.136), the trapezium methods of integration and suitable data given respectively in Tables 5.5–5.7, are

$$M(1) = \int_0^\infty \Re^{(2/3)}(t,1)\mathrm{d}t \cong \frac{0.01}{2}\left[\Re^{(2/3)}(0.00,1) + 2\left[\Re^{(2/3)}(0.01,1)\right.\right.$$
$$\left.\left. + \Re^{(2/3)}(0.02,1) + \cdots + \Re^{(2/3)}(0.19,1)\right] + \Re^{(2/3)}(0.20,1)\right]$$
$$= \frac{0.01}{2}[0.9991 + 2[0.9976 + 0.9958 + \cdots + 0.0006] + 0.0000]$$
$$\cong 0.1015 \tag{5.143}$$

$$M(2) = \int_0^\infty \Re^{(2/3)}(t,2)\mathrm{d}t \cong \frac{0.005}{2}\left[\Re^{(2/3)}(0.000,2) + 2\left[\Re^{(2/3)}(0.005,2)\right.\right.$$
$$\left.\left. + \Re^{(2/3)}(0.010,2) + \cdots + \Re^{(2/3)}(0.095,2)\right] + \Re^{(2/3)}(0.100,2)\right]$$
$$= \frac{0.005}{2}[0.9991 + 2[0.9976 + 0.9958 + \cdots + 0.0006] + 0.0000]$$
$$\cong 0.0507 \tag{5.144}$$

$$M(3) = \int_0^\infty \Re^{(2/3)}(t,3)\mathrm{d}t \cong \frac{0.003}{2}\left[\Re^{(2/3)}(0.000,3) + 2\left[\Re^{(2/3)}(0.003,3)\right.\right.$$
$$\left.\left. + \Re^{(2/3)}(0.006,3) + \cdots + \Re^{(2/3)}(0.066,3)\right] + \Re^{(2/3)}(0.069,3)\right]$$
$$= \frac{0.003}{2}[0.9991 + 2[0.9976 + 0.9952 + \cdots + 0.0002] + 0.0000]$$
$$\cong 0.0338 \tag{5.145}$$

Hence, applying Eq. (3.17), the mean values of the system lifetimes in the particular reliability states are

$$\overline{M}(1) \cong 0.11015 - 0.0507 = 0.0508,$$
$$\overline{M}(2) \cong 0.0507 - 0.0338 = 0.0169, \quad \overline{M}(3) \cong 0.0338 \tag{5.146}$$

Since the critical reliability state is $r = 2$, then the approximate system risk function, according to Eq. (3.18), is given by

$$r(t) = 1 - \Re^{(2/3)}(t, 2)$$
$$= 1 - \left[1 - \frac{1}{\sqrt{2\pi}} \int_{-\infty}^{12\sqrt{17.5}t + 2\sqrt{17.5}\log(2/3)} e^{-x^2/2}\,\mathrm{d}x\right]^3 \quad \text{for } t \geq 0 \tag{5.147}$$

Hence, the moment when the system risk function exceeds a permitted level, for instance $\delta = 0.025$, by Eq. (3.19), is

$$\tau = \boldsymbol{r}^{-1}(\delta) \cong 0.020 \tag{5.148}$$

6 Reliability Evaluation of Port and Shipyard Transportation Systems

6.1 Auxiliary Results

Assuming in Corollary 5.4

$$a_n(u) = [\bar{d}(t,u)\beta(u)n]^{-1/\alpha(u)} = \left[n \sum_{i:\alpha_i(u)=\alpha(u)} q_i\beta_i(u) \right]^{-1/\alpha(u)},$$
$$b_n(u) = 0 \quad \text{for } u = 1, 2, \ldots, z$$

after considering the justification given in Chapter 1, prior to Definition 1.6, we obtain the following result.

Corollary 6.1

If components of the non-homogeneous multi-state series system have Weibull reliability functions

$$R^{(i)}(t,\cdot) = [1, R^{(i)}(t,1), \ldots, R^{(i)}(t,z)], \quad t \in (-\infty, \infty)$$

where

$$R^{(i)}(t,u) = 1 \text{ for } t<0,\ R^{(i)}(t,u) = \exp[-\beta_i(u)t^{\alpha_i(u)}] \text{ for } t \geq 0,\ \alpha_i(u)>0,\ \beta_i(u)>0,$$
$$i = 1,2,\ldots,a,\ u = 1,2,\ldots,z \tag{6.1}$$

and

$$a_n(u) = [\beta(u)n]^{-1/\alpha(u)}, \quad b_n(u) = 0 \quad \text{for } u = 1, 2, \ldots, z \tag{6.2}$$

where

$$\alpha(u) = \min_{1 \leq i \leq a} \{\alpha_i(u)\}, \quad \beta(u) = \sum_{i:\alpha_i(u)=\alpha(u)} q_i\beta_i(u) \quad \text{for } u = 1, 2, \ldots, z \tag{6.3}$$

then

$$\overline{\mathfrak{R}}'_2(t,\cdot) = [1, \overline{\mathfrak{R}}'_2(t,1), \ldots, \overline{\mathfrak{R}}'_2(t,z)]$$

Reliability of Large Systems. DOI: http://dx.doi.org/10.1016/B978-0-08-099949-4.00006-4

where

$$\overline{\mathfrak{R}}'_2(t,u) = 1 \quad \text{for } t<0, \quad \overline{\mathfrak{R}}'_2(t,u) = \exp[-t^{\alpha(u)}] \quad \text{for } t \geq 0, \quad u = 1,2,\ldots,z$$

is its limit reliability function.

Corollary 6.2

If components of the homogeneous regular multi-state parallel–series system have Weibull reliability functions

$$R(t,\cdot) = [1, R(t,1),\ldots,R(t,z)], \quad t \in (-\infty, \infty)$$

where

$$R(t,u) = 1 \ \text{ for } t<0, \ R(t,u) = \exp[-\beta(u)t^{\alpha(u)}] \ \text{ for } t \geq 0, \ \alpha(u)>0, \\ \beta(u)>0, \ u = 1,2,\ldots,z \tag{6.4}$$

and

$$k_n = n, \quad l_n - c\log n \gg s, \quad c>0, \quad s>0 \tag{6.5}$$

$$a_n(u) = b_n(u)/(a(u)b(u)(b_n(u))^{a(u)}\log n), \quad b_n(u) = [(1/b(u))\log(l_n/\log n)]^{1/a(u)}, \\ u = 1,2,\ldots,z \tag{6.6}$$

then

$$\overline{\mathfrak{R}}_3(t,\cdot) = [1, \overline{\mathfrak{R}}_3(t,1),\ldots,\overline{\mathfrak{R}}_3(t,z)], \quad t \in (-\infty, \infty)$$

where

$$\overline{\mathfrak{R}}_3(t,u) = \exp[-\exp[t]] \quad \text{for } t \in (-\infty, \infty), \quad u = 1,2,\ldots,z$$

is its limit reliability function.

Motivation: Since for each fixed u, sufficiently large n and all $t \in (-\infty, \infty)$, from Eqs. (6.5) and (6.6), we have

$$a_n(u)t + b_n(u) > 0 \ \text{ and } \ a_n(u)/b_n(u) \to 0 \quad \text{as } n \to \infty$$

then according to Eq. (6.4), we get

$$\begin{aligned} F(a_n(u)t + b_n(u), u) &= 1 - \exp[-\beta(u)(a_n(u)t + b_n(u))^{\alpha(u)}] \\ &= 1 - \exp[-\beta(u)(b_n(u))^{a(u)}(1 + (a_n(u)/b_n(u))t)^{\alpha(u)}] \\ &= 1 - \exp[-\beta(u)(b_n(u))^{a(u)}(1 + \alpha(u)(a_n(u)/b_n(u))t) \\ &\qquad + o(a_n(u)/b_n(u))] \end{aligned}$$

Moreover, from the condition

$$a(u)\beta(u)(b_n(u))^{a(u)}a_n(u)/b_n(u) = 1/\log n \to 0 \quad \text{as } n \to \infty$$

it follows that

$$\begin{aligned} F(a_n(u)t + b_n(u), u) &= 1 - \exp[-\log(l_n/\log n) - t/\log n + o(1/\log n)] \\ &= 1 - (\log n)/l_n(1 - t/\log n) + o(1/\log n) \\ &= 1 - (\log n)/l_n + t/l_n + o(1/l_n) \quad \text{for } t \in (-\infty, \infty) \end{aligned}$$

and according to Eq. (5.65)

$$\begin{aligned} \overline{V}(t,u) &= \lim_{n\to\infty} k_n[F(a_n(u)t + b_n(u), u)]^{l_n} \\ &= \lim_{n\to\infty} n[1 - (\log n)/l_n + t/l_n + o(1/l_n)]^{l_n} \\ &= \lim_{n\to\infty} \exp[\log n - \log n + t + l_n o(1/l_n)] \\ &= \exp[t] \quad \text{for } t \in (-\infty, \infty) \end{aligned}$$

which from Lemma 5.12 completes the proof.

6.2 Reliability of a Port Grain Transportation System

The grain elevator, presented in Figure 6.1, is the basic structure in the Baltic Grain Terminal of the Port of Gdynia assigned to handle the clearing of exported

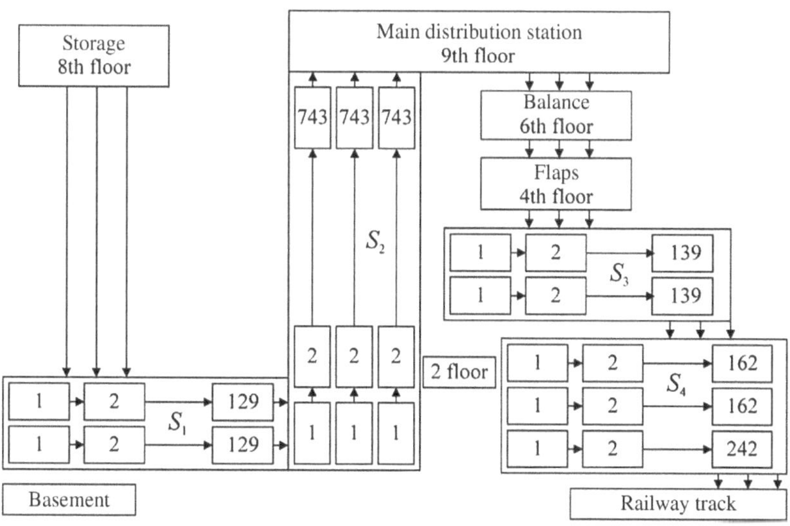

Figure 6.1 The scheme of the grain transportation system.

and imported grain. Elevator technological potentialities allow us to join different loading and unloading relations of ships, cars and railway trucks. Its output in the grain unloading process is 400 tons/h and in grain loading is 360 tons/h. The whole technological process is controlled electronically. A computer station delivers full visual information about the grain stream (flow) and its balance and the elevator's working state.

One of the basic elevator functions is loading railway trucks with grain. The railway truck loading is performed in the following successive elevator operation steps [79]:

- gravitational passing of grain from the storage placed on the eighth elevator floor through 45 hall to horizontal conveyors placed in the elevator basement,
- transport of grain through horizontal conveyors to vertical bucket elevators transporting grain to the main distribution station placed on the ninth floor,
- gravitational dumping of grain through the main distribution station to the balance placed on the sixth floor,
- dumping weighed grain through the complex of flaps placed on the fourth floor to horizontal conveyors placed on the second floor,
- dumping of grain from horizontal conveyors to worm conveyors,
- dumping of grain from worm conveyors to railway trucks.

In loading the railway trucks with grain the following elevator transportation subsystems take part:

S_1: horizontal conveyors of the first type,
S_2: vertical bucket elevators,
S_3: horizontal conveyors of the second type,
S_4: worm conveyors,
the main distribution station and the balance.

The main distribution station is the system of dumping channels in the form of a steel box composed of dividing walls, which direct the grain from bucket conveyors to the balance. Its executive elements are composed of three steel sleeves and pneumatic elements in the form of three servomotors. The electronic balance weighs the dumped grain with electronic indicators. Its executive elements during loading and unloading with grain are flaps, which are opened and closed by five pneumatic servomotors.

The transporting subsystems have steel covers and they are provided with drives in the form of electrical engines with gears. In their reliability analysis, we omit their drives as they are different types of mechanisms. We also omit their covers as they have a high reliability and, practically, do not fail.

Taking into account the efficiency of the considered transportation system we distinguish the following three reliability states of its components [79,146]:

- state 2: the state ensuring the largest efficiency of the conveyor,
- state 1: the state ensuring less efficiency of the conveyor caused by throwing grain off the belt,
- state 0: the state involving failure of the conveyor.

Subsystem S_1 is composed of two identical belt conveyors of the first type, each composed of a ribbon belt, a drum driving the belt, a reversible drum, 117 channelled rollers and nine rollers supporting the belt. Thus subsystem S_1 consists of $k_n = 2$ conveyors, each composed of $l_n = 129$ components. In each conveyor there is one belt with reliability functions

$$R^{(1,1)}(t,1) = \exp[-0.0125t^2],\ \ R^{(1,1)}(t,2) = \exp[-0.022t^2] \quad \text{for } t \geq 0$$

two drums with reliability functions

$$R^{(1,2)}(t,1) = \exp[-0.0015t^2],\ \ R^{(1,2)}(t,2) = \exp[-0.0018t^2] \quad \text{for } t \geq 0$$

117 channelled rollers with reliability functions

$$R^{(1,3)}(t,1) = \exp[-0.005t^2],\ \ R^{(1,3)}(t,2) = \exp[-0.0075t^2] \quad \text{for } t \geq 0$$

and nine supporting rollers with reliability functions

$$R^{(1,4)}(t,1) = \exp[-0.004t^2],\ \ R^{(1,4)}(t,2) = \exp[-0.005t^2] \quad \text{for } t \geq 0$$

Thus it is a non-homogeneous regular multi-state series–parallel system where, according to Definition 3.20, we have

$$k_n = k = 2,\ \ l_n = 129,\ \ a = 1,\ \ q_1 = 1,$$
$$e_1 = 4,\ \ p_{11} = 1/129,\ \ p_{12} = 2/129,\ \ p_{13} = 117/129,\ \ p_{14} = 9/129$$

and according to Eqs. (3.43) – (3.45) its exact reliability function is

$$\boldsymbol{R}'_{2,129}(t,\cdot) = [1, 1-[1-\exp[-0.6365t^2]]^2, 1-[1-\exp[-0.9481t^2]]^2] \quad \text{for } t \geq 0$$

Further, applying Corollary 5.10, according to Eq. (5.62), we have

$$\alpha_1(1) = \min\{2,2,2,\ 2\} = 2,$$
$$\beta_1(1) = \frac{1}{129}0.0125 + \frac{2}{129}0.0015 + \frac{117}{129}0.005 + \frac{9}{129}0.004 = 0.004934108,$$

$$\alpha_1(2) = \min\{2,2,2,\ 2\} = 2,$$
$$\beta_1(2) = \frac{1}{129}0.022 + \frac{2}{129}0.0018 + \frac{117}{129}0.0075 + \frac{9}{129}0.005 = 0.007349612$$

and according to Eq. (5.63)

$$\alpha(1) = \max\{2\} = 2,\ \ \beta(1) = \min\{0.004934108\} = 0.004934108,$$
$$\alpha(2) = \max\{2\} = 2,\ \ \beta(2) = \min\{0.007349612\} = 0.007349612$$

and according to Eq. (5.61)

$$a_n(1) = (0.004934108 \cdot 129)^{-1/2} = 1.253432119, \quad b_n(1) = 0,$$
$$a_n(2) = (0.007349612 \cdot 129)^{-1/2} = 1.027005872, \quad b_n(2) = 0$$

and we conclude that the limit reliability function of subsystem S_1 is

$$\mathfrak{R}'_9(t, \cdot) = [1, 1 - [1 - \exp[-t^2]]^2, 1 - [1 - \exp[-t^2]]^2] \quad \text{for } t \geq 0$$

Hence, from Eq. (3.49), we get the approximate subsystem reliability function (the formula is exact in this case) given by

$$\boldsymbol{R}'_{2,129}(t, \cdot) \cong \mathfrak{R}'_9((t - b_n(u))/a_n(u), \cdot)$$
$$= [1, 1 - [1 - \exp[-0.6365t^2]]^2, 1 - [1 - \exp[-0.9481t^2]]^2] \quad \text{for } t \geq 0$$

The expected values of the subsystem lifetimes in the state subsets and their standard deviations, from Eqs. (3.13)–(3.15), are

$$M(1) \cong 1.44 \text{ years}, \quad M(2) \cong 1.18 \text{ years},$$
$$\sigma(1) \cong 0.53 \text{ years}, \quad \sigma(2) \cong 0.44 \text{ years}$$

Hence and from Eq. (3.17), the subsystem mean lifetimes in particular states are

$$\overline{M}(1) \cong 0.26 \text{ years}, \quad \overline{M}(2) \cong 1.18 \text{ years}$$

If the critical reliability state of the subsystem is $r = 2$, then according to Eq. (3.18) its risk function is given by

$$\boldsymbol{r}(t) \cong [1 - \exp[-0.9481t^2]]^2$$

Hence and from Eq. (3.19), the moment when the subsystem risk function exceeds an admissible level $\delta = 0.05$ is

$$\tau = \boldsymbol{r}^{-1}(\delta) = [-(1/0.9481)\log(1 - \sqrt{\delta})]^{1/2} \cong 0.52 \text{ years}$$

Subsystem S_2 is composed of three identical bucket elevators [103], each composed of a buttress belt, a drum driving the belt, a reversible drum and 740 buckets. Thus subsystem S_2 consists of $k_n = 3$ elevators, each composed of $l_n = 743$ components. In each elevator there is one belt with reliability functions

$$R^{(1,1)}(t, 1) = \exp[-0.025t^2], \quad R^{(1,1)}(t, 2) = \exp[-0.026t^2] \quad \text{for } t \geq 0$$

two drums with reliability functions

$$R^{(1,2)}(t, 1) = \exp[-0.0015t^2], \quad R^{(1,2)}(t, 2) = \exp[-0.0018t^2] \quad \text{for } t \geq 0$$

and 740 buckets with reliability functions

$$R^{(1,3)}(t,1) = \exp[-0.03t^2],\ \ R^{(1,3)}(t,2) = \exp[-0.06t^2] \quad \text{for } t \geq 0$$

Thus it is a non-homogeneous regular multi-state series–parallel system where, according to Definition 3.20, we have

$$k_n = k = 3,\ \ l_n = 743,\ \ a = 1,\ \ q_1 = 1,$$
$$e_1 = 3,\ \ p_{11} = 1/743,\ \ p_{12} = 2/743,\ \ p_{13} = 740/743$$

and according to Eqs. (3.43)–(3.45) its exact reliability function is given by

$$\boldsymbol{R}'_{3,743}(t,\cdot) = [1, 1-[1-\exp[-22.228t^2]]^3, 1-[1-\exp[-44.4296t^2]]^3] \ \text{ for } t \geq 0$$

Further, applying Corollary 5.10, according to Eq. (5.62), we have

$$\alpha_1(1) = \min\{2,2,2\} = 2,$$
$$\beta_1(1) = \frac{1}{743}0.025 + \frac{2}{743}0.0015 + \frac{740}{743}0.03 = 0.029916554,$$

$$\alpha_1(2) = \min\{2,2,2\} = 2,$$
$$\beta_1(2) = \frac{1}{743}0.026 + \frac{2}{743}0.0018 + \frac{740}{743}0.06 = 0.059797577$$

and according to Eq. (5.63)

$$\alpha(1) = \max\{2\} = 2,\ \ \beta(1) = \min\{0.029916554\} = 0.029916554,$$
$$\alpha(2) = \max\{2\} = 2,\ \ \beta(2) = \min\{0.059797577\} = 0.059797577$$

and according to Eq. (5.61)

$$a_n(1) = (0.029916554 \cdot 743)^{-1/2} = 0.212104464,\ \ b_n(1) = 0,$$
$$a_n(2) = (0.059797577 \cdot 743)^{-1/2} = 0.0150025056,\ \ b_n(2) = 0$$

and we conclude that the limit reliability function of subsystem S_2 is

$$\mathfrak{R}'_9(t,\cdot) = [1, 1-[1-\exp[-t^2]]^3, 1-[1-\exp[-t^2]]^3] \quad \text{for } t \geq 0$$

Thus, from Eq. (3.49), the approximate formula (it is exact in this case) takes the form

$$\boldsymbol{R}'_{3,743}(t,\cdot) \cong \mathfrak{R}'_9((t-b_n(u))/a_n(u),\cdot)$$
$$= [1, 1-[1-\exp[-22.228t^2]]^3, 1-[1-\exp[-44.4296t^2]]^3] \ \text{ for } t \geq 0$$

The expected values and standard deviations of the subsystem lifetimes in the state subsets, according to Eqs. (3.13)–(3.15), are

$$M(1) \cong 0.27 \text{ years}, \quad M(2) \cong 0.19 \text{ years},$$
$$\sigma(1) \cong 0.10 \text{ years}, \quad \sigma(2) \cong 0.07 \text{ years}$$

Hence and from Eq. (3.17), the subsystem mean lifetimes in particular states are

$$\overline{M}(1) \cong 0.08 \text{ years}, \quad \overline{M}(2) \cong 0.19 \text{ years}$$

If the critical reliability state of the subsystem is $r = 2$, then from Eq. (3.18) its risk function has the form

$$\boldsymbol{r}(t) \cong [1 - \exp[-44.2296t^2]]^3$$

Hence, and from Eq. (3.19), the moment when the subsystem risk function exceeds an admissible level $\delta = 0.05$ is

$$\tau = \boldsymbol{r}^{-1}(\delta) = [-(1/44.2296)\log(1 - \sqrt[3]{\delta})]^{1/2} \cong 0.10 \text{ years}$$

Subsystem S_3 is composed of two identical belt conveyors of the second type, each composed of a buttress belt, a drum driving the belt, a reversible drum, 117 channelled rollers and 19 rollers supporting the belt. Thus subsystem S_3 consists of $k_n = 2$ conveyors, each composed of $l_n = 139$ components. In each conveyor there is one belt with reliability functions

$$R^{(1,1)}(t,1) = \exp[-0.0125t^2], \quad R^{(1,1)}(t,2) = \exp[-0.022t^2] \quad \text{for } t \geq 0$$

two drums with reliability functions

$$R^{(1,2)}(t,1) = \exp[-0.0015t^2], \quad R^{(1,2)}(t,2) = \exp[-0.0018t^2] \quad \text{for } t \geq 0$$

117 channelled rollers with reliability functions

$$R^{(1,3)}(t,1) = \exp[-0.0015t^2], \quad R^{(1,3)}(t,2) = \exp[-0.0075t^2] \quad \text{for } t \geq 0$$

and 19 supporting rollers with reliability functions

$$R^{(1,4)}(t,1) = \exp[-0.004t^2], \quad R^{(1,4)}(t,2) = \exp[-0.005t^2] \quad \text{for } t \geq 0$$

Thus it is a non-homogeneous regular multi-state series–parallel system where, according to Definition 3.20, we have

$$k_n = k = 2, \quad l_n = 139, \quad a = 1, \quad q_1 = 1,$$
$$e_1 = 4, \quad p_{11} = 1/139, \quad p_{12} = 2/139, \quad p_{13} = 117/139, \quad p_{14} = 19/139$$

and according to Eqs. (3.43)–(3.45) its exact reliability function is

$$\boldsymbol{R}'_{2,139}(t,\cdot) = [1, 1-[1-\exp[-0.6765t^2]]^2, 1-[1-\exp[-0.9981t^2]]^2] \quad \text{for } t \geq 0$$

Further, applying Corollary 5.10, according to Eq. (5.62), we have

$$\alpha_1(1) = \min\{2, 2, 2, \ 2\} = 2,$$
$$\beta_1(1) = \frac{1}{139}0.0125 + \frac{2}{139}0.0015 + \frac{117}{139}0.005 + \frac{19}{139}0.004 = 0.004866906,$$
$$\alpha_1(2) = \min\{2, 2, 2, \ 2\} = 2,$$
$$\beta_1(2) = \frac{1}{139}0.022 + \frac{2}{139}0.0018 + \frac{117}{139}0.0075 + \frac{19}{139}0.005 = 0.007180575$$

and according to Eq. (5.63)

$$\alpha(1) = \max\{2\} = 2, \ \ \beta(1) = \min\{0.004866906\} = 0.004866906,$$
$$\alpha(2) = \max\{2\} = 2, \ \ \beta(2) = \min\{0.007180575\} = 0.007180575$$

and according to Eq. (5.61)

$$a_n(1) = (0.004866906 \cdot 139)^{-1/2} = 1.215811147, \ \ b_n(1) = 0,$$
$$a_n(2) = (0.007180575 \cdot 139)^{-1/2} = 1.000951394, \ \ b_n(2) = 0$$

and we conclude that the limit reliability function of subsystem S_3 is

$$\mathfrak{R}'_9(t,\cdot) = [1, 1-[1-\exp[-t^2]]^2, 1-[1-\exp[-t^2]]^2] \quad \text{for } t \geq 0$$

Thus, from Eq. (3.49), the approximate formula (it is exact in this case) for the subsystem reliability function takes the form

$$\boldsymbol{R}'_{2,139}(t,\cdot) \cong \mathfrak{R}'_9((t-b_n(u))/a_n(u),\cdot)$$
$$= [1, 1-[1-\exp[-0.6765t^2]]^2, 1-[1-\exp[-0.9981t^2]]^2] \quad \text{for } t \geq 0$$

The expected values and standard deviations of the subsystem sojourn times in the state subsets, according to Eqs. (3.13)–(3.15), are

$$M(1) \cong 1.39 \text{ years}, \ \ M(2) \cong 1.15 \text{ years},$$
$$\sigma(1) \cong 0.53 \text{ years}, \ \ \sigma(2) \cong 0.42 \text{ years}$$

Hence and from Eq. (3.17), the mean lifetimes of the subsystem in particular states are

$$\overline{M}(1) \cong 0.24 \text{ years}, \ \ \overline{M}(2) \cong 1.15 \text{ years}$$

If the critical reliability state of the subsystem is $r = 2$, then according to Eq. (3.18) its risk function takes the form

$$\boldsymbol{r}(t) \cong [1 - \exp[-0.9981t^2]]^2$$

Hence, and from Eq. (3.19), the moment when the subsystem risk exceeds an admissible level $\delta = 0.05$ is

$$\tau = \boldsymbol{r}^{-1}(\delta) = [-(1/0.9981)\log(1-\sqrt{\delta})]^{1/2} \cong 0.50 \text{ years}$$

Subsystem S_4 is composed of three chain conveyors, each composed of a wheel driving the belt, a reversible wheel and 160, 160 and 240 links respectively. The subsystem consists of three conveyors. Two of these are composed of 162 components and the remaining one is composed of 242 components. Thus it is a non-homogeneous non-regular multi-state series–parallel system. In order to make it a regular system we conventionally complete two first conveyors having 162 components with 80 components that do not fail. After this supplement subsystem S_4 consists of $k_n = 3$ conveyors, each composed of $l_n = 242$ components. In two of them there are two driving wheels with reliability functions

$$R^{(1,1)}(t,1) = \exp[-0.005t^2], \quad R^{(1,1)}(t,2) = \exp[-0.008t^2] \quad \text{for } t \geq 0$$

160 links with reliability functions

$$R^{(1,2)}(t,1) = \exp[-0.012t^2], \quad R^{(1,2)}(t,2) = \exp[-0.018t^2] \quad \text{for } t \geq 0$$

and 80 components with 'reliability functions'

$$R^{(1,3)}(t,1) = \exp[-\beta_1 t^2], \quad R^{(1,3)}(t,2) = \exp[-\beta_2 t^2] \quad \text{for } t \geq 0$$

where $\beta_1 = \beta_2 = 0$.

The third conveyer is composed of two driving wheels with reliability functions

$$R^{(2,1)}(t,1) = \exp[-0.022t^{3/2}], \quad R^{(2,1)}(t,2) = \exp[-0.026t^{3/2}] \quad \text{for } t \geq 0$$

and 240 links with reliability functions

$$R^{(2,2)}(t,1) = \exp[-0.034t^{3/2}], \quad R^{(2,2)}(t,2) = \exp[-0.042t^{3/2}] \quad \text{for } t \geq 0$$

Thus the subsystem is a non-homogeneous regular multi-state series–parallel system where, according to Definition 3.20, we have

$$k_n = k = 3, \quad l_n = 242, \quad a = 2, \quad q_1 = 2/3, \quad q_2 = 1/3,$$
$$e_1 = 3, \quad p_{11} = 2/242, \quad p_{12} = 160/242, \quad p_{13} = 80/242,$$
$$e_2 = 2, \quad p_{21} = 2/242, \quad p_{22} = 240/242$$

and according to Eqs. (3.43)–(3.45) its exact reliability function is given by

$$\boldsymbol{R}'_{3,242}(t,\cdot) = [1, 1-[1-\exp[-1.93t^2]]^2 \cdot [1-\exp[-8.204t^{3/2}]] \\ 1-[1-\exp[-2.896t^2]]^2 \cdot [1-\exp[-10.132t^{3/2}]]] \quad \text{for } t \geq 0$$

Further, applying Corollary 5.10, from Eq. (5.62), we have

$$\alpha_1(1) = \min\{2, 2, 2\} = 2,$$
$$\beta_1(1) = \frac{2}{242}0.005 + \frac{160}{242}0.012 + \frac{80}{242}0 = 0.007975206,$$

$$\alpha_2(1) = \min\{3/2, 3/2\} = 3/2,$$
$$\beta_2(1) = \frac{2}{242}0.022 + \frac{240}{242}0.034 = 0.033900826,$$

$$\alpha_1(2) = \min\{2, 2, 2\} = 2,$$
$$\beta_1(2) = \frac{2}{242}0.008 + \frac{160}{242}0.018 + \frac{80}{242}0 = 0.011966942,$$

$$\alpha_2(2) = \min\{3/2, 3/2\} = 3/2,$$
$$\beta_2(2) = \frac{2}{242}0.026 + \frac{240}{242}0.042 = 0.041867768$$

and from Eq. (5.63)

$$\alpha(1) = \max\{2, 3/2\} = 2, \quad \beta(1) = \min\{0.007975206\} = 0.007975206, \\ \alpha(2) = \max\{2, 3/2\} = 2, \quad \beta(2) = \min\{0.011966942\} = 0.011966942$$

and from Eq. (5.61)

$$a_n(1) = (0.007975206 \cdot 242)^{-1/2} = 0.719815778, \quad b_n(1) = 0, \\ a_n(2) = (0.011966942 \cdot 242)^{-1/2} = 0.587625622, \quad b_n(2) = 0$$

and we conclude that the limit reliability function of subsystem S_4 is

$$\mathfrak{R}'_9(t,\cdot) = [1, 1-[1-\exp[-t]]^2, \quad 1-[1-\exp[-t]]^2] \quad \text{for } t \geq 0$$

Thus, from Eq. (3.49), we obtain the following approximate formula:

$$\boldsymbol{R}'_{3,242}(t,\cdot) \cong \mathfrak{R}'_9((t-b_n(u))/a_n(u),\cdot) \\ = [1, 1-[1-\exp[-1.93t^2]]^2, 1-[1-\exp[-2.896t^2]]^2] \quad \text{for } t \geq 0$$

The approximate mean values and standard deviations of the subsystem lifetimes in the state subsets, according to Eqs. (3.13)–(3.15), are

$$M(1) \cong 0.82 \text{ years}, \quad M(2) \cong 0.67 \text{ years}, \quad \sigma(1) \cong 0.32 \text{ years}, \quad \sigma(2) \cong 0.26 \text{ years}$$

Hence, and from Eq. (3.17), the subsystem mean lifetimes in particular states are

$$\overline{M}(1) \cong 0.15 \text{ years}, \quad \overline{M}(2) \cong 0.67 \text{ years}$$

If the critical reliability state of the subsystem is $r = 2$, then due to Eq. (3.18) its risk function is given by

$$\boldsymbol{r}(t) \cong [1 - \exp[-2.896t^2]]^2$$

Hence, and from Eq. (3.19), it follows that the moment when the subsystem risk exceeds an admissible level $\delta = 0.05$ is

$$\tau = \boldsymbol{r}^{-1}(\delta) = [-(1/2.896)\log(1-\sqrt{\delta})]^{1/2} \cong 0.30 \text{ years}$$

Since the considered subsystems create a series structure in a reliability sense, then according to Definition 3.17 and formulae (3.33) and (3.34), the reliability function of the whole transportation system is given by

$$\overline{\boldsymbol{R}}'(t,\cdot) \cong [1, \overline{\boldsymbol{R}}'(t,1), \overline{\boldsymbol{R}}'(t,2)], \quad t \geq 0$$

where

$$\begin{aligned}
\overline{\boldsymbol{R}}'(t,1) \cong\ & 24\exp[-25.471t^2] - 12\exp[-27.401t^2] - 12\exp[-26.1475t^2] \\
& + 6\exp[-28.0775t^2] - 24\exp[-47.699t^2] + 12\exp[-49.629t^2] \\
& + 12\exp[-48.3755t^2] - 6\exp[-50.3055t^2] + 8\exp[-69.927t^2] \\
& - 4\exp[-71.857t^2] - 4\exp[-70.6035t^2] + 2\exp[-72.5335t^2] \\
& - 12\exp[-26.1075t^2] + 6\exp[-28.0375t^2] + 6\exp[-26.784t^2] \\
& - 3\exp[-28.714t^2] + 12\exp[-48.3355t^2] - 6\exp[-50.2655t^2] \\
& - 6\exp[-49.012t^2] + 3\exp[-50.942t^2] - 4\exp[-70.5635t^2] \\
& + 2\exp[-72.4935t^2] + 2\exp[-71.24t^2] - \exp[-73.17t^2]
\end{aligned}$$

$$\begin{aligned}
\overline{\boldsymbol{R}}'(t,2) \cong\ & 24\exp[-49.2718t^2] - 12\exp[-52.1678t^2] - 12\exp[-50.2699t^2] \\
& + 6\exp[-53.1659t^2] - 24\exp[-93.7014t^2] + 12\exp[-96.5974t^2] \\
& + 12\exp[-94.6995t^2] - 6\exp[-97.5955t^2] + 8\exp[-138.131t^2] \\
& - 4\exp[-141.027t^2] - 4\exp[-139.1291t^2] + 2\exp[-142.0251t^2] \\
& - 12\exp[-50.2199t^2] + 6\exp[-53.1159t^2] + 6\exp[-51.218t^2] \\
& - 3\exp[-54.114t^2] + 12\exp[-94.6495t^2] - 6\exp[-97.5455t^2] \\
& - 6\exp[-95.6476t^2] + 3\exp[-98.5436t^2] - 4\exp[-139.0791t^2] \\
& + 2\exp[-141.9751t^2] + 2\exp[-140.0772t^2] - \exp[-142.9732t^2]
\end{aligned}$$

The approximate mean values and standard deviations of the system lifetimes in the state subsets, from Eqs. (3.13)–(3.15), are

$$M(1) \cong 0.27 \text{ years}, \quad M(2) \cong 0.19 \text{ years},$$
$$\sigma(1) \cong 0.09 \text{ years}, \quad \sigma(2) \cong 0.07 \text{ years}$$

Hence, from Eq. (3.17), the mean values of the system lifetime in particular states are

$$\overline{M}(1) \cong 0.08 \text{ years}, \quad \overline{M}(2) \cong 0.19 \text{ years}$$

If a critical reliability state of the system is $r = 2$, then from Eq. (3.18) its risk function takes the form

$$\boldsymbol{r}(t) \cong 1 - \overline{\boldsymbol{R}}'(t, 2)$$

Hence, from Eq. (3.19), the moment when the risk exceeds the critical level $\delta = 0.05$ is

$$\tau = r^{-1}(\delta) \cong 0.10 \text{ years}$$

The behaviour of the exact and approximate reliability functions and the risk function of the considered system are given in Table 6.1 and illustrated in Figure 6.2. Moreover, in Table 6.1 the differences between the exact and approximate values of the components of system multi-state reliability functions marked by $\Delta(1)$ and $\Delta(2)$ are given. These differences show that replacement of the system's exact reliability function by its approximate form does not result in large mistakes in this evaluation.

Table 6.1 The Values of the Components of Multi-State Reliability Functions and the Risk Function of the Port Grain Transportation System

t	$\overline{R}'(t,1)$		$\Delta(1)$	$\overline{R}'(t,2)$		$\Delta(2)$	$r(t)$
	Exact Reliability Function	**Approximate Reliability Function**		**Exact Reliability Function**	**Approximate Reliability Function**		
0.00	1.00000	1.00000	0.0000	1.00000	1.00000	0.0000	0.0000
0.05	0.99984	0.99981	0.0000	0.99882	0.99877	0.0001	0.0012
0.10	0.99191	0.99164	0.0003	0.95345	0.95288	0.0006	0.0471
0.15	0.93800	0.93695	0.0011	0.74554	0.74389	0.0017	0.2561
0.20	0.79233	0.79023	0.0021	0.42211	0.42005	**0.0021**	0.5779
0.25	0.57025	0.56759	**0.0027**	0.17068	0.16933	0.0014	0.8307
0.30	0.34445	0.34213	0.0023	0.05099	0.05046	0.0005	0.9495
0.35	0.17559	0.17411	0.0015	0.01162	0.01148	0.0001	0.9885
0.40	0.07652	0.07579	0.0007	0.00205	0.00203	0.0000	0.9980
0.45	0.02885	0.02857	0.0003	0.00028	0.00028	0.0000	0.9997
0.50	0.00948	0.00940	0.0001	0.00003	0.00003	0.0000	1.0000
0.55	0.00273	0.00271	0.0000	0.00000	0.00000	0.0000	1.0000
0.60	0.00069	0.00069	0.0000	0.00000	0.00000	0.0000	1.0000

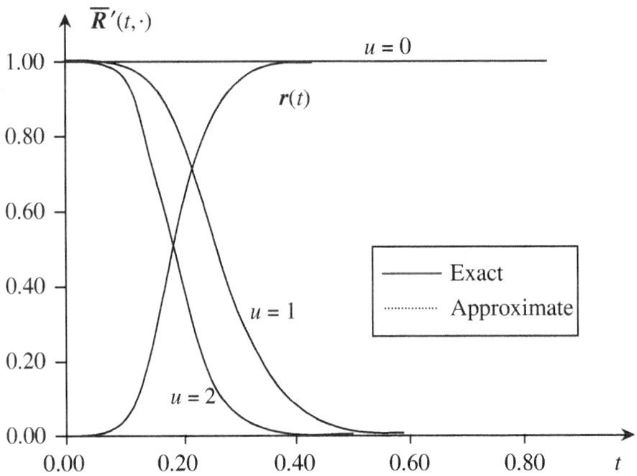

Figure 6.2 The graphs of the components of multi-state reliability functions and the risk function of the port grain transportation system.

6.3 Reliability of a Port Oil Transportation System

Oil Terminal No. 21 in Dębogórze, presented in Figure 6.3, is set up to receive from ships, store and send by carriages or cars oil products, such as petrol, driving oil and fuel oil. Three terminal parts, *A*, *B* and *C*, fulfil these purposes. They are linked by the piping transportation systems. The unloading of tankers is performed at the pier located in the Port of Gdynia. The pier is connected to terminal part *A* through the transportation subsystem S_1 built of two piping lines composed of steel pipe segments with a diameter of 600 mm. In part *A* there is a supporting station fortifying tankers' pumps and making possible further transport of oil by means of subsystem S_2 to terminal part *B*. Subsystem S_2 is built of two piping lines composed of steel pipe segments of 600 mm diameter. Terminal part *B* is connected to terminal part *C* by subsystem S_3. Subsystem S_3 is built of one piping line composed of steel pipe segments of 500 mm diameter and two piping lines composed of steel pipe segments of 350 mm diameter. Terminal part *C* is set up for loading the rail cisterns with oil products and for the wagon carrying these to the railway station of the Port of Gdynia and further into the country.

Considering the safety of the operation of the piping in the system, the following three reliability states of the piping components are distinguished:

- state 2: the piping operation is fully safe,
- state 1: the piping operation is less safe and dangerous because of the possibility of environment pollution,
- state 0: the piping is destroyed.

Subsystem S_1 consists of two identical piping lines, each composed of 176 pipe segments of length 12 m and 2 valves. It means that it is built of $k_n = 2$ piping lines, each composed of $l_n = 178$ components. In each of the lines there are 176 pipe segments with reliability functions

$$R^{(1,1)}(t,1) = \exp[-0.000000001t^4],\ R^{(1,1)}(t,2) = \exp[-0.000000004t^4] \quad \text{for } t \geq 0$$

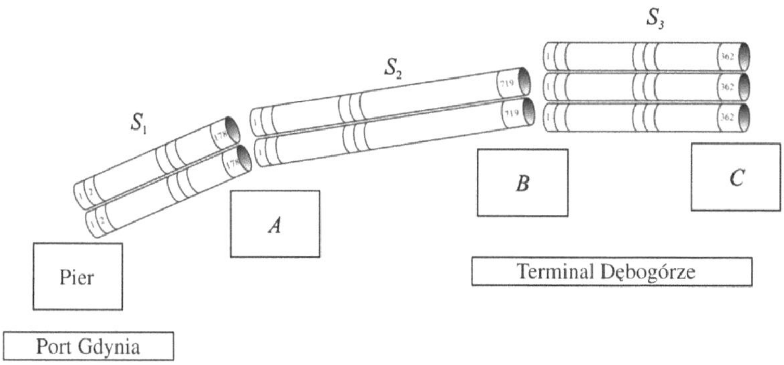

Figure 6.3 The scheme of the oil transportation system.

and two valves with reliability functions

$$R^{(1,2)}(t,1) = \exp[-0.000000052t^4],\ R^{(1,2)}(t,2) = \exp[-0.000000107t^4] \quad \text{for } t \geq 0$$

Thus subsystem S_1 is a non-homogeneous regular multi-state series–parallel system, where, according to Definition 3.20,

$$k_n = k = 2,\ \ l_n = 178,\ \ a = 1,\ \ q_1 = 1,$$
$$e_1 = 2,\ \ p_{11} = 176/178,\ \ p_{12} = 2/178$$

The exact subsystem reliability function for $t \geq 0$, according to Eqs. (3.43)–(3.45) is given by

$$\boldsymbol{R}'_{2,178}(t,\cdot) = [1, 1 - [1 - \exp[-0.00000028t^4]]^2, 1 - [1 - \exp[-0.000000918t^4]]^2]$$

Applying Corollary 5.10, according to Eq. (5.62), we have

$$\alpha_1(1) = \min\{4,4\} = 4,$$
$$\beta_1(1) = \frac{176}{178}0.000000001 + \frac{2}{178}0.000000052 = 0.000000001573,$$
$$\alpha_1(2) = \min\{4,4\} = 4,$$
$$\beta_1(2) = \frac{176}{178}0.000000004 + \frac{2}{178}0.000000107 = 0.0000000051573$$

and from Eq. (5.63)

$$\alpha(1) = \max\{4\} = 4,$$
$$\beta(1) = \min\{0.000000001573\} = 0.000000001573,$$
$$\alpha(2) = \max\{4\} = 4,$$
$$\beta(2) = \min\{0.000000005157\} = 0.0000000051573$$

and from Eq. (5.61)

$$a_n(1) = (0.000000001573 \cdot 178)^{-1/4} = 43.47208719, \;\; b_n(1) = 0,$$
$$a_n(2) = (0.000000005157 \cdot 178)^{-1/4} = 32.30645681, \;\; b_n(2) = 0$$

and we conclude that the subsystem limit reliability function is given by

$$\mathfrak{R}'_9(t, \cdot) = [1, 1 - [1 - \exp[-t^4]]^2, 1 - [1 - \exp[-t^4]]^2] \quad \text{for } t \geq 0$$

Thus, from Eq. (3.49), for $t \geq 0$, we have

$$\boldsymbol{R}'_{2,178}(t, \cdot) \cong [1, 1 - [1 - \exp[-0.00000028t^4]]^2, 1 - [1 - \exp[-0.000000918t^4]]^2]$$

The mean values and standard deviations of subsystem S_1 sojourn times in the state subsets, from Eqs. (3.13)–(3.15), are

$$M(1) \cong 45.67 \text{ years}, \;\; M(2) \cong 33.94 \text{ years},$$
$$\sigma(1) \cong 8.92 \text{ years}, \;\; \sigma(2) \cong 6.63 \text{ years}$$

Hence, from Eq. (3.17), the mean lifetimes of the subsystem in the particular states are

$$\overline{M}(1) \cong 11.73 \text{ years}, \;\; \overline{M}(2) \cong 33.94 \text{ years}$$

If a critical state is $r = 2$, then according to Eq. (3.18) the system risk function takes the form

$$\boldsymbol{r}(t) \cong [1 - \exp[-0.000000918t^4]]^2$$

and from Eq. (3.19) the moment when the risk exceeds the critical level $\delta = 0.05$ is

$$\tau = \boldsymbol{r}^{-1}(\delta) = [-(1/0.000000918)\log(1 - \sqrt{\delta})]^{1/4} \cong 22.91 \text{ years}$$

Subsystem S_2 consists of two identical piping lines, each composed of 717 pipe segments of length 12 m and 2 valves. It means that it is built of $k_n = 2$ piping lines, each composed of $l_n = 719$ components. In each of the lines there are 717 pipe segments with reliability functions

$$R^{(1,1)}(t, 1) = \exp[-0.000000001t^4], \; R^{(1,1)}(t, 2) = \exp[-0.000000004t^4] \quad \text{for } t \geq 0$$

and two valves with reliability function

$$R^{(1,2)}(t, 1) = \exp[-0.000000052t^4], \; R^{(1,2)}(t, 2) = \exp[-0.000000107t^4] \quad \text{for } t \geq 0$$

Thus subsystem S_2 is a non-homogeneous regular multi-state series−parallel system, where, according to Definition 3.20,

$$k_n = k = 2,\ \ l_n = 719,\ \ a = 1,\ \ q_1 = 1,$$
$$e_1 = 2,\ \ p_{11} = 717/719,\ \ p_{12} = 2/719$$

The exact subsystem reliability function for $t \geq 0$, according to Eqs. (3.43)−(3.45), is given by

$$\boldsymbol{R}'_{2,719}(t, \cdot) = [1, 1 - [1 - \exp[-0.000000821t^4]]^2, 1 - [1 - \exp[-0.000003082t^4]]^2$$

Applying Corollary 5.10, according to Eq. (5.62), we have

$$\alpha_1(1) = \min\{4, 4\} = 4,$$
$$\beta_1(1) = \frac{717}{719}0.000000001 + \frac{2}{719}0.000000052 = 0.0000000011418,$$

$$\alpha_1(2) = \min\{4, 4\} = 4,$$
$$\beta_1(2) = \frac{717}{719}0.000000001 + \frac{2}{719}0.000000052 = 0.0000000042865$$

and from Eq. (5.63)

$$\alpha(1) = \max\{4\} = 4,$$
$$\beta(1) = \min\{0.0000000011418\} = 0.0000000011418,$$
$$\alpha(2) = \max\{4\} = 4,$$
$$\beta(2) = \min\{0.0000000042865\} = 0.0000000042865$$

and from Eq. (5.61)

$$a_n(1) = (0.0000000011418 \cdot 719)^{-1/4} = 33.22111547,\ \ b_n(1) = 0,$$
$$a_n(2) = (0.0000000042865 \cdot 719)^{-1/4} = 23.86667072,\ \ b_n(2) = 0$$

and we conclude that the subsystem S_2 limit reliability is given by

$$\mathfrak{R}'_9(t, \cdot) = [1, 1 - [1 - \exp[-t^4]]^2, 1 - [1 - \exp[-t^4]]^2] \quad \text{for } t \geq 0$$

Hence, from Eq. (3.49), for $t \geq 0$, we get

$$\boldsymbol{R}'_{2,719}(t, \cdot) \cong [1, 1 - [1 - \exp[-0.000000821t^4]]^2, 1 - [1 - \exp[-0.000003082t^4]]^2]$$

The mean values and standard deviations of the subsystem sojourn times in the state subsets, according to Eqs. (3.13)−(3.15), are

$$M(1) \cong 34.90 \text{ years},\ \ M(2) \cong 25.07 \text{ years},$$
$$\sigma(1) \cong 6.82 \text{ years},\ \ \ \sigma(2) \cong 4.92 \text{ years}$$

and from Eq. (3.17), the mean values of the subsystem lifetimes in the particular states are

$$\overline{M}(1) \cong 9.83 \text{ years}, \quad \overline{M}(2) \cong 25.07 \text{ years}$$

If the critical state is $r = 2$, then from Eq. (3.18), the subsystem risk function takes the form

$$\boldsymbol{r}(t) \cong [1 - \exp[-0.000003082t^4]]^2$$

Hence, from Eq. (3.19), the moment when the subsystem risk exceeds the critical level $\delta = 0.05$ is

$$\tau = \boldsymbol{r}^{-1}(\delta) = [-(1/0.000003082)\log(1 - \sqrt{\delta})]^{1/4} \cong 16.93 \text{ years}$$

Subsystem S_3 consists of three different piping lines, each composed of 360 pipe segments of either 10 m or 7.5 m length and two valves. It means that it is built of $k_n = 3$ piping lines, each composed of $l_n = 362$ components. In two lines there are 60 pipe segments with reliability functions

$$R^{(1,1)}(t,1) = \exp[-0.0000000008t^4],\ R^{(1,1)}(t,2) = \exp[-0.000000002t^4] \quad \text{for } t \geq 0$$

and two valves with reliability functions

$$R^{(1,2)}(t,1) = \exp[-0.000000052t^4],\ R^{(1,2)}(t,2) = \exp[-0.000000107t^4] \quad \text{for } t \geq 0$$

In the third line there are 360 pipe segments with reliability functions

$$R^{(2,1)}(t,1) = \exp[-0.00000022t^3],\ R^{(2,1)}(t,2) = \exp[-0.0000003t^3] \quad \text{for } t \geq 0$$

and two valves with reliability functions

$$R^{(2,2)}(t,1) = \exp[-0.000000052t^4],\ R^{(2,2)}(t,2) = \exp[-0.000000107t^4] \quad \text{for } t \geq 0$$

Thus subsystem S_3 is a non-homogeneous regular multi-state series–parallel system, where, according to Definition 3.20,

$$k_n = k = 3,\ l_n = 362,\ a = 2,\ q_1 = 2/3,\ q_2 = 1/3,$$
$$e_1 = 2,\ p_{11} = 360/362,\ p_{12} = 2/362,$$
$$e_2 = 2,\ p_{21} = 360/362,\ p_{22} = 2/362$$

The exact subsystem reliability function for $t \geq 0$, according to Eqs. (3.43)–(3.45), is given by

$$R'_{3,362}(t,\cdot) = [1, 1-[1-\exp[-0.000000392t^4]]^2 \cdot [1-\exp[-0.0000792t^3 - 0.000000104t^4]]$$
$$1-[1-\exp[-0.000000934t^4]]^2 \cdot [1-\exp[-0.000108t^3 - 0.000001294t^4]]]$$

Applying Corollary 5.10, according to Eq. (5.62), we have

$$\alpha_1(1) = \min\{4,4\} = 4,$$
$$\beta_1(1) = \frac{360}{362}0.0000000008 + \frac{2}{362}0.000000052 = 0.0000000010828,$$

$$\alpha_2(1) = \min\{3,4\} = 3,$$
$$\beta_2(1) = \frac{360}{362}0.00000022 + \frac{2}{362}0.000000052 = 0.000000218,$$

$$\alpha_1(2) = \min\{4,4\} = 4,$$
$$\beta_1(2) = \frac{360}{362}0.000000002 + \frac{2}{362}0.000000107 = 0.0000000025801,$$

$$\alpha_2(2) = \min\{3,4\} = 3,$$
$$\beta_2(2) = \frac{360}{362}0.0000003 + \frac{2}{362}0.000000107 = 0.000000298$$

next from Eq. (5.63)

$$\alpha(1) = \max\{4,3\} = 4, \quad \beta(1) = \min\{0.0000000010828\} = 0.0000000010828,$$
$$\alpha(2) = \max\{4,3\} = 4, \quad \beta(2) = \min\{0.0000000025801\} = 0.0000000025801$$

and from Eq. (5.61)

$$a_n(1) = (0.0000000010828 \cdot 362)^{-1/4} = 39.96487724, \quad b_n(1) = 0,$$
$$a_n(2) = (0.0000000025801 \cdot 362)^{-1/4} = 32.1672016, \quad b_n(2) = 0$$

and we conclude that the subsystem S_3 limit reliability function is given by

$$\mathfrak{R}'_9(t,\cdot) = [1, 1-[1-\exp[-t^4]]^2, 1-[1-\exp[-t^4]]^2] \quad \text{for } t \geq 0$$

Hence, applying Eq. (3.49), for $t \geq 0$, we get

$$\boldsymbol{R}'_{3,362}(t,\cdot) \cong [1, 1-[1-\exp[-0.000000392t^4]]^2, 1-[1-\exp[-0.000000934t^4]]^2]$$

The mean values and standard deviations of the subsystem sojourn times in the state subsets, according to Eqs. (3.13)–(3.15), are

$$M(1) \cong 41.99 \text{ years}, \quad M(2) \cong 33.80 \text{ years},$$
$$\sigma(1) \cong 8.17 \text{ years}, \quad \sigma(2) \cong 6.57 \text{ years}$$

Hence, from Eq. (3.17), the mean values of the subsystem lifetimes in the particular states are

$$\overline{M}(1) \cong 8.19 \text{ years}, \quad \overline{M}(2) \cong 33.80 \text{ years}$$

If the critical state is $r = 2$, then from Eq. (3.18), the subsystem risk function takes the form

$$r(t) \cong [1 - \exp[-0.000000934t^4]]^2$$

and, from Eq. (3.19), the moment when the risk exceeds the critical level $\delta = 0.05$ is

$$\tau = \boldsymbol{r}^{-1}(\delta) = [-(1/0.000000934)\log(1-\sqrt{\delta})]^{1/4} \cong 22.82 \text{ years}$$

Since the subsystems create a series reliability structure, then by Eqs. (3.33) and (3.34) the multi-state reliability function of the whole oil transportation system is given by

$$\overline{\boldsymbol{R}}'(t,\cdot) \cong [1, \overline{\boldsymbol{R}}'(t,1), \overline{\boldsymbol{R}}'(t,2)]$$

where

$$\begin{aligned}\overline{\boldsymbol{R}}'(t,1) \cong\ & 8\exp[-0.000001493t^4] - 4\exp[-0.000001885t^4] - 4\exp[-0.000002314t^4] \\ & + 2\exp[-0.000002706t^4] - 4\exp[-0.000001773t^4] + 2\exp[-0.000002165t^4] \\ & + 2\exp[-0.000002594t^4] - \exp[-0.000002986t^4]\end{aligned}$$

$$\begin{aligned}\overline{\boldsymbol{R}}'(t,2) \cong\ & 8\exp[-0.000004934t^4] - 4\exp[-0.000005868t^4] - 4\exp[-0.000008016t^4] \\ & + 2\exp[-0.00000895t^4] - 4\exp[-0.000005852t^4] + 2\exp[-0.000006786t^4] \\ & + 2\exp[-0.000008934t^4] - \exp[-0.000009868t^4].\end{aligned}$$

From Eqs. (3.13)–(3.15), the approximate mean values and standard deviations of the system sojourn times in the state subsets are

$$\begin{aligned} M(1) &\cong 32.60 \text{ years}, & M(2) &\cong 23.98 \text{ years}, \\ \sigma(1) &\cong 5.90 \text{ years}, & \sigma(2) &\cong 4.43 \text{ years}\end{aligned}$$

while, according to Eq. (3.17), the mean values of the system lifetimes in particular states are

$$\overline{M}(1) \cong 8.62 \text{ years}, \quad \overline{M}(2) \cong 23.98 \text{ years}$$

If the critical state is $r = 2$, then from Eq. (3.18) the system risk function is given by

$$\boldsymbol{r}(t) \cong 1 - \overline{\boldsymbol{R}}'(t,2)$$

Hence, from Eq. (3.19), the moment when the risk exceeds the admissible level $\delta = 0.05$ is

$$\tau = \boldsymbol{r}^{-1}(\delta) \cong 16.50 \text{ years}$$

The behaviour of the system's exact and approximate reliability functions, the differences between them and its risk function are given in Table 6.2 and illustrated in Figure 6.4.

The considered reliability structure of the oil transportation system is adequate in the case when each of the oil products may be transported by each of three system pipelines. However, in such transportation, some parts of different products are mixed and they have to be refined. If mixing of different oil products is not tolerated, the system structure should be considered as a non-homogeneous series system composed of $n = 2880$ components. In this case the system is composed of 1786 pipe segments with reliability functions

$$R^{(1)}(t,1) = \exp[-0.000000001t^4],\ R^{(1)}(t,2) = \exp[-0.000000004t^4] \quad \text{for } t \geq 0$$

720 pipe segments with reliability functions

$$R^{(2)}(t,1) = \exp[-0.0000000008t^4],\ R^{(2)}(t,2) = \exp[-0.000000002t^4] \quad \text{for } t \geq 0$$

Table 6.2 The Values of the Multi-State Reliability Functions Components and the Risk Function of the Port Oil Transportation System

t	$\overline{\boldsymbol{R}}'(t,1)$		$\Delta(1)$	$\overline{\boldsymbol{R}}'(t,2)$		$\Delta(2)$	$r(t)$
	Exact Reliability Function	**Approximate Reliability Function**		**Exact Reliability Function**	**Approximate Reliability Function**		
0	1.00000	1.00000	0.0000	1.00000	1.00000	0.0000	0.0000
4	1.00000	1.00000	0.0000	1.00000	1.00000	0.0000	0.0000
8	0.99999	0.99998	0.0000	0.99983	0.99981	0.0000	0.0002
12	0.99967	0.99962	0.0001	0.99574	0.99545	0.0003	0.0046
16	0.99674	0.99628	0.0005	0.96187	0.95986	0.0020	0.0401
20	0.98122	0.97932	0.0019	0.82208	0.81658	0.0055	0.1834
24	0.92630	0.92181	0.0045	0.51602	0.51032	**0.0057**	0.4897
28	0.79042	0.78421	**0.0062**	0.18509	0.18326	0.0018	0.8167
32	0.55756	0.55282	0.0047	0.02933	0.02919	0.0001	0.9708
36	0.29212	0.29028	0.0018	0.00161	0.00161	0.0000	0.9984
40	0.10179	0.10146	0.0003	0.00002	0.00002	0.0000	1.0000
44	0.02121	0.02118	0.0000	0.00000	0.00000	0.0000	1.0000
48	0.00239	0.00239	0.0000	0.00000	0.00000	0.0000	1.0000
52	0.00013	0.00013	0.0000	0.00000	0.00000	0.0000	1.0000
56	0.00000	0.00000	0.0000	0.00000	0.00000	0.0000	1.0000
60	0.00000	0.00000	0.0000	0.00000	0.00000	0.0000	1.0000

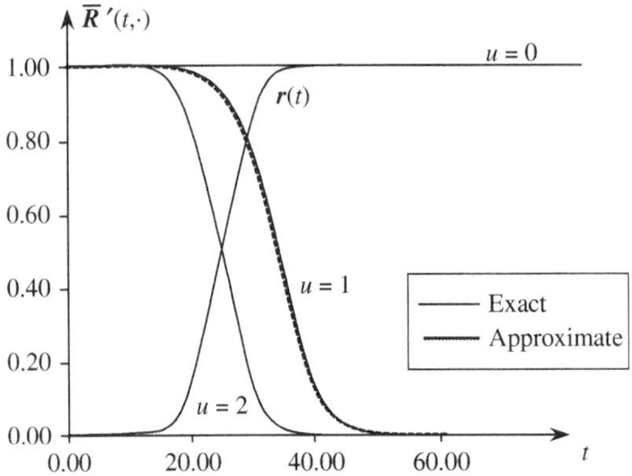

Figure 6.4 The graphs of the multi-state reliability functions and the risk function of the port oil transportation system.

360 pipe segments with reliability functions

$$R^{(3)}(t,1) = \exp[-0.00000022t^3],\ \ R^{(3)}(t,2) = \exp[-0.0000003t^3] \quad \text{for } t \geq 0$$

and 14 valves with reliability functions

$$R^{(4)}(t,1) = \exp[-0.000000052t^4], R^{(4)}(t,2) = \exp[-0.000000107t^4] \quad \text{for } t \geq 0$$

According to Definition 3.17, the considered system is a non-homogeneous multi-state series system with parameters

$$n = 2880,\ \ a = 4,$$
$$q_1 = 1786/2880,\ \ q_2 = 720/2880,\ \ q_3 = 360/2880,\ \ q_4 = 14/2880$$

and according to Eqs. (3.33) and (3.34) its exact reliability function is given by

$$\overline{\boldsymbol{R}}'_{2880}(t,\cdot) = [1, \exp[-000000309t^4 - 00000792t^3]$$
$$\exp[-0000010082t^4 - 0000108t^3]] \quad \text{for } t \geq 0$$

The expected values of the system sojourn times in the state subset and their standard deviations, according to Eqs. (3.13) − (3.15), are

$$M(1) \cong 17.30 \text{ years},\ \ M(2) \cong 14.00 \text{ years},$$
$$\sigma(1) \cong 5.60 \text{ years},\ \ \ \sigma(2) \cong 4.40 \text{ years}$$

Hence, from Eq. (3.17), the system mean lifetimes in the states are

$$\overline{M}(1) \cong 3.30 \text{ years},\ \ \overline{M}(2) \cong 14.00 \text{ years}$$

If a critical system reliability state is $r = 2$, then according to Eq. (3.18) its risk function takes the form

$$\boldsymbol{r}(t) \cong 1 - \exp[-0000010082t^4 - 0000108t^3]$$

Further, from Eq. (3.19), the moment when the system risk exceeds a permitted level $d = 0.05$ is

$$\tau \cong \boldsymbol{r}^{-1}(\delta) \cong 6.20 \text{ years}$$

6.4 Reliability of a Port Bulk Transportation System

The bulk conveyor system is the part of the Baltic Bulk Terminal of the Port of Gdynia assigned to load ships with bulk cargo from Terminal Storage. Its scheme is given in Figure 6.5. Three self-acting loading machines, the transportation system composed of belt conveyors and the coastal loading system carry out the loading of the ships.

In the conveyor loading system we distinguish the following transportation subsystems:

S_1: the dosage conveyor,
S_2: the horizontal conveyor,
S_3: the horizontal conveyor,
S_4: the sloping conveyors,
S_5: the dosage conveyor with buffer,
S_6: the loading system.

The transporting subsystems have steel covers and they are provided with drives in the form of electrical engines with gears. In their reliability analysis we omit their drives, as they are mechanisms of different types. We also omit their covers as they have a high reliability and, practically, do not fail.

Taking into account the efficiency of the considered transportation system, we distinguish the following three reliability states of its components:

- state 3: the state ensuring the largest efficiency of the conveyor,
- state 2: the state ensuring less efficiency of the conveyor caused by throwing material off the belt,
- state 1: the state ensuring least efficiency of the conveyor caused by throwing material off the belt and needing human assistance,
- state 0: the state involving failure of the conveyer.

Subsystem S_1 is composed of three identical belt conveyers, which consist of a ribbon belt, a drum driving the belt, a reversible drum, 12 channelled rollers and

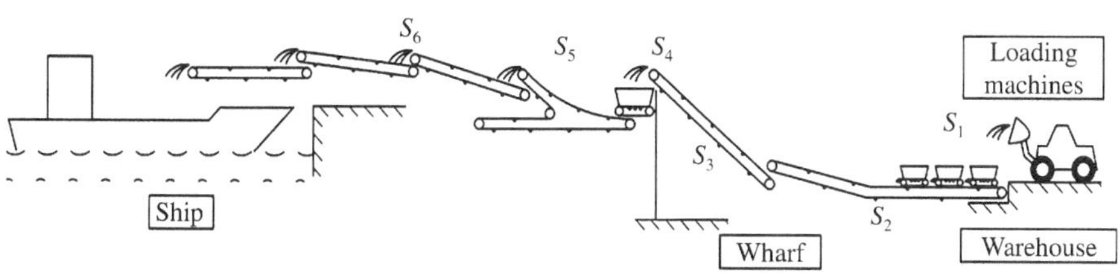

Figure 6.5 The scheme of the bulk cargo transportation system.

three rollers supporting the belt. It means that subsystem S_1 consists of $k_n = 3$ conveyors, each composed of $l_n = 18$ components. In each conveyer there is one belt with reliability functions

$$R^{(1,1)}(t,1) = \exp[-0.012t^2],\ R^{(1,1)}(t,2) = \exp[-0.022t^2],\ R^{(1,1)}(t,3) = \exp[-0.049t^2] \quad \text{for } t \geq 0$$

two drums driving the belt with reliability functions

$$R^{(1,2)}(t,1) = \exp[-0.0019t^2],\ R^{(1,2)}(t,2) = \exp[-0.0024t^2],\ R^{(1,2)}(t,3) = \exp[-0.0029t^2] \quad \text{for } t \geq 0$$

12 channelled rollers with reliability functions

$$R^{(1,3)}(t,1) = \exp[-0.028t^2],\ R^{(1,3)}(t,2) = \exp[-0.03t^2],\ R^{(1,3)}(t,3) = \exp[-0.032t^2] \quad \text{for } t \geq 0$$

and three supporting rollers with reliability functions

$$R^{(1,4)}(t,1) = \exp[-0.0075t^2],\ R^{(1,4)}(t,2) = \exp[-0.01t^2],\ R^{(1,4)}(t,3) = \exp[-0.02t^2] \quad \text{for } t \geq 0$$

Thus, according to Definition 3.20, it is a non-homogeneous regular multi-state series–parallel system with parameters

$$k_n = k = 3,\ l_n = 18,\ a = 1,\ q_1 = 1,$$
$$e_1 = 4,\ p_{11} = 1/18,\ p_{12} = 2/18,\ p_{13} = 12/18,\ p_{14} = 3/18$$

and from Eqs. (3.43)–(3.45) its exact reliability function is given by

$$\boldsymbol{R}'_{3,18}(t,\cdot) = [1, 1-[1-\exp[-0.3743t^2]]^3, 1-[1-\exp[-0.4168t^2]]^3$$
$$1-[1-\exp[-0.4988t^2]]^3] \quad \text{for } t \geq 0$$

Applying Corollary 5.10, according to Eq. (5.62), we have

$$\alpha_1(1) = \min\{2,2,2,2\} = 2,$$
$$\beta_1(1) = \frac{1}{18}0.012 + \frac{2}{18}0.0019 + \frac{12}{18}0.028 + \frac{3}{18}0.0075 = 0.020794444,$$

$$\alpha_1(2) = \min\{2,2,2,2\} = 2,$$
$$\beta_1(2) = \frac{1}{18}0.022 + \frac{2}{18}0.0024 + \frac{12}{18}0.03 + \frac{3}{18}0.01 = 0.023155555,$$

$$\alpha_1(3) = \min\{2,2,2,2\} = 2,$$
$$\beta_1(3) = \frac{1}{18}0.049 + \frac{2}{18}0.0029 + \frac{12}{18}0.032 + \frac{3}{18}0.02 = 0.027711111$$

and according to Eq. (5.63)

$$\alpha(1) = \max\{2\} = 2, \quad \beta(1) = \min\{0.020794444\} = 0.020794444,$$
$$\alpha(2) = \max\{2\} = 2, \quad \beta(2) = \min\{0.023155555\} = 0.023155555,$$
$$\alpha(3) = \max\{2\} = 2, \quad \beta(3) = \min\{0.027711111\} = 0.027711111$$

and according to Eq. (5.61)

$$a_n(1) = (0.020794444 \cdot 18)^{-1/2} = 1.634519443, \quad b_n(1) = 0,$$
$$a_n(2) = (0.023155555 \cdot 18)^{-1/2} = 1.548945546, \quad b_n(2) = 0,$$
$$a_n(3) = (0.027711111 \cdot 18)^{-1/2} = 1.415913682, \quad b_n(3) = 0$$

and we conclude that the limit reliability function of the subsystem S_1 is given by

$$\mathfrak{R}'_9(t,\cdot) = [1, 1-[1-\exp[-t^2]]^3, 1-[1-\exp[-t^2]]^3, 1-[1-\exp[-t^2]]^3] \text{ for } t \geq 0$$

Therefore, from Eq. (3.49), the approximate reliability function (the formula is exact in this case) takes the form

$$\begin{aligned} \boldsymbol{R}'_{3,18}(t,\cdot) &\cong \boldsymbol{R}'_9((t-b_n(u))/a_n(u)) \\ &= [1, 1-[1-\exp[-0.3743t^2]]^3, 1-[1-\exp[-0.4168t^2]]^3, \\ &\quad 1-[1-\exp[-0.4988t^2]]^3] \quad \text{for } t \geq 0 \end{aligned}$$

The expected values of the subsystem lifetimes in the state subsets and their standard deviations, according to Eqs. (3.13)–(3.15), are

$$M(1) \cong 2.11 \text{ years}, \quad M(2) \cong 2.00 \text{ years}, \quad M(3) \cong 1.83 \text{ years},$$
$$\sigma(1) \cong 0.67 \text{ years}, \quad \sigma(2) \cong 0.63 \text{ years}, \quad \sigma(3) \cong 0.57 \text{ years}$$

Hence, from Eq. (3.17), the subsystem mean lifetimes in the states are

$$\overline{M}(1) \cong 0.11 \text{ years}, \quad \overline{M}(2) \cong 0.17 \text{ years}, \quad \overline{M}(3) \cong 1.83 \text{ years}$$

If a critical state is $r = 2$, then from Eq. (3.18) the subsystem risk function is given by

$$\boldsymbol{r}(t) \cong [1-\exp[-0.4168t^2]]^3$$

Hence, from Eq. (3.19), the moment when the subsystem risk exceeds the permitted level $\delta = 0.05$ is

$$\tau = \boldsymbol{r}^{-1}(\delta) = [-(1/0.4168)\log(1-\sqrt[3]{\delta})]^{1/2} \cong 1.05 \text{ years}$$

Subsystem S_2 is composed of one belt conveyor that consists of a ribbon belt, a drum driving the belt, a reversible drum, 125 channelled rollers and 45 rollers supporting the belt. It means that subsystem S_2 consists of one conveyor composed of $n = 173$ components. In the conveyor there is one belt with reliability functions

$$R^{(1)}(t,1) = \exp[-0.012t^2],\ R^{(1)}(t,2) = \exp[-0.022t^2],\ R^{(1)}(t,3) = \exp[-0.049t^2] \quad \text{for } t \geq 0$$

two drums driving the belt with reliability functions

$$R^{(2)}(t,1) = \exp[-0.0019t^2],\ R^{(2)}(t,2) = \exp[-0.0024t^2],\ R^{(2)}(t,3) = \exp[-0.0029t^2] \quad \text{for } t \geq 0$$

125 channelled rollers with reliability functions

$$R^{(3)}(t,1) = \exp[-0.0074t^2],\ R^{(3)}(t,2) = \exp[-0.012t^2],\ R^{(3)}(t,3) = \exp[-0.021t^2] \quad \text{for } t \geq 0$$

and 45 supporting rollers with reliability functions

$$R^{(4)}(t,1) = \exp[-0.002t^2],\ R^{(4)}(t,2) = \exp[-0.0025t^2],\ R^{(4)}(t,3) = \exp[-0.003t^2] \quad \text{for } t \geq 0$$

Thus, according to Definition 3.17, it is a non-homogeneous multi-state series system with parameters

$$n = 173,\ a = 4,$$
$$q_1 = 1/173,\ q_2 = 2/173,\ q_3 = 125/173,\ q_4 = 45/173$$

and according to Eqs. (3.33) and (3.34) its exact multi-state reliability function is given by

$$\overline{\boldsymbol{R}}'_{173}(t,\cdot) = [1, \exp[-1.0308t^2], \exp[-1.6393t^2], \exp[-2.8148t^2]] \quad \text{for } t \geq 0$$

Next, applying Corollary 5.4, we have

$$\alpha(1) = \min\{2,2,2,2\} = 2,$$
$$\beta(1) = \max\{0.012, 0.0019, 0.0074, 0.002\} = 0.012,$$
$$\alpha(2) = \min\{2,2,2,2\} = 2,$$
$$\beta(2) = \max\{0.022, 0.0024, 0.012, 0.0025\} = 0.022,$$
$$\alpha(3) = \min\{2,2,2,2\} = 2,$$
$$\beta(3) = \max\{0.049, 0.0029, 0.021, 0.003\} = 0.049,$$
$$a_n(1) = (0.012 \cdot 173)^{-1/2} = 0.694042915,\ b_n(1) = 0,$$
$$a_n(2) = (0.022 \cdot 173)^{-1/2} = 0.512584663,\ b_n(2) = 0,$$
$$a_n(3) = (0.049 \cdot 173)^{-1/2} = 0.343462169,\ b_n(3) = 0$$

and moreover

$$\bar{d}(t,1) = \frac{1}{173}\cdot\frac{0.012}{0.012} + \frac{2}{173}\cdot\frac{0.0019}{0.012} + \frac{125}{173}\cdot\frac{0.0074}{0.012} + \frac{45}{173}\cdot\frac{0.002}{0.012} = 0.49653175,$$

$$\bar{d}(t,2) = \frac{1}{173}\cdot\frac{0.022}{0.022} + \frac{2}{173}\cdot\frac{0.0024}{0.022} + \frac{125}{173}\cdot\frac{0.012}{0.022} + \frac{45}{173}\cdot\frac{0.0025}{0.022} = 0.430714636,$$

$$\bar{d}(t,3) = \frac{1}{173}\cdot\frac{0.049}{0.049} + \frac{2}{173}\cdot\frac{0.0029}{0.049} + \frac{125}{173}\cdot\frac{0.021}{0.049} + \frac{45}{173}\cdot\frac{0.003}{0.049} = 0.332051428$$

and we conclude that the subsystem limit reliability function is given by

$$\overline{\mathfrak{R}}_2'(t,\cdot) = [1, \exp[-0.49653175t^2],\ \exp[-0.430714636t^2],\ \exp[-0.332051428t^2]],\quad t \geq 0$$

Therefore, from Eq. (3.49), the approximate reliability function (the formula is exact in this case) takes the form

$$\overline{\boldsymbol{R}}'_{173}(t,\cdot) \cong \overline{\mathfrak{R}}_2'((t - b_n(u))/a_n(u)) = [1, \exp[-1.0308t^2], \exp[-1.6393t^2],\ \exp[-2.8148t^2]] \quad \text{for } t \geq 0$$

The expected values of the subsystem lifetimes in the state subsets and their standard deviations, according to Eqs. (3.13)–(3.15),

$$M(1) \cong 0.87 \text{ years},\quad M(2) \cong 0.69 \text{ years},\quad M(2) \cong 0.53 \text{ years},$$
$$\sigma(1) \cong 0.46 \text{ years},\quad \sigma(2) \cong 0.36 \text{ years},\quad \sigma(3) \cong 0.28 \text{ years}$$

Hence, from Eq. (3.17), the subsystem mean lifetimes in the states are

$$\overline{M}(1) \cong 0.18 \text{ years},\quad \overline{M}(2) \cong 0.16 \text{ years},\quad \overline{M}(3) \cong 0.53 \text{ years}$$

If a critical state is $r = 2$, then from Eq. (3.18) the subsystem risk function is given by

$$\boldsymbol{r}(t) \cong 1 - \exp[-1.6393t^2]$$

Hence, from Eq. (3.19), the moment when the subsystem risk exceeds the permitted level $\delta = 0.05$ is

$$\tau = \boldsymbol{r}^{-1}(\delta) = [-(1/1.6393)\log(1-\delta)]^{1/2} \cong 0.18 \text{ years}$$

We will perform the reliability evaluation of the remaining subsystems using Corollary 6.1, which is a simplified modification of Corollary 5.4 and is much easier to use.

Subsystem S_3 is composed of one belt conveyor that consists of a ribbon belt, a drum driving the belt, a reversible drum, 65 channelled rollers and 20 rollers supporting the belt. It means that subsystem S_3 consists of one conveyor composed of $n = 88$ components. In the conveyor there is one belt with reliability functions

$$R^{(1)}(t,1) = \exp[-0.012t^2],\ R^{(1)}(t,2) = \exp[-0.022t^2],\ R^{(1)}(t,3) = \exp[-0.049t^2] \quad \text{for } t \geq 0$$

two drums driving the belt with reliability functions

$$R^{(2)}(t,1) = \exp[-0.0019t^2],\ R^{(2)}(t,2) = \exp[-0.0024t^2],\ R^{(2)}(t,3) = \exp[-0.0029t^2] \quad \text{for } t \geq 0$$

65 channelled rollers with reliability functions

$$R^{(3)}(t,1) = \exp[-0.0074t^2],\ R^{(3)}(t,2) = \exp[-0.012t^2],\ R^{(3)}(t,3) = \exp[-0.021t^2] \quad \text{for } t \geq 0$$

and 20 supporting rollers with reliability functions

$$R^{(4)}(t,1) = \exp[-0.002t^2],\ R^{(4)}(t,2) = \exp[-0.0025t^2],\ R^{(4)}(t,3) = \exp[-0.003t^2] \quad \text{for } t \geq 0$$

Thus, according to Definition 3.17, it is a non-homogeneous multi-state series system with parameters

$$n = 88,\ a = 4,$$
$$q_1 = 1/88,\ q_2 = 2/88,\ q_3 = 65/88,\ q_4 = 20/88$$

and according to Eqs. (3.33) and (3.34) its exact multi-state reliability function is given by

$$\overline{\boldsymbol{R}}'_{88}(t,\cdot) = [1, \exp[-0.5368t^2], \exp[-0.8568t^2], \exp[-1.4798t^2]] \quad \text{for } t \geq 0$$

Next, applying Corollary 6.1, according to Eq. (6.3), we have

$$\alpha(1) = \min\{2,2,2,2\} = 2,$$
$$\beta(1) = \frac{1}{88}0.012 + \frac{2}{88}0.0019 + \frac{65}{88}0.0074 + \frac{20}{88}0.002 = 0.0061,$$

$$\alpha(2) = \min\{2,2,2,2\} = 2,$$
$$\beta(2) = \frac{1}{88}0.022 + \frac{2}{88}0.0024 + \frac{65}{88}0.012 + \frac{20}{88}0.0025 = 0.009736363,$$

$$\alpha(3) = \min\{2,2,2,2\} = 2,$$
$$\beta(3) = \frac{1}{88}0.049 + \frac{2}{88}0.0029 + \frac{65}{88}0.021 + \frac{20}{88}0.003 = 0.016815909$$

and according to Eq. (6.2)

$$a_n(1) = (0.0061 \cdot 88)^{-1/2} = 1.364877726, \quad b_n(1) = 0,$$
$$a_n(2) = (0.009736363 \cdot 88)^{-1/2} = 1.080339575, \quad b_n(2) = 0,$$
$$a_n(3) = (0.016815909 \cdot 88)^{-1/2} = 0.822050484, \quad b_n(3) = 0$$

and we conclude that the subsystem limit reliability function is given by

$$\overline{\mathfrak{R}}_2'(t, \cdot) = [1, \exp[-t^2], \exp[-t^2], \exp[-t^2]] \quad \text{for } t \geq 0$$

Therefore, from Eq. (3.49), the approximate reliability function (the formula is exact in this case) takes the form

$$\overline{\boldsymbol{R}}_{88}'(t, \cdot) \cong \overline{\mathfrak{R}}_2'((t - b_n(u))/a_n(u))$$
$$= [1, \exp[-0.5368t^2], \exp[-0.8568t^2], \exp[-1.4798t^2]] \quad \text{for } t \geq 0$$

The expected values of the subsystem lifetimes in the state subsets and their standard deviations, according to Eqs. (3.13)–(3.15), are

$$M(1) \cong 1.21 \text{ years}, \quad M(2) \cong 0.96 \text{ years}, \quad M(2) \cong 0.73 \text{ years},$$
$$\sigma(1) \cong 0.63 \text{ years}, \quad \sigma(2) \cong 0.50 \text{ years}, \quad \sigma(3) \cong 0.38 \text{ years}$$

Hence, from Eq. (3.17), the subsystem mean lifetimes in the states are

$$\overline{M}(1) \cong 0.25 \text{ years}, \quad \overline{M}(2) \cong 0.23 \text{ years}, \quad \overline{M}(3) \cong 0.73 \text{ years}$$

If a critical state is $r = 2$, then from Eq. (3.18) the subsystem risk function is given by

$$\boldsymbol{r}(t) \cong 1 - \exp[-0.8568t^2]$$

Hence, from Eq. (3.19), the moment when the subsystem risk exceeds the permitted level $\delta = 0.05$ is

$$\tau = \boldsymbol{r}^{-1}(\delta) = [-(1/0.8568)\log(1-\delta)]^{1/2} \cong 0.24 \text{ years}$$

Subsystem S_4 is composed of one belt conveyor that consists of a ribbon belt, a drum driving the belt, a reversible drum, 12 channelled rollers and three rollers supporting the belt. It means that subsystem S_4 consists of one conveyor composed of $n = 18$ components. In the conveyor there is one belt with reliability functions

$$R^{(1)}(t, 1) = \exp[-0.012t^2], \quad R^{(1)}(t, 2) = \exp[-0.022t^2], \quad R^{(1)}(t, 3)$$
$$= \exp[-0.049t^2] \quad \text{for } t \geq 0$$

two drums driving the belt with reliability functions

$$R^{(2)}(t,1)=\exp[-0.0019t^2],\ R^{(2)}(t,2)=\exp[-0.0024t^2],$$
$$R^{(2)}(t,3)=\exp[-0.0029t^2] \quad \text{for } t\geq 0$$

12 channelled rollers with reliability functions

$$R^{(3)}(t,1)=\exp[-0.028t^2],\ R^{(3)}(t,2)=\exp[-0.03t^2],$$
$$R^{(3)}(t,3)=\exp[-0.032t^2] \quad \text{for } t\geq 0$$

and three supporting rollers with reliability functions

$$R^{(4)}(t,1)=\exp[-0.0075t^2],\ R^{(4)}(t,2)=\exp[-0.01t^2],$$
$$R^{(4)}(t,3)=\exp[-0.02t^2] \quad \text{for } t\geq 0$$

Thus, according to Definition 3.17, it is a non-homogeneous multi-state series system with parameters

$$n=18,\ a=1,$$
$$q_1=1/18,\ q_2=2/18,\ q_3=12/18,\ q_4=3/18$$

and according to Eqs. (3.33) and (3.34) its exact multi-state reliability function is given by

$$\overline{\boldsymbol{R}}'_{18}(t,\cdot)=[1,\exp[-0.3743t^2],\exp[-0.4168t^2],\exp[-0.4988t^2]] \quad \text{for } t\geq 0$$

Next, applying Corollary 6.1, according to Eq. (6.3), we have

$$\alpha(1)=\min\{2,2,2,2\}=2,$$
$$\beta(1)=\frac{1}{18}0.012+\frac{2}{18}0.0019+\frac{12}{18}0.028+\frac{3}{18}0.0075=0.020794444,$$

$$\alpha(2)=\min\{2,2,2,2\}=2,$$
$$\beta(2)=\frac{1}{18}0.022+\frac{2}{18}0.0024+\frac{12}{18}0.03+\frac{3}{18}0.01=0.023155555,$$

$$\alpha(3)=\min\{2,2,2,2\}=2,$$
$$\beta(3)=\frac{1}{18}0.049+\frac{2}{18}0.0029+\frac{12}{18}0.032+\frac{3}{18}0.02=0.027711111$$

and according to Eq. (6.2)

$$a_n(1)=(0.020794444\cdot 18)^{-1/2}=1.634519443,\ b_n(1)=0,$$
$$a_n(2)=(0.023155555\cdot 18)^{-1/2}=1.548945546,\ b_n(2)=0,$$
$$a_n(3)=(0.027711111\cdot 18)^{-1/2}=1.415913682,\ b_n(3)=0$$

and we conclude that the subsystem limit reliability function is given by

$$\overline{\Re}'_2(t,\cdot) = [1, \exp[-t^2], \exp[-t^2], \exp[-t^2]] \quad \text{for } t \geq 0$$

Therefore, from Eq. (3.49), the approximate reliability function (the formula is exact in this case) takes the form

$$\begin{aligned}\overline{\boldsymbol{R}}'_{18}(t,\cdot) &\cong \overline{\boldsymbol{R}}'_2((t-b_n(u))/a_n(u)) \\ &= [1, \exp[-0.3743t^2], \exp[-0.4168t^2], \exp[-0.4988t^2]] \quad \text{for } t \geq 0\end{aligned}$$

The expected values of the subsystem lifetimes in the state subsets and their standard deviations, according to Eqs. (3.13)–(3.15), are

$$\begin{aligned}&M(1) \cong 1.45 \text{ years}, \quad M(2) \cong 1.37 \text{ years}, \quad M(3) \cong 1.25 \text{ years}, \\ &\sigma(1) \cong 0.76 \text{ years}, \quad \sigma(2) \cong 0.72 \text{ years}, \quad \sigma(3) \cong 0.66 \text{ years}\end{aligned}$$

Hence, from Eq. (3.17), the subsystem mean lifetimes in the states are

$$\overline{M}(1) \cong 0.08 \text{ years}, \quad \overline{M}(2) \cong 0.12 \text{ years}, \quad \overline{M}(3) \cong 1.25 \text{ years}$$

If a critical state is $r = 2$, then from Eq. (3.18) the subsystem risk function is given by

$$\boldsymbol{r}(t) \cong 1 - \exp[-0.4168t^2]$$

Hence, from Eq. (3.19), the moment when the subsystem risk exceeds the permitted level $\delta = 0.05$ is

$$\tau = \boldsymbol{r}^{-1}(\delta) = [-(1/04168)\log(1-\delta)]^{1/2} \cong 0.35 \text{ years}$$

Subsystem S_5 is composed of one belt conveyor, which consists of a ribbon belt, a drum driving the belt, a reversible drum, 162 channelled rollers and 53 rollers supporting the belt. It means that subsystem S_5 consists of one conveyor composed of $n = 218$ components. In the conveyor there is one belt with reliability functions

$$\begin{aligned}&R^{(1)}(t,1) = \exp[-0.012t^2], \quad R^{(1)}(t,2) = \exp[-0.022t^2], \\ &R^{(1)}(t,3) = \exp[-0.049t^2] \quad \text{for } t \geq 0\end{aligned}$$

two drums driving the belt with reliability functions

$$\begin{aligned}&R^{(2)}(t,1) = \exp[-0.0019t^2], \quad R^{(2)}(t,2) = \exp[-0.0024t^2], \\ &R^{(2)}(t,3) = \exp[-0.0029t^2] \quad \text{for } t \geq 0\end{aligned}$$

162 channelled rollers with reliability functions

$$R^{(3)}(t,1) = \exp[-0.0074t^2],\ \ R^{(3)}(t,2) = \exp[-0.012t^2],$$
$$R^{(3)}(t,3) = \exp[-0.021t^2] \quad \text{for } t \ge 0$$

and 53 supporting rollers with reliability functions

$$R^{(4)}(t,1) = \exp[-0.002t^2],\ \ R^{(4)}(t,2) = \exp[-0.0025t^2],$$
$$R^{(4)}(t,3) = \exp[-0.003t^2] \quad \text{for } t \ge 0$$

Thus, according to Definition 3.17, it is a non-homogeneous multi-state series system with parameters

$$n = 218,\ \ a = 4,$$
$$q_1 = 1/218,\ \ q_2 = 2/218,\ \ q_3 = 162/218,\ \ q_4 = 53/218$$

and according to Eqs. (3.33) and (3.34) its exact multi-state reliability function is given by

$$\overline{\boldsymbol{R}}'_{218}(t,\cdot) = [1, \exp[-1.3206t^2], \exp[-2.1033t^2], \exp[-3.6158t^2]] \quad \text{for } t \ge 0$$

Next, applying Corollary 6.1, according to Eq. (6.3), we have

$$\alpha(1) = \min\{2,2,2,2\} = 2,$$
$$\beta(1) = \frac{1}{218}0.012 + \frac{2}{218}0.0019 + \frac{162}{218}0.0074 + \frac{53}{218}0.002 = 0.006057798,$$

$$\alpha(2) = \min\{2,2,2,2\} = 2,$$
$$\beta(2) = \frac{1}{218}0.022 + \frac{2}{218}0.0024 + \frac{162}{218}0.012 + \frac{53}{218}0.0025 = 0.009648165,$$

$$\alpha(3) = \min\{2,2,2,2\} = 2,$$
$$\beta(3) = \frac{1}{218}0.049 + \frac{2}{218}0.0029 + \frac{162}{218}0.021 + \frac{53}{218}0.003 = 0.016586238$$

and according to Eq. (6.2)

$$a_n(1) = (0.006057798 \cdot 218)^{-1/2} = 0.870190543,\ \ b_n(1) = 0,$$
$$a_n(2) = (0.009648165 \cdot 218)^{-1/2} = 0.689524008,\ \ b_n(2) = 0,$$
$$a_n(3) = (0.016586238 \cdot 218)^{-1/2} = 0.525893504,\ \ b_n(3) = 0$$

and we conclude that the subsystem limit reliability function is given by

$$\mathfrak{R}'_2(t,\cdot) = [1, \exp[-t^2], \exp[-t^2], \exp[-t^2]] \quad \text{for } t \ge 0$$

Therefore, from Eq. (3.49), the approximate reliability function (the formula is exact in this case) takes the form

$$\overline{\boldsymbol{R}}'_{218}(t,\cdot) \cong \overline{\mathfrak{R}}'_2((t-b_n(u))/a_n(u))$$
$$= [1, \exp[-1.3206t^2], \exp[-2.1033t^2], \exp[-3.6158t^2]] \quad \text{for } t \geq 0$$

The expected values of the subsystem lifetimes in the state subsets and their standard deviations, according to Eqs. (3.13)–(3.15), are

$$M(1) \cong 0.77 \text{ years}, \quad M(2) \cong 0.61 \text{ years}, \quad M(2) \cong 0.47 \text{ years},$$
$$\sigma(1) \cong 0.40 \text{ years}, \quad \sigma(2) \cong 0.32 \text{ years}, \quad \sigma(3) \cong 0.24 \text{ years}$$

Hence, from Eq. (3.17), the subsystem mean lifetimes in the states are

$$\overline{M}(1) \cong 0.16 \text{ years}, \quad \overline{M}(2) \cong 0.14 \text{ years}, \quad \overline{M}(3) \cong 0.47 \text{ years}$$

If a critical state is $r = 2$, then from Eq. (3.18) the subsystem risk function is given by

$$\boldsymbol{r}(t) \cong 1 - \exp[-2.1033t^2]$$

Hence, from Eq. (3.19), the moment when the subsystem risk exceeds permitted level $\delta = 0.05$ is

$$\tau = \boldsymbol{r}^{-1}(\delta) = [-(1/2.1033)\log(1-\delta)]^{1/2} \cong 0.16 \text{ years}$$

Subsystem S_6 is a set of belt conveyors placed on the moving tracks and having the possibility of moving in relation to each other. It is composed of a rotary mechanism, two stable belt conveyors and one moving belt conveyer. One of the stable conveyors consists of one ribbon belt, a drum driving the belt, a reversible drum, 34 channelled rollers and 10 rollers supporting the belt. The second stable conveyor consists of one ribbon belt, a drum driving the belt, a reversible drum, 15 channelled rollers and five rollers supporting the belt. The moving conveyor consists of one ribbon belt, a drum driving the belt, a reversible drum, 15 channelled rollers and five rollers supporting the belt. Thus subsystem S_6 is a non-homogeneous multi-state series system composed of $n = 93$ components of four types. In the subsystem there are three belts with reliability functions

$$R^{(1)}(t,1) = \exp[-0.012t^2], \quad R^{(1)}(t,2) = \exp[-0.022t^2],$$
$$R^{(1)}(t,3) = \exp[-0.049t^2] \quad \text{for } t \geq 0$$

six drums driving the belt with reliability functions

$$R^{(2)}(t,1) = \exp[-0.0019t^2], \quad R^{(2)}(t,2) = \exp[-0.0024t^2],$$
$$R^{(2)}(t,3) = \exp[-0.0029t^2] \quad \text{for } t \geq 0$$

64 channelled rollers with reliability functions

$$R^{(3)}(t,1) = \exp[-0.0046t^2], \quad R^{(3)}(t,2) = \exp[-0.0075t^2],$$
$$R^{(3)}(t,3) = \exp[-0.012t^2] \quad \text{for } t \geq 0$$

and 20 supporting rollers with reliability functions

$$R^{(4)}(t,1) = \exp[-0.0012t^2], \quad R^{(4)}(t,2) = \exp[-0.0018t^2],$$
$$R^{(4)}(t,3) = \exp[-0.0024t^2] \quad \text{for } t \geq 0$$

Thus, according to Definition 3.17, it is a non-homogeneous multi-state series system with parameters

$$n = 93, \quad a = 4,$$
$$q_1 = 3/93, \quad q_2 = 6/93, \quad q_3 = 64/93, \quad q_4 = 20/93$$

and according to Eqs. (3.33) and (3.34) its exact multi-state reliability function is given by

$$\overline{\boldsymbol{R}}'_{93}(t,\cdot) = [1, \exp[-0.3658t^2], \exp[-0.5964t^2], \exp[-0.9804t^2]] \quad \text{for } t \geq 0$$

Next, applying Corollary 6.1, according to Eq. (6.3), we have

$$\alpha(1) = \min\{2,2,2,2\} = 2,$$
$$\beta(1) = \frac{3}{93}0.012 + \frac{6}{93}0.0019 + \frac{64}{93}0.0046 + \frac{20}{93}0.0012 = 0.003933333,$$

$$\alpha(2) = \min\{2,2,2,2\} = 2,$$
$$\beta(2) = \frac{3}{93}0.022 + \frac{6}{93}0.0024 + \frac{64}{93}0.0075 + \frac{20}{93}0.0018 = 0.006412903,$$

$$\alpha(3) = \min\{2,2,2,2\} = 2,$$
$$\beta(3) = \frac{3}{93}0.049 + \frac{6}{93}0.0029 + \frac{64}{93}0.012 + \frac{20}{93}0.0024 = 0.010541935$$

and according to Eq. (6.2)

$$a_n(1) = (0.003933333 \cdot 93)^{-1/2} = 1.6534000893, \quad b_n(1) = 0,$$
$$a_n(2) = (0.006412903 \cdot 93)^{-1/2} = 1.294884971, \quad b_n(2) = 0,$$
$$a_n(3) = (0.010541935 \cdot 93)^{-1/2} = 1.009946454, \quad b_n(3) = 0$$

and we conclude that the subsystem limit reliability function is given by

$$\mathfrak{R}'_2(t,\cdot) = [1, \exp[-t^2], \exp[-t^2], \exp[-t^2]] \quad \text{for } t \geq 0$$

Therefore, from Eq. (3.49), the approximate reliability function (the formula is exact in this case) takes the form

$$\overline{\boldsymbol{R}}'_{93}(t,\cdot) \cong \overline{\mathfrak{R}}'_2((t - b_n(u))/a_n(u))$$
$$= [1, \exp[-0.3658t^2], \exp[-0.5964t^2], \exp[-0.9804t^2]] \quad \text{for } t \geq 0$$

The expected values of the subsystem lifetimes in the state subsets and their standard deviations, according to Eqs. (3.13)–(3.15), are

$$M(1) \cong 1.47 \text{ years}, \quad M(2) \cong 1.15 \text{ years}, \quad M(3) \cong 0.90 \text{ years},$$
$$\sigma(1) \cong 0.77 \text{ years}, \quad \sigma(2) \cong 0.60 \text{ years}, \quad \sigma(3) \cong 0.47 \text{ years}$$

Hence, from Eq. (3.17), the subsystem mean lifetimes in the states are

$$\overline{M}(1) \cong 0.32 \text{ years}, \quad \overline{M}(2) \cong 0.25 \text{ years}, \quad \overline{M}(3) \cong 0.90 \text{ years}$$

If a critical state is $r = 2$, then from Eq. (3.18) the subsystem risk function is given by

$$\boldsymbol{r}(t) \cong 1 - \exp[-0.5964t^2]$$

Hence, from Eq. (3.19), the moment when the subsystem risk exceeds the permitted level $\delta = 0.05$ is

$$\tau = \boldsymbol{r}^{-1}(\delta) = [-(1/05964)\log(1-\delta)]^{1/2} \cong 0.29 \text{ years}$$

Since, according to Definition 3.17, all subsystems create a series reliability structure, then from Eqs. (3.33) and (3.34) the reliability function of the whole transportation system is given by

$$\overline{\boldsymbol{R}}'(t,\cdot) \cong [1, \overline{\boldsymbol{R}}'(t,1), \overline{\boldsymbol{R}}'(t,2), \overline{\boldsymbol{R}}'(t,3)]$$

where

$$\overline{\boldsymbol{R}}'(t,1) = 3\exp[-4.0026t^2] - 3\exp[-4.3769t^2] + \exp[-4.7512t^2],$$
$$\overline{\boldsymbol{R}}'(t,2) = 3\exp[-6.0294t^2] - 3\exp[-6.4462t^2] + \exp[-6.8630t^2],$$
$$\overline{\boldsymbol{R}}'(t,3) = 3\exp[-9.8884t^2] - 3\exp[-10.3872t^2] + \exp[-10.8860t^2]$$

Table 6.3 The Values of the Multi-State Reliability Function and the Risk Function of the Port Bulk Transportation System

t	$\overline{R}'(t,1)$	$\overline{R}'(t,2)$	$\overline{R}'(t,3)$	$r(t)$
0.00	1.00000	1.00000	1.00000	0.00000
0.10	0.96437	0.94542	0.91038	0.05458
0.20	0.86490	0.79891	0.68683	0.20109
0.30	0.72138	0.60339	0.42949	0.39661
0.40	0.55949	0.40727	0.22251	0.59273
0.50	0.40342	0.24558	0.09546	0.75442
0.60	0.27031	0.13223	0.03389	0.86777
0.70	0.16820	0.06351	0.00994	0.93649
0.80	0.09712	0.02719	0.00241	0.97281
0.90	0.05198	0.01036	0.00048	0.98964
1.00	0.02575	0.00351	0.00008	0.99649
1.10	0.01180	0.00105	0.00001	0.99895
1.20	0.00499	0.00028	0.00000	0.99972
1.30	0.00195	0.00007	0.00000	0.99993
1.40	0.00070	0.00001	0.00000	0.99999
1.50	0.00023	0.00000	0.00000	1.00000

The expected values of the subsystem lifetimes in the state subsets and their standard deviations, according to Eqs. (3.13)–(3.15), are

$$M(1) \cong 0.45 \text{ years}, \quad M(2) \cong 0.37 \text{ years}, \quad M(3) \cong 0.29 \text{ years},$$
$$\sigma(1) \cong 0.27 \text{ years}, \quad \sigma(2) \cong 0.20 \text{ years}, \quad \sigma(3) \cong 0.15 \text{ years}$$

Hence, from Eq. (3.17), the subsystem mean lifetimes in the states are

$$\overline{M}(1) \cong 0.08 \text{ years}, \quad \overline{M}(2) \cong 0.08 \text{ years}, \quad \overline{M}(3) \cong 0.29 \text{ years}$$

If a critical state is $r = 2$, then from Eq. (3.18) the subsystem risk function is given by

$$\boldsymbol{r}(t) \cong 1 - \overline{\boldsymbol{R}}'(t, 2)$$

Hence, by Eq. (3.19), the moment when the subsystem risk exceeds the permitted level $\delta = 0.05$ is

$$\tau = \boldsymbol{r}^{-1}(\delta) \cong 0.10 \text{ years}$$

The system exact and approximate reliability functions are identical. Their behaviour and the system risk function are presented in Table 6.3 and Figure 6.6.

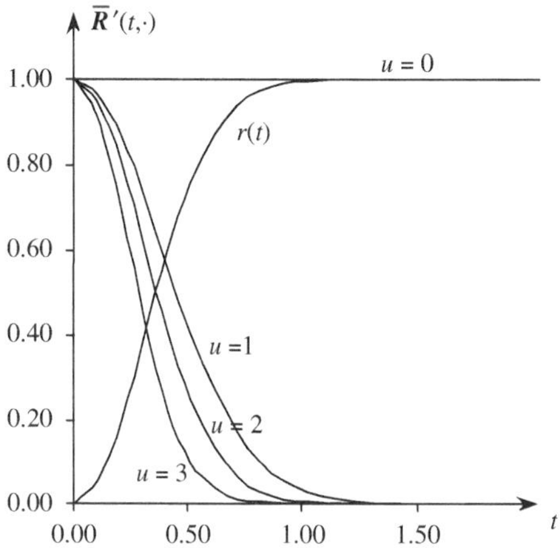

Figure 6.6 Graphs of the multi-state reliability function and the risk function of the port bulk cargo transportation system.

6.5 Reliability of a Shipyard Rope Transportation System

Ship-rope elevators are used to dock and undock ships coming to shipyards for repairs. The elevator utilized in the Naval Shipyard in Gdynia, with the scheme presented in Figure 6.7, is composed of a steel platform carriage placed in its syncline (hutch). The platform is moved vertically with 10 rope-hoisting winches fed by separate electric motors, each rope having a maximum load of 300 tonnes. During ship docking the platform, with the ship settled in special supporting carriages on the platform, is raised to the wharf level (upper position). During undocking, the operation is reversed. While the ship is moving into or out of the syncline, and while stopped in the upper position, the platform is held on hooks and the loads in the ropes are relieved. Since the platform carriage and electric motors are highly reliable in comparison to the ropes, which work in extremely aggressive conditions (salt water, wind, pollution and so on), in our further analysis we will discuss the reliability of the rope system only.

The system under consideration is in order if all its ropes do not fail. Thus we may assume that it is a series system composed of 10 components. Each of the ropes is composed of 22 strands: 10 outer and 12 inner. The outer strands of ropes are composed of 26 steel wires. Their inner strands are composed of 19 steel wires and they form the rope's steel core, which is covered by a plastic layer. The cross section of the rope is shown in Figure 6.8.

Thus, considering the strands as basic components of the system, according to Definitions 3.14–3.16 we conclude that the rope elevator is a parallel–series system composed of $k_n = 10$ series-linked subsystems (ropes) with $l_n = 22$ parallel-linked components (strands).

According to safety standards and to the requirements for approximately comparable reliability (durability of outer and inner strands, after considering technical

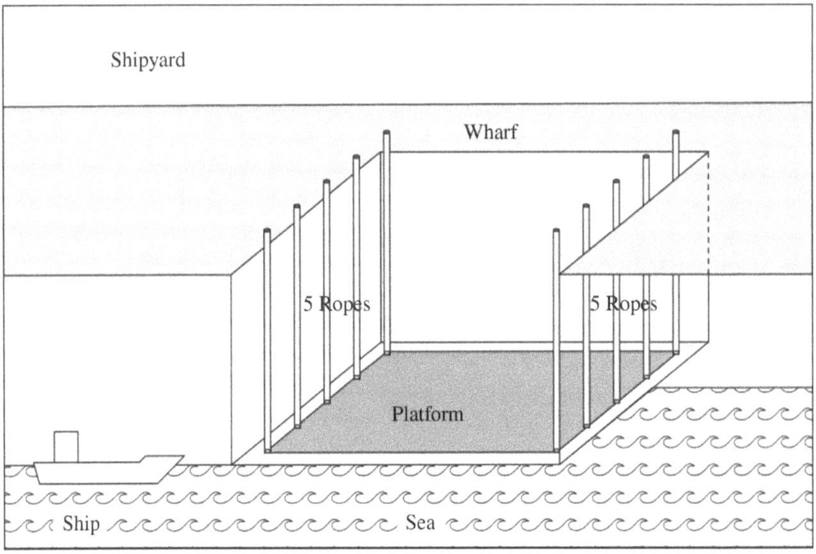

Figure 6.7 The scheme of the ship-rope transportation system.

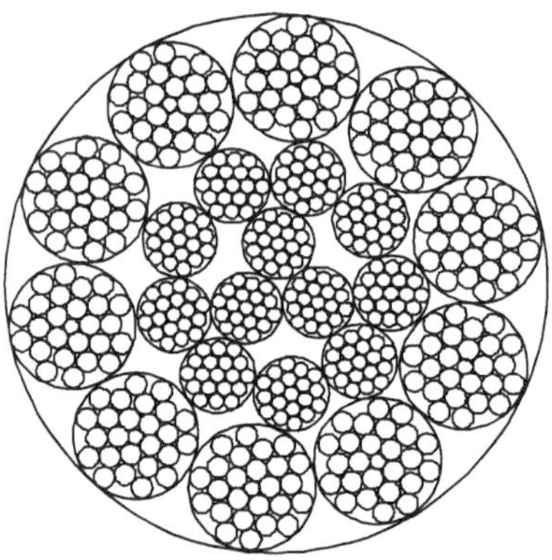

Figure 6.8 The cross section of the rope.

norms [24,145] and expert opinion [117], the following reliability states of all strands have been distinguished:

- state 3: a strand is new, without any defects,
- state 2: the number of broken wires in the strand is >0% and <25% of all its wires, or corrosion of wires is >0% and <25%,
- state 1: the number of broken wires in the strand is ≥25% and <50% of all its wires, or corrosion of wires is ≥25% and <50%,
- state 0: otherwise (a strand is failed).

Considering these component reliability states, according to Definitions 3.14–3.16, we conclude that the rope system is a homogeneous regular four-state

Table 6.4 The Rope Elevator Loading State Characteristics

i	n_i	$t_i(h)$	$T_i(h)$
4	20	8	160
3	80	6	480
2	40	3	120
1	20	2	40
0	–	–	7960
Total	160	–	8760

parallel–series system. Thus, from Eqs. (3.30) and (3.31), its reliability function is defined by

$$\overline{\boldsymbol{R}}_{10,22}(t,\cdot) = [1, \overline{\boldsymbol{R}}_{10,22}(t,1), \overline{\boldsymbol{R}}_{10,22}(t,2), \overline{\boldsymbol{R}}_{10,22}(t,3)]$$

where

$$\overline{\boldsymbol{R}}_{10,22}(t,u) = [1-[F(t,u)]^{22}]^{10}, \quad t \in (-\infty, \infty), \quad u = 1,2,3 \tag{6.7}$$

According to experts' opinions [79] the mean numbers of ship docking and undocking are equal to 80 per year. It means that the elevator is active 160 times per year. Moreover, the rope elevator system reliability depends strongly on the tonnage of the docking ships and therefore the following states of the elevator system loading have been distinguished:

- state 4: loading from 2250 to 1750 tonnes,
- state 3: loading from 1750 to 1250 tonnes,
- state 2: loading from 1250 to 750 tonnes,
- state 1: loading from 750 to 250 tonnes,
- state 0: without loading.

The loading states i, the frequencies n_i of loading states per year, the duration times n_i of the loading states and the total duration times T_i of the loading states per year of the rope elevator system are given in Table 6.4.

Using these data and the formula

$$p_i = \frac{T_i}{8760}, \quad i = 0, 1, \ldots, 4$$

it is possible to evaluate the probabilities of the elevator loading states. They are as follows:

$$p_0 = 0.9087, \quad p_1 = 0.0046, \quad p_2 = 0.0137, \quad p_3 = 0.0548, \quad p_4 = 0.0182 \tag{6.8}$$

Considering these particular elevator loading states, we obtain the following formula for its reliability function

$$\overline{\boldsymbol{R}}'_{10,22}(t,\cdot) = [1, \overline{\boldsymbol{R}}'_{10,22}(t,1), \overline{\boldsymbol{R}}'_{10,22}(t,2), \overline{\boldsymbol{R}}'_{10,22}(t,3)] \tag{6.9}$$

where

$$\overline{\boldsymbol{R}}'_{10,22}(t,u) = p_0\overline{\boldsymbol{R}}^{(0)}_{10,22}(t,u) + p_1\overline{\boldsymbol{R}}^{(1)}_{10,22}(t,u) + p_2\overline{\boldsymbol{R}}^{(2)}_{10,22}(t,u) + p_3\overline{\boldsymbol{R}}^{(3)}_{10,22}(t,u) + p_4\overline{\boldsymbol{R}}^{(4)}_{10,22}(t,u), t\in(-\infty,\infty), \quad u=1,2,3 \tag{6.10}$$

and according to Eq. (6.7)

$$\overline{\boldsymbol{R}}^{(i)}_{10,22}(t,u) = [1-[F^{(i)}(t,u)]^{22}]^{10}, \quad i=0,1,\ldots,4 \tag{6.11}$$

are its multi-state reliability functions in particular loading states, while

$$R^{(i)}(t,u) = 1 - F^{(i)}(t,u), \quad t\in(-\infty,\infty), \quad u=1,2,3, \quad i=0,1,\ldots,4$$

are component (strand) multi-state reliability functions in particular elevator loading states.

According to rope reliability data given in their technical certificates [24] and experts' opinions [117] based on the nature of strand failures, it has been assumed that the strands have multi-state Weibull reliability functions

$$R^{(i)}(t,u) = 1 \quad \text{for } t<0, \quad R^{(i)}(t,u) = \exp[-\beta_i(u)t^{\alpha_i(u)}] \quad \text{for } t\geq 0, \quad u=1,2,3, \quad i=0,1,\ldots,4$$

with the following parameters

$$\alpha_i(u),\ \beta_i(u), \quad i=0,1,\ldots,4$$

in the particular elevator loading states:

$$\begin{aligned}
&\alpha_0(u)=3, \ \beta_0(u)=0.003u,\\
&\alpha_1(u)=3, \ \beta_1(u)=0.006u,\\
&\alpha_2(u)=3, \ \beta_2(u)=0.008u,\\
&\alpha_3(u)=2, \ \beta_3(u)=0.090u,\\
&\alpha_4(u)=1.5, \ \beta_4(u)=0.250u, \quad u=1,2,3
\end{aligned}$$

Hence, after considering Eqs. (6.8), (6.10) and (6.11), the exact elevator reliability function is given by formula (6.9) with

$$\begin{aligned}
\overline{\boldsymbol{R}}'_{10,22}(t,u) = {} & 0.9087[1-[1-\exp[-0.003ut^3]]^{22}]^{10}\\
& + 0.0046[1-[1-\exp[-0.006ut^3]]^{22}]^{10}\\
& + 0.0137[1-[1-\exp[-0.008ut^3]]^{22}]^{10}\\
& + 0.0548[1-[1-\exp[-0.09ut^2]]^{22}]^{10}\\
& + 0.0182[1-[1-\exp[-0.25ut^{3/2}]]^{22}]^{10}, \quad u=1,2,3
\end{aligned} \tag{6.12}$$

Since the number of parallel subsystems in the system is $k_n = 10$ and the number of components in each subsystem is $l_n = 22$, then considering that

$$l_n = 22 \gg \log k_n = \log 10 \cong 2.3$$

it seems reasonable to apply, in finding the limit reliability function of the rope elevator system, either Corollary 6.2 or Corollary 5.11. We will use both of them separately, first at different loading states for estimating the system reliability functions given by Eq. (6.11), and approximating the reliability function of the system by combining the results according to Eq. (6.12). Next, we will compare and comment on the achieved results.

6.5.1 Application of Corollary 6.2

According to Eq. (6.6), assuming normalizing constants

$$b_n^{(i)}(u) = \left[\frac{1}{\beta_i(u)}\log\frac{l_n}{\log k_n}\right]^{1/\alpha_i(u)}$$

$$a_n^{(i)}(u) = (b_n^{(i)}(u))^{1-\alpha_i(u)}/(\alpha_i(u)\beta_i(u)\log k_n), \quad u = 1,2,3, \quad i = 0,1,\ldots,4 \qquad (6.13)$$

we conclude that

$$\overline{\Re}_3^{(i)}(t,\cdot) = [1, \overline{\Re}_3^{(i)}(t,1), \overline{\Re}_3^{(i)}(t,2), \overline{\Re}_3^{(i)}(t,3)], \quad t \in (-\infty, \infty)$$

where

$$\overline{\Re}_3^{(i)}(t,u) = \exp[-\exp[t]] \quad \text{for } t \in (-\infty, \infty), \quad u = 1,2,3, \quad i = 0,1,\ldots,4$$

is the multi-state reliability function of the rope elevator system in the ith loading state.

According to Eqs. (6.13) and (3.49) for particular elevator loading states we have the following:

- loading state 0

$$b_n^{(0)}(1) = 9.0950, \quad a_n^{(0)}(1) = 0.5834,$$
$$b_n^{(0)}(2) = 7.2187, \quad a_n^{(0)}(2) = 0.4630,$$
$$b_n^{(0)}(3) = 6.3061, \quad a_n^{(0)}(3) = 0.4045$$

$$\overline{\boldsymbol{R}}_{10,22}^{(0)}(t,1) \cong \exp[-\exp[1.7141t - 15.5909]],$$
$$\overline{\boldsymbol{R}}_{10,22}^{(0)}(t,2) \cong \exp[-\exp[2.1598t - 15.5909]],$$
$$\overline{\boldsymbol{R}}_{10,22}^{(0)}(t,3) \cong \exp[-\exp[2.4722t - 15.5909]]$$

- loading state 1

$$b_n^{(1)}(1) = 7.2187, \quad a_n^{(1)}(1) = 0.4630,$$
$$b_n^{(1)}(2) = 5.7295, \quad a_n^{(1)}(2) = 0.3675,$$
$$b_n^{(1)}(3) = 5.0052, \quad a_n^{(1)}(3) = 0.3210$$

$$\overline{\boldsymbol{R}}_{10,22}^{(1)}(t,1) \cong \exp[-\exp[2.1598t - 15.5909]],$$
$$\overline{\boldsymbol{R}}_{10,22}^{(1)}(t,2) \cong \exp[-\exp[2.7211t - 15.5909]],$$
$$\overline{\boldsymbol{R}}_{10,22}^{(1)}(t,3) \cong \exp[-\exp[3.1153t - 15.5909]]$$

- loading state 2

$$b_n^{(2)}(1) = 6.5587, \quad a_n^{(2)}(1) = 0.4207,$$
$$b_n^{(2)}(2) = 5.2056, \quad a_n^{(2)}(2) = 0.3339,$$
$$b_n^{(2)}(3) = 4.5475, \quad a_n^{(2)}(3) = 0.2917$$

$$\overline{\boldsymbol{R}}_{10,22}^{(2)}(t,1) \cong \exp[-\exp[2.3771t - 15.5909]],$$
$$\overline{\boldsymbol{R}}_{10,22}^{(2)}(t,2) \cong \exp[-\exp[2.9950t - 15.5909]],$$
$$\overline{\boldsymbol{R}}_{10,22}^{(2)}(t,3) \cong \exp[-\exp[3.4282t - 15.5909]]$$

- loading state 3

$$b_n^{(3)}(1) = 5.0078, \quad a_n^{(3)}(1) = 0.4818,$$
$$b_n^{(3)}(2) = 3.5410, \quad a_n^{(3)}(2) = 0.3407,$$
$$b_n^{(3)}(3) = 2.8912, \quad a_n^{(3)}(3) = 0.2782$$

$$\overline{\boldsymbol{R}}_{10,22}^{(3)}(t,1) \cong \exp[-\exp[2.0756t - 10.3939]],$$
$$\overline{\boldsymbol{R}}_{10,22}^{(3)}(t,2) \cong \exp[-\exp[2.9351t - 10.3939]],$$
$$\overline{\boldsymbol{R}}_{10,22}^{(3)}(t,3) \cong \exp[-\exp[3.5945t - 10.3939]]$$

- loading state 4

$$b_n^{(4)}(1) = 4.3357, \quad a_n^{(4)}(1) = 0.5562,$$
$$b_n^{(4)}(2) = 2.7313, \quad a_n^{(4)}(2) = 0.3504,$$
$$b_n^{(4)}(3) = 2.0844, \quad a_n^{(4)}(3) = 0.2674$$

$$\overline{\boldsymbol{R}}_{10,22}^{(4)}(t,1) \cong \exp[-\exp[1.7979t - 7.7954]],$$
$$\overline{\boldsymbol{R}}_{10,22}^{(4)}(t,2) \cong \exp[-\exp[2.8539t - 7.7954]],$$
$$\overline{\boldsymbol{R}}_{10,22}^{(4)}(t,3) \cong \exp[-\exp[3.7397t - 7.7954]]$$

Combining the achieved results of reliability approximation, according to Eqs. (6.9) and (6.10), we get the approximate reliability function of the rope elevator system given by

$$\overline{\boldsymbol{R}}'_{10,22}(t,\cdot) = [1, \overline{\boldsymbol{R}}'_{10,22}(t,1), \overline{\boldsymbol{R}}'_{10,22}(t,2), \overline{\boldsymbol{R}}'_{10,22}(t,3)], \quad t \in (-\infty, \infty) \tag{6.14}$$

where

$$\begin{aligned}\overline{\boldsymbol{R}}'_{10,22}(t,1) \cong\ & 0.9087 \exp[-\exp[1.7141t - 15.5909]] \\ & + 0.0046 \exp[-\exp[2.1598t - 15.5909]] \\ & + 0.0137 \exp[-\exp[2.3771t - 15.5909]] \\ & + 0.0548 \exp[-\exp[2.0756t - 10.3939]] \\ & + 0.0182 \exp[-\exp[1.7979t - 7.7954]]\end{aligned}$$

$$\begin{aligned}\overline{\boldsymbol{R}}'_{10,22}(t,2) \cong\ & 0.9087 \exp[-\exp[2.1598t - 15.5909]] \\ & + 0.0046 \exp[-\exp[2.7211t - 15.5909]] \\ & + 0.0137 \exp[-\exp[2.9950t - 15.5909]] \\ & + 0.0548 \exp[-\exp[2.9351t - 10.3939]] \\ & + 0.0182 \exp[-\exp[2.8539t - 7.7954]]\end{aligned}$$

$$\begin{aligned}\overline{\boldsymbol{R}}'_{10,22}(t,3) \cong\ & 0.9087 \exp[-\exp[2.4722t - 15.5909]] \\ & + 0.0046 \exp[-\exp[3.1153t - 15.5909]] \\ & + 0.0137 \exp[-\exp[3.4282t - 15.5909]] \\ & + 0.0548 \exp[-\exp[3.5945t - 10.3939]] \\ & + 0.0182 \exp[-\exp[3.7397t - 7.7954]]\end{aligned}$$

The approximate expected values of the system lifetimes $T(u)$ in the state subsets, according to Eq. (3.13), are [80]

$$M(1) \cong 8.40 \text{ years}, \quad M(2) \cong 6.70 \text{ years}, \quad M(3) \cong 6.30 \text{ years}$$

Hence, according to Eq. (3.17), the mean lifetimes of the system in the states are

$$\overline{M}(1) \cong 1.70 \text{ years}, \quad \overline{M}(2) \cong 0.40 \text{ years}, \quad \overline{M}(3) \cong 6.30 \text{ years}$$

If a critical state is $r = 2$, then a system risk, by Eq. (3.18), is given by

$$\begin{aligned}\boldsymbol{r}(t) = 1 - \overline{\boldsymbol{R}}'_{10,22}(t,2) \cong\ & 1 - 0.9087 \exp[-\exp[2.1598t - 15.5909]] \\ & - 0.0046 \exp[-\exp[2.7211t - 15.5909]] \\ & - 0.0137 \exp[-\exp[2.9950t - 15.5909]] \\ & - 0.0548 \exp[-\exp[2.9351t - 10.3939]] \\ & - 0.0182 \exp[-\exp[2.8539t - 7.7954]]\end{aligned} \tag{6.15}$$

The time when the risk exceeds a permitted level $\delta = 0.05$, according to Eq. (3.19), is $\tau = \boldsymbol{r}^{-1}(\delta) \cong 3.50$ years.

6.5.2 Application of Corollary 5.11

According to Eq. (5.75), letting

$$b_n^{(i)}(u) = \left[\frac{1}{\beta_i(u)} \log l_n\right]^{1/\alpha_i(u)}$$

$$a_n^{(i)}(u) = (b_n^{(i)}(u))^{1-\alpha_i(u)}/(\alpha_i(u)\beta_i(u)), \quad u = 1, 2, 3, \quad i = 0, 1, \ldots, 4 \tag{6.16}$$

we conclude that

$$\overline{\mathfrak{R}}_{10}^{(i)}(t, \cdot) = [1, \overline{\mathfrak{R}}_{10}^{(i)}(t, 1), \overline{\mathfrak{R}}_{10}^{(i)}(t, 2), \overline{\mathfrak{R}}_{10}^{(i)}(t, 3)], \quad t \in (-\infty, \infty)$$

where

$$\overline{\mathfrak{R}}_{10}^{(i)}(t, u) = [1 - \exp[-\exp[-t]]^k \quad \text{for } t \in (-\infty, \infty), \; u = 1, 2, 3, \; i = 0, 1, \ldots, 4$$

is the multi-state limit reliability function of the rope elevator system in the ith loading state.

Hence, considering Eqs. (6.16) and (3.4), for the particular states we have the following:

- loading state 0

$$b_n^{(0)}(1) = 10.1002, \quad a_n^{(0)}(1) = 1.0892,$$
$$b_n^{(0)}(2) = 8.0165, \quad a_n^{(0)}(2) = 0.8645,$$
$$b_n^{(0)}(3) = 7.0031, \quad a_n^{(0)}(3) = 0.7552$$

$$\overline{\boldsymbol{R}}_{10,22}^{(0)}(t, 1) \cong [1 - \exp[-\exp[-0.9181t + 9.2731]]]^{10},$$
$$\overline{\boldsymbol{R}}_{10,22}^{(0)}(t, 2) \cong [1 - \exp[-\exp[-1.1568t + 9.2731]]]^{10},$$
$$\overline{\boldsymbol{R}}_{10,22}^{(0)}(t, 3) \cong [1 - \exp[-\exp[-1.3242t + 9.2731]]]^{10}$$

- loading state 1

$$b_n^{(1)}(1) = 8.01650, \quad a_n^{(1)}(1) = 0.8645,$$
$$b_n^{(1)}(2) = 6.3627, \quad a_n^{(1)}(2) = 0.6861,$$
$$b_n^{(1)}(3) = 5.5583, \quad a_n^{(1)}(3) = 0.5994$$

$$\overline{R}^{(1)}_{10,22}(t,1) \cong [1-\exp[-\exp[-1.1568t+9.2731]]]^{10},$$
$$\overline{R}^{(1)}_{10,22}(t,2) \cong [1-\exp[-\exp[-1.4574t+9.2731]]]^{10},$$
$$\overline{R}^{(1)}_{10,22}(t,1) \cong [1-\exp[-\exp[-1.6683t+9.2731]]]^{10}$$

- loading state 2

$$b^{(2)}_n(1) = 7.2835, \quad a^{(2)}_n(1) = 0.7854,$$
$$b^{(2)}_n(2) = 5.7809, \quad a^{(2)}_n(2) = 0.6234,$$
$$b^{(2)}_n(3) = 5.0501, \quad a^{(2)}_n(3) = 0.5446$$

$$\overline{R}^{(2)}_{10,22}(t,1) \cong [1-\exp[-\exp[-1.2732t+9.2731]]]^{10},$$
$$\overline{R}^{(1)}_{10,22}(t,1) \cong [1-\exp[-\exp[-1.6041t+9.2731]]]^{10},$$
$$\overline{R}^{(1)}_{10,22}(t,1) \cong [1-\exp[-\exp[-1.8362t+9.2731]]]^{10}$$

- loading state 3

$$b^{(3)}_n(1) = 5.8605, \quad a^{(3)}_n(1) = 0.9480,$$
$$b^{(3)}_n(2) = 4.1440, \quad a^{(3)}_n(2) = 0.6703,$$
$$b^{(3)}_n(3) = 3.3835, \quad a^{(3)}_n(3) = 0.5473$$

$$\overline{R}^{(3)}_{10,22}(t,1) \cong [1-\exp[-\exp[-1.0549t+6.1821]]]^{10},$$
$$\overline{R}^{(3)}_{10,22}(t,2) \cong [1-\exp[-\exp[-1.4918t+6.1821]]]^{10},$$
$$\overline{R}^{(3)}_{10,22}(t,3) \cong [1-\exp[-\exp[-1.8271t+6.1821]]]^{10}$$

- loading state 4

$$b^{(4)}_n(1) = 5.3470, \quad a^{(4)}_n(1) = 1.1532,$$
$$b^{(4)}_n(2) = 3.3684, \quad a^{(4)}_n(2) = 0.7265,$$
$$b^{(4)}_n(3) = 2.5706, \quad a^{(4)}_n(3) = 0.5544$$

$$\overline{R}^{(4)}_{10,22}(t,1) \cong [1-\exp[-\exp[-0.8671t+4.6366]]]^{10},$$
$$\overline{R}^{(4)}_{10,22}(t,2) \cong [1-\exp[-\exp[-1.3765t+4.6366]]]^{10},$$
$$\overline{R}^{(4)}_{10,22}(t,3) \cong [1-\exp[-\exp[-1.8037t+4.6366]]]^{10}$$

Combining the achieved results of reliability approximation, from Eqs. (6.9) and (6.10), we get the approximate reliability function of the rope elevator system given by

$$\overline{R}'_{10,22}(t,\cdot) = [1, \overline{R}'_{10,22}(t,1), \overline{R}'_{10,22}(t,2), \overline{R}'_{10,22}(t,3)] \tag{6.17}$$

where

$$\overline{\boldsymbol{R}}'_{10,22}(t,1) \cong 0.9087[1-\exp[-\exp[-0.9181t+9.2731]]]^{10}$$
$$+0.0046[1-\exp[-\exp[-1.1568t+9.2731]]]^{10}$$
$$+0.0137[1-\exp[-\exp[-1.2732t+9.2731]]]^{10}$$
$$+0.0548[1-\exp[-\exp[-1.0549t+6.1821]]]^{10}$$
$$+0.0182[1-\exp[-\exp[-0.8671t+4.6366]]]^{10}$$

$$\overline{\boldsymbol{R}}'_{10,22}(t,2) \cong 0.9087[1-\exp[-\exp[-1.1568t+9.2731]]]^{10}$$
$$+0.0046[1-\exp[-\exp[-1.4574t+9.2731]]]^{10}$$
$$+0.0137[1-\exp[-\exp[-1.6041t+9.2731]]]^{10}$$
$$+0.0548[1-\exp[-\exp[-1.4918t+6.1821]]]^{10}$$
$$+0.0182[1-\exp[-\exp[-1.3765t+4.6366]]]^{10}$$

$$\overline{\boldsymbol{R}}'_{10,22}(t,3) \cong 0.9087[1-\exp[-\exp[-1.3242t+9.2731]]]^{10}$$
$$+0.0046[1-\exp[-\exp[-1.6683t+9.2731]]]^{10}$$
$$+0.0137[1-\exp[-\exp[-1.8362t+9.2731]]]^{10}$$
$$+0.0548[1-\exp[-\exp[-1.8271t+6.1821]]]^{10}$$
$$+0.0182[1-\exp[-\exp[-1.8037t+4.6366]]]^{10}$$

The approximate expected values of the system lifetimes $T(u)$ in the state subsets, according to Eq. (3.13), are [80]

$$M(1) \cong 8.70 \text{ years}, \quad M(2) \cong 6.90 \text{ years}, \quad M(3) \cong 6.40 \text{ years}$$

Hence, according to Eq. (3.17), the mean lifetimes of the system in the states are

$$\overline{M}(1) \cong 1.80 \text{ years}, \quad \overline{M}(2) \cong 0.50 \text{ years}, \quad \overline{M}(3) \cong 6.40 \text{ years}$$

If a critical state is $r = 2$, then a system risk, according to Eq. (3.18), is given by

$$\begin{aligned} \boldsymbol{r}(t) \cong 1-\overline{\boldsymbol{R}}'_{10,22}(t,2) \cong 1 &- 0.9087[1-\exp[-\exp[-1.1568t+9.2731]]]^{10} \\ &- 0.0046[1-\exp[-\exp[-1.4574t+9.2731]]]^{10} \\ &- 0.0137[1-\exp[-\exp[-1.6041t+9.2731]]]^{10} \\ &- 0.0548[1-\exp[-\exp[-1.4918t+6.1821]]]^{10} \\ &- 0.0182[1-\exp[-\exp[-1.3765t+4.6366]]]^{10} \end{aligned} \tag{6.18}$$

The time when the risk exceeds a permitted level $\delta = 0.05$, according to Eq. (3.19), is $\tau = \boldsymbol{r}^{-1}(\delta) \cong 3.60$ years.

6.5.3 Results Comparison

The behaviour of the rope elevator's exact reliability function given by Eq. (6.12) and its approximate reliability functions given by Eqs. (6.14) and (6.17) in the state subsets is shown in Tables 6.5–6.7 and Figures 6.9–6.11. The behaviour of the approximate risk functions determined by Eqs. (6.15) and (6.18) is given in Table 6.8 and illustrated in Figure 6.12. Moreover, in Tables 6.5–6.7 there are differences between exact and approximate system reliability functions in the state subsets marked by $\Delta_1(u)$ in the case where approximate formula (6.14) applies and by $\Delta_2(u)$ where approximate formula (6.17) applies. From data presented in the tables and the graphs it can be noticed that the evaluation of the elevator system reliability given by Eq. (6.14) is pessimistic and that the evaluation given by Eq. (6.17) is optimistic. Therefore, in fact, the elevator reliability characteristics are better than those calculated from Eq. (6.14) and worse than those calculated from Eq. (6.17). From these it is possible to find the following upper and lower bounds for the elevator mean sojourn times $T(u)$ in the state subsets [80]:

$$8.40 < M(1) < 8.70\,\text{years},\ 6.70 < M(2) < 6.90\,\text{years},\ 6.30 < M(3) < 6.40\,\text{years}$$

and in particular states as well:

$$1.70 < \overline{M}(1) < 1.80\,\text{years},\ 0.40 < \overline{M}(2) < 0.50\,\text{years},\ 6.30 < \overline{M}(3) < 6.40\,\text{years}$$

The moment when the risk exceeds a permitted level $\delta = 0.05$ falls in the interval $3.50 < \tau < 3.60$ years.

Table 6.5 The Values of the Rope Elevator Exact and Approximate Reliability Functions in the State Subset $u \geq 1$

t	$\overline{R}'_{10,22}(t, 1)$				
	Formula (6.12)	**Formula (6.14)**	**Formula (6.17)**	$\Delta_1(1)$	$\Delta_2(1)$
0.00	1.00000	0.99999	1.00000	+0.00001	0.00000
1.00	1.00000	0.99994	1.00000	+0.00006	0.00000
2.00	1.00000	0.99962	1.00000	+0.00038	0.00000
3.00	0.99983	0.99755	0.99991	+0.00228	−0.00008
4.00	0.99239	0.98579	0.99344	+0.00660	−0.00105
5.00	0.95025	0.94697	0.95073	+0.00328	−0.00048
6.00	0.92547	0.91902	0.92644	+0.00645	−0.00097
7.00	0.91191	0.88727	0.91268	+0.02464	−0.00077
8.00	0.86579	0.77990	0.89943	**+0.08589**	−0.03364
9.00	0.42655	0.38888	0.46813	+0.03767	**−0.04158**
10.00	0.01781	0.00817	0.01559	+0.00964	+0.00222
11.00	0.00002	0.00000	0.00003	+0.00002	−0.00001
12.00	0.00000	0.00000	0.00000	0.00000	0.00000

Table 6.6 The Values of the Rope Elevator Exact and Approximate Reliability Functions in the State Subset $u \geq 2$

t	$\overline{R}'_{10,22}(t, 2)$				
	Formula (6.12)	**Formula (6.14)**	**Formula (6.17)**	$\Delta_1(2)$	$\Delta_2(2)$
0.00	1.00000	0.99999	1.00000	+0.00001	0.00000
1.00	1.00000	0.99984	1.00000	+0.00016	0.00000
2.00	0.99961	0.99727	0.99975	+0.00234	−0.00014
3.00	0.98009	0.97367	0.98184	+0.00642	−0.00175
4.00	0.92903	0.92690	0.92880	+0.00213	+0.00023
5.00	0.92215	0.91318	0.92335	+0.00897	−0.00120
6.00	0.90136	0.84621	0.90902	+0.05515	−0.00766
7.00	0.54698	0.48709	0.60936	**+0.05989**	**−0.06238**
8.00	0.01183	0.00408	0.01031	+0.00775	+0.00152
9.00	0.00000	0.00000	0.00000	0.00000	0.00000
10.00	0.00000	0.00000	0.00000	0.00000	0.00000
11.00	0.00000	0.00000	0.00000	0.00000	0.00000
12.00	0.00000	0.00000	0.00000	0.00000	0.00000

Table 6.7 The Values of the Rope Elevator Exact and Approximate Reliability Functions in the State $u = 3$

t	$\overline{R}'_{10,22}(t, 3)$				
	Formula (6.12)	**Formula (6.14)**	**Formula (6.17)**	$\Delta_1(3)$	$\Delta_2(3)$
0.00	1.00000	0.99999	1.00000	+0.00001	0.00000
1.00	1.00000	0.99962	1.00000	+0.00038	0.00000
2.00	0.99152	0.98838	0.99151	+0.00314	0.00001
3.00	0.94035	0.93919	0.94011	+0.00116	+0.00024
4.00	0.92634	0.92183	0.92686	+0.00451	−0.00052
5.00	0.90922	0.87529	0.91095	+0.03393	−0.00173
6.00	0.64678	0.56865	0.72020	**+0.07813**	**−0.07342**
7.00	0.01084	0.00352	0.00946	+0.00732	+0.00138
8.00	0.00000	0.00000	0.00000	0.00000	0.00000
9.00	0.00000	0.00000	0.00000	0.00000	0.00000
10.00	0.00000	0.00000	0.00000	0.00000	0.00000
11.00	0.00000	0.00000	0.00000	0.00000	0.00000
12.00	0.00000	0.00000	0.00000	0.00000	0.00000

The values calculated from Eqs. (6.12), (6.14) and (6.17) and the graphs illustrated in the figures allow us to state that the differences between exact and approximate reliability functions of the rope elevator are insignificant in practice. Thus replacing the exact reliability function with one of asymptotic form does not lead to large mistakes in the evaluation of system reliability characteristics.

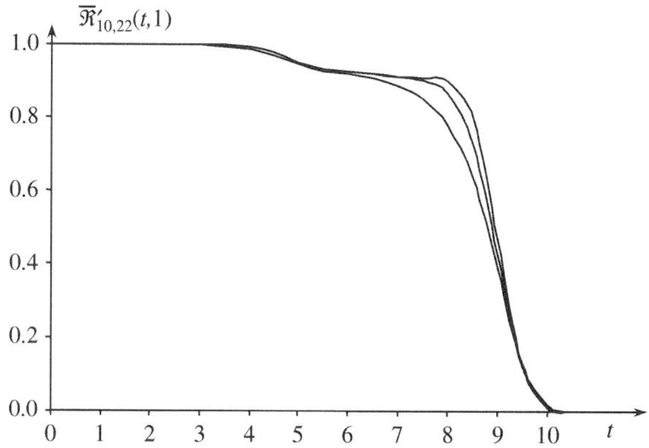

Figure 6.9 Graph of the rope elevator exact and approximate reliability functions in the state subset $u \geq 1$.

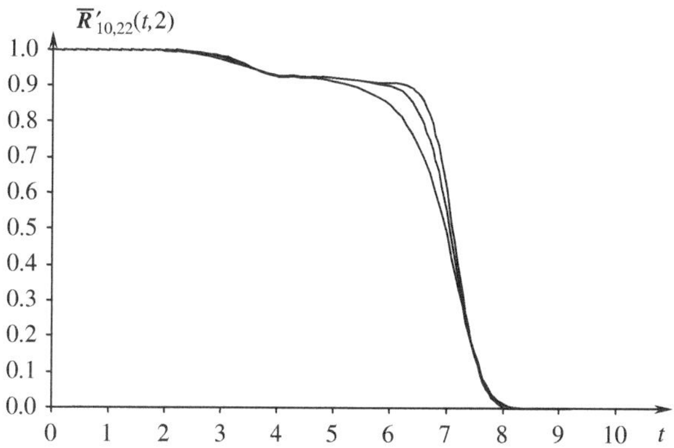

Figure 6.10 Graphs of the rope elevator exact and approximate reliability functions in the state subset $u \geq 2$.

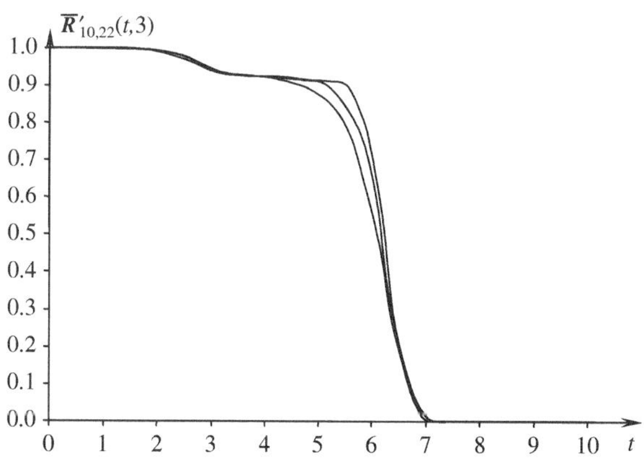

Figure 6.11 Graphs of the rope elevator exact and approximate reliability functions in the state $u = 3$.

Table 6.8. The Approximate Values of the Rope Elevator Risk Functions

t	$r(t)$	
	Formula (6.15)	**Formula (6.18)**
0.00	0.00001	0.00000
1.00	0.00016	0.00000
2.00	0.00273	0.00025
3.00	0.02633	0.01816
4.00	0.07310	0.07120
5.00	0.08682	0.07665
6.00	0.15379	0.09098
7.00	0.51291	0.39064
8.00	0.99592	0.98969
9.00	1.00000	1.00000
10.00	1.00000	1.00000

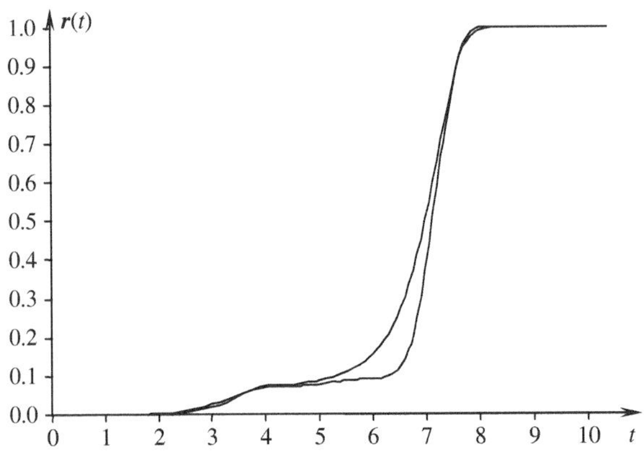

Figure 6.12 Graphs of the approximate rope elevator risk functions.

The accuracy of the rope elevator reliability evaluation testifies that the asymptotic approach to system reliability evaluation is sensible even in cases where the numbers of the system's components are not large enough.

Limit reliability functions of the regular homogeneous parallel–series Weibull systems have been applied here to the reliability evaluation of the ship-rope elevator. The application of the asymptotic approach gives approximate formulae, which allow us to evaluate lower and upper limits for the system exact reliability function. The results of evaluation are also approximate because of the lack of exact reliability data for components. In cases when data are more exact it is possible to develop methods to non-homogeneous systems assuming different reliability functions of the system components (strands). It is also possible to consider the system more particularly and try to approximate its reliability by analysing its basic components, i.e. particular wires of the strands. In this case it seems to be possible to construct an 'm out of n'-series reliability model of the system.

7 Reliability of Large Multi-State Exponential Systems

7.1 Auxiliary Theorems

Results presented in this chapter are established and proved in Ref. [82].

Proposition 7.1

If components of the non-homogeneous multi-state series system have exponential reliability functions

$$R^{(i)}(t,\cdot) = [R^{(i)}(t,1),\ldots,R^{(i)}(t,z)], \quad t \in (-\infty,\infty)$$

where

$$R^{(i)}(t,u) = 1 \quad \text{for } t<0, R^{(i)}(t,u) = \exp[-\lambda_i(u)t]$$
$$\text{for } t \geq 0,\ i = 1,2,\ldots,a,\ u = 1,2,\ldots,z,$$
$$a_n(u) = \frac{1}{\lambda(u)n},\ b_n(u) = 0,\ u = 1,2,\ldots,z$$

where

$$\lambda(u) = \max_{1 \leq i \leq a}\{\lambda_i(u)\},\ u = 1,2,\ldots,z$$

then

$$\overline{\mathfrak{R}}'_2(t,\cdot) = [1, \overline{\mathfrak{R}}'_2(t,1),\ldots,\overline{\mathfrak{R}}'_2(t,z)], \quad t \in (-\infty,\infty)$$

where

$$\overline{\mathfrak{R}}'_2(t,u) = 1 \quad \text{for } t<0, \overline{\mathfrak{R}}'_2(t,u) = \exp[-\overline{d}(t,u)t] \quad \text{for } t \geq 0,\ u = 1,2,\ldots,z$$

and

$$\overline{d}(t,u) = \sum_{i=1}^{a} q_i \frac{\lambda_i(u)}{\lambda(u)} \quad \text{for } t \geq 0,\ u = 1,2,\ldots,z$$

is its limit reliability function.

Reliability of Large Systems. DOI: http://dx.doi.org/10.1016/B978-0-08-099949-4.00007-6

Proposition 7.2

If components of the non-homogeneous multi-state parallel system have exponential reliability functions

$$R^{(i)}(t,\cdot) = [R^{(i)}(t,1),\ldots,R^{(i)}(t,z)], \quad t \in (-\infty, \infty)$$

where

$$R^{(i)}(t,u) = 1 \quad \text{for } t < 0, R^{(i)}(t,u) = \exp[-\lambda_i(u)t]$$
$$\text{for } t \geq 0,\ i = 1,2,\ldots,a,\ u = 1,2,\ldots,z,$$
$$a_n(u) = \frac{1}{\lambda(u)},\ b_n(u) = \frac{1}{\lambda(u)}\log n,\ u = 1,2,\ldots,z$$

where

$$\lambda(u) = \min_{1 \leq i \leq a} \{\lambda_i(u)\},\ u = 1,2,\ldots,z$$

then

$$\mathfrak{R}_3'(t,\cdot) = [1, \mathfrak{R}_3'(t,1),\ldots,\mathfrak{R}_3'(t,z)], \quad t \in (-\infty, \infty)$$

where

$$\mathfrak{R}_3'(t,u) = 1 - \exp[-d(t,u)\exp[-t]] \quad \text{for } t \in (-\infty, \infty),\ u = 1,2,\ldots,z$$

and

$$d(t,u) = \sum_{(i:\lambda_i(u) = \lambda(u))} q_i \quad \text{for } t \in (-\infty, \infty),\ u = 1,2,\ldots,z$$

is its limit reliability function.

Proposition 7.3

If components of the homogeneous multi-state 'm out of n' system have exponential reliability functions

$$R(t,\cdot) = [R(t,1),\ldots,R(t,z)], \quad t \in (-\infty, \infty)$$

where

$$R(t,u) = 1 \quad \text{for } t < 0, R(t,u) = \exp[-\lambda(u)t] \quad \text{for } t \geq 0,\ u = 1,2,\ldots,z$$

and

Case 1. $m = \text{constant}$ ($m/n \to 0$ as $n \to \infty$),

$$a_n(u) = \frac{1}{\lambda(u)},\ b_n(u) = \frac{1}{\lambda(u)}\log n,\ u = 1, 2, \ldots, z$$

then

$$\mathfrak{R}_3^{(0)}(t, \cdot) = [1, \mathfrak{R}_3^{(0)}(t, 1), \ldots, \mathfrak{R}_3^{(0)}(t, z)], \quad t \in (-\infty, \infty)$$

where

$$\mathfrak{R}_3^{(0)}(t, u) = 1 - \sum_{i=0}^{m-1} \frac{[\exp[-it]}{i!} \exp[-\exp[-t]] \quad \text{for } t \in (-\infty, \infty),\ u = 1, 2, \ldots, z$$

is its limit reliability function.

Case 2. $m/n \to \mu$, $0 < \mu < 1$, as $n \to \infty$,

$$a_n(u) = \frac{1}{\lambda(u)}\sqrt{\frac{n - m + 1}{(n+1)m}},\ b_n(u) = \frac{1}{\lambda(u)}\log\frac{n+1}{m}, \quad u = 1, 2, \ldots, z$$

then

$$\mathfrak{R}_6^{(\mu)}(t, \cdot) = [1, \mathfrak{R}_6^{(\mu)}(t, 1), \ldots, \mathfrak{R}_6^{(\mu)}(t, z)], \quad t \in (-\infty, \infty)$$

where

$$\mathfrak{R}_6^{(\mu)}(t, u) = 1 - \frac{1}{\sqrt{2\pi}} \int_{-\infty}^{t} e^{-\frac{x^2}{2}} \mathrm{d}x \quad \text{for } t \in (-\infty, \infty),\ u = 1, 2, \ldots, z$$

is its limit reliability function.

Case 3. $n - m = \overline{m} = \text{constant}$ ($m/n \to 1$, as $n \to \infty$),

$$a_n(u) = \frac{1}{\lambda n},\ b_n(u) = 0, \quad u = 1, 2, \ldots, z$$

then

$$\overline{\mathfrak{R}}_9^{(1)}(t, \cdot) = [1, \mathfrak{R}_9^{(1)}(t, 1), \ldots, \overline{\mathfrak{R}}_9^{(1)}(t, z)], \quad t \in (-\infty, \infty)$$

where

$$\overline{\mathfrak{R}}_9^{(1)}(t,u)=1 \quad \text{for } t<0, \overline{\mathfrak{R}}_9^{(1)}(t,u)=\sum_{i=0\overline{m}}\frac{t^i}{i!}\exp[-t] \quad \text{for } t\geq 0,\ u=1,2,\ldots,z$$

is its limit reliability function.

Proposition 7.4

If components of the non-homogeneous regular multi-state series–parallel system have exponential reliability functions

$$R^{(i,j)}(t,\cdot)=[R^{(i,j)}(t,1),\ldots,R^{(i,j)}(t,z)], \quad t\in(-\infty,\infty)$$

where

$$R^{(i,j)}(t,u)=1 \quad \text{for } t<0, R^{(i,j)}(t,u)=\exp[-\lambda_{ij}(u)t]$$
$$\text{for } t\geq 0,\ i=1,2,\ldots,a,\ j=1,2,\ldots,e_i, \quad u=1,2,\ldots,z$$

and

Case 1. $k_n=n,\ l_n>0,$

$$a_n(u)=\frac{1}{\lambda(u)l_n},\ b_n(u)=\frac{\log n}{\lambda(u)l_n}, \quad u=1,2,\ldots,z$$

where

$$\lambda_i(u)=\sum_{j=1}^{e_i}p_{ij}\lambda_{ij}(u),\ \lambda(u)=\min_{1\leq i\leq a}\{\lambda_i(u)\}, \quad u=1,2,\ldots,z$$

then

$$\mathfrak{R}_3'(t,\cdot)=[1,\mathfrak{R}_3'(t,1),\ldots,\mathfrak{R}_3'(t,z)], \quad t\in(-\infty,\infty)$$

where

$$\mathfrak{R}_3'(t,u)=1-\exp[-d(t,u)\exp[-t]] \quad \text{for} \quad t\in(-\infty,\infty), \quad u=1,2,\ldots,z$$

and

$$d(t,u)=\sum_{(i:\lambda_i(u)=\lambda(u))}q_i \quad \text{for } t\in(-\infty,\infty), \quad u=1,2,\ldots,z$$

is its limit reliability function.

Case 2. $k_n \to k,\ l_n \to \infty$,

$$a_n(u) = \frac{1}{\lambda(u) l_n},\ b_n(u) = 0, \quad u = 1, 2, \ldots, z$$

where

$$\lambda_i(u) = \sum_{j=1}^{e_i} p_{ij} \lambda_{ij}(u),\ \lambda(u) = \min_{1 \le i \le a} \{\lambda_i(u)\}, \quad u = 1, 2, \ldots, z$$

then

$$\mathfrak{R}'_9(t, \cdot) = [1, \mathfrak{R}'_9(t, 1), \ldots, \mathfrak{R}'_9(t, z)], \quad t \in (-\infty, \infty)$$

where

$$\mathfrak{R}'_9(t, u) = 1 \text{ for } t < 0$$
$$\mathfrak{R}'_9(t, u) = 1 - \prod_{i=1}^{a} [1 - d_i(t, u) \exp[-t]]^{q_i k} \quad \text{for} \quad t \ge 0, \quad u = 1, 2, \ldots, z$$

and

$$d_i(t, u) = \exp\left[-\left(\frac{\lambda_i(u)}{\lambda(u)} - 1 \right) t \right] \quad \text{for } t \ge 0,\ i = 1, 2, \ldots, a, \quad u = 1, 2, \ldots, z$$

is its limit reliability function.

Proposition 7.5

If components of the non-homogeneous regular multi-state parallel−series system have exponential reliability function

$$R^{(i,j)}(t, \cdot) = [R^{(i,j)}(t, 1), \ldots, R^{(i,j)}(t, z)], \quad t \in (-\infty, \infty)$$

where

$$R^{(i,j)}(t, u) = 1 \quad \text{for } t < 0, R^{(i,j)}(t, u) = \exp[-\lambda_{ij}(u) t]$$
$$\text{for } t \ge 0,\ i = 1, 2, \ldots, a,\ j = 1, 2, \ldots, e_i, \quad u = 1, 2, \ldots, z$$

and

Case 1. $k_n = n,\ l_n \to l,\ l \in (0, \infty)$,

$$a_n(u) = \frac{1}{\lambda(u) n^{1/l_n}},\ b_n(u) = 0, \quad u = 1, 2, \ldots, z$$

where

$$\lambda_i(u) = \prod_{i=1}^{e_i} \lambda_{ij}^{p_{ij}}(u),\ \lambda(u) = \max_{1 \le i \le a} \{\lambda_i(u)\}, \quad u = 1, 2, \ldots, z$$

then

$$\overline{\mathfrak{R}}_2'(t, \cdot) = [1, \overline{\mathfrak{R}}_2'(t, 1), \ldots, \overline{\mathfrak{R}}_2'(t, z)], \quad t \in (-\infty, \infty)$$

where

$$\overline{\mathfrak{R}}_2'(t, u) = 1 \text{ for } t < 0,\ \overline{\mathfrak{R}}_2'(t, u) = \exp[-\overline{d}(t, u)t^l] \quad \text{for } t \ge 0, \quad u = 1, 2, \ldots, z$$

and

$$\overline{d}(t, u) = \sum_{i=1}^{a} q_i \left(\frac{\lambda_i(u)}{\lambda(u)}\right)^l \quad \text{for } t \ge 0, \quad u = 1, 2, \ldots, z$$

is its limit reliability function.

Case 2. $k_n = n,\ c \ll l_n,\ c \log n - l_n \gg s,\ c > 0,\ s > 0$

$$a_n(u) = \frac{1}{(n^{1/l_n} - 1)\lambda(u) l_n},\ b_n(u) = -\frac{1}{\lambda(u)} \log(1 - n^{-1/l_n}), \quad u = 1, 2, \ldots, z$$

where

$$\lambda_i(u) = \prod_{i=1}^{e_i} \lambda_{ij}^{p_{ij}}(u),\ \lambda(u) = \max_{1 \le i \le a} \{\lambda_i(u)\}, \quad u = 1, 2, \ldots, z$$

then

$$\overline{\mathfrak{R}}_3'(t, \cdot) = [1, \overline{\mathfrak{R}}_3'(t, 1), \ldots, \overline{\mathfrak{R}}_3'(t, z)], \quad t \in (-\infty, \infty)$$

where

$$\overline{\mathfrak{R}}_3'(t, u) = \exp[-\overline{d}(t, u)\exp[t]] \quad \text{for } t \in (-\infty, \infty), \quad u = 1, 2, \ldots, z$$

and

$$\overline{d}(t, u) = \sum_{(i:\, \lambda_i(u) = \lambda(u))} q_i \quad \text{for } t \in (-\infty, \infty), \quad u = 1, 2, \ldots, z$$

is its limit reliability function.

Case 3. $k_n = n,\ l_n - c \log n \sim s,\ c > 0, \quad s \in (-\infty, \infty)$

$$a_n(u) = \frac{1}{\alpha(u) l_n},\ n\beta(u)^{l_n} = 1, \quad u = 1, 2, \ldots, z$$

where

$$\beta i(u) = \prod_{j=1}^{e_i} [1 - \exp[-\lambda_{ij}(u) b_n(u)]]^{p_{ij}},\ \alpha_i(u) = \sum_{j=1}^{e_i} \frac{\exp[-\lambda_{ij}(u) b_n(u)]}{1 - \exp[-\lambda_{ij}(u) b_n(u)]} p_{ij} \lambda_{ij}(u)$$

$$\beta(u) = \max_{1 \le i \le a} \{\beta_i(u)\}, \alpha(u) = \max\{\alpha_i(u) : \beta_i(u) = \beta(u)\}, \quad u = 1, 2, \ldots, z$$

then

$$\overline{\mathfrak{R}}_3'(t, \cdot) = [1, \overline{\mathfrak{R}}_3'(t, 1), \ldots, \overline{\mathfrak{R}}_3'(t, z)], \quad t \in (-\infty, \infty)$$

where

$$\overline{\mathfrak{R}}_3'(t, u) = \exp[-\overline{d}(t, u) \exp[t]] \quad \text{for } t \in (-\infty, \infty),\ u = 1, 2, \ldots, z$$

and

$$d(t, u) = \sum_{(i: \beta_i(u) = \beta(u))} q_i \exp\left[\frac{\alpha_i(u) - \alpha(u)}{\alpha(u)} t\right] \quad \text{for } t \in (-\infty, \infty), \quad u = 1, 2, \ldots, z$$

is its limit reliability function.

Case 3'. (For a homogeneous system)
$k_n = n,\ l_n - c \log n \sim s,\ c > 0,\ s \in (-\infty, \infty)$,

$$a_n(u) = \frac{1}{\lambda(u) l_n (n^{1/l_n} - 1)},\ b_n(u) = -\frac{1}{\lambda(u)} \log(1 - n^{-1/l_n}), \quad u = 1, 2, \ldots, z$$

then

$$\overline{\mathfrak{R}}_3(t, \cdot) = [1, \overline{\mathfrak{R}}_3(t, 1), \ldots, \overline{\mathfrak{R}}_3(t, z)], \quad t \in (-\infty, \infty)$$

where

$$\overline{\mathfrak{R}}_3(t, u) = \exp[-\exp[t]] \text{ for } t \in (-\infty, \infty), \quad u = 1, 2, \ldots, z$$

is its limit reliability function

Case 4. $k_n = n,\ l_n - \log n \gg s,\ c > 0,\ s > 0,$

$$a_n(u) = \frac{1}{\lambda(u)\log n},\ b_n(u) = \frac{1}{\lambda(u)}\log\left(\frac{l_n}{\log n}\right),\quad u = 1, 2, \ldots, z$$

where

$$\lambda_i(u) = \max_{1 \le j \le e_i}\{\lambda_{ij}(u)\},\ \lambda(u) = \max_{1 \le i \le a}\{\lambda_i(u)\},\quad u = 1, 2, \ldots, z$$

then

$$\overline{\mathfrak{R}}'_3(t, \cdot) = [1, \overline{\mathfrak{R}}'_3(t, 1), \ldots, \overline{\mathfrak{R}}'_3(t, z)],\quad t \in (-\infty, \infty)$$

where

$$\overline{\mathfrak{R}}'_3(t, u) = \exp[-\bar{d}(t, u)\exp[t]] \quad \text{for } t \in (-\infty, \infty),\quad u = 1, 2, \ldots, z$$

and

$$\bar{d}(t, u) = \sum_{(i: \lambda_i(u) = \lambda(u))} q_i \quad \text{for } t \in (-\infty, \infty),\quad u = 1, 2, \ldots, z$$

is its limit reliability function.

Case 5. $k_n \to k,\ \ l_n \to \infty,$

$$a_n(u) = \frac{1}{\lambda(u)},\ b_n(u) = \frac{1}{\lambda(u)}\log l_n,\quad u = 1, 2, \ldots, z$$

where

$$\lambda_i(u) = \max_{1 \le j \le e_i}\{\lambda_{ij}(u)\},\ \lambda(u) = \max_{1 \le j \le a}\{\lambda_i(u)\},\quad u = 1, 2, \ldots, z$$

then

$$\overline{\mathfrak{R}}'_{10}(t, \cdot) = [1, \overline{\mathfrak{R}}'_{10}(t, 1), \ldots, \overline{\mathfrak{R}}'_{10}(t, z)],\quad t \in (-\infty, \infty)$$

where

$$\overline{\mathfrak{R}}'_{10}(t, u) = \prod_{(i: \lambda_i(u) = \lambda(u))} [1 - \exp[-\exp[-t]]]^{q_i k} \quad \text{for } t \in (-\infty, \infty),\quad u = 1, 2, \ldots, z$$

is its limit reliability function.

7.2 Algorithms for Reliability Evaluation of Multi-state Exponential Systems

The results of Section 7.1 are presented here in the form of tables containing exact algorithms of the procedure, while evaluating reliability characteristics of these systems' reliability in order to deliver reliability practitioners a simple and convenient tool for everyday practice. Tables 7.1–7.5 are composed of three parts containing reliability data of the evaluated system, necessary calculations and results of the system reliability evaluation. The method of using the algorithms is illustrated by several examples. These algorithms are also the basis of the computer program [93], allowing us to accelerate the system reliability evaluation.

Table 7.1 Algorithm of Reliability Evaluation of a Series System

System Type	A Series System
Data	Number of components N
	Number of component kinds a
	Fractions of different component kinds $q_1, \ldots, q_a$
	Number of component and system states z
	Transition rates between components' state subsets
	$\lambda_1(u), \ldots, \lambda_a(u),\ u = 1, 2, \ldots, z$
	System critical reliability state r
	Admissible system risk level δ
Calculations	$A(u) = n \sum_{i=1}^{a} q_i \lambda_i(u), u = 1, 2, \ldots, z$
Results	System reliability function
	$\overline{\boldsymbol{R}}'_n(t, \cdot) = \overline{\mathfrak{R}}'_2((t - b_n(u))/a_n(u), \cdot) = [1, \exp[-A(1)t], \ldots, \exp[-A(z)t]]$
	System mean lifetimes in state subsets
	$M(u) = 1/A(u), u = 1, 2, \ldots, z$
	Standard deviations of system lifetimes in state subsets
	$(u) = 1/A(u), u = 1, 2, \ldots, z$
	System mean lifetimes in particular states
	$\overline{M}(u) = M(u) - M(u+1),\ u = 1, 2, \ldots, z-1,\ \overline{M}(z) = M(z)$
	System risk function
	$\boldsymbol{r}(t) = 1 - \exp[-A(r)t]$
	Exceeding moment of admissible system risk level δ
	$\tau = -(\log(1 - \delta))/A(r)$

Table 7.2 Algorithm of Reliability Evaluation of a Parallel System

System Type	**A Parallel System**
Data	Number of components n
	Number of component kinds a
	Fractions of different component kinds $q_1, \ldots, q_a$
	Number of component and system states z
	Transition rates between components' state subsets
	$\lambda_1(u), \ldots, \lambda_a(u),\ u = 1, 2, \ldots, z$
	System critical reliability state r
	Admissible system risk level δ
Calculations	$\lambda(u) = \min_{1 \le i \le a}\{\lambda_i(u)\}, u = 1, 2, \ldots, z$ $d(t, u) = \sum q_{i,(i:\lambda_i(u) = \lambda(u))}\ u = 1, 2, \ldots, z$ $A(u) = \lambda(u), B(u) = \log(nd(t, u)),\ u = 1, 2, \ldots, z$
Results	System reliability function
	$\boldsymbol{R}'_n(t, \cdot) \cong \Re'_3((t - b_n(u))/a_n(u), \cdot)$ $= [1, 1 - \exp[-\exp[-A(1)t + B(1)]], \ldots$ $1 - \exp[-\exp[-A(z)t + B(z)]]]$
	System mean lifetimes in state subsets
	$M(u) = [05772 + B(u)]/A(u), u = 1, 2, \ldots, z$
	Standard deviations of system lifetimes in state subsets
	$\sigma(u) = \pi/(A(u)\sqrt{6}),\ u = 1, 2, \ldots, z$
	System mean lifetimes in particular states
	$\overline{M}(u) = M(u) - M(u+1),\ u = 1, 2, \ldots, z-1,\ \overline{M}(z) = M(z)$
	System risk function
	$\boldsymbol{r}(t) = \exp[-\exp[-A(r)t + B(r)]]$
	Exceeding moment of admissible system risk level δ
	$\tau = [B(r) - \log(-\log\delta)]/A(r)$

Table 7.3 Algorithm of Reliability Evaluation of an 'm out of n' System

System Type	**An 'm out of n' System**
Data	Number of components n
	Threshold number m
	Number of component and system states z
	Transition rates between components' state subsets $\lambda(u),\ u = 1, 2, \ldots, z$
	System critical reliability state r
	Admissible system risk level δ

(Continued)

Table 7.3 (Continued)

System Type	**An '*m* out of *n*' System**
Calculations	***Case 1.*** $m=$ constant ($m/n \to 0$ as $n \to \infty$)
	$A(u)=\lambda(u),\ B(u)=\log n,\ u=1,2,\ldots,z$
	Case 2. ($m/n \to \mu,\ 0<\mu<1,$ as $n \to \infty$)
	$A(u)=\lambda(u)\sqrt{(n+1)m/(n-m+1},\ B(u)=\sqrt{(n+1)m/(n-m+1}\log n+1/m$
	$u=1,2,\ldots,z$
	Case 3. $n-m=\overline{m}=$ constant ($m/n \to 1$ as $n \to \infty$)
	$A(u)=n\lambda(u),\ u=1,2,\ldots,z$
Results	***Case 1.*** $m=$ constant ($m/n \to 0$ as $n \to \infty$)
	System reliability function
	$\boldsymbol{R}_n^{(m)}(t,\cdot) \cong \mathfrak{R}_3^{(0)}((t-b_n(u))/a_n(u),\cdot)$ $= \left[1, 1-\sum_{i=0}^{m-1}\frac{\exp[-iA(1)t+iB(1)]}{i!}\exp[\exp[-A(1)t+B(1)]], \ldots\right.$ $\left. 1-\sum_{i=0}^{m-1}\frac{\exp[-iA(z)t+iB(z)]}{i!}\exp[\exp[-A(1)t+B(1)]]\right]$
	System mean lifetimes in state subsets
	$M(u)=$ (a geometric integration), $u=1,2,\ldots,z$
	Standard deviations of system lifetimes in state subsets
	$\sigma(u)=$ (a geometric integration), $u=1,2,\ldots,z$
	System mean lifetimes in particular states
	$\overline{M}(u)=M(u)-M(u+1),\ u=1,2,\ldots,z-1,\ \overline{M}(z)=M(z)$
	System risk function
	$\boldsymbol{r}(t)=\sum_{i=0}^{m-1}\frac{\exp[-iA(r)t+iB(r)]}{i!}\exp[\exp[-A(r)t+B(r)]]$
	Exceeding moment of admissible system risk level δ
	$\tau=\boldsymbol{r}^{-1}(t)$
Results	***Case 2.*** $m/n \to \mu$ ($0<\mu<1$ as $n \to \infty$)
	System reliability function
	$\boldsymbol{R}_n^{(m)}(t,\cdot) \cong \mathfrak{R}_6^{(\mu)}((t-b_n(u))/a_n(u),\cdot)$ $= \left[1, 1-\frac{1}{\sqrt{2\pi}}\int_{-\infty}^{A(1)t-B(1)} e^{-\frac{x^2}{2}}dx, \ldots, 1-\frac{1}{\sqrt{2\pi}}\int_{-\infty}^{A(z)t-B(z)} e^{-\frac{x^2}{2}}dx\right]$
	System mean lifetimes in state subsets
	$M(u)=B(u)/A(u),\ u=1,2,\ldots,z,$
	System mean lifetimes in particular states
	$\overline{M}(u)=M(u)-M(u+1),\ u=1,2,\ldots,z-1,\ \overline{M}(z)=M(z)$

(Continued)

Table 7.3 (Continued)

System Type	An '*m* out of *n*' System
	System risk function
	$\boldsymbol{r}(t) = 1/\sqrt{2\pi}\int_{-\infty}^{A(r)t-B(r)} e^{-x^2/2}\,dx$
	Exceeding moment of admissible system risk level δ
	$\tau = \boldsymbol{r}^1(t)$
Results	***Case 3***. $n - m = \overline{m} = \text{constant}(m/n \to 1 \text{ as } n \to \infty)$
	System reliability function
	$\overline{\boldsymbol{R}}_n^{(\overline{m})}(t,\cdot) \cong \overline{\mathfrak{R}}_9^{(1)}((t - b_n(u))/a_n(u),\cdot) = \left[1, \sum_{i=0\overline{m}} \frac{[A(1)]^i t^i}{i!}\exp[-A(1)t] \quad \sum_{i=0\overline{m}} \frac{[A(2)]^i t^i}{i!}\exp[-A(2)t], \ldots \sum_{i=0\overline{m}} \frac{[A(z)]^i t^i}{i!}\exp[-A(z)t] \quad \text{for } t \ge 0\right.$
	System mean lifetimes in state subsets
	$M(u) = (\text{a geometric integration}),\ u = 1, 2, \ldots, z$
	Standard deviations of system lifetimes in state subsets
	$\sigma(u) = (\text{a geometric integration}),\ u = 1, 2, \ldots, z$
	System mean lifetimes in particular states
	$\overline{M}(u) = M(u) - M(u+1),\ u = 1, 2, \ldots, z-1,\ \overline{M}(z) = M(z)$
	System risk function
	$\boldsymbol{r}(t) = 1 - \sum_{i=0\overline{m}} \frac{[A(r)]^i t^i}{i!}\exp[-A(r)t]$
	Exceeding moment of admissible system risk level δ
	$\tau = \boldsymbol{r}^1(t)$

Table 7.4 Algorithm of Reliability Evaluation of a Series–Parallel System

System Type	A Series–Parallel System
Data	Number of series subsystems linked in parallel k_n
	Number of components in series subsystems l_n
	Number of types of series subsystems a
	Fractions of series subsystems of particular types $q_1,\ldots,q_a$
	Numbers of types of components in series subsystems $e_1,\ldots,e_a$
	Fractions of components of particular types in series subsystems
	$p_{11},\ldots,p_{1e_1}$
	...
	$p_{a1},\ldots,p_{ae_a}$

(Continued)

Table 7.4 (Continued)

System Type	A Series–Parallel System
	Number of component and system states z
	Component transition rates between state subsets
	$\lambda_{11}(u),\ldots,\lambda_{1e_1}(u),$
	...
	$\lambda_{a1}(u),\ldots,\lambda_{ae_a}(u),\ u=1,2,\ldots,z$
	System critical state r
	System risk admissible level δ
Calculations	***Case 1.*** System shape: $k_n \to \infty,\ l_n > 0$
	$\lambda_i(u)=\sum_{j=1}^{e_i} p_{ij}\lambda_{ij}(u),\ i=1,2,\ldots,a$
	$\lambda(u)=\min_{1\le i\le a}\{\lambda_i(u)\},\ u=1,2,\ldots,z$
	$d(t,u)=\sum q_{i,(i:\lambda_i(u)=\lambda(u))}\ u=1,2,\ldots,z$
	$A(u)=\lambda(u)l_n,\ B(u)=\log(nd(t,u)),\ u=1,2,\ldots,z$
Calculations	***Case 2.*** System shape: $k_n \to k,\ l_n \to \infty$
	$\lambda_i(u)=\sum_{j=1}^{e_i} p_{ij}\lambda_{ij}(u),\ i=1,2,\ldots,a$
	$A_i(u)=\lambda_i(u)l_n,\ i=1,2,\ldots a,\ u=1,2,\ldots,z$
Results	***Case 1.*** System shape: $k_n \to \infty,\ l_n > 0$
	System reliability function
	$\boldsymbol{R}'_{k_n l_n}(t,\cdot) \cong \mathfrak{R}'_3(t-b_n(u))/a_n(u),\cdot)$ $=[1,1-\exp[-\exp[-A(1)t+B(1)]],\ldots$ $1-\exp[-\exp[-A(z)t+B(z)]]]$
	System mean lifetimes in state subsets
	$M(u)=[05772+B(u)]/A(u),\ u=1,2,\ldots,z$
	Standard deviations of system lifetimes in state subsystems
	$\sigma(u)=\pi/(A(u)\sqrt{6}),\ u=1,2,\ldots,z$
	System mean lifetimes in particular states
	$\overline{M}(u)=M(u)-M(u+1),\ u=1,2,\ldots,z-1,\ \overline{M}(z)=M(z)$
	System risk function
	$\boldsymbol{r}(t)=\exp[-\exp[-A(r)t+B(r)]]$
	Exceeding moment of admissible system risk level δ
	$\tau=[B(r)-\log(-\log\delta)]/A(r)$
Results	***Case 2.*** System shape: $k_n \to k,\ l_n \to \infty$
	System reliability function
	$\boldsymbol{R}'_{k_n l_n}(t,\cdot)=\mathfrak{R}'_9((t-b_n(u))/a_n(u),\cdot)$ $=\lfloor 1,1-\prod_{i=1}^{u}[1-\exp[-A_i(1)t]]^{q_i k},\ldots$ $1-\prod_{i=1}^{a}[1-\exp[-A_i(z)t]]^{q_i k}]$

(*Continued*)

Table 7.4 (Continued)

System Type	A Series–Parallel System
	System mean lifetimes in state subsystems $M(u) = \int_0^\infty \left[1 - \prod_{i=1}^{a}[1-\exp[-A_i(u)t]]^{q_i k}\right] dt,\ u = 1, 2, \ldots, z$ Standard deviations of system lifetimes in state subsystems $\sigma(u) = \sqrt{N(u) - M^2(u)},\ N(u) = 2\int_0^\infty t\left[1 - \prod_{i=1}^{a}[1-\exp[-A_i(u)t]]^{q_i k}\right] dt,$ $u = 1, 2, \ldots, z$ System mean lifetimes in particular states $\overline{M}(u) = M(u) - M(u+1),\ u = 1, 2, \ldots, z-1,\ \overline{M}(z) = M(z)$ System risk function $\boldsymbol{r}(t) = \prod_{i=1}^{a}[1-\exp[-A_i(r)t]]^{q_i k}$ Exceeding moment of admissible system risk level δ $\tau = \boldsymbol{r}^{1}(\delta)$

Table 7.5 Algorithm of Reliability Evaluation of a Parallel–Series System

System Type	A Parallel–Series System
Data	Number of parallel subsystems linked in series k_n Number of components in parallel subsystems l_n Number of types of parallel subsystems a Fractions of parallel subsystems of particular types $q_1,\ldots,q_a$ Numbers of types of components in parallel subsystems $e_1,\ldots,e_a$ Fractions of components of particular types in parallel subsystems $p_{11}, \ldots, p_{1e_1}$ $\vdots$ $p_{a1}, \ldots, p_{ae_a}$ Number of component and system states z Component transition rates between state subsets $\lambda_{11}(u), \ldots, \lambda_{1e_1}(u)$ $\vdots$ $\lambda_{a1}(u), \ldots, \lambda_{ae_a}(u),\ u = 1, 2, \ldots, z$ System critical state r System risk admissible level δ
Calculations	***Case 1.*** System shape: $k_n \to \infty,\ l_n \to l$ $\lambda_i(u) = \prod_{j=1}^{e_i} \lambda_{ij}^{p_{ij}}(u),\ i = 1, 2, \ldots, a,\ u = 1, 2, \ldots, z$ $A(u) = n\sum_{i=1}^{a} q_i[\lambda_i(u)]^l,\ u = 1, 2, \ldots, z$

(Continued)

Table 7.5 (Continued)

System Type	A Parallel–Series System
Calculations	***Case 2.*** System shape: $k_n \to \infty, c \ll l_n,\ c\log n - l_n \gg s,\ c>0,\ s>0$ $\lambda_i(u) = \sum_{j=1}^{e_i} \lambda_{ij}^{p_{ij}}(u),\ i=1,2,\ldots,a,\ \lambda(u) = \max_{1\le i\le a}\{\lambda_i(u)\}$ $\overline{d}(t,u) = \sum_{(i:\lambda_i(u)=\lambda(u))} q_i, u=1,2,\ldots,z$ $A(u) = (n^{1/l_n}-1)\lambda(u)l_n$ $B(u) = (n^{1/l_n}-1)l_n \log(1-n^{-1/l_n}) + \log \overline{d}(t,u)$
Calculations	***Case 3.*** System shape: $k_n = n, l_n - c\log n \sim s,\ c>0,\ s\in(-\infty,\infty)$ $\beta_i(u) = \prod_{j=1}^{e_i}[1-\exp[-\lambda_{ij}(u)b_n(u)]]^{p_{ij}}$ $\alpha_i(u) = \sum_{j=1}^{e_i} \frac{\exp[-\lambda_{ij}(u)b_n(u)]}{1-\exp[-\lambda_{ij}(u)b_n(u)]} p_{ij}\lambda_{ij}(u)$ $\beta(u) = \max_{1\le i\le a}\{\beta_i(u)\}$ $n\beta(u)^{l_n} = 1, \alpha(u) = \max\{\alpha_i(u):\beta_i(u)=\beta(u)\}$ $A_i(u) = \alpha_i(u)l_n,\ B_i(u) = -b_n(u)\alpha_i(u)l_n,\ i=1,2,\ldots,a$
Calculations	***Case 3′.*** (A homogeneous system) system shape: $k_n = n,\ l_n - c\log n \sim s,\ c>0,\ s\in(-\infty,\infty)$ $A(u) = (n^{1/l_n}-1)\lambda(u)l_n,\ B(u) = (n^{1/l}-1)l_n \log(1-n^{-1/l_n})$
Calculations	***Case 4.*** System shape: $k_n \to \infty,\ l_n - c\log n \gg s,\ c>0,\ s>0$ $\lambda_i(u) = \max_{1\le j\le e_i}\{\lambda_{ij}(u)\}, \lambda(u) = \max_{1\le i\le a}\{\lambda_i(u)\}$ $\overline{d}(t,u) = \sum_{(i:\lambda_i(u)=\lambda(u))} q_i$ $A(u) = \lambda(u)\cdot\log n,\ B(u) = -\log n \log(l_n/\log n) + \log\overline{d}(t,u)$
Calculations	***Case 5.*** System shape: $k_n \to k,\ l_n \to \infty$ $\lambda_i(u) = \max_{1\le j\le e_i}\{\lambda_{ij}(u)\}, \lambda(u) = \max_{1\le i\le a}\{\lambda_i(u)\}$ $A(u) = \lambda(u),\ B(u) = \log l_n$
Results	***Case 1.*** System shape: $k_n \to \infty,\ l_n \to l$ System reliability function $\overline{\boldsymbol{R}}'_{k_n l_n}(t,\cdot) \cong \overline{\mathfrak{R}}'_2((t-b_n(u))/a_n(u),\cdot) = [1, \exp[-A(1)t^l], \ldots, \exp[-A(z)t^l]]$ Mean system sojourn times in state subsets $M(u) = \Gamma((l+1)/l)[A(u)]^{-1/l},\ u=1,2,\ldots,z$ Standard deviations of system sojourn times in state subsets $\sigma(u) = \sqrt{\Gamma((l+2)/l) - \Gamma^2((l+1)/l)}[A(u)]^{-1/l},\ u=1,2,\ldots,z$ Mean system sojourn times in particular states $\overline{M}(u) = M(u) - M(u+1),\ u=1,2,\ldots,z-1,\ \overline{M}(z) = M(z)$

(Continued)

Table 7.5 (Continued)

System Type	**A Parallel–Series System**
	System risk function $r(t) \cong 1 - \exp[-A(r)t^l]$ Exceeding moment of admissible system risk level δ $\tau \cong [-(\log(1-\delta))/A(r)]^{1/l}$
Results	***Case 2.*** System shape: $k_n \to \infty,\ c \ll l_n,\ c \log n - l_n \gg s,\ c>0,\ s>0$ System reliability function $\overline{\boldsymbol{R}}'_{k_n l_n}(t,\cdot) = \overline{\Re}'_3((t-b_n(u))/a_n(u),\cdot)$ $= [1, \exp[-\exp[A(1)t + B(1)]], \ldots$ $\exp[-\exp[A(z)t + B(z)]]]$ Mean system sojourn times in state subsets $M(u) = [-05772 - B(u)]/A(u),\ u = 1,2,\ldots,z$ Standard deviations of system sojourn times in state subsets $\sigma(u) = \pi/(A(u)\sqrt{6}),\ u = 1,2,\ldots,z$ Mean system sojourn times in particular states $\overline{M}(u) = M(u) - M(u+1),\ u = 1,2,\ldots,z-1,\ \overline{M}(z) = M(z)$ System risk function $r(t) \cong 1 - \exp[-\exp[A(r)t + B(r)]$ Exceeding moment of admissible system risk level δ $\tau \cong [\log(-\log(1-\delta)) - B(r)]/A(r)$
Results	***Case 3.*** System shape: $k_n = n,\ l_n - c\log n \sim s,\ c>0,\ s \in (-\infty,\infty)$ System reliability function $\overline{\boldsymbol{R}}'_{k_n l_n}(t,\cdot) = \overline{\Re}'_3((t-b_n(u))/a_n(u),\cdot)$ $= \left[1, \exp\left[-\sum_{(i:\beta_i(1)=\beta(1))} q_i \exp[A_i(1)t + B_i(1)]\right], \ldots\right.$ $\left.\exp\left[-\sum_{(i:\beta_i(z1)=\beta(z))} q_i \exp[A_i(z)t + B_i(z)]\right]\right]$ Mean system sojourn times in state subsets $M(u) = \int_0^\infty \exp\left[-\sum_{(i:\beta_i(u)=\beta(u))} q_i \exp[A_i(u)t + B_i(u)]\right] dt,\ u = 1,2,\ldots,z$ Standard deviations of system sojourn times in state subsets $\sigma(u) = \sqrt{N(u) - M^2(u)}$ $N(u) = 2\int_0^\infty t \exp\left[-\sum_{(i:\beta_i(u)=\beta(u))} q_i \exp[A_i(u)t + B_i(u)]\right] dt,\ u = 1,2,\ldots,z$ Mean system sojourn times in particular states $\overline{M}(u) = M(u) - M(u+1),\ u = 1,2,\ldots,z-1,\ \overline{M}(z) = M(z)$

(Continued)

Table 7.5 (Continued)

System Type	A Parallel–Series System
	S system risk function
	$r(t) = 1 - \exp\left[-\sum_{(i:\beta_i(r)=\beta(r))}\exp[A_i(r)t + B_i(r)]\right],\ u = 1,2,\ldots,z$
	Exceeding moment of admissible system risk level δ
	$\tau = r^1(\delta)$
Results	***Case 3′.*** (A homogeneous system) system shape: $k_n = n$, $l_n - c\log n \sim s,\ c > 0,\ s \in (-\infty, \infty)$
	System reliability function
	$\overline{R}_{k_n l_n}(t,\cdot) \cong \overline{\mathfrak{R}}_3((t - b_n(u))/a_n(u),\cdot) = [1, \exp[-\exp[A(1)t + B(1)]], \ldots \exp[-\exp[A(z)t + B(z)]]]$
	Mean system sojourn times in state subsets
	$M(u) = [-05772 - B(u)]/A(u),\ u = 1,2,\ldots,z$
	Standard deviations of system sojourn times in state subsets
	$\sigma(u) = \pi/(A(u)\sqrt{6}),\ u = 1,2,\ldots,z$
	Mean system sojourn times in particular states
	$\overline{M}(u) = M(u) - M(u+1),\ u = 1,2,\ldots,z-1,\ \overline{M}(z) = M(z)$
	System risk function
	$r(t) \cong 1 - \exp[-\exp[A(r)t + B(r)]]$
	Exceeding moment of admissible system risk level δ
	$\tau \cong [\log(-\log(1-\delta)) - B(r)]/A(r)$
Results	***Case 4.*** System shape: $k_n \to \infty$, $l_n - c\log n \gg s,\ c > 0,\ s > 0$
	System reliability function
	$\overline{R}'_{k_n l_n}(t,\cdot) = \overline{\mathfrak{R}}'_3((t - b_n(u))/a_n(u),\cdot) = [1, \exp[-\exp[A(1)t + B(1)]], \ldots \exp[-\exp[A(z)t + B(z)]]]$
	Mean system sojourn times in state subsets
	$M(u) = [-05772 - B(u)]/A(u),\ u = 1,2,\ldots,z$
	Standard deviations of system sojourn times in state subsets
	$\sigma(u) = \pi/(A(u)\sqrt{6}),\ u = 1,2,\ldots,z$
	Mean system sojourn times in particular states
	$\overline{M}(u) = M(u) - M(u+1),\ u = 1,2,\ldots,z-1,\ \overline{M}(z) = M(z)$
	System risk function
	$r(t) \cong 1 - \exp[-\exp[A(r)t + B(r)]$
	Exceeding moment of admissible system risk level δ
	$\tau \cong [\log(-\log(1-\delta))\quad B(r)]/A(r)$

(Continued)

Table 7.5 (Continued)

System Type	A Parallel–Series System
Results	*Case 5.* System shape: $k_n \to k,\ l_n \to \infty$ System reliability function $\overline{\boldsymbol{R}}'_{k_n l_n}(t,\cdot) = \overline{\mathfrak{R}}'_{10}((t - b_n(u))/a_n(u),\cdot)$ $= \left[1, \prod_{(i:\lambda_i(1)=\lambda(1))} [1-\exp[-\exp[-A(1)t+B(1)]]]^{q_i k}, \ldots \prod_{(i:\lambda_i(z)=\lambda(z))} [1-\exp[-\exp[-A(z)t+B(z)]]]^{q_i k}\right]$ System mean sojourn times in state subsets $M(u) = \int_0^\infty \left[\prod_{(i:\lambda_i(1)=\lambda(1))} [1-\exp[-\exp[-A(u)t+B(u)]]]^{q_i k}\right] \mathrm{d}t,\ u=1,2,\ldots,z$ Standard deviations of system sojourn times in state subsets $\sigma(u) = \sqrt{N(u) - M^2(u)}$ $N(u) = 2\int_0^\infty t\left[\prod_{(i:\lambda_i(u)=\lambda(u))} [1 - \exp[-\exp[A(u)t+B(u)]]]^{q_i k}\right] \mathrm{d}t,\ u=1,2,\ldots,z$ Mean system sojourn times in particular states $\overline{M}(u) = M(u) - M(u+1),\ u = 1,2,\ldots,z-1,\ \overline{M}(z) = M(z)$ System risk function $r(t) = 1 - \prod_{(i:\lambda_i(r)=\lambda(r))} [1-\exp[-\exp[-A(r)t+B(r)]]]^{q_i k}$ Exceeding moment of admissible system risk level δ $\tau = \boldsymbol{r}^1(\delta)$

7.3 Algorithms Application to Reliability Evaluation of Exponential Systems

The examples of the system reliability evaluation presented here are an illustration of the usage of the algorithms, addressed directly to reliability practitioners.

Example 7.1. (A Piping System)
Let us consider a piping system composed of $n = 80$ four-state pipe segments of four types such that 20 pipe segments have exponential reliability functions with transition rates between state subsets

$$\lambda_1(u) = 4^{u-4}, \quad u = 1,2,3$$

20 pipe segments have exponential reliability functions with transition rates between state subsets

$$\lambda_2(u) = (4.5)^{u-4}, \quad u = 1,2,3$$

10 pipe segments have exponential reliability functions with transition rates between state subsets

$$\lambda_3(u) = (6.5)^{u-4}, \quad u = 1, 2, 3$$

and 30 pipe segments have exponential reliability functions with transition rates between state subsets

$$\lambda_4(u) = 8^{u-4}, \quad u = 1, 2, 3$$

According to Definition 3.17, the considered piping is a non-homogeneous multi-state series system with parameters

$$n = 80,\ a = 4,\ q_1 = 2/8,\ q_2 = 2/8,\ q_3 = 1/8,\ q_4 = 3/8$$

Its reliability evaluation may be performed using the algorithm presented in Table 7.1. Sequential steps of the procedure are given in Table 7.6.

The behaviour of the multi-state reliability function and risk function of the piping system is illustrated in Table 7.7 and Figure 7.1.

Example 7.2. (A Model Parallel System)
Let us consider a homogeneous six-state parallel system such that

$$n = 30,\ \lambda(u) = 10^{u-6}h^{-1}, \quad u = 1, 2, 3, 4, 5$$

We will perform its reliability evaluation based on Table 7.2. The procedure is given in Table 7.8.

Example 7.3. (A Piping System)
Let us consider a piping system composed of $k_n = 3$ pipelines linked in parallel, each of them composed of $l_n = 100$ six-state pipe segments of two types linked in series. Two of the pipelines consist of 40 pipe segments that have exponential reliability functions with transition rates between state subsets

$$\lambda_{11}(u) = (25)^{u-6}, \quad u = 1, 2, 3, 4, 5$$

and 60 pipe segments that have exponential reliability functions with transition rates between state subsets

$$\lambda_{12}(u) = 2^{u-6}, \quad u = 1, 2, 3, 4, 5$$

The third pipeline consists of 50 pipe segments that have exponential reliability functions with transition rates between state subsets

$$\lambda_{21}(u) - (1.9)^{u-6}, \quad u = 1, 2, 3, 4, 5$$

Table 7.6 Reliability Evaluation of the Piping System

System Type	**A Non-homogeneous Series System**
Data	Number of system components $n = 80$ Number of component types $a = 4$ Fractions of different component types $q_1 = 2/8,\ q_2 = 2/8,\ q_3 = 1/8,\ q_4 = 3/8$ Number of component and system states $z = 3$ Component transition rates between state subsets $\lambda_1(1) = 4^{-3},\ \lambda_2(1) = (4.5)^{-3},\ \lambda_3(1) = (6.5)^{-3},\ \lambda_4(1) = 8^{-3}$ $\lambda_1(2) = 4^{-2},\ \lambda_2(2) = (4.5)^{-2},\ \lambda_3(2) = (6.5)^{-2},\ \lambda_4(2) = 8^{-2}$ $\lambda_1(3) = 4^{-1},\ \lambda_2(3) = (4.5)^{-1},\ \lambda_3(3) = (6.5)^{-1},\ \lambda_4(3) = 8^{-1}$ System reliability critical state $r = 2$ Admissible level of system risk $\delta = 0.05$
Calculations	$A(1) = 80[(2/8)4^{-3} + (2/8)(4.5)^{-3} + (1/8)(6.5)^{-3} + (3/8)8^{-3}] = 0.627$ $A(2) = 80[(2/8)4^{-2} + (2/8)(4.5)^{-2} + (1/8)(6.5)^{-2} + (3/8)8^{-2}] = 2.943$ $A(3) = 80[(2/8)4^{-1} + (2/8)(4.5)^{-1} + (1/8)(6.5)^{-1} + (3/8)8^{-1}] = 14.733$
Results	System approximate reliability function $\overline{\boldsymbol{R}}'_{80}(t,\cdot) = \overline{\Re}'_2((t - b_n(u))/a_n(u),\cdot) = [1, \exp[-0.627t], \exp[-2.943t], \exp[-14.733t]]$ for $t \geq 0$ System mean lifetimes in state subsets $M(1) = 1/0.627 \cong 1.59, M(2) = 1/2.943 \cong 0.34$ $M(3) = 1/14.733 \cong 0.07$ Standard deviations of system lifetimes in state subsets $\sigma(1) \cong 1.59,\ \sigma(2) \cong 0.34,\ \sigma(3) \cong 0.07$ System mean lifetimes in particular states $\overline{M}(1) \cong 1.59 - 0.34 \cong 1.25,\ \overline{M}(2) \cong 0.34 - 0.07 \cong 0.27$ $\overline{M}(3) \cong 0.07$ System risk function $\boldsymbol{r}(t) = 1 - \exp[-A(2)t] = 1 - \exp[-2.943t]$ Exceeding moment of admissible system risk level δ $\tau = -(\log(1 - 005))/294 \cong 0.017$

Table 7.7 The Values of the Multi-state Reliability Function and the Risk Function of the Piping System

t	$\overline{\Re}'_2(t - b_n(1)/a_n(1), 1)$	$\overline{\Re}'_2(t - b_n(2)/a_n(2), 2)$	$\overline{\Re}'_2(t - b_n(3)/a_n(3), 3)$	$r(t)$
0.0	1.00000	1.00000	1.00000	0.00000
0.2	0.88214	0.55510	0.05252	0.44456
0.4	0.77818	0.30814	0.00276	0.69149
0.6	0.68647	0.17105	0.00014	0.82864
0.8	0.60556	0.09495	0.00001	0.90482
1.0	0.53419	0.05271	0.00000	0.94713
1.2	0.47123	0.02926	0.00000	0.97064
1.4	0.41570	0.01624	0.00000	0.98369
1.6	0.36670	0.00902	0.00000	0.99094
1.8	0.32349	0.00500	0.00000	0.99497
2.0	0.28536	0.00278	0.00000	0.99721
2.2	0.25173	0.00154	0.00000	0.99845
2.4	0.22206	0.00086	0.00000	0.99914
2.6	0.19589	0.00048	0.00000	0.99952
2.8	0.17280	0.00026	0.00000	0.99974

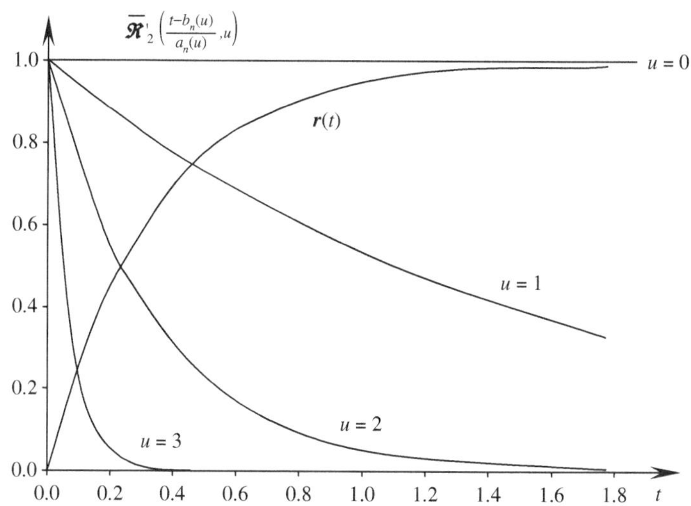

Figure 7.1 Graphs of the multi-state reliability function and the risk function of the piping system.

Table 7.8 Reliability Evaluation of the Homogeneous Parallel System

System Type	**A Homogeneous Parallel System**
Data	Number of system components $n = 30$ Number of different component types $a = 1$ Fractions of different component types $q_1 = 1$ Number of component and system states $z = 5$ Component transition rates between state subsets $\lambda_1(1) = 10^{-5},\ \lambda_1(2) = 10^{-4},\ \lambda_1(3) = 10^{-3},\ \lambda_1(4) = 10^{-2},\ \lambda_1(5) = 10^{-1}$ System reliability critical state $r = 2$ Admissible critical level of system risk $\delta = 0.05$
Calculations	$\lambda(1) = 10^{-5}, \lambda(2) = 10^{-4}, \lambda(3) = 10^{-3}, \lambda(4) = 10^{-2}$ $\lambda(5) = 10^{-1}$ $d(u) = 1,\ u = 1, 2, 3, 4, 5$ $A(1) = 10^{-5},\ A(2) = 10^{-4},\ A(3) = 10^{-3},\ A(4) = 10^{-2},\ A(5) = 10^{-1}$ $B(u) = \log 80 \cong 3.4012, u = 1, 2, \ldots, z$
Results	System reliability function $\boldsymbol{R}_{30}(t,\cdot) \cong \Re_3((t - b_n(u))/a_n(u), \cdot)$ $= [1, 1 - \exp[-\exp[-000001t + 34012]]$ $1 - \exp[-\exp[-0.0001t + 3.4012]]$ $1 - \exp[-\exp[-0.001t + 3.4012]]$ $1 - \exp[-\exp[-0.01t + 3.4012]]$ $1 - \exp[-\exp[-0.1t + 3.4012]]]$ for $t \in (-\infty, \infty)$ System mean lifetimes in state subsets $M(1) \cong 397840,\ M(2) \cong 39784,\ M(3) \cong 3978.4,\ M(4) \cong 397.84$ $M(5) \cong 39.784$ Standard deviations of system lifetimes in state subsets $\sigma(1) \cong 128190,\ \sigma(2) \cong 12819,\ \sigma(3) \cong 1281.9,\ \sigma(4) \cong 128.19$ $\sigma(5) \cong 12.819$ System mean lifetimes in particular states $\overline{M}(1) \cong 358056,\ \overline{M}(2) \cong 35806,\ \overline{M}(3) \cong 3581,\ \overline{M}(4) \cong 358$ $\overline{M}(5) \cong 40$ System risk function $\boldsymbol{r}(t) \cong \exp[-\exp[-00001t + 34012]],$ Exceeding moment of admissible system risk level δ $\tau = [3.4012 - \log(-\log 0.05)]/0.0001 \cong 23040$

and 50 pipe segments that have exponential reliability functions with transition rates between state subsets

$$\lambda_{22}(u) = (2.1)^{u-6}, \quad u = 1, 2, 3, 4, 5$$

Thus the considered piping system is a non-homogeneous regular six-state series–parallel system with parameters

$$k_n = k = 3,\ l_n = 100,\ a = 2,\ q_1 = 2/3,\ q_2 = 1/3$$
$$q_1 = 2,\ p_{11} = 0.4,\ p_{12} = 0.6$$
$$e_2 = 2,\ p_{21} = 0.5,\ p_{22} = 0.5$$

We will perform its reliability evaluation according to the procedure given in Table 7.4 (Case 2). This procedure is presented in Table 7.9.

$$\sigma = \sqrt{5/[4A_1]^2] + 1/[A_2]^2 + 1/[2A_1 + A_2]^2 + 4/[(A_1 + A_2)(2A_1 + A_2)] - 8/[A_1 + A_2]^2}$$

The behaviour of the system multi-state reliability function and risk function is illustrated in Table 7.10 and Figure 7.2.

Table 7.9 Reliability Evaluation of the Piping System

System Type	A Non-homogeneous Regular Series–Parallel System
Data	System shape $k_n \to k,\ l_n \to \infty$ Number of series subsystems linked in parallel $k_n = 3$ Number of components in series subsystems $l_n = 100$ Number of subsystem types $a = 2$ Fractions of different subsystem types $q_1 = 2/3,\ q_2 = 1/3$ Numbers of component types in series subsystems $e_1 = 2,\ e_2 = 2$ Fractions of different component types in series subsystems $p_{11} = 0.4,\ p_{12} = 0.6$ $p_{21} = 0.5,\ p_{22} = 0.5$ Number of component and system states $z = 5$ Component transition rates between state subsets $\lambda_{11}(u) = (2.5)^{u-6},\ \lambda_{12}(u) = 2^{u-6}$ $\lambda_{21}(u) = (1.9)^{u-6},\ \lambda_{22}(u) = (2.1)^{u-6},\ u = 1, 2, 3, 4, 5$

(Continued)

Table 7.9 (Continued)

System Type	**A Non-homogeneous Regular Series–Parallel System**
	System reliability critical state $r = 2$ Admissible level of system risk $\delta = 0.05$
Calculations	***Case 2***. $k_n \to k,\ l_n \to \infty$ $\lambda_1(u) = 0.4 \cdot (2.5)^{u-6} + 0.6 \cdot 2^{u-6}$ $\lambda_2(u) = 0.5 \cdot (1.9)^{u-6} + 0.5 \cdot (2.1)^{u-6}$ $A_1(u) = 40 \cdot (2.5)^{u-6} + 60 \cdot 2^{u-6}$ $A_2(u) = 50 \cdot (1.9)^{u-6} + 50 \cdot (2.1)^{u-6},\ u = 1, 2, 3, 4, 5$
Results	***Case 2***. $k_n \to k,\ l_n \to \infty$ System reliability function $\boldsymbol{R}'_{3,100}(t,\cdot) = \Re'_9((t - b_n(u))/a_n(u), \cdot)$ $= [1, 1 - [1 - \exp[-2.285t]]^2[1 - \exp[-3.244t]]$ $1 - [1 - \exp[-4.774t]]^2[1 - \exp[-6.408t]]$ $1 - [1 - \exp[-10.060t]]^2[1 - \exp[-19.096t]]$ $1 - [1 - \exp[-21.400t]]^2[1 - \exp[-25.188t]]$ $1 - [1 - \exp[-46.000t]]^2[1 - \exp[-50.125t]]]$ for $t \geq 0$ System mean lifetimes in state subsets $M(u) = \int_0^{\infty} [1 - [1 - \exp[-A_1(u)t]]^2[1 - \exp[-A_2(u)]]]dt$ $= 3/[2A_1(u)] + 1/A_2(u) + 1/[2A_1(u) + A_2(u)] - 2/[A_1(u) + A_2(u)]$ i.e. $M(1) = 0.74,\ M(2) = 0.35,\ M(3) = 0.16,\ M(4) = 0.05,\ M(5) = 0.04$ Standard deviations of system lifetimes in state subsets $\sigma(u) = \sqrt{N(u) - M^2(u)}\ (*)$ $N(u) = 2\int_0^{\infty} t[1 - [1 - \exp[-A_1(u)t]]^2[1 - \exp[-A_2(u)t]]]dt$ i.e. $\sigma(1) = 0.43,\ \sigma(2) = 0.20,\ \sigma(3) = 0.10,\ \sigma(4) = 0.05,\ \sigma(5) = 0.02$ System mean lifetimes in particular states $\overline{M}(1) = 0.39,\ \overline{M}(2) = 0.19,\ \overline{M}(3) = 0.11,\ \overline{M}(4) = 0.01$ $\overline{M}(5) = 0.04$ System risk function $\boldsymbol{r}(t) = [1 - \exp[-4.774t]]^2[1 - \exp[-6.408t]]$ Exceeding moment of admissible system risk level δ $\tau = \boldsymbol{r}^{-1}(\delta) = 0.09$ years

Table 7.10 The Values of the Piping System Multi-state Reliability Function and its Risk Function

	$\Re'_9((t-b_n(u))/a_n(u),u)$					
t	$u=1$	$u=2$	$u=3$	$u=4$	$u=5$	$r(t)$
0.00	1.0000	1.0000	1.0000	1.0000	1.0000	0.0000
0.05	0.9983	0.9876	0.9039	0.6909	0.2565	0.0124
0.10	0.9884	0.9318	0.6572	0.2842	0.0265	0.0682
0.20	0.9358	0.7267	0.2660	0.0338	0.0002	0.2733
0.25	0.8948	0.6123	0.1623	0.0113	0.0000	0.3877
0.30	0.8468	0.5053	0.0983	0.0038	0.0000	0.4947
0.35	0.7943	0.4108	0.0594	0.0013	0.0000	0.5892
0.40	0.7391	0.3303	0.0359	0.0004	0.0000	0.6697
0.45	0.6832	0.2634	0.0217	0.0001	0.0000	0.7366
0.50	0.6279	0.2088	0.0131	0.0000	0.0000	0.7912
0.55	0.5741	0.1649	0.0079	0.0000	0.0000	0.8351
0.60	0.5228	0.1298	0.0048	0.0000	0.0000	0.8702
0.65	0.4743	0.1020	0.0029	0.0000	0.0000	0.8980
0.70	0.4289	0.0800	0.0017	0.0000	0.0000	0.9200

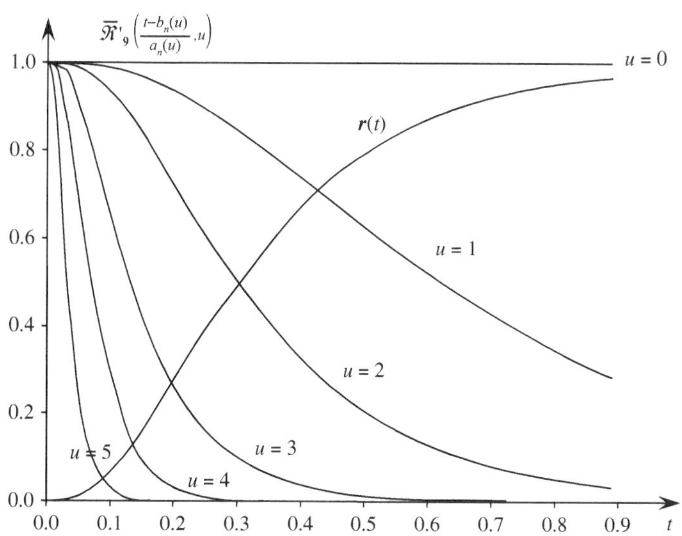

Figure 7.2 The graphs of the multi-state reliability function of the piping system and its risk function.

Example 7.4. (A Bus Transportation System)
A bus transportation company has $k_n = 100$ transportation lines. In each of the lines passengers have at their disposal $l_n = 3$ buses on which they may travel. Buses have six-state reliability functions, i.e. $z = 5$. Forty of the lines in the considered system have two buses that have exponential reliability functions with transition rates between state subsets

$$\lambda_{11}(u) = (2.5)^{(2u-12)/3}, \quad u = 1, 2, 3, 4, 5$$

and one bus that has exponential reliability functions with transition rates between state subsets

$$\lambda_{12}(u) = 2^{(2u-12)/3}, \quad u = 1, 2, 3, 4, 5$$

The remaining 60 lines have two buses that have exponential reliability functions with transition rates between state subsets

$$\lambda_{21}(u) = (1.9)^{(2u-12)/3}, \quad u = 1, 2, 3, 4, 5$$

and one bus that has exponential reliability functions with transition rates between state subsets

$$\lambda_{22}(u) = (2.1)^{(2u-12)/3}, \quad u = 1, 2, 3, 4, 5$$

The bus transportation system is able to perform its transportation tasks if at least one bus on each of the lines is not failed. Thus, according to Definition 3.21, it is a multi-state non-homogeneous regular parallel–series system with parameters

$$k_n = n = 100,\ l_n = l = 3,\ a = 2,\ q_1 = 0.4,\ q_2 = 0.6$$
$$e_1 = 2,\ p_{11} = 2/3,\ p_{12} = 1/3,\ e_2 = 2,\ p_{21} = 2/3,\ p_{22} = 1/3$$

Its reliability evaluation is performed in Table 7.11, according to the algorithm given in Table 7.5 (Case 1).

Table 7.11 Reliability Evaluation of the Bus Transportation System

System Type	A Non-homogeneous Regular Parallel–Series System
Data	System shape:
	Number of parallel subsystems linked in series $k_n = 100$
	Number of components in parallel subsystems $l_n = 3$
	Number of types of parallel subsystems $a = 2$

(*Continued*)

Table 7.11 (Continued)

System Type	**A Non-homogeneous Regular Parallel–Series System**
	Fractions of parallel subsystems of particular types $q_1 = 0.4,\ q_2 = 0.6$ Numbers of component types in parallel subsystems $e_1 = 2,\ e_2 = 2$ Fractions of components of particular types in parallel subsystems $p_{11} = 2/3,\ p_{12} = 1/3$ $p_{21} = 2/3,\ p_{22} = 1/3$ Number of component and system states $z = 5$ Component transitions rates between state subsets $\lambda_{11}(u) = (2.5)^{(2u-12)/3},\ \lambda_{12}(u) = 2^{(2u-12)/3}$ $\lambda_{21}(u) = (1.9)^{(2u-12)/3},\ \lambda_{22}(u) = (2.1)^{(2u-12)/3},\ u = 1,2,3,4,5$ A system critical state $r = 2$ System risk admissible level $\delta = 0.05$
Calculations	***Case 1.*** System shape: $k_n \to \infty,\ l_n \to l$ $\lambda_1(u) = (2.5)^{(2u-12)2/9} \cdot 2^{(2u-12)1/9}$ $\lambda_2(u) = (1.9)^{(2u-12)2/9} \cdot (2.1)^{(2u-12)1/9}$ $A(u) = 40(2.5)^{(2u-12)2/3} \cdot 2^{(2u-12)/3} + 60(1.9)^{(2u-12)2/3} \cdot (2.1)^{(2u-12)/3}$
Results	***Case 1.*** System shape: $k_n \to \infty,\ l_n \to l$ System reliability function $\overline{\boldsymbol{R}}'_{100,3}(t,\cdot) \cong \overline{\mathfrak{R}}'_2((t-b_n(u))/a_n(u),\cdot)$ $= [1, \exp[-0.079t^3], \exp[-0.318t^3], \exp[-1.300t^3]$ $\exp[-5.408t^3], \exp[-22.974t^3]]$ for $t \geq 0$ Mean system sojourn times in state subsets $M(u) = \Gamma(4/3)[A(u)]^{1/3},\ u = 1,2,\ldots,5$, i.e. $M(1) = 2.08,\ M(2) = 1.31,\ M(3) = 0.82,\ M(4) = 0.51,\ M(5) = 0.31$ Standard deviations of system sojourn times in state subsets $\sigma(u) = [\Gamma(5/3) - \Gamma^2(4/3)]^{1/2}[A(u)]^{-1/3},\ u = 1,2,\ldots,5$, i.e. $\sigma(1) = 0.57,\ \sigma(2) = 0.23,\ \sigma(3) = 0.09,\ \sigma(4) = 0.03,\ \sigma(5) = 0.01$ Mean system sojourn times in particular states $\overline{M}(1) = 0.77,\ \overline{M}(2) = 0.49,\ \overline{M}(3) = 0.31,\ \overline{M}(4) = 0.20,$ $\overline{M}(5) = 0.31$ System risk function $r(t) = 1 - \exp[-0.318t^3]$ Exceeding moment of admissible system risk level δ $\tau = r^{-1}(\delta) = [(-\log(1-\delta))/0.318]^{1/3} = 0.54$ years

The behaviour of the approximate multi-state reliability function components and the risk function of the bus transportation system are presented in Table 7.12 and Figure 7.3.

Table 7.12 Values of the Multi-state Reliability Function and the Risk Function of the Bus Transportation System

	$\overline{\mathfrak{R}}'_2((t-b_n(u))/a_n(u),\cdot)$					
t	$\overline{u}=1$	$u=2$	$u=3$	$u=4$	$u=5$	$r(t)$
0.0	1.0000	1.0000	1.0000	1.0000	1.0000	0.0000
0.4	0.9950	0.9799	0.9222	0.7092	0.8317	0.0191
0.6	0.9811	0.9336	0.7609	0.3136	0.2289	0.0629
0.8	0.9604	0.8497	0.5233	0.0640	0.0069	0.1428
1.0	0.9240	0.7276	0.2822	0.0047	0.0000	0.2599
1.2	0.8723	0.5772	0.1124	0.0001	0.0000	0.4056
1.4	0.8051	0.4179	0.0311	0.0000	0.0000	0.5622
1.6	0.7236	0.2718	0.0056	0.0000	0.0000	0.7086
1.8	0.6308	0.1565	0.0006	0.0000	0.0000	0.8272
2.0	0.5315	0.0786	0.0000	0.0000	0.0000	0.9100
2.2	0.4312	0.0338	0.0000	0.0000	0.0000	0.9594
2.4	0.3355	0.0123	0.0000	0.0000	0.0000	0.9844
2.6	0.2494	0.0037	0.0000	0.0000	0.0000	0.9950

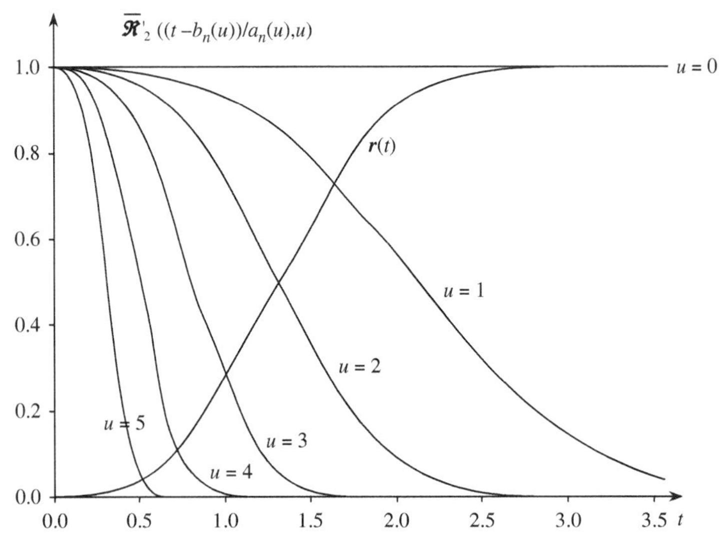

Figure 7.3 Graphs of the multi-state reliability function and the risk function of the bus transportation system.

8 Large Systems Related Problems

8.1 Domains of Attraction for System Limit Reliability Functions

The problem of domains of attraction for the limit reliability functions of two-state systems considered in this book is solved completely in Refs. [25,93,95,119,120]. We will illustrate this problem partly for two-state series homogeneous systems only. From Theorem 4.1 it follows that the class of limit reliability functions for a homogeneous series system is composed of three functions, $\overline{\mathfrak{R}}_i(t)$, $i = 1, 2, 3$, defined by Eqs. (4.3)–(4.5). Now we will determine domains of attraction $D_{\overline{\mathfrak{R}}_i}$ for these fixed functions, that is we will determine the conditions which the reliability functions $R(t)$ of the particular components of the homogeneous series system have to satisfy in order that the system limit reliability function is one of the reliability functions $\overline{\mathfrak{R}}_i(t)$, $i = 1,2,3$.

Proposition 8.1
If $R(t)$ is a reliability function of the homogeneous series system components, then

$$R(t) \in D_{\overline{\mathfrak{R}}_1}$$

if and only if

$$\lim_{r \to -\infty} \frac{1 - R(r)}{1 - R(rt)} = t^{\alpha} \quad \text{for } t > 0$$

Proposition 8.2
If $R(t)$ is a reliability function of the homogeneous series system components, then

$$R(t) \in D_{\overline{\mathfrak{R}}_2}$$

if and only if

i. $\exists y \in (-\infty, \infty) \quad R(y) = 1 \text{ and } R(y + \varepsilon) < 1 \quad \text{for } \varepsilon > 0$

Reliability of Large Systems. DOI: http://dx.doi.org/10.1016/B978-0-08-099949-4.00008-8

ii. $$\lim_{r \to 0^+} \frac{1 - R(rt + y)}{1 - R(r + y)} = t^{\alpha} \quad \text{for } t > 0$$

Proposition 8.3
If $R(t)$ is a reliability function of the homogeneous series system components, then

$$R(t) \in D_{\overline{\mathfrak{R}}_3}$$

if and only if

$$\lim_{n \to \infty} n[1 - R(a_n t + b_n)] = e^t \quad \text{for } t \in (-\infty, \infty)$$

with

$$b_n = \inf\left\{ t{:}R(t+0) \leq 1 - \frac{1}{n} \leq R(t-0) \right\}$$

$$a_n = \inf\left\{ t{:}R(t(1+0) + b_n) \leq 1 - \frac{e}{n} \leq R(t(1-0) + b_n) \right\}$$

Example 8.1
If components of the homogeneous series system have reliability functions

$$R(t) = \begin{cases} 1, & t < 0 \\ 1 - t, & 0 \leq t < 1 \\ 0, & t \geq 1 \end{cases}$$

then

$$R(t) \in D_{\overline{\mathfrak{R}}_2}$$

Motivation: Since

i. $R(0) = 1$ and $R(\varepsilon) < 1$ for each $\varepsilon > 0$

ii. $$\lim_{r \to 0^+} \frac{1 - R(rt + y)}{1 - R(r + y)} = \lim_{r \to 0^+} \frac{1 - (1 - rt)}{1 - (1 - r)} = t \quad \text{for } t > 0$$

then by Proposition 8.2, $R(t) \in D_{\overline{\Re}_2}$, where

$$\overline{\Re}_2(t) = \begin{cases} 1, & t < 0 \\ \exp[-t], & t \geq 0 \end{cases}$$

The results of the analysis on domains of attraction for limit reliability functions of two-state systems may automatically be transmitted to multi-state systems. To do this, it is sufficient to apply theorems about two-state systems such as the ones presented here to each vector coordinate of the multi-state reliability functions [93].

8.2 Speed of Convergence of System Reliability Function Sequences

A practically important problem of accuracy of the asymptotic approach to large systems reliability evaluation is concerned with the speed of convergence of the system reliability sequence to its limit reliability function. This problem is progressively solved for two-state systems in Refs. [31,93,151–153], and we will illustrate it by analysing the speed of convergence of the homogeneous series–parallel system reliability sequence to its limit reliability function [93].

Proposition 8.4
If

$$\lim_{n \to \infty} k_n = \infty, \quad \lim_{n \to \infty} l_n = l$$

and a component reliability function $R(t)$ of the regular homogeneous two-state series–parallel system satisfies the inequality

$$0 < [R(a_n t + b_n)]^l < \frac{1}{2}$$

then for the limit reliability function $\Re_3(t)$ defined by Eq. (4.54) and for the system exact reliability function $R_{k_n,l_n}(t)$ given by Eq. (2.5), we have

$$\left| \Re_3(t) - \boldsymbol{R}_{k_n,l_n}(a_n t + b_n) + \frac{1}{2}[k_n R^l(a_n t + b_n)]^2 \exp\left[-k_n R^l(a_n t + b_n)\right] \right|$$
$$\leq \left| R^l n(a_n t + b_n) - R^l(a_n t + b_n) \right| \cdot k_n + \frac{1}{k_n^2} C(k_n, R^l(a_n t + b_n)) + \left| \int_{k_n R^l(a_n t + b_n)}^{V_{3(t)}} e^{-x} \, dx \right|$$

where

$$C(k_n, R^l(a_n t + b_n))$$
$$= \frac{1}{2}\pi \exp\left[2k_n R^l(a_n t + b_n)\right] \cdot [k_n R^l(a_n t + b_n)]^3$$
$$\cdot \left[\frac{4}{3[1 - 2R^l(a_n t + b_n)]} + \left[\frac{16}{9}\frac{n[R^l(a_n t + b_n)]^3}{[1 - 2R^l(at + b_n)]^2} + \frac{8}{3}\frac{n[R^l(a_n t + b_n)]^2}{[1 - 2R^l(at + b_n)]}\right.\right.$$
$$\left.\left. + k_n R^l(a_n t + b_n)] \cdot \exp\left[2k_n[R^l(a_n t + b_n)]^2 + \frac{8}{3}\frac{n[R^l(a_n t + b_n)]^3}{[1 - 2R^l(at + b_n)]}\right]\right]\right]$$

Example 8.2

A homogeneous series–parallel system is composed of $k_n = 100$ series subsystems linked in parallel. Each series subsystem is composed of $l_n = 4$ components with Rayleigh reliability function

$$R(t) = \begin{cases} 1, & t \leq 0 \\ \exp[-t^2], & t > 0 \end{cases} \tag{8.1}$$

Assuming

$$a_n = \frac{1}{2\sqrt{l_n \log k_n}}, \quad b_n = \sqrt{\frac{1}{l_n}\log k_n}$$

and considering the relationships

$$\lim_{n\to\infty} k_n = \infty, \quad \lim_{n\to\infty} l_n = l$$

for $t \in (-\infty, \infty)$ and sufficiently large n, we have

$$a_n t + b_n = \frac{t}{2\sqrt{l_n \log k_n}} + \sqrt{\frac{1}{l_n}\log k_n} > 0$$

Hence and from Eq. (8.1), we have

$$V(t) = \lim_{n\to\infty} k_n[R(a_n t + b_n)]^{l_n}$$
$$= \lim_{n\to\infty} k_n \exp\left[-l_n\left(\frac{t}{2\sqrt{l_n \log k_n}} + \sqrt{\frac{1}{l_n}\log k_n}\right)^2\right]$$
$$= \lim_{n\to\infty} k_n \exp\left[-l_n\left(\frac{1}{4l_n \log k_n}t^2 + \frac{1}{l_n}t + \frac{1}{l_n}\log k_n\right)\right]$$
$$= \lim_{n\to\infty} k_n \exp\left[-\frac{1}{4\log k_n}t^2 - t - \log k_n\right]$$
$$= \lim_{n\to\infty} \exp\left[-\frac{1}{4\log k_n}t^2 - t\right] = \exp[-t] \quad \text{for } t \in (-\infty, \infty)$$

Thus, considering Lemma 4.12, the system limit reliability function is given by

$$\mathfrak{R}_3(t) = 1 - \exp[-\exp[-t]] \quad \text{for } t \in (-\infty, \infty)$$

Moreover, from Proposition 8.4, we have

$$\left| \mathfrak{R}_3(t) - R_{100,4}(a_{100}t + b_{100}) + \frac{1}{2}[100R^4(a_{100}t + b_{100})]^2 \exp\left[-100R^4(a_{100}t + b_{100})\right] \right|$$
$$\leq \frac{1}{10{,}000} C(100, R^4(a_{100}t + b_{100})) + \left| \int_{100R^4(a_{100}t+b_{100})}^{V_3(t)} e^{-x} dx \right|$$

where for $t > 0$, we have

$$C(100, R^4(a_{100}t + b_{100}))$$
$$= \frac{1}{2}\pi \exp\left[2\exp\left[-\frac{1}{4\log 100}t^2 - t\right]\right]\left[\exp\left[-\frac{1}{4\log 100}t^2 - t\right]\right]^3$$
$$\cdot \left[\frac{200}{3[50 - \exp[-(1/4\log 100)t^2 - t]]} + \left[\frac{16}{9}\frac{50\left[\exp\left[-(1/4\log 100)t^2 - t\right]\right]^3}{\left[50 - \exp\left[-(1/4\log 100)t^2 - t\right]\right]^2}\right]\right.$$
$$+ \frac{8}{3}\frac{50\left[\exp\left[-(1/4\log 100)t^2 - t\right]\right]^2}{50 - \exp\left[-(1/4\log 100)t^2 - t\right]} + \exp\left[-\frac{1}{4\log 100}t^2 - t\right]\Bigg]$$
$$\cdot \exp\left[2\left[\exp\left[-\frac{1}{4\log 100}t^2 - t\right]\right]^2 + \frac{8}{3}\frac{50[\exp[-1/4\log 100\, t^2 - t]]^3}{50 - \exp[-1/4\log 100\, t^2 - t]}\right]$$

These evaluations, i.e. the system limit reliability function and its lower and upper bounds for $k_n = 100$ and $l_n = 4$, are presented in Table 8.1 and Figure 8.1. Additionally, to illustrate the speed of convergence, the evaluations for $k_n = 10$ and $l_n = 4$ and for $k_n = 50$ and $l_n = 4$ are presented in Tables 8.2 and 8.3 and in Figures 8.2 and 8.3.

The results of the analysis of the speed of reliability sequences' convergence to limit reliability functions for two-state systems may automatically be transferred to multi-state systems. To do this, it is sufficient to apply theorems on two-state systems like the one presented here to each vector coordinate of the multi-state reliability functions [93].

8.3 Reliability of Large Series–'*m* Out of *n*' Systems

Definition 8.1
A two-state system is called a series–'m out of k_n' system if its lifetime T is given by

$$T = T_{(k_n - m + 1)}, \quad m = 1, 2, \ldots, k_n$$

Table 8.1 The Evaluation of the Speed of Convergence of Reliability Function Sequences for a Homogeneous Series–Parallel System ($k_n = 100$, $l_n = 4$)

t	$\Re_3(t)$	Δ_1	Δ_2	$\Re_3(t) + \Delta_1$	$\Re_3(t) + \Delta_2$	$R_{100}(a_{100}t + b_{100})$
0.00	0.6321	−0.0028	0.0028	0.6293	0.6349	0.6340
0.25	0.5410	−0.0019	0.0020	0.5391	0.5430	0.5412
0.50	0.4548	−0.0047	0.0047	0.4501	0.4595	0.4513
0.75	0.3765	−0.0090	0.0090	0.3675	0.3855	0.3682
1.00	0.3078	−0.0136	0.0136	0.2942	0.3214	0.2946
1.25	0.2491	−0.0177	0.0177	0.2314	0.2668	0.2317
1.50	0.2000	−0.0208	0.0208	0.1792	0.2208	0.1794
1.75	0.1595	−0.0227	0.0227	0.1368	0.1822	0.1369
2.00	0.1266	−0.0234	0.0234	0.1032	0.1500	0.1033
2.25	0.1000	−0.0231	0.0231	0.0770	0.1231	0.0770
2.50	0.0788	−0.0220	0.0220	0.0568	0.1008	0.0568
2.75	0.0619	−0.0204	0.0204	0.0415	0.0823	0.0415
3.00	0.0486	−0.0185	0.0185	0.0301	0.0671	0.0301
3.25	0.0380	−0.0164	0.0164	0.0216	0.0544	0.0216
3.50	0.0297	−0.0143	0.0143	0.0154	0.0441	0.0154
3.75	0.0232	−0.0123	0.0123	0.0109	0.0356	0.0109

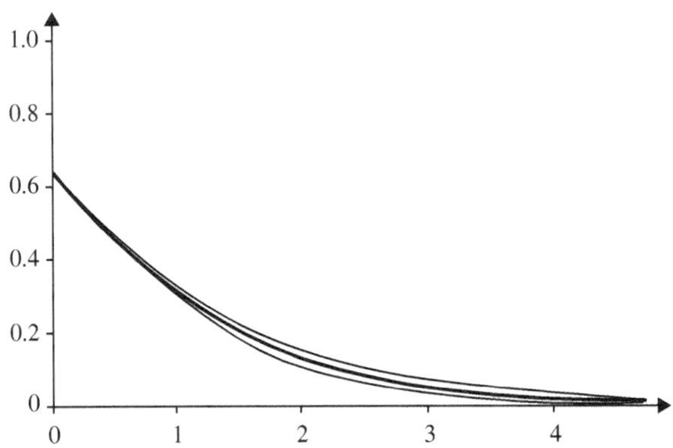

Figure 8.1 The graphs of the limit reliability function and their lower and upper evaluations for a homogeneous series–parallel system ($k_n = 100$, $l_n = 4$).

where $T_{(k_n-m+1)}$ is the mth maximal order statistic in the set of random variables

$$T_i = \min_{1 \le j \le l_i} \{T_{ij}\}, \quad i = 1, 2, \ldots, k_n$$

The above definition means that the series–'m out of k_n' system is composed of k_n series subsystems and it is not failed if and only if at least m out of its k_n series subsystems are not failed.

Table 8.2 The Evaluation of the Speed of Convergence of Reliability Function Sequences for a Homogeneous Series–Parallel System ($k_n = 10$, $l_n = 4$)

t	$\mathfrak{R}_3(t)$	Δ_I	Δ_2	$\mathfrak{R}_3(t) + \Delta_I$	$\mathfrak{R}_3(t) + \Delta_2$	$R_{10}(a_{10}t + b_{10})$
0.00	0.6321	−0.3911	0.3948	0.2410	1.0269	0.6513
0.25	0.5410	−0.0930	0.0957	0.4481	0.6368	0.5530
0.50	0.4548	−0.0318	0.0337	0.4229	0.4885	0.4558
0.75	0.3765	−0.0237	0.0250	0.3527	0.4015	0.3653
1.00	0.3078	−0.0283	0.0291	0.2795	0.3369	0.2851
1.25	0.2491	−0.0347	0.0352	0.2144	0.2843	0.2171
1.50	0.2000	−0.0397	0.0400	0.1603	0.2399	0.1616
1.75	0.1595	−0.0423	0.0425	0.1172	0.2020	0.1179
2.00	0.1266	−0.0426	0.0427	0.0839	0.1693	0.0843
2.25	0.1000	−0.0410	0.0410	0.0590	0.1411	0.0592
2.50	0.0788	−0.0380	0.0380	0.0408	0.1168	0.0409
2.75	0.0619	−0.0342	0.0342	0.0277	0.0961	0.0278
3.00	0.0486	−0.0300	0.0300	0.0186	0.0786	0.0186
3.25	0.0380	−0.0258	0.0258	0.0122	0.0638	0.0122
3.50	0.0297	−0.0218	0.0218	0.0080	0.0515	0.0080
3.75	0.0232	−0.0181	0.0181	0.0051	0.0414	0.0051

Table 8.3 The Evaluation of the Speed of Convergence of Reliability Function Sequences for a Homogeneous Series–Parallel System ($k_n = 50$, $l_n = 4$)

t	$\mathfrak{R}_3(t)$	Δ_1	Δ_2	$\mathfrak{R}_3(t) + \Delta_1$	$\mathfrak{R}_3(t) + \Delta_2$	$R_{50}(a_{50}t + b_{50})$
0.00	0.6321	−0.0115	0.0116	0.6206	0.6438	0.6358
0.25	0.5410	−0.0044	0.0045	0.5366	0.5456	0.5424
0.50	0.4548	−0.0061	0.0062	0.4487	0.4609	0.4515
0.75	0.3765	−0.0107	0.0108	0.3657	0.3873	0.3673
1.00	0.3078	−0.0160	0.0160	0.2918	0.3238	0.2927
1.25	0.2491	−0.0207	0.0208	0.2284	0.2699	0.2289
1.50	0.2000	−0.0243	0.0243	0.1757	0.2243	0.1760
1.75	0.1595	−0.0264	0.0264	0.1331	0.1859	0.1333
2.00	0.1266	−0.0271	0.0271	0.0995	0.1537	0.0996
2.25	0.1000	−0.0266	0.0266	0.0734	0.1266	0.0735
2.50	0.0788	−0.0252	0.0252	0.0536	0.1040	0.0536
2.75	0.0619	−0.0233	0.0233	0.0387	0.0852	0.0387
3.00	0.0486	−0.0209	0.0209	0.0276	0.0695	0.0276
3.25	0.0380	−0.0185	0.0185	0.0195	0.0565	0.0196
3.50	0.0297	−0.0160	0.0160	0.0137	0.0458	0.0137
3.75	0.0232	−0.0137	0.0137	0.0095	0.0370	0.0095

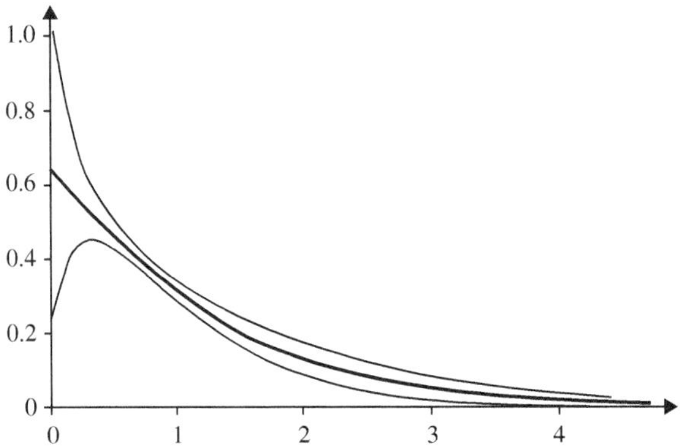

Figure 8.2 The graphs of the limit reliability function and their lower and upper evaluations for a homogeneous series–parallel system ($k_n = 10$, $l_n = 4$).

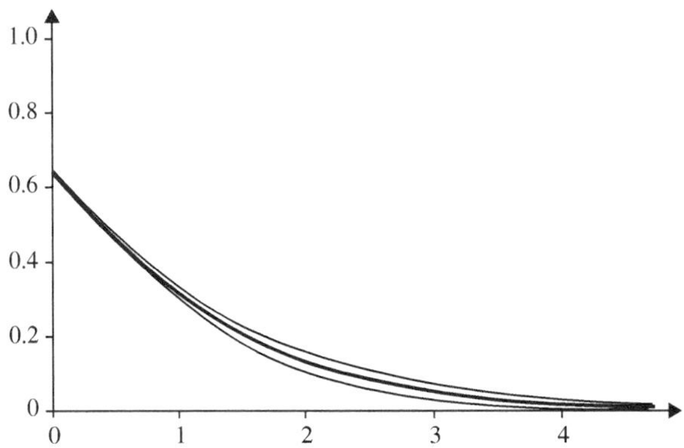

Figure 8.3 The graphs of the limit reliability function and their lower and upper evaluations for a homogeneous series–parallel system ($k_n = 50$, $l_n = 4$).

The series–'m out of k_n' system is a series–parallel for $m = 1$ and it becomes a series system for $m = k_n$.

The diagram of a series–'m out of k_n' system is shown in Figure 8.4, where $i_1, i_2, \ldots, i_{k_n} \in \{1, 2, \ldots, k_n\}$ and $i_j \neq i_k$ for $j \neq k$.

The reliability function of the two-state series–'m out of k_n' system is given either by

$$\boldsymbol{R}^{(m)}_{k_n,l_1,l_2,\ldots,l_{k_n}}(t) = 1 - \sum_{\substack{r_1,r_2,\ldots,r_{k_n}=0 \\ r_1+r_2+\cdots+r_{k_n} \leq m-1}}^{1} \prod_{i=1}^{k_n} \left[\prod_{j=1}^{l_i} R_{ij}(t)\right]^{r_i} \left[1 - \prod_{j=1}^{l_i} R_{ij}(t)\right]^{1-r_i}, t \in (-\infty, \infty)$$

or by

$$\overline{\boldsymbol{R}}^{(\overline{m})}_{k_n,l_1,l_2,\ldots,l_{k_n}}(t) = \sum_{\substack{r_1,r_2,\ldots,r_{k_n}=0 \\ r_1+r_2+\cdots+r_{k_n} \leq \overline{m}}}^{1} \prod_{i=1}^{k_n} \left[1 - \prod_{j=1}^{l_i} R_{ij}(t)\right]^{r_i} \left[\prod_{j=1}^{l_i} R_{ij}(t)\right]^{1-r_i}, \ t \in (-\infty, \infty)$$

where $\overline{m} = k_n - m$.

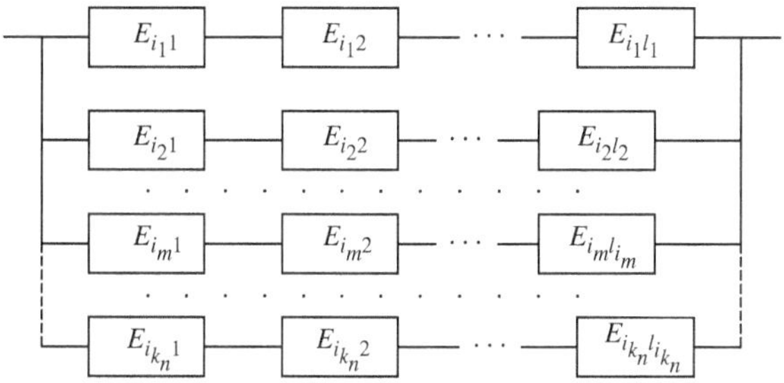

Figure 8.4 The scheme of a series–'m out of k_n' system.

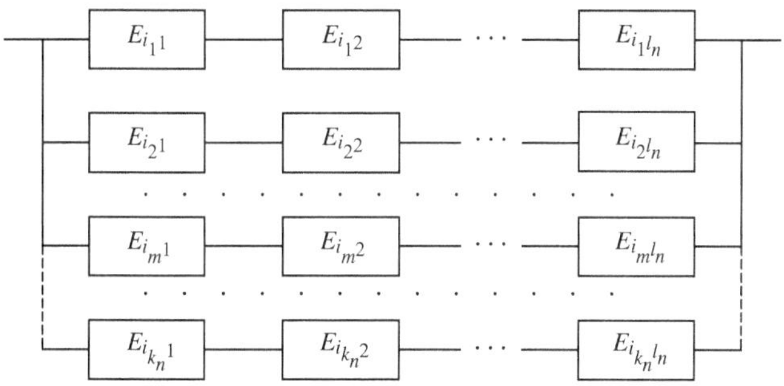

Figure 8.5 The scheme of a regular series–'m out of k_n' system.

Definition 8.2
The series–'m out of k_n' system is called regular if

$$l_1 = l_2 = \cdots = l_{k_n} = l_n, \quad l_n \in N$$

The diagram of a regular series –'m out of k_n' system is shown in Figure 8.5.

Definition 8.3
The series –'m out of k_n' system is called homogeneous if its component lifetimes T_{ij} have an identical distribution function

$$F(t) = P(T_{ij} \le t), \quad t \in (-\infty, \infty), \ i = 1, 2, \ldots, k_n, \ j = 1, 2, \ldots, l_i$$

i.e. if its components E_{ij} have the same reliability function

$$R(t) = 1 - F(t), \quad t \in (-\infty, \infty)$$

From the above definitions it follows that the reliability function of the homogeneous and regular series –'m out of k_n' system is given either by

$$\boldsymbol{R}^{(m)}_{k_n, l_n}(t) = 1 - \sum_{i=0}^{m-1} \binom{k_n}{i} [R^{l_n}(t)]^i [1 - R^{l_n}(t)]^{k_n - i}, \quad t \in (-\infty, \infty)$$

or by

$$\overline{\boldsymbol{R}}^{(\overline{m})}_{k_n,l_n}(t) = \sum_{i=0}^{\overline{m}} \binom{k_n}{i} [1 - R^{l_n}(t)]^i [R^{l_n}(t)]^{k_n - i}, \quad t \in (-\infty, \infty), \; \overline{m} = k_n - m$$

where k_n is the number of series subsystems in the 'm out of k_n' system and l_n is the number of components of the series subsystems.

Corollary 8.1
If components of the homogeneous and regular two-state series–'m out of k_n' system have Weibull reliability function

$$R(t) = 1 \quad \text{for } t < 0, \; R(t) = \exp[-\beta t^{\alpha}] \quad \text{for } t \geq 0, \; \alpha > 0, \; \beta > 0$$

then its reliability function is given either by

$$\boldsymbol{R}^{(m)}_{k_n,l_n}(t) = 1 \quad \text{for } t < 0$$

$$\boldsymbol{R}^{(m)}_{k_n,l_n}(t) = 1 - \sum_{i=0}^{m-1} \binom{k_n}{i} [\exp[-i l_n \beta t^{\alpha}]][1 - \exp[-l_n \beta t^{\alpha}]]^{k_n - i} \quad \text{for } t \geq 0$$

or by

$$\overline{\boldsymbol{R}}^{(\overline{m})}_{k_n,l_n}(t) = 1 \quad \text{for } t < 0$$

$$\overline{\boldsymbol{R}}^{(\overline{m})}_{k_n,l_n}(t) = \sum_{i=0}^{\overline{m}} \binom{k_n}{i} [1 - \exp[-l_n \beta t^{\alpha}]]^i [\exp[-(k_n - i) l_n \beta t^{\alpha}]], \quad t \geq 0, \; \overline{m} = k_n - m \tag{8.2}$$

Proposition 8.5
If components of the two-state homogeneous and regular series–'m out of k_n' system have Weibull reliability function

$$R(t) = 1 \quad \text{for } t < 0, \; R(t) = \exp[-\beta t^{\alpha}] \quad \text{for } t \geq 0, \; \alpha > 0, \; \beta > 0$$

and

$$k_n = n, \; n - m = \overline{m} = \text{constant} \quad (m/n \to 1 \text{ as } n \to \infty)$$
$$a_n = (\beta l_n n)^{-1/\alpha}, \; b_n = 0$$

then

$$\overline{\mathfrak{R}}^{(1)}_9(t) = \begin{cases} 1 & \text{for } t < 0 \\ \sum_{i=0}^{\overline{m}} \frac{t^{i\alpha}}{i!} \exp[-t^{\alpha}] & \text{for } t \geq 0, \; \alpha > 0 \end{cases} \tag{8.3}$$

is its limit reliability function.
Motivation: We will use Lemma 4.11 with condition (4.34) modified into the form

$$\overline{V}(t) = \lim_{n\to\infty} k_n[1 - R^{l_n}(a_n t + b_n)] \quad \text{for } t \in C_{\overline{V}}$$

Since

$$a_n t + b_n = (\beta l_n n)^{-1/\alpha} \quad \text{for } t < 0$$

and

$$a_n t + b_n = (\beta l_n n)^{-1/\alpha} \quad \text{for } t \geq 0$$

then

$$R(a_n t + b_n) = 1 \quad \text{for } t < 0 \text{ and } R(a_n t + b_n) = \exp[-\beta[(\beta l_n n)^{-1/\alpha} t]^\alpha] \quad \text{for } t \geq 0$$

Hence

$$\overline{V}(t) = \lim_{n\to\infty} k_n[1 - R^{l_n}(a_n t + b_n)] = 0 \quad \text{for } t < 0$$

and

$$\begin{aligned}\overline{V}(t) &= \lim_{n\to\infty} k_n[1 - R^{l_n}(a_n t + b_n)]\\ &= \lim_{n\to\infty} n[1 - \exp[-l_n\beta(a_n t + b_n)^\alpha]]\\ &= \lim_{n\to\infty} n\left[1 - \exp\left[-\frac{1}{n}t^\alpha\right]\right]\\ &= \lim_{n\to\infty} n\left[\frac{1}{n}t^\alpha + o\left(\frac{1}{n}\right)\right] = t^\alpha \quad \text{for } t \geq 0\end{aligned}$$

Thus, from Lemma 4.11, $\overline{\Re}_9^{(1)}(t)$ given by Eq. (8.3) is the system limit reliability function.

Example 8.3
An illumination of a sports hall is constructed of $k_n = 30$ rows of glow lamps. Each row is composed of $l_n = 10$ series-connected glow lamps. The lighting is good enough if at least 25 of the rows work. This lighting may be considered a series–'25 out of 30' system. If the reliability functions of system components are Weibull with parameters $\alpha = 2$, $\beta = 0.0001$, i.e. if they are given by the formula

$$R(t) = \begin{cases} 1, & t < 0 \\ \exp[-0.0001t^2, & t \geq 0 \end{cases}$$

then according to Eq. (8.2), the reliability function of the system is given by

$$\overline{\boldsymbol{R}}^{(5)}_{30,10}(t) = \sum_{i=0}^{5} \binom{30}{i} [1 - R^{10}(t)]^i [R^{10}(t)]^{30-i}$$
$$= \sum_{i=0}^{5} \binom{30}{i} (1 - \exp[-10^{-3}t^2])^i \exp[-10^{-3}t^2(30-i)]$$

Assuming

$$a_n = \frac{1}{(\beta l_n k_n)^{1/\alpha}} = 10\sqrt{3^{-1}}, \quad b_n = 0$$

from Proposition 8.5 and using Eq. (1.1), we get

$$\overline{\boldsymbol{R}}^{(5)}_{30,10}(t) \cong \overline{\mathfrak{R}}^{(1)}_{9}(10^{-1}\sqrt{3}t) = \begin{cases} 1 & \text{for } t < 0 \\ \sum_{i=0}^{5} \frac{3^i t^{2i}}{10^{2i} i!} \exp\left[-10^{-2} 3t^2\right] & \text{for } t \geq 0 \end{cases}$$

The graphs of the system exact reliability function $\overline{\boldsymbol{R}}^{(5)}_{30,10}(t)$ and its approximate reliability function $\overline{\mathfrak{R}}^{(1)}_{9}(10^{-1}\sqrt{3}t)$ are shown in Figure 8.6 [95].

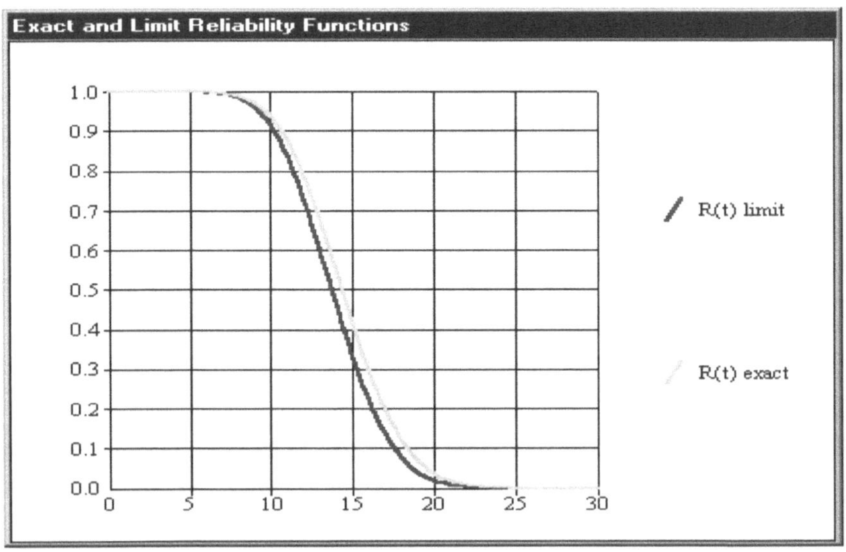

Figure 8.6 The graphs of the reliability functions $\overline{\boldsymbol{R}}^{(5)}_{30,10}(t)$ and $\overline{\mathfrak{R}}^{(1)}_{9}(10^{-1}\sqrt{3}t)$.

8.4 Reliability of Large 'm Out of n'-Series Systems

Definition 8.4

A two-state system is called an 'm_i out of l_i'−series system if its lifetime T is given by

$$T = \min_{1 \le i \le k_n} T_{(l_i - m_i + 1)}, \quad m_i = 1, 2, \ldots, l_i$$

where $T_{(l_i - m_i + 1)}$ is the m_ith maximal order statistic in the set of random variables

$$T_{i1}, T_{i2}, \ldots, T_{il_i}, \quad i = 1, 2, \ldots, k_n$$

The above definition means that the 'm_i out of l_i'−series system is composed of k_n subsystems that are 'm_i out of l_i' systems and it is not failed if all its 'm_i out of l_i' subsystems are not failed.

The 'm_i out of l_i'−series system is a parallel−series system if $m_1 = m_2 = \cdots = m_{k_n} = 1$ and it becomes a series system if $m_i = l_i$ for all $i = 1, 2, \ldots, k_n$.

The diagram of an 'm_i out of l_i'−series system is shown in Figure 8.7, where $j_1, j_2, \ldots, j_{l_i} \in \{1, 2, \ldots, l_i\}$, for $i = 1, 2, ..., k_n$ and $i_j \neq i_k$ for $j \neq k$.

The reliability function of the two-state 'm_i out of l_i'−series system is given either by

$$\overline{\boldsymbol{R}^{(m_1, m_2, \ldots, m_{k_n})}_{k_n, l_1, l_2, \ldots, l_{k_n}}}(t) = \prod_{i=1}^{k_n} \left[1 - \sum_{\substack{r_1, r_2, \ldots, r_{l_i} = 0 \\ r_1 + r_2 + \cdots + r_{l_i} \le m_i - 1}}^{1} \left[\prod_{j=1}^{l_i} R_{ij}(t) \right]^{r_i} \left[1 - \prod_{j=1}^{l_i} R_{ij}(t) \right]^{1 - r_i} \right], t \in (-\infty, \infty)$$

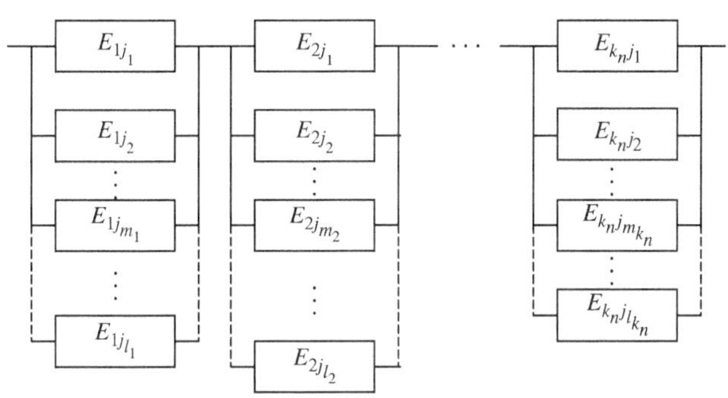

Figure 8.7 The scheme of an 'm_i out of l_i'−series system.

or by

$$\overline{\boldsymbol{R}}^{(\overline{m}_1,\overline{m}_2,\ldots,\overline{m}_{k_n})}_{k_n,l_1,l_2,\ldots,l_{k_n}}(t) = \prod_{i=1}^{k_n}\left[\sum_{\substack{r_1,r_2,\ldots,r_{l_i}=0 \\ r_1+r_2+\cdots+r_{l_i}\leq\overline{m}_i}}^{1}\left[1-\prod_{j=1}^{l_i}R_{ij}(t)\right]^{r_i}\left[\prod_{j=1}^{l_i}R_{ij}(t)\right]^{1-r_i}\right]$$

where $\overline{m}_i = l_i - m_i$, $i = 1, 2, \ldots, k_n$.

Definition 8.5
The two-state 'm_i out of l_i'$-$series system is called homogeneous if its component lifetimes T_{ij} have an identical distribution function

$$F(t) = P(T_{ij} \leq t), \quad t\in(-\infty,\infty), \; i = 1, 2, \ldots, k_n, \; j = 1, 2, \ldots l_i$$

i.e. if its components E_{ij} have the same reliability function

$$R(t) = 1 - F(t), \quad t\in(-\infty,\infty)$$

Definition 8.6
The 'm_i out of l_i'$-$series system is called regular if

$$l_1 = l_2 = \cdots = l_{k_n} = l_n \text{ and } m_1 = m_2 = \cdots = m_{k_n} = m, \quad \text{where } l_n,\ m\in N,\ m\leq l_n$$

The diagram of a regular 'm out of l_n'$-$series system is shown in Figure 8.8.
The reliability function of the two-state homogeneous and regular 'm out of l_n'$-$series system is given either by

$$\overline{\boldsymbol{R}}^{(m)}_{k_n,l_n}(t) = \left[1-\sum_{i=0}^{m-1}\binom{l_n}{i}[R(t)]^i[1-R(t)]^{l_n-i}\right]^{k_n}, \quad t\in(-\infty,\infty)$$

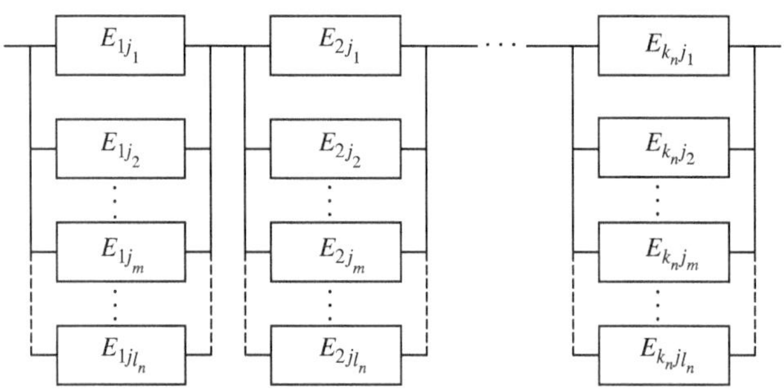

Figure 8.8 The scheme of a regular 'm out of l_n'$-$series system.

or by

$$\overline{\boldsymbol{R}^{(\overline{m})}_{k_n,l_n}}(t) = \left[\sum_{i=0}^{\overline{m}} \binom{l_n}{i}[1-R(t)]^i[R(t)]^{l_n-i}\right]^{k_n}, \quad t\in(-\infty,\infty), \ \overline{m}=l_n-m$$

where k_n is the number of 'm out of l_n' subsystems linked in series and l_n is the number of components in the 'm out of l_n' subsystems.

Corollary 8.2
If the components of the two-state homogeneous and regular 'm out of l_n'–series system have Weibull reliability function

$$R(t)=1 \quad \text{for } t<0, \ R(t)=\exp[-\beta t^{\alpha}] \quad \text{for } t\geq 0, \ \alpha>0, \ \beta>0$$

then its reliability function is given either by

$$\overline{\boldsymbol{R}^{(m)}_{k_n,l_n}}(t)=1 \quad \text{for } t<0$$

$$\overline{\boldsymbol{R}^{(m)}_{k_n,l_n}}(t) = \left[1-\sum_{i=0}^{m-1}\binom{l_n}{i}\exp[-i\beta t^{\alpha}][1-\exp[-\beta t^{\alpha}]]^{l_n-i}\right]^{k_n} \quad \text{for } t\geq 0 \tag{8.4}$$

or by

$$\overline{\boldsymbol{R}^{(\overline{m})}_{k_n,l_n}}(t)=1 \quad \text{for } t<0$$

$$\overline{\boldsymbol{R}^{(\overline{m})}_{k_n,l_n}}(t) = \left[\sum_{i=0}^{\overline{m}}\binom{l_n}{i}[1-\exp[-\beta t^{\alpha}]]^i\exp[-(l_n-i)\beta t^{\alpha}]\right]^{k_n} \quad \text{for } t\geq 0, \ \overline{m}=l_n-m$$

Proposition 8.6
If components of the two-state homogeneous and regular 'm out of l_n'–series system have Weibull reliability function

$$R(t)=1 \quad \text{for } t<0, \ R(t)=\exp[-\beta t^{\alpha}] \quad \text{for } t\geq 0, \ \alpha>0, \ \beta>0$$

and

$$k_n \to k, \ l_n=n, \ m=\text{constant} \quad (m/n\to 0 \text{ as } n\to\infty),$$

$$a_n = \left[\frac{b_n}{\alpha \log n}\right], \quad b_n = \left[\frac{\log n}{\beta}\right]^{1/\alpha}$$

then

$$[\overline{\mathfrak{R}}_3^{(0)}(t)]^k = \left[1-\exp[-\exp[-t]]\sum_{i=0}^{m-1}\frac{\exp[-it]}{i!}\right]^k, \quad t\in(-\infty,\infty) \tag{8.5}$$

is its limit reliability function.

Motivation: We will use Lemma 4.9 with unmodified condition (4.30) in the form

$$V(t) = \lim_{n\to\infty} n[R(a_n t + b_n)] \quad \text{for } t \in C_V$$

Since for sufficiently large n, we have

$$a_n t + b_n = \left[\frac{\log n}{\beta}\right]^{1/\alpha}\left[1 + \frac{t}{\alpha \log n}\right] > 0 \quad \text{for } t \in (-\infty, \infty)$$

then

$$\begin{aligned} R(a_n t + b_n) &= \exp[-\beta(a_n t + b_n)^{\alpha}] \\ &= \exp\left[-\log n\left[1 + \frac{t}{\alpha \log n}\right]^{\alpha}\right] \\ &= \exp[-\log n - t - o(1)] \quad \text{for } t \in (-\infty, \infty) \end{aligned}$$

Hence

$$\begin{aligned} V(t) &= \lim_{n\to\infty} n[R(a_n t + b_n)] \\ &= \lim_{n\to\infty} n \exp[-\log n - t - o(1)] \\ &= \lim_{n\to\infty} \exp[-t - o(1)] = \exp[-t] \quad \text{for } t \in (-\infty, \infty) \end{aligned}$$

Thus, from Lemma 4.9, $[\overline{\mathfrak{R}}_3^{(0)}(t)]^k$ given by Eq. (8.5) is the system limit reliability function.

Example 8.4

Let us consider the ship-rope elevator used to dock and undock ships coming into shipyards for repairs. The elevator is composed of a steel platform-carriage placed in its syncline (hutch). The platform is moved vertically with 10 rope hoisting winches fed by separate electric motors. During ship docking the platform, with the ship settled in special supporting carriages on the platform, is raised to the wharf level (upper position). During undocking, the operation is reversed. While the ship is moving into or out of the syncline, and while stopped in the upper position, the platform is held on hooks and the loads in the ropes are relieved. In our further analysis we will discuss the reliability of the rope system only. The system under consideration is in order if all its ropes do not fail. Thus we may assume that it is a series system composed of 10 components (ropes). Each of the ropes is composed of 22 strands. Thus, considering the strands as basic components of the system and assuming that each of the ropes is not failed if at least $m = 5$ of its strands are not failed, according to Definitions 8.4 and 8.5, we conclude that the rope elevator is the two-state homogeneous and regular ‘5 out of 22’−series system. It is composed

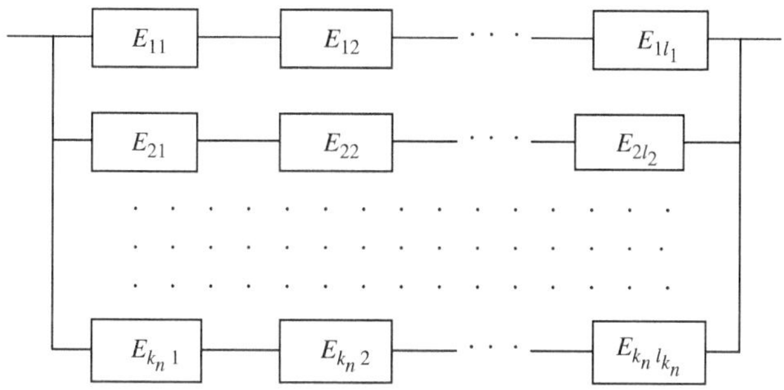

Figure 8.9 The scheme of a series–parallel system of order 1.

of $k_n = 10$ series-linked '5 out of 22' subsystems (ropes) with $l_n = 22$ components (strands). Assuming additionally that strands have Weibull reliability functions with parameters $\alpha = 2$, $\beta = 0.05$, i.e.

$$R(t) = \exp[-0.05t^2] \quad \text{for } t \geq 0$$

from Eq. (8.4), we conclude that the elevator reliability function is given by

$$\overline{\boldsymbol{R}^{(5)}_{10,22}}(t) = \left[1 - \sum_{i=0}^{4} \binom{22}{i} \exp[-i0.05t^2][1 - \exp[-0.05t^2]]^{22-i}\right]^{10} \quad \text{for } t \geq 0$$

Next, applying Proposition 8.6 with

$$a_n = \frac{7.8626}{2 \log 22} \cong 1.2718, \quad b_n = \left[\frac{\log 22}{0.05}\right]^{1/2} \cong 7.8626$$

and Eq. (1.1), we get the following approximate formula for the elevator reliability function

$$\overline{\boldsymbol{R}^{(5)}_{10,22}}(t) \cong [\overline{\Re}_3^{(0)}(0.7863t - 6.1821)]^{10}$$

$$= \left[1 - \exp[-\exp[-0.7863t + 6.1821]] \sum_{i=0}^{4} \frac{\exp[-0.7863it + 6.1821i]}{i!}\right]^{10} \quad \text{for } t \in (-\infty, \infty)$$

8.5 Reliability of Large Hierarchical Systems

Prior to defining the hierarchical systems of any order we once again consider a series–parallel system, presented in Figure 8.9. This system here is called a series–parallel system of order 1.

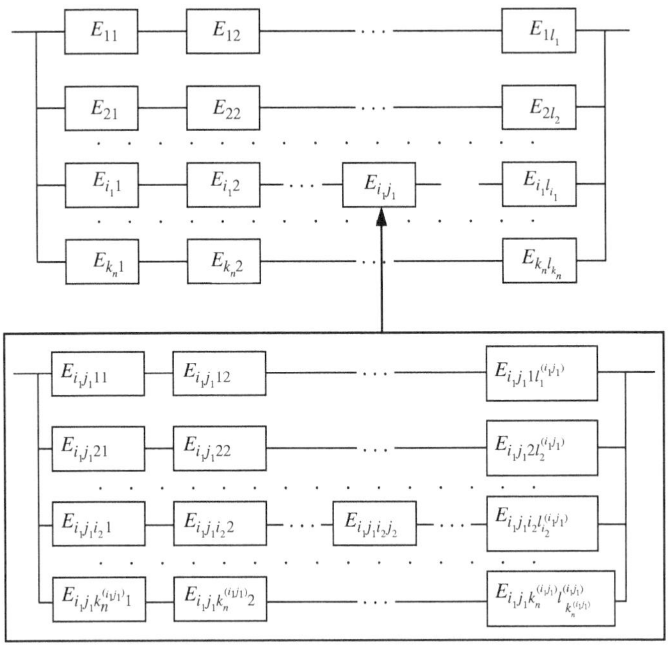

Figure 8.10 The scheme of a series–parallel system of order 2.

It is made up of components

$$E_{i_1 j_1},\quad i_1 = 1, 2, \ldots, k_n,\ \ j_1 = 1, 2, \ldots, l_{i_1}$$

with the lifetimes respectively

$$T_{i_1 j_1},\quad i_1 = 1, 2, \ldots, k_n,\ \ j_1 = 1, 2, \ldots, l_{i_1}$$

Its lifetime is given by

$$T = \max_{1 \le i_1 \le k_n} \left\{ \min_{1 \le j_1 \le l_{i_1}} \{T_{i_1 j_1}\} \right\} \tag{8.6}$$

Now we assume that each component

$$E_{i_1 j_1},\quad i_1 = 1, 2, \ldots, k_n,\ \ j_1 = 1, 2, \ldots, l_{i_1}$$

of the series–parallel system of order 1 is a subsystem composed of components (Figure 8.10)

$$E_{i_1 j_1 i_2 j_2},\quad i_2 = 1, 2, \ldots, k_n^{(i_1 j_1)},\ \ j_2 = 1, 2, \ldots, l_{i_2}^{(i_1 j_1)}$$

and has a series–parallel structure.

This means that each subsystem lifetime $T_{i_1j_1}$ is given by

$$T_{i_1j_1} = \max_{1 \le i_2 \le k_n^{(i_1j_1)}} \left\{ \min_{1 \le j_2 \le l_{i_2}^{(i_1j_1)}} \{T_{i_1j_1i_2j_2}\} \right\}, \quad i_1 = 1, 2, \ldots, k_n, \; j_1 = 1, 2, \ldots, l_{i_1} \tag{8.7}$$

where

$$T_{i_1j_1i_2j_2}, \quad i_2 = 1, 2, \ldots, k_n^{(i_1j_1)}, \; j_2 = 1, 2, \ldots, l_{i_2}^{(i_1j_1)}$$

are the lifetimes of the subsystem components $E_{i_1j_1i_2j_2}$.

The system defined this way is called a hierarchical series–parallel system of order 2. Its lifetime, from Eqs. (8.6) and (8.7), is given by the formula

$$T = \max_{1 \le i_1 \le k_n} \left\{ \min_{1 \le j_1 \le l_{i_1}} \left[\max_{1 \le i_2 \le k_n^{(i_1j_1)}} \left(\min_{1 \le j_2 \le l_{i_2}^{(i_1j_1)}} T_{i_1j_1i_2j_2} \right) \right] \right\}$$

where k_n is the number of series systems linked in parallel and composed of series–parallel subsystems $E_{i_1j_1}$, l_{i_1} are the numbers of series–parallel subsystems $E_{i_1j_1}$ in these series systems, $k_n^{(i_1j_1)}$ are the numbers of series systems in the series–parallel subsystems $E_{i_1j_1}$ linked in parallel and $l_{i_2}^{(i_1j_1)}$ are the numbers of components in these series systems of the series–parallel subsystems $E_{i_1j_1}$.

In an analogous way it is possible to define two-state parallel–series systems of order 2.

Generally, in order to define hierarchical series–parallel and parallel–series systems of any order r, $r \ge 1$, we assume that

$$E_{i_1j_1\ldots i_rj_r}$$

where

$$i_1 = 1, 2, \ldots, k_n, \; j_1 = 1, 2, \ldots, l_{i_1}, \; i_2 = 1, 2, \ldots, k_n{}^{(i_1j_1)}, \; j_2 = 1, 2, \ldots, l_{i_2}^{(i_1j_1)}, \ldots,$$
$$i_r = 1, 2, \ldots, k_n^{(i_1j_1\ldots i_{r-1}j_{r-1})}, \; j_r = 1, 2, \ldots, l_{i_r}^{(i_1j_1\ldots i_{r-1}j_{r-1})}$$

and

$$k_n, \; l_{i_1}, \; k_n^{(i_1j_1)}, \; l_{i_2}^{(i_1j_1)}, \ldots, \; k_n^{(i_1j_1\ldots i_{r-1}j_{r-1})}, \; l_{i_r}^{(i_1j_1\ldots i_{r-1}j_{r-1})} \in N$$

are two-state components having reliability functions

$$R_{i_1j_1\ldots i_rj_r}(t) = P(T_{i_1j_1\ldots i_rj_r} > t), \quad t \in (-\infty, \infty)$$

and random variables

$$T_{i_1j_1\ldots i_rj_r}$$

and where

$$i_1 = 1,2,\ldots,k_n,\ \ j_1 = 1,2,\ldots,l_{i_1},\ \ i_2 = 1,2,\ldots,k_n{}^{(i_1j_1)},\ \ j_2 = 1,2,\ldots,l_{i_2}^{(i_1j_1)},\ldots$$

$$i_r = 1,2,\ldots,k_n^{(i_1j_1\ldots i_{r-1}j_{r-1})},\ \ j_r = 1,2,\ldots,l_{i_r}^{(i_1j_1\ldots i_{r-1}j_{r-1})}$$

are independent random variables with distribution functions

$$F_{i_1j_1\ldots i_rj_r}(t) = P(T_{i_1j_1\ldots i_rj_r} \leq t),\quad t \in (-\infty, \infty)$$

representing the lifetimes of the components $E_{i_1j_1\ldots i_rj_r}$.

Definition 8.7
A two-state system is called a series–parallel system of order r if its lifetime T is given by

$$T = \max_{1\leq i_1\leq k_n}\left\{\min_{1\leq j_1\leq l_{i_1}}\left\{\max_{1\leq i_2\leq k_n{}^{(i_1j_1)}}\left\{\min_{1\leq j_2\leq l_{i_2}^{(i_1j_1)}}\cdots\left[\max_{1\leq i_r\leq k_n{}^{(i_1j_1\ldots i_{r-1}j_{r-1})}}\left(\min_{1\leq j_r\leq l_{i_r}^{(i_1j_1\ldots i_{r-1}j_{r-1})}} T_{i_1j_1\ldots i_rj_r}\right)\right]\cdots\right\}\right\}\right\}$$

where $k_n, k_n{}^{(i_1j_1)},\ldots,k_n{}^{(i_1j_1i_2j_2\ldots i_{r-1}j_{r-1})}$ are the numbers of suitable series systems of the system composed of series–parallel subsystems and linked in parallel, $l_{i_1}, l_{i_2}^{(i_1j_1)},\ldots,l_{i_{r-1}}^{(i_1j_1i_2j_2\ldots i_{r-2}j_{r-2})}$ are the numbers of suitable series–parallel subsystems in these series systems and $l_{i_r}^{(i_1j_1i_2j_2\ldots i_{r-1}j_{r-1})}$ are the numbers of components in the series systems of the series–parallel subsystems.

Definition 8.8
A two-state series–parallel system of order r is called homogeneous if its component lifetimes $T_{i_1j_1\ldots i_rj_r}$ have an identical distribution function

$$F(t) = P(T_{i_1j_1\ldots i_rj_r} \leq t),\quad t \in (-\infty, \infty)$$

where

$$i_1 = 1,2,\ldots,k_n,\ \ j_1 = 1,2,\ldots,l_{i_1},\ \ i_2 = 1,2,\ldots,k_n{}^{(i_1j_1)},\ \ j_2 = 1,2,\ldots,l_{i_2}^{(i_1j_1)},\ldots$$

$$i_r = 1,2,\ldots,k_n^{(i_1j_1\ldots i_{r-1}j_{r-1})},\ \ j_r = 1,2,\ldots,l_{i_r}^{(i_1j_1\ldots i_{r-1}j_{r-1})}$$

i.e. if its components $E_{i_1j_1\ldots i_rj_r}$ have the same reliability function

$$R(t) = 1 - F(t), \quad t \in (-\infty, \infty)$$

Definition 8.9
A two-state series–parallel system of order r is called regular if

$$l_{i_1} = l_{i_2}^{(i_1j_1)} = \ldots = l_{i_r}^{(i_1j_1\ldots i_{r-1}j_{r-1})} = l_n \text{ and } k_n^{(i_1j_1)} = \ldots = k_n^{(i_1j_1\ldots i_{r-1}j_{r-1})} = k_n$$

where k_n is the number of series systems in the series–parallel subsystems and l_n are the numbers of series–parallel subsystems or, respectively, the numbers of components in these series systems.

Using mathematical induction it is possible to prove that the reliability function of the homogeneous and regular two-state hierarchical series–parallel system of order r is given by

$$\boldsymbol{R}_{k,k_n,l_n}(t) = 1 - [1 - [\boldsymbol{R}_{k-1,k_n,l_n}(t)]^{l_n}]^{k_n} \quad \text{for } k = 2, 3, \ldots, r$$

and

$$R_{1,k_n,l_n}(t) = 1 - [1 - [R(t)]^{l_n}]^{k_n}, \quad t \in (-\infty, \infty)$$

where k_n and l_n are the numbers defined in Definition 8.9.

Corollary 8.3
If components of the homogeneous and regular two-state hierarchical series–parallel system of order r have an exponential reliability function

$$R(t) = 1 \quad \text{for } t < 0, \quad R(t) = \exp[-\lambda t] \quad \text{for } t \geq 0, \quad \lambda > 0$$

then its reliability function is given by

$$\boldsymbol{R}_{k,k_n,l_n}(t) = 1 \quad \text{for } t < 0, \quad \boldsymbol{R}_{k,k_n,l_n}(t) = 1 - [1 - [\boldsymbol{R}_{k-1,k_n,l_n}(t)]^{l_n}]^{k_n} \quad \text{for } t \geq 0$$

for $k = 2, 3, \ldots, r$ and

$$\boldsymbol{R}_{1,k_n,l_n}(t) = 1 - [1 - \exp[-\lambda l_n t]]^{k_n}, \quad t \in (-\infty, \infty)$$

Theorem 8.1 [26]
If

i. $\mathfrak{R}(t) = 1 - \exp[-V(t)], \ t \in (-\infty, \infty)$, is a non-degenerate reliability function,

ii.

$$\lim_{n\to\infty} l_n^{r-1} k_n^{-1/l_n} = 0 \quad \text{for } r \geq 1$$

iii.

$$\lim_{n\to\infty} k_n^{l_n^{r-1}+\cdots+1}[R(a_n t + b_n)]^{l_n^r} = V(t) \quad \text{for } t \in C_V,\ r \geq 1,\ t \in (-\infty, \infty)$$

then

$$\lim_{n\to\infty} \boldsymbol{R}_{r,k_n,l_n}(a_n t + b_n) = \Re(t) \quad \text{for } t \in C_{\Re},\ r \geq 1,\ t \in (-\infty, \infty)$$

Proposition 8.7

If components of the homogeneous and regular two-state hierarchical series–parallel system of order r have an exponential reliability function

$$R(t) = 1 \quad \text{for } t < 0,\ R(t) = \exp[-\lambda t] \quad \text{for } t \geq 0,\ \lambda > 0$$

$$\lim_{n\to\infty} l_n^{r-1} k_n^{-1/l_n} = 0 \quad \text{for } r \geq 1$$

and

$$a_n = \frac{1}{\lambda l_n^r},\ b_n = \frac{1}{\lambda}\left(\frac{1}{l_n} + \frac{1}{l_n^2} + \cdots + \frac{1}{l_n^r}\right)\log k_n$$

then

$$\Re_3(t) = 1 - \exp[-\exp[-t]] \quad \text{for } t \in (-\infty, \infty) \tag{8.8}$$

is its limit reliability function.

Motivation: Since for sufficiently large n, we have

$$a_n t + b_n = \frac{t + (l_n^{r-1} + \cdots + 1)\log k_n}{\lambda l_n^r} > 0 \quad \text{for } t \in (-\infty, \infty)$$

then

$$R(a_n t + b_n) = \exp\left[-\frac{t + (l_n^{r-1} + \cdots + 1)\log k_n}{l_n^r}\right] \quad \text{for } t \in (-\infty, \infty)$$

and further

$$\begin{aligned} V(t) &= \lim_{n\to\infty} k_n^{l_n^{r-1}+\cdots+1} R^{l_n^r}(a_n t + b_n) \\ &= \lim_{n\to\infty} k_n^{l_n^{r-1}+\cdots+1} \exp\left[-l_n^r \frac{t + (l_n^{r-1} + \cdots + 1)\log k_n}{l_n^r}\right] = \exp[-t] \ \text{ for } t \in (-\infty, \infty) \end{aligned}$$

Hence from Theorem 8.1, $\mathfrak{R}_3(t)$ given by Eq. (8.8) is the system limit reliability function.

Example 8.5

A hierarchical regular series–parallel homogeneous system of order $r = 2$ is such that $k_n = 200$, $l_n = 3$. The system components have identical exponential reliability functions with the failure rate $\lambda = 0.01$.

Under these assumptions its exact reliability function, according to Corollary 8.3, is given by

$$\boldsymbol{R}_{2,200,3}(t) = 1, \quad t < 0, \quad \boldsymbol{R}_{2,200,3}(t) = 1 - [1 - [1 - [1 - \exp[-0.01 \cdot 3t]]^{200}]^3]^{200}, \quad t \geq 0$$

Next, applying Proposition 8.7 with normalizing constants

$$a_n = \frac{1}{0.01 \cdot 9} = 11.1, \quad b_n = \frac{1}{0.01}\left(\frac{1}{3} + \frac{1}{9}\right)\log 200 = 235.5$$

we conclude that the system limit reliability function is given by

$$\mathfrak{R}_3(t) = 1 - \exp[-\exp[-t]] \quad \text{for } t \in (-\infty, \infty)$$

and from Eq. (1.1), the following approximate formula is valid

$$\boldsymbol{R}_{2,200,3}(t) \cong \mathfrak{R}_3(0.09t - 21.2) = 1 - \exp[-\exp[-0.09t + 21.2]] \quad \text{for } t \in (-\infty, \infty)$$

The accuracy of this approximation is illustrated in Table 8.4 and Figure 8.11.

Definition 8.10

A two-state system is called a parallel–series system of order r if its lifetime T is given by

$$T = \min_{1 \leq i_1 \leq k_n} \left\{ \max_{1 \leq j_1 \leq l_{i_1}} \left\{ \min_{1 \leq i_2 \leq k_n^{(i_1 j_1)}} \left\{ \max_{1 \leq j_2 \leq l_{i_2}^{(i_1 j_1)}} \cdots \left[\min_{1 \leq i_r \leq k_n^{(i_1 j_1 \ldots i_{r-1} j_{r-1})}} \left(\max_{1 \leq j_r \leq l_{i_r}^{(i_1 j_1 \ldots i_{r-1} j_{r-1})}} T_{i_1 j_1 \ldots i_r j_r} \right) \right] \cdots \right\} \right\} \right\}$$

where $k_n, k_n^{(i_1 j_1)}, \ldots, k_n^{(i_1 j_1 i_2 j_2 \ldots i_{r-1} j_{r-1})}$ are the numbers of suitable parallel systems of the system composed of parallel–series subsystems and linked in series, $l_{i_1}, l_{i_2}^{(i_1 j_1)}, \ldots, l_{i_{r-1}}^{(i_1 j_1 i_2 j_2 \ldots i_{r-2} j_{r-2})}$ are the numbers of suitable parallel–series subsystems in these parallel systems and $l_{i_r}^{(i_1 j_1 i_2 j_2 \ldots i_{r-1} j_{r-1})}$ are the numbers of components in the parallel systems of the parallel–series subsystems.

Table 8.4 Values of Exact and Approximate Reliability Functions of a Hierarchical Regular Series–Parallel Homogeneous System of Order 2

t	$\boldsymbol{R_{2,200,3}(t)}$	$\boldsymbol{\Re_3(0.09t-21.2)}$	$\boldsymbol{\Delta = R_{2,200,3}(t) - \Re_3(0.09t-21.2)}$
200	0.999996	1.000000	−0.000004
210	0.997281	0.999953	−0.002672
220	0.934547	0.982668	−0.048121
230	0.705338	0.807704	−0.102366
240	0.414313	0.488455	**−0.074142**
250	0.205450	0.238551	−0.033101
260	0.092891	0.104885	−0.011994
270	0.040069	0.044050	−0.003981
280	0.016873	0.018149	−0.001276
290	0.007014	0.007419	−0.000405
300	0.002895	0.003023	−0.000128
310	0.001189	0.001230	−0.000041
320	0.000487	0.000500	−0.000013
330	0.000199	0.000203	−0.000116
340	0.000081	0.000083	−0.000002

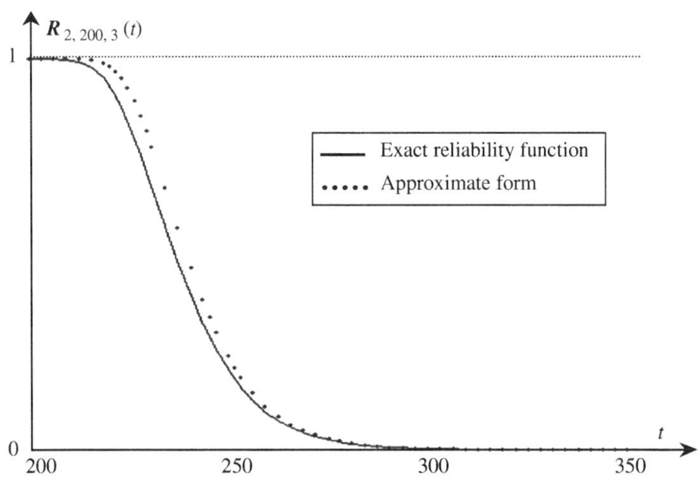

Figure 8.11 Graphs of exact and approximate reliability functions of a hierarchical regular series–parallel homogeneous system of order 2.

Definition 8.11

A two-state parallel–series system of order r is called homogeneous if its component lifetimes $T_{i_1 j_1 \ldots i_r j_r}$ have an identical distribution function

$$F(t) = P(T_{i_1 j_1 \ldots i_r j_r} \le t)$$

where

$$i_1 = 1,2,\ldots,k_n,\ j_1 = 1,2,\ldots,l_{i_1},\ i_2 = 1,2,\ldots,k_n^{(i_1j_1)},\ j_2 = 1,2,\ldots,l_{i_2}^{(i_1j_1)},\ldots$$

$$i_r = 1,2,\ldots,k_n^{(i_1j_1\ldots i_{r-1}j_{r-1})},\ j_r = 1,2,\ldots,l_{i_r}^{(i_1j_1\ldots i_{r-1}j_{r-1})}$$

i.e. if its components $E_{i_1j_1\ldots i_rj_r}$ have the same reliability function

$$R(t) = 1 - F(t),\quad t\in(-\infty,\infty)$$

Definition 8.12
A two-state parallel–series system of order r is called regular if

$$l_{i_1} = l_{i_2}^{(i_1j_1)} = \cdots = l_{i_r}^{(i_1j_1\ldots i_{r-1}j_{r-1})} = l_n \text{ and } k_n^{(i_1j_1)} = \cdots = k_n^{(i_1j_1\ldots i_{r-1}j_{r-1})} = k_n$$

where k_n is the number of parallel systems in the parallel–series subsystems and l_n are the numbers of parallel–series subsystems or, respectively, the numbers of components in these parallel systems.

Using mathematical induction it is possible to prove that the reliability function of the homogeneous and regular two-state hierarchical parallel–series system of order r is given by

$$\overline{\mathbf{R}}_{k,k_n,l_n}(t) = 1 - [1-[\overline{\mathbf{R}}_{k-1,k_n,l_n}(t)]^{l_n}]^{k_n}\quad \text{for } k = 2,3,...,r$$

and

$$\overline{\mathbf{R}}_{1,k_n,l_n}(t) = [1-[F(t)]^{l_n}]^{k_n},\quad t\in(-\infty,\infty)$$

where k_n and l_n are the numbers defined in Definition 8.12.

Corollary 8.4
If components of the homogeneous and regular two-state hierarchical parallel–-series system of order r have an exponential reliability function

$$R(t) = 1\quad \text{for } t<0,\ R(t) = \exp[-\lambda t]\quad \text{for } t\geq 0,\ \lambda>0$$

then its reliability function is given by

$$\overline{\boldsymbol{R}}_{k,k_n,l_n}(t) = 1 - [1-[\overline{\boldsymbol{R}}_{k-1,k_n,l_n}(t)]^{l_n}]^{k_n}\quad \text{for } k = 2,3,\ldots,r$$

and

$$\overline{\boldsymbol{R}}_{1,k_n,l_n}(t) = [1-[1-\exp[-\lambda t]]^{l_n}]^{k_n},\quad t\in(-\infty,\infty)$$

Theorem 8.2 [26]
If

i. $\overline{\Re}(t) = \exp[-\overline{V}(t)],\ t \in (-\infty, \infty)$, is a non-degenerate reliability function,

ii. $\lim_{n\to\infty} l_n^{r-1} k_n^{-1/l_n} = 0 \quad \text{for } r \geq 1$

iii. $\lim_{n\to\infty} k_n^{l_n^{r-1}+\cdots+1}[F(a_n t + b_n)]^{l_n^r} = \overline{V}(t) \quad \text{for } t \in C_V,\ r \geq 1,\ t \in (-\infty, \infty)$

then

$$\lim_{n\to\infty} \overline{\mathbf{R}}_{r,k_n,l_n}(a_n t + b_n) = \overline{\Re}(t) \quad \text{for } t \in C_{\overline{\Re}},\ r \geq 1,\ t \in (-\infty, \infty) \tag{8.9}$$

Proposition 8.8
If components of the homogeneous and regular two-state hierarchical parallel–series system of order r have an exponential reliability function

$$R(t) = 1 \quad \text{for } t < 0,\ R(t) = \exp[-\lambda t] \quad \text{for } t \geq 0,\ \lambda > 0$$

$$\lim_{n\to\infty} l_n^{r-1} k_n^{-1/l_n} = 0 \quad \text{for } r \geq 1, \quad \lim_{n\to\infty} l_n = l,\ l \in N$$

and

$$a_n = \frac{1}{\lambda} \frac{1}{k_n^{(1/l_n)+\cdots+(1/l_n^r)}},\quad b_n = 0$$

then

$$\Re_2(t) = 1 \quad \text{for } t < 0 \text{ and } \Re_2(t) = \exp[-t^{l^r}] \quad \text{for } t \geq 0 \tag{8.10}$$

is its limit reliability function.
Motivation: Since for sufficiently large n, we have

$$a_n t + b_n = \frac{1}{\lambda} \frac{t}{k_n^{1/l_n+\cdots+1/l_n^r}} < 0 \text{ for } t < 0$$

and

$$a_n t + b_n = \frac{1}{\lambda} \frac{t}{k_n^{1/l_n+\cdots+1/l_n^r}} \geq 0 \quad \text{for } t \geq 0$$

then

$$F(a_n t + b_n) = 0 \quad \text{for } t < 0$$

and

$$F(a_nt+b_n)=1-\exp\left[-\frac{t}{k_n^{(1/l_n)+\cdots+(1/l_n^r)}}\right]\quad \text{for } t\geq 0$$

and further

$$\overline{V}(t)=\lim_{n\to\infty} k_n^{l_n^{r-1}+\cdots+1}F^{l_n^r}(a_nt+b_n)=0\quad \text{for } t<0$$

and

$$\begin{aligned}\overline{V}(t)&=\lim_{n\to\infty} k_n^{l_n^{r-1}+\cdots+1}F^{l_n^r}(a_nt+b_n)\\&=\lim_{n\to\infty} k_n^{l_n^{r-1}+\cdots+1}\left[1-\exp\left[-\frac{t}{k_n^{(1/l_n)+\cdots+(1/l_n^r)}}\right]\right]^{l_n^r}\\&=\lim_{n\to\infty} k_n^{l_n^{r-1}+\cdots+1}\left[\frac{t^{l_n^r}}{k_n^{l_n^{r-1}+\cdots+1}}+o\left(\frac{1}{k_n^{l_n^{r-1}+\cdots+1}}\right)\right]=t^{l^r}\quad \text{for } t\geq 0\end{aligned}$$

Hence from Theorem 8.2, $\overline{\mathfrak{R}}_2(t)$ given by Eq. (8.10) is the system limit reliability function.

Example 8.6

We consider a hierarchical regular parallel−series homogeneous system of order $r=2$ such that $k_n=200$, $l_n=3$, whose components have identical exponential reliability functions with the failure rate $\lambda=0.01$.

Its exact reliability function, according to Corollary 8.4, is given by

$$\overline{\mathbf{R}}_{2,200,3}(t)=1\quad \text{for } t<0$$

and

$$\overline{\mathbf{R}}_{2,200,3}(t)=[1-[1-[1-[1-\exp[-0.01t]]^3]^{200}]^3]^{200}\quad \text{for } t\geq 0$$

Next, applying Proposition 8.8 with normalizing constants

$$a_n=\frac{1}{0.01}\cdot\frac{1}{200^{1/3+1/9}}=9.4912,\quad b_n=0$$

we conclude that

$$\overline{\mathfrak{R}}_2(t)=1\quad \text{for } t<0 \text{ and } \overline{\mathfrak{R}}_2(t)=\exp[-t^9]\quad \text{for } t\geq 0$$

Table 8.5 Values of Exact and Approximate Reliability Functions of a Hierarchical Regular Parallel–Series Homogeneous System of Order 2

t	$\overline{\boldsymbol{R}}_{2,200,3}(t)$	$\overline{\mathfrak{R}}_2(\mathbf{0.1054}t)$	$\Delta = \overline{\mathbf{R}}_{2,200,3}(t) - \overline{\mathfrak{R}}_2(\mathbf{0.1054}t)$
0	1.000000	1.000000	0.000000
2	0.999999	0.999999	0.000000
4	0.999656	0.999579	0.000077
6	0.988445	0.983952	0.004493
8	0.876948	0.806167	0.070781
10	0.451036	0.200822	**0.250214**
12	0.040806	0.000253	0.040553
14	0.000070	0.000000	0.000070
16	0.000000	0.000000	0.000000

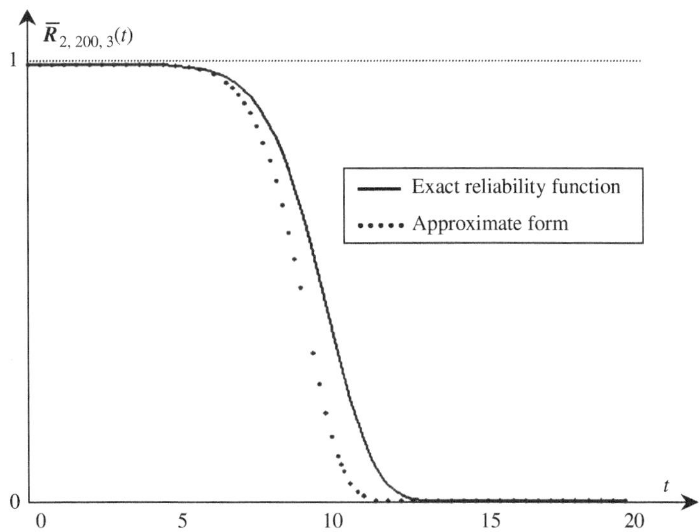

Figure 8.12 Graphs of exact and approximate reliability functions of a hierarchical regular parallel–series homogeneous system of order 2.

is the system limit reliability function, and from Eq. (1.1), the following approximate formula is valid

$$\overline{\mathbf{R}}_{2,200,3}(t) \cong \overline{\mathfrak{R}}_2(0.1054t) = \exp[-(0.1054t)^9] \quad \text{for } t \geq 0$$

The accuracy of this approximation is illustrated in Table 8.5 and Figure 8.12.

8.6 Asymptotic Approach to Systems Reliability Improvement

We first consider the homogeneous series system illustrated in Figure 8.13.

It is composed of n components E_{i1}, $i = 1, 2, \ldots, n$, having lifetimes T_{i1}, $i = 1, 2, \ldots, n$, and exponential reliability functions

$$R(t) = 1 \quad \text{for } t < 0, \quad R(t) = \exp[-\lambda t] \quad \text{for } t \geq 0, \quad \lambda > 0$$

Its lifetime and its reliability function respectively are given by

$$T = \min_{1 \leq i \leq n}\{T_{i1}\}$$
$$\mathbf{R}_n(t) = [R(t)]^n = \exp[-\lambda n t], \quad t \geq 0$$

In order to improve the reliability of this series system, the following exemplary methods can be used:

- improving the reliability of its components by reducing their failure rates by a factor ρ, $0 < \rho < 1$,
- a warm duplication (single reservation) of system components,
- a cold duplication of system components,
- a mixed duplication of system components,
- a hot system duplication,
- a cold system duplication.

It is supposed here that the reserve components are identical to the basic ones.

The results of these methods of system reliability improvement are briefly presented below, giving the system schemes, lifetimes and reliability functions.

Case 1. Improving the reliability of the system components by reducing their failure rates by a factor ρ, $0 < \rho < 1$,

$$T = \min_{1 \leq i \leq n}\{T_{i1}\}$$
$$\boldsymbol{R}_n^{(1)}(t) = [R(\rho t)]^n = \exp[-\rho \lambda n t], \quad t \geq 0$$

Case 2. A hot reservation of the system components (Figure 8.14)

$$T = \min_{1 \leq i \leq n}\{\max_{1 \leq j \leq 2}\{T_{ij}\}\}$$
$$\boldsymbol{R}_n^{(2)}(t) = [1 - [F(t)]^2]^n = [1 - [1 - \exp[-\lambda t]]^2]^n, t \geq 0$$

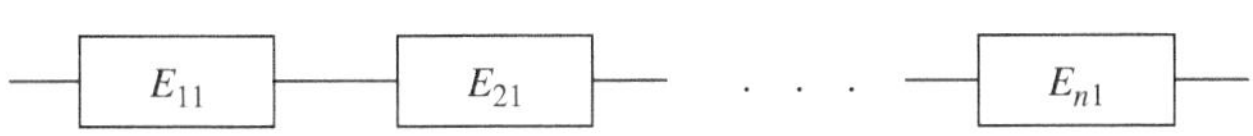

Figure 8.13 The scheme of a series system.

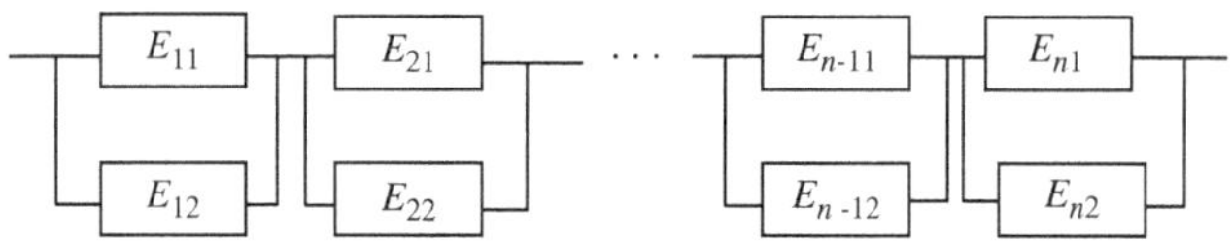

Figure 8.14 The scheme of a series system with components having hot reservation.

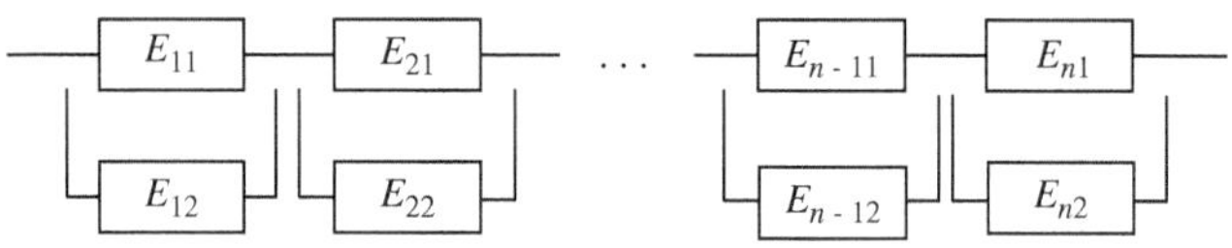

Figure 8.15 The scheme of a series system with components having cold reservation.

Case 3. A cold reservation of the system components (Figure 8.15)

$$T = \min_{1 \le i \le n} \left\{ \sum_{j=1}^{2} T_{ij} \right\}$$

$$\begin{aligned} \boldsymbol{R}_n^{(3)}(t) &= [1-[F(t)] * [F(t)]]^n = [[1+\lambda t]\exp[-\lambda t]]^n \\ &= [1+\lambda t]^n \exp[-n\lambda t], t \ge 0 \end{aligned}$$

Case 4. A mixed reservation of the system components (Figure 8.16)

$$T = \min\left\{ \min_{1 \le i \le m} \left\{ \sum_{j=1}^{2} T_{ij} \right\}, \min_{m+1 \le i \le n} \left\{ \max_{1 \le j \le 2} \{T_{ij}\} \right\} \right\}$$

$$\begin{aligned} \boldsymbol{R}_n^{(4)}(t) &= [1-[F(t)] * [F(t)]]^m [1-R^2(t)]^{n-m} \\ &= [[1+\lambda t]\exp[-\lambda t]]^m [1-[1-\exp[-\lambda t]]^2]^{n-m} \\ &= [1+\lambda t]^m \exp[-\lambda m t][2\exp[-\lambda t]-\exp[-2\lambda t]]^{n-m} \\ &= [1+\lambda t]^m \exp[-\lambda n t][2-\exp[-\lambda t]]^{n-m}, \quad t \ge 0 \end{aligned}$$

Case 5. A hot system reservation (Figure 8.17)

$$T = \max_{1 \le j \le 2} \left\{ \min_{1 \le j \le n} \{T_{ij}\} \right\}$$

$$\boldsymbol{R}_n^{(5)}(t) = 1 - [1-[R(t)]^n]^2 = 1 - [1-\exp[-n\lambda t]]^2, \quad t \ge 0$$

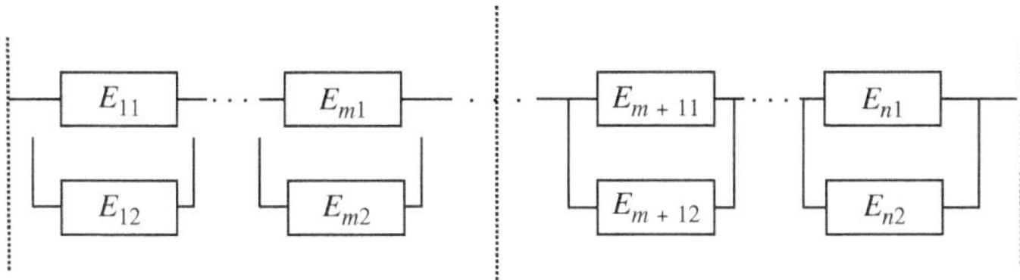

Figure 8.16 The scheme of a series system with components having mixed reservation.

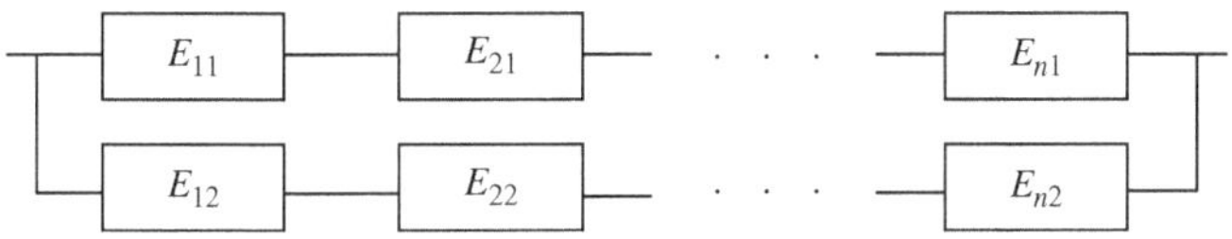

Figure 8.17 The scheme of a series system with hot reservation.

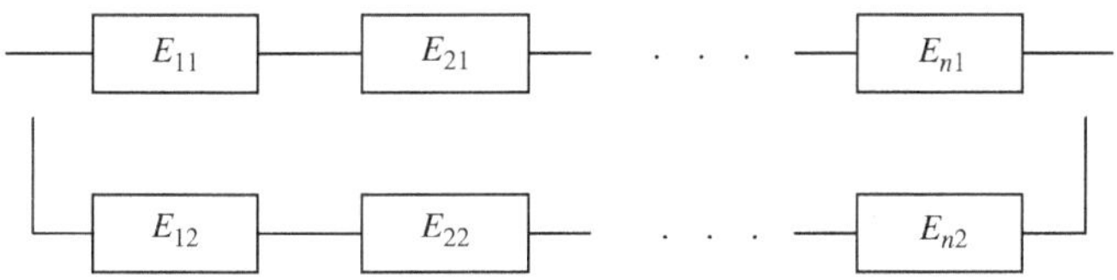

Figure 8.18 The scheme of a series system with cold reservation.

Case 6. A cold system reservation (Figure 8.18)

$$T = \sum_{j=1}^{2} \min_{1 \le i \le n} \{T_{ij}\}$$

$$\boldsymbol{R}_n^{(6)}(t) = 1 - [1 - [R(t)]^n] * [1 - [R(t)]^n] = [1 + n\lambda t]\exp[-n\lambda t], \quad t \ge 0$$

The difficulty arises when selecting the right method of improvement of reliability for a large system. This problem may be simplified and approximately solved by the application of the asymptotic approach. Comparisons of the limit reliability functions of the systems with different types of reserve and such systems with improved components allow us to find the value of the components' decreasing failure rate factor ρ, which gives rise to an equivalent effect on the system reliability improvement. Similar results are obtained under comparison of the system lifetime mean values. As an example we will present the asymptotic approach to the above methods of improving reliability for homogeneous two-state series systems.

Proposition 8.9
Case 1. If

$$a_n = 1/\lambda\rho n, \quad b_n = 0$$

then

$$\Re^{(1)}(t) = 1 \quad \text{for } t < 0 \text{ and } \Re^{(1)}(t) = \exp[-t] \quad \text{for } t \geq 0$$

is the limit reliability function of the homogeneous exponential series system with reduced failure rates of its components, i.e.

$$\boldsymbol{R}_n^{(1)}(t) = \Re^{(1)}(\lambda\rho n t) = \exp[-\lambda\rho n t] \quad \text{for } t \geq 0$$

and

$$\boldsymbol{T}^{(1)} = E[T] = \frac{1}{\lambda\rho n}$$

Case 2. If

$$a_n = 1/\lambda\sqrt{n}, \quad b_n = 0$$

then

$$\Re^{(2)}(t) = 1 \quad \text{for } t < 0 \text{ and } \Re^{(2)}(t) = \exp[-t^2] \quad \text{for } t \geq 0$$

is the limit reliability function of the homogeneous exponential series system with hot reservation of its components, i.e.

$$\boldsymbol{R}_n^{(2)}(t) \cong \Re^{(2)}(\lambda\sqrt{n}t) = \exp[-\lambda^2 n t^2] \quad \text{for } t \geq 0$$

and

$$\boldsymbol{T}^{(2)} = E[T] \cong \Gamma\left(\frac{3}{2}\right)[\lambda^2 n]^{-1/2} = \Gamma\left(\frac{3}{2}\right)\frac{1}{\lambda\sqrt{n}}$$

Case 3. If

$$a_n = \sqrt{2}/\lambda\sqrt{n}, \quad b_n = 0$$

then

$$\Re^{(3)}(t) = 1 \quad \text{for } t < 0 \text{ and } \Re^{(3)}(t) = \exp[-t^2] \quad \text{for } t \geq 0$$

is the limit reliability function of the homogeneous exponential series system with cold reservation of its components, i.e.

$$\boldsymbol{R}_n^{(3)}(t) \cong \mathfrak{R}^{(3)}\left(\lambda\sqrt{\frac{nt}{2}}\right) = \exp\left[-\frac{1}{2}\lambda^2 nt^2\right] \quad \text{for } t \geq 0$$

and

$$\boldsymbol{T}^{(3)} = E[T] \cong \Gamma\left(\frac{3}{2}\right)\left[\frac{1}{2}\lambda^2 n\right]^{-1/2} = \Gamma\left(\frac{3}{2}\right)\frac{1}{\lambda}\sqrt{\frac{2}{n}}$$

Case 4. If

$$a_n = \frac{1}{\lambda}\sqrt{\frac{2}{2n-m}}, \quad b_n = 0$$

then

$$\mathfrak{R}^{(4)}(t) = 1 \quad \text{for } t < 0 \text{ and } \mathfrak{R}^{(4)}(t) = \exp[-t^2] \quad \text{for } t \geq 0$$

is the limit reliability function of the homogeneous exponential series system with mixed reservation of its components, i.e.

$$\boldsymbol{R}_n^{(4)}(t) \cong \mathfrak{R}^{(4)}\left(\lambda\sqrt{\frac{2n-m}{2}}\right) = \exp\left[-\frac{2n-m}{2}\lambda^2 t^2\right] \quad \text{for } t \geq 0$$

and

$$\boldsymbol{T}^{(4)} = E[T] \cong \Gamma\left(\frac{3}{2}\right)\left[\frac{2n-m}{2}\lambda^2\right]^{-1/2} = \Gamma\left(\frac{3}{2}\right)\frac{1}{\lambda}\sqrt{\frac{2}{2n-m}}$$

Case 5. If

$$a_n = \frac{1}{\lambda n}, \quad b_n = 0$$

then

$$\mathfrak{R}^{(5)}(t) = 1 \quad \text{for } t < 0 \text{ and } \mathfrak{R}^{(5)}(t) = 1 - [1 - \exp[-t]]^2 \quad \text{for } t \geq 0$$

is the limit reliability function of the homogeneous exponential series system with hot reservation, i.e.

$$\boldsymbol{R}_n^{(5)}(t) = \mathfrak{R}^{(5)}(\lambda nt) = 1 - [1 - \exp[-\lambda nt]]^2 \quad \text{for } t \geq 0$$

and

$$\boldsymbol{T}^{(5)} = E[T] = \frac{3}{2\lambda n}$$

Case 6. If

$$a_n = \frac{1}{\lambda n}, \quad b_n = 0$$

then

$$\mathfrak{R}^{(6)}(t) = 1 \quad \text{for } t < 0 \text{ and } \mathfrak{R}^{(6)}(t) = [1 + t]\exp[-t] \quad \text{for } t \geq 0$$

is the limit reliability function of the homogeneous exponential series system with cold reservation, i.e.

$$\boldsymbol{R}_n^{(6)}(t) = \mathfrak{R}^{(6)}(\lambda n t) = [1 + \lambda n t]\exp[-\lambda n t] \quad \text{for } t \geq 0$$

and

$$\boldsymbol{T}^{(6)} = E[T] = \frac{2}{\lambda n}$$

Motivation: We will prove the proposition parts concerned with the considered system limit reliability functions only. The approximate formulae for the system reliability functions follow directly from Eq. (1.1) and the expressions for the system lifetimes are easy to obtain by direct integration of the system approximate reliability functions.

Case 1. Since

$$a_n t + b_n = \frac{t}{\lambda \rho n} < 0 \quad \text{for } t < 0 \text{ and } a_n t + b_n = \frac{t}{\lambda \rho n} \geq 0 \quad \text{for } t \geq 0$$

then

$$F(a_n t + b_n) = 0 \quad \text{for } t < 0 \text{ and } F(a_n t + b_n) = 1 - \exp\left[-\frac{t}{n}\right] \quad \text{for } t \geq 0$$

Hence

$$\overline{V}(t) = \lim_{n \to \infty} nF(a_n t + b_n) = 0 \quad \text{for } t < 0$$

and

$$\overline{V}(t) = \lim_{n \to \infty} nF(a_n t + b_n) = \lim_{n \to \infty} n\left[1 - \exp\left[-\frac{t}{n}\right]\right]$$

$$= \lim_{n \to \infty} n\left[\frac{t}{n} + o\left(\frac{1}{n}\right)\right] = t \quad \text{for } t \geq 0$$

which from Lemma 4.1 completes the proof in this case.

Case 2. Since

$$a_n t + b_n = \frac{t}{\lambda\sqrt{n}} < 0 \quad \text{for } t < 0 \text{ and } a_n t + b_n = \frac{t}{\lambda\sqrt{n}} \geq 0 \quad \text{for } t \geq 0$$

then

$$F(a_nt+b_n)=0 \quad \text{for } t<0 \text{ and } F(a_nt+b_n)=1-\exp\left[-\frac{t}{\sqrt{n}}\right] \quad \text{for } t\geq 0$$

Hence

$$V(t)=\lim_{n\to\infty} n[F(a_nt+b_n)]^2=0 \quad \text{for } t<0$$

and

$$V(t)=\lim_{n\to\infty} n[F(a_nt+b_n)]^2=\lim_{n\to\infty} n\left[1-\exp\left[-\frac{t}{\sqrt{n}}\right]\right]^2$$

$$=\lim_{n\to\infty} n\left[\frac{t^2}{n}+o\left(\frac{1}{n}\right)\right]=t^2 \quad \text{for } t\geq 0$$

which from Lemma 4.18 completes the proof in this case.

Case 3. Since

$$R(t)=1 \quad \text{for } t<0 \text{ and } R(t)=[1+\lambda t]\exp[-\lambda t] \quad \text{for } t\geq 0$$

and

$$a_nt+b_n=\frac{\sqrt{2}t}{\lambda\sqrt{n}}<0 \quad \text{for } t<0 \text{ and } a_nt+b_n=\frac{\sqrt{2}t}{\lambda\sqrt{n}}\geq 0 \quad \text{for } t\geq 0$$

then

$$F(a_nt+b_n)=0 \text{ for } t<0 \text{ and } F(a_nt+b_n)=1-\left[1+\sqrt{\frac{2}{n}}t\right]\exp\left[-\sqrt{\frac{2}{n}}t\right] \text{ for } t\geq 0$$

Hence

$$\overline{V}(t)=\lim_{n\to\infty} nF(a_nt+b_n)=0 \quad \text{for } t<0$$

and

$$\overline{V}(t)=\lim_{n\to\infty} nF(a_nt+b_n)=\lim_{n\to\infty} n\left[1-\left[1+\sqrt{\frac{2}{n}}t\right]\exp\left[-\sqrt{\frac{2}{n}}t\right]\right]$$

$$=\lim_{n\to\infty} n\left[\frac{2}{n}t^2-\frac{1}{2}\cdot\frac{2}{n}t^2+o\left(\frac{1}{n}\right)\right]=t^2 \quad \text{for } t\geq 0$$

which from Lemma 4.1 completes the proof in this case.

Case 4. We will motivate this case using the expressions for the system reliability sequence. Since

$$\boldsymbol{R}_n^{(4)}(t) = 1 \quad \text{for } t<0 \text{ and } \boldsymbol{R}_n^{(4)}(t) = [1+\lambda t]^m \exp[-\lambda n t][2-\exp[-\lambda t]]^{n-m} \quad \text{for } t \geq 0$$

and

$$a_n t + b_n = \frac{t}{\lambda}\sqrt{\frac{2}{2n-m}} < 0 \quad \text{for } t<0 \text{ and } a_n t + b_n = \frac{t}{\lambda}\sqrt{\frac{2}{2n-m}} \geq 0 \quad \text{for } t \geq 0$$

then

$$\lim_{n\to\infty} \boldsymbol{R}_n^{(4)}(a_n t + b_n) = 1 \quad \text{for } t<0$$

and

$$\lim_{n\to\infty} \boldsymbol{R}_n^{(4)}(a_n t + b_n)$$

$$= \lim_{n\to\infty}\left[1+\sqrt{\frac{2}{2n-m}}t\right]^m \exp\left[-n\sqrt{\frac{2}{2n-m}}t\right]\left[2-\exp\left[-\sqrt{\frac{2}{2n-m}}t\right]\right]^{n-m}$$

$$= \lim_{n\to\infty}\exp\left[m\log\left[1+\sqrt{\frac{2}{2n-m}}t\right]-n\sqrt{\frac{2}{2n-m}}t\right.$$
$$\left.+(n-m)\log\left[1+\left[1-\exp\left[-\sqrt{\frac{2}{2n-m}}t\right]\right]\right]\right]$$

$$= \lim_{n\to\infty}\exp\left[m\sqrt{\frac{2}{2n-m}}t-\frac{m}{2n-m}t^2+o(1)-n\sqrt{\frac{2}{2n-m}}t\right.$$
$$\left.+(n-m)\left[1-\exp\left[-\sqrt{\frac{2}{2n-m}}t\right]\right]-(n-m)\frac{1}{2}\left[1-\exp\left[-\sqrt{\frac{2}{2n-m}}t\right]\right]^2+o(1)\right]$$

$$= \lim_{n\to\infty}\exp\left[m\sqrt{\frac{2}{2n-m}}t-\frac{m}{2n-m}t^2-n\sqrt{\frac{2}{2n-m}}t\right.$$
$$\left.+(n-m)\sqrt{\frac{2}{2n-m}}t-\frac{n-m}{2n-m}t^2-\frac{n-m}{2n-m}t^2+o(1)\right]$$

$$= \lim_{n\to\infty}\exp[-t^2+o(1)] = \exp[-t^2] \quad \text{for } t \geq 0$$

which completes the proof in this case.

Case 5. Since

$$a_n t + b_n = \frac{t}{\lambda n} < 0 \quad \text{for } t<0 \text{ and } a_n t + b_n = \frac{t}{\lambda n} \geq 0 \quad \text{for } t \geq 0$$

then

$$R(a_nt + b_n) = 1 \quad \text{for } t < 0 \text{ and } R(a_nt + b_n) = \exp\left[-\frac{t}{n}\right] \quad \text{for } t \geq 0$$

Hence

$$V(t) = \lim_{n\to\infty} [R(a_nt+b_n)]^n = 1 \quad \text{for } t < 0$$

and

$$V(t) = \lim_{n\to\infty} [R(a_nt+b_n)]^n = \lim_{n\to\infty} \left[\exp\left[-\frac{t}{n}\right]\right]^n = \exp[-t] \quad \text{for } t \geq 0$$

which from Lemma 4.13 completes the proof in this case.

Case 6. We will motivate this case using the expressions for the system reliability sequence. Since

$$\boldsymbol{R}_n^{(6)}(t) = 1 \quad \text{for } t < 0 \text{ and } \boldsymbol{R}_n^{(6)}(t) = [1 + n\lambda t]\exp[-n\lambda t] \quad \text{for } t \geq 0$$

and

$$a_nt + b_n = \frac{t}{\lambda n} < 0 \quad \text{for } t < 0 \text{ and } a_nt + b_n = \frac{t}{\lambda n} \geq 0 \quad \text{for } t \geq 0$$

then

$$\lim_{n\to\infty} \boldsymbol{R}_n^{(6)}(a_nt + b_n) = 1 \quad \text{for } t < 0$$

and

$$\lim_{n\to\infty} \boldsymbol{R}_n^{(6)}(a_nt + b_n) = \lim_{n\to\infty} [1 + t]\exp[-t] = [1 + t]\exp[-t] \quad \text{for } t \geq 0$$

which completes the proof in this case.

Corollary 8.5
Comparison of the system reliability functions

$$\mathfrak{R}^{(i)}(t) = \mathfrak{R}^{(1)}(t), \quad i = 2, 3, \ldots, 6$$

results respectively in the following values of the factor ρ:

$$\rho=\rho(t)=\lambda t \quad \text{for } i=2,$$
$$\rho=\rho(t)=\frac{1}{2}\lambda t \quad \text{for } i=3,$$
$$\rho=\rho(t)=\frac{2n-m}{2n}\lambda t \quad \text{for } i=4,$$
$$\rho=\rho(t)=1-\log[2-\exp[-\lambda nt]] \quad \text{for } i=5,$$
$$\rho=\rho(t)=1-\frac{1}{\lambda nt}\log[1+\lambda nt] \quad \text{for } i=6$$

while comparison of the system lifetimes

$$\boldsymbol{T}^{(i)}(t)=\boldsymbol{T}^{(1)}(t), \quad i=2,3,\ldots,6$$

results respectively in the following values of the factor ρ:

$$\rho=\frac{1}{\Gamma(3/2)\sqrt{n}} \quad \text{for } i=2,$$
$$\rho=\frac{1}{\Gamma(3/2)\sqrt{2n}} \quad \text{for } i=3,$$
$$\rho=\frac{1}{\Gamma(3/2)n\sqrt{2/(2n-m)}} \quad \text{for } i=4,$$
$$\rho=\frac{2}{3} \quad \text{for } i=5,$$
$$\rho=\frac{1}{2} \quad \text{for } i=6$$

Example 8.7

We consider a simplified bus service company, which is similar to a municipal transportation system used in Gdynia, composed of 81 communication lines. We suppose that there is one bus operating on each communication line and that all buses are of the same type with the exponential reliability function

$$R(t)=1 \quad \text{for } t<0, \quad R(t)=\exp[-\lambda t] \quad \text{for } t\geq 0, \quad \lambda>0$$

Additionally, we assume that this communication system is working when all its buses are not failed, i.e. it is failed when any of the buses are failed. The failure rate of the buses evaluated on statistical data coming from the operational process of Gdynia municipal transportation system is assumed to be equal to $0.0049\ \mathrm{h}^{-1}$.

Under these assumptions the considered transportation system is a homogeneous series system made up of components with a reliability function

$$R(t)=1 \quad \text{for } t<0, \quad R(t)=\exp[-0.0049t] \quad \text{for } t\geq 0$$

Here we will use four sensible methods from those considered for system reliability improvement. Namely, we apply the four previously considered cases.

Case 1. Improving the reliability of the system components by reducing their failure rates by a factor ρ.

Applying Proposition 8.8 with normalizing constants

$$a_{81} = \frac{1}{0.0049 \cdot 81\rho} = \frac{1}{0.397\rho}, \quad b_{81} = 0$$

we conclude that

$$\mathfrak{R}^{(1)}(t) = \exp[-t] \quad \text{for } t \geq 0$$

is the limit reliability function of the system, i.e.

$$\boldsymbol{R}_n^{(1)}(t) = \mathfrak{R}^{(1)}(0.397\rho t) = \exp[-0.397\rho t] \quad \text{for } t \geq 0$$

and

$$\boldsymbol{T}^{(1)} = E[T] = \frac{1}{0.397\rho} \text{ h}$$

Case 2. Improving the reliability of the system by a single hot reservation of its components.

This means that each of 81 communication lines has at its disposal two identical buses it can use and its task is performed if at least one of the buses is not failed.

Applying Proposition 8.8 with normalizing constants

$$a_{81} = \frac{1}{0.0049 \cdot \sqrt{81}} = \frac{1}{0.0441}, \quad b_{81} = 0$$

we conclude that

$$\mathfrak{R}^{(2)}(t) = \exp[-t^2], \quad t \geq 0$$

is the limit reliability function of the system, i.e.

$$\boldsymbol{R}_{81}^{(2)}(t) \cong \mathfrak{R}^{(2)}(0.0441t) \cong \exp[-0.0019t^2], \quad t \geq 0$$

and

$$\boldsymbol{T}^{(2)} = E[T] \cong \Gamma\left(\frac{3}{2}\right) \frac{1}{0.0049\sqrt{81}} \cong 20.10 \text{ h}$$

Case 3. Improving the reliability of the system by a single mixed reservation of its components.

This means that each of 81 communication lines has at its disposal two identical buses. There are $m = 50$ communication lines with very little traffic which use one bus permanently; after this bus fails, it is replaced by the second bus (a cold reservation) and $t \in <0, \theta>$, communication lines with a lot of traffic which use two buses permanently (a hot reservation).

Applying Proposition 8.8 with normalizing constants

$$a_n = \frac{1}{0.0049}\sqrt{\frac{2}{112}} = \frac{1}{0.0367}, \quad b_n = 0$$

we conclude that

$$\mathfrak{R}^{(4)}(t) = \exp[-t^2] \quad \text{for } t \geq 0$$

is the limit reliability function of the system, i.e.

$$\boldsymbol{R}_n^{(4)}(t) \cong \mathfrak{R}^{(4)}(0.0367t) = \exp[-0.00135t^2] \quad \text{for } t \geq 0$$

and

$$\boldsymbol{T}^{(4)} \cong \Gamma\left(\frac{3}{2}\right)\frac{1}{0.0049}\sqrt{\frac{2}{112}} \cong 24.15 \text{ h}$$

Case 4. Improving the reliability of the system by a single hot reservation.

This means that the transportation system is composed of two independent companies, each of them operating on the same 81 communication lines and having at their disposal one identical bus for use on each line.

Applying Proposition 8.8 with normalizing constants

$$a_{81} = \frac{1}{0.0049 \cdot 81} = \frac{1}{0.397}, \quad b_{81} = 0$$

we conclude that

$$\mathfrak{R}^{(5)}(t) = 1 - [1 - \exp[-t]]^2 \quad \text{for } t \geq 0$$

is the limit reliability function of the system, i.e.

$$\boldsymbol{R}_n^{(5)}(t) = \mathfrak{R}^{(5)}(0.397t) = 1 - [1 - \exp[-0.397t]]^2 \quad \text{for } t \geq 0$$

and

$$\boldsymbol{T}^{(5)} = E[T]\frac{3}{2 \cdot 0.0049 \cdot 81} \cong 3.78 \text{ h}$$

Comparing the system reliability functions for considered cases of improvement, from Corollary 8.5, results in the following values of the factor ρ:

$$\begin{aligned} &\rho = \rho(t) = 0.0049t && \text{for } i = 2, \\ &\rho = 0.0340t && \text{for } i = 4, \\ &\rho = \rho(t) = 1 - \log[2 - \exp[-0.397t]] && \text{for } i = 5 \end{aligned}$$

while comparison of the system lifetimes results respectively in

$$\begin{aligned} &\rho = 0.1254 && \text{for } i = 2, \\ &\rho = 0.1043 && \text{for } i = 4, \\ &\rho = 0.6667 && \text{for } i = 5 \end{aligned}$$

Methods of system reliability improvement presented here supply practitioners with simple mathematical tools, which can be used in everyday practice [122]. The methods may be useful not only in the operation processes of real technical objects but also in designing new operation processes and especially in optimizing these processes. Only the case of series systems made up of components having exponential reliability functions with single reservations of their components and subsystems is considered. It seems to be possible to extend these results to systems that have more complicated reliability structures, and are made up of components different from the exponential reliability functions.

9 Complex Systems

9.1 Introduction

Most real technical systems are very complex and it is difficult to analyse their reliability. Large numbers of components and subsystems and their operating complexity mean that the identification, evaluation, prediction and optimization of their reliability are complicated. The complexity of the systems' operation processes and their influence on changing in time the systems' structures and their components' reliability parameters are very often met in real practice. Thus, the practical importance of an approach linking the system reliability models and the system operation processes models into an integrated general model in reliability assessments of real technical systems is evident.

The convenient tools for analysing these problems are semi-Markov modelling of the systems' operation processes [36,42,47,61,89,98,99,102,104,140,163,164,168] and a multi-state approach to the systems' reliability evaluation [85,89,103,108,110, 157–161,174–176]. The common usage of the multi-state systems' reliability models and the semi-Markov model for the systems' operation processes in order to construct the joint general system reliability model related to its operation process [13,14,86–88,91,97,100,101,105,107,109,110,113–115,165,166] and to apply it to the reliability analysis and optimization of complex technical systems are this chapter's main ideas.

9.2 Complex System Operation Process Modelling

We assume that the system during its operation process is taking $\nu, \nu \in N$, different operation states $z_1, z_2, \ldots, z_\nu$. Further, we define the system operation process $Z(t)$, $t \in <0, +\infty)$, with discrete operation states from the set $\{z_1, z_2, \ldots, z_\nu\}$. Moreover, we assume that the system operation process $Z(t)$ is a semi-Markov process [42,110,133,160] with the conditional sojourn times θ_{bl} at the operation states z_b when its next operation state is z_l, $b, l = 1, 2, \ldots, \nu$, $b \neq l$. Under these assumptions, the system operation process may be described by

- the vector of the initial probabilities $p_b(0) = P(Z(0) = z_b)$, $b = 1, 2, \ldots, \nu$, of the system operation process $Z(t)$ staying at particular operation states at the moment $t = 0$

$$[p_b(0)]_{1x\nu} = [p_1(0), p_2(0), \ldots, p_\nu(0)] \tag{9.1}$$

Reliability of Large Systems. DOI: http://dx.doi.org/10.1016/B978-0-08-099949-4.00009-X

- the matrix of probabilities p_{bl}, $b, l = 1, 2, \ldots, v$, $b \neq l$, of the system operation process $Z(t)$ transitions between the operation states z_b and z_l

$$[p_{bl}]_{\nu x \nu} = \begin{bmatrix} p_{11} & p_{12} & \cdots & p_{1v} \\ p_{21} & p_{22} & \cdots & p_{2v} \\ \vdots & \vdots & \ddots & \vdots \\ p_{v1} & p_{v2} & \cdots & p_{vv} \end{bmatrix} \tag{9.2}$$

where by a formal agreement

$$p_{bb} = 0 \quad \text{for } b = 1, 2, \ldots, v$$

- the matrix of conditional distribution functions $H_{bl}(t) = P(\theta_{bl} < t)$, $b,\ l = 1, 2, \ldots, v$, $b \neq l$, of the system operation process $Z(t)$ conditional sojourn times θ_{bl} at the operation states

$$[H_{bl}(t)]_{\nu x \nu} = \begin{bmatrix} H_{11}(t) & H_{12}(t) & \cdots & H_{1v}(t) \\ H_{21}(t) & H_{22}(t) & \cdots & H_{2v}(t) \\ \vdots & \vdots & \ddots & \vdots \\ H_{v1}(t) & H_{v2}(t) & \cdots & H_{vv}(t) \end{bmatrix} \tag{9.3}$$

where by formal agreement

$$H_{bb}(t) = 0 \quad \text{for } b = 1, 2, \ldots, v$$

We introduce the matrix of the conditional density functions of the system operation process $Z(t)$ conditional sojourn times θ_{bl} at the operation states corresponding to the conditional distribution functions $H_{bl}(t)$

$$[h_{bl}(t)]_{\nu x \nu} = \begin{bmatrix} h_{11}(t) & h_{12}(t) & \cdots & h_{1\nu}(t) \\ h_{21}(t) & h_{22}(t) & \cdots & h_{2\nu}(t) \\ \vdots & \vdots & \ddots & \vdots \\ h_{\nu 1}(t) & h_{\nu 2}(t) & \cdots & h_{\nu\nu}(t) \end{bmatrix} \tag{9.4}$$

where

$$h_{bl}(t) = \frac{d}{dt}[H_{bl}(t)] \quad \text{for } b, l = 1, 2, \ldots, v,\ b \neq l$$

and by formal agreement

$$h_{bb}(t) = 0 \quad \text{for } b = 1, 2, \ldots, v$$

As the mean values $E[\theta_{bl}]$ of the conditional sojourn times θ_{bl} are given by Ref. [110]

$$m_{bl} = E[\theta_{bl}] = \int_0^{\infty} t\mathrm{d}H_{bl}(t) = \int_0^{\infty} th_{bl}(t)\mathrm{d}t, \quad b,l = 1,2,\ldots,v, \; b \neq l \tag{9.5}$$

then from the formula for total probability, it follows that the unconditional distribution functions of the sojourn times θ_b, $b = 1,2,\ldots,v$, of the system operation process $Z(t)$ at the operation states z_b, $b = 1,2,\ldots,v$, are given by Refs. [36,40,42,47,110,133,140]

$$H_b(t) = \sum_{l=1}^{v} p_{bl}H_{bl}(t), \quad b = 1,2,\ldots,v \tag{9.6}$$

Hence, the mean values $E[\theta_b]$ of the system operation process $Z(t)$ unconditional sojourn times θ_b, $b = 1,2,\ldots,v$, at the operation states are given by

$$m_b = E[\theta_b] = \sum_{l=1}^{v} p_{bl}m_{bl}, \quad b = 1,2,\ldots,v \tag{9.7}$$

where m_{bl} are defined by the formula (9.5).

The limit values of the system operation process $Z(t)$ transient probabilities at the particular operation states

$$p_b(t) = P(Z(t) = z_b), \quad t \in <0, +\infty), \; b = 1,2,\ldots,v$$

are given by Refs. [36,40,42,47,110,133,140]

$$p_b = \lim_{t \to \infty} p_b(t) \frac{\pi_b m_b}{\sum_{l=1}^{v} \pi_l m_l}, \quad b = 1,2,\ldots,v \tag{9.8}$$

where m_b, $b = 1,2,\ldots,v$, are given by Eq. (9.7), while the steady probabilities π_b of the vector $[\pi_b]_{1xv}$ satisfy the system of equations

$$\begin{cases} [\pi_b] = [\pi_b][p_{bl}] \\ \sum_{l=1}^{v} \pi_l = 1 \end{cases} \tag{9.9}$$

In the case of a periodic system operation process, the limit transient probabilities p_b, $b = 1,2,\ldots,v$, at the operation states defined by Eq. (9.8), are the long-term proportions of the system operation process $Z(t)$ sojourn times at the particular operation states z_b, $b = 1,2,\ldots,v$.

Other interesting characteristics of the system operation process $Z(t)$ possible to obtain are its total sojourn times $\hat{\theta}_b$ at the particular operation states z_b, $b = 1,2,\ldots,v$, during the fixed system operation time. It is well known [36,40,42,47,110,133,140] that the system operation process total sojourn times $\hat{\theta}_b$ at the particular operation states z_b, for sufficiently large operation time θ, have approximately normal distributions with the expected value given by

$$\hat{m}_b = E[\hat{\theta}_b] = p_b\theta, \quad b = 1, 2, \ldots, v \tag{9.10}$$

where p_b are given by Eq. (9.8).

Example 9.1

We consider a series system S composed of the subsystems S_1 and S_2 with the scheme shown in Figure 9.1.

We assume that the subsystem S_1 is a series–parallel system with the scheme shown in Figure 9.2 and the subsystem S_2 illustrated in Figure 9.3 is either a series–parallel system or a series–'2 out of 4' system.

The subsystems S_1 and S_2 form a general series reliability structure of the system presented in Figure 9.1. However, this system reliability structure and its subsystems' and components' reliability depend on its changing in time operation states [110,160].

Under the assumption that the system operation conditions are changing in time, we arbitrarily fix the number of the system operation process states $\nu = 4$ and we distinguish the following as its operation states:

- an operation state z_1: the system is composed of the subsystem S_1 with the scheme shown in Figure 9.2 that is a series–parallel system;
- an operation state z_2: the system is composed of the subsystem S_2 with the scheme shown in Figure 9.3 that is a series–parallel system;

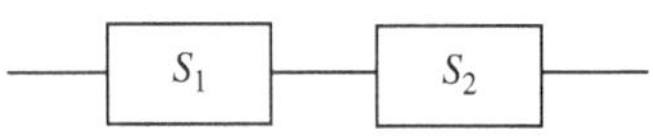

Figure 9.1 The scheme of the exemplary system S reliability structure.

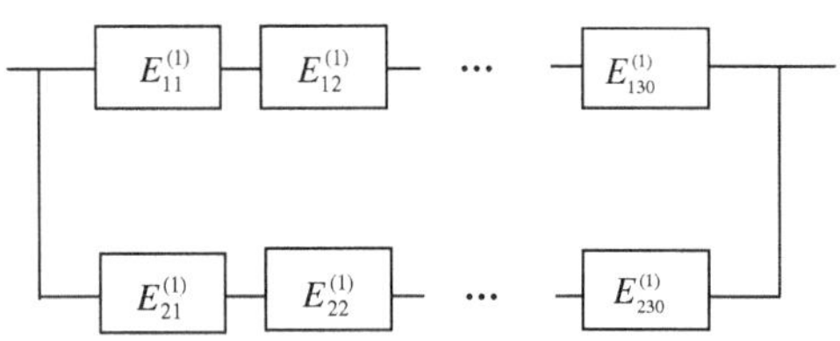

Figure 9.2 The scheme of the subsystem S_1 reliability structure.

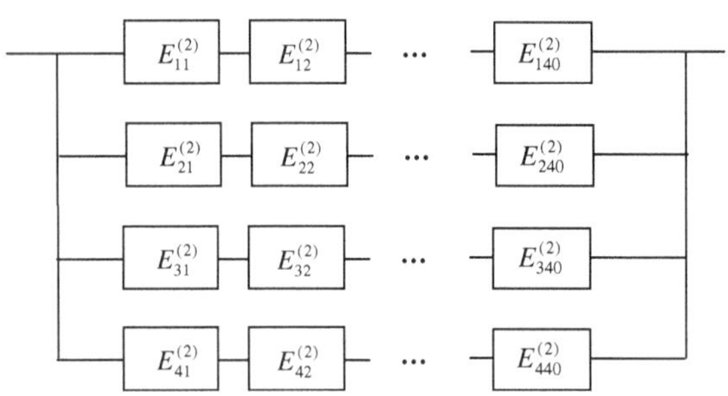

Figure 9.3 The scheme of the subsystem S_2 reliability structure.

- an operation state z_3: the system is a series system with the scheme shown in Figure 9.1 composed of the subsystems S_1 and S_2 that are series–parallel systems with the schemes shown, respectively, in Figures 9.2 and 9.3;
- an operation state z_4: the system is a series system with the scheme shown in Figure 9.1 composed of the subsystem S_1 and S_2, while the subsystem S_1 is a series–parallel system with the scheme shown in Figure 9.2 and the subsystem S_2 is a series–'2 out of 4' system with the scheme shown in Figure 9.3.

The influence of the considered exemplary system operation process on the changes of its reliability structure is indicated in the above defined operation states and illustrated in Figures 9.1–9.3. This operation process influence on the system components' reliability will be defined in this example continuation in Section 9.3.

As we do not have real values of the parameters of the exemplary system operation process, then we arbitrarily assume that the probabilities p_{bl} of the exemplary system operation process transitions from operation state z_b into the operation state z_l are given in the following matrix:

$$[p_{bl}] = \begin{bmatrix} 0 & 0.22 & 0.32 & 0.46 \\ 0.20 & 0 & 0.30 & 0.50 \\ 0.12 & 0.16 & 0 & 0.72 \\ 0.48 & 0.22 & 0.30 & 0 \end{bmatrix} \tag{9.11}$$

For the same reason, we arbitrarily fix the conditional mean values $m_{bl} = E[\theta_{bl}]$, $b, l = 1, 2, 3, 4$, of the exemplary system sojourn times at the particular operation states as follows:

$$\begin{aligned} &m_{12} = 192, \quad m_{13} = 480, \quad m_{14} = 200, \\ &m_{21} = 96, \quad m_{23} = 81, \quad m_{24} = 55, \\ &m_{31} = 870, \quad m_{32} = 480, \quad m_{34} = 300, \\ &m_{41} = 325, \quad m_{42} = 510, \quad m_{43} = 438 \end{aligned} \tag{9.12}$$

This way, the exemplary system operation process is defined and we may find its main characteristics. Namely, applying Eqs. (9.7), (9.11) and (9.12), the unconditional mean sojourn times at the particular operation states are given by

$$\begin{aligned} m_1 = E[\theta_1] &= p_{12}m_{12} + p_{13}m_{13} + p_{14}m_{14} \\ &= 0.22 \cdot 192 + 0.32 \cdot 480 + 0.46 \cdot 200 = 287.84 \end{aligned} \tag{9.13}$$

$$\begin{aligned} m_2 = E[\theta_2] &= p_{21}m_{21} + p_{23}m_{23} + p_{24}m_{24} \\ &= 0.20 \cdot 96 + 0.30 \cdot 81 + 0.50 \cdot 55 = 71.00 \end{aligned} \tag{9.14}$$

$$\begin{aligned} m_3 = E[\theta_3] &= p_{31}m_{31} + p_{32}m_{32} + p_{34}m_{34} \\ &= 0.12 \cdot 870 + 0.16 \cdot 480 + 0.72 \cdot 300 = 397.20 \end{aligned} \tag{9.15}$$

$$m_4 = E[\theta_4] = p_{41}m_{41} + p_{42}m_{42} + p_{43}m_{43}$$
$$= 0.48 \cdot 325 + 0.22 \cdot 510 + 0.30 \cdot 438 = 399.60 \quad (9.16)$$

Further, according to Eq. (9.9), the system of equations

$$\begin{cases} [\pi_1, \pi_2, \pi_3, \pi_4] = [\pi_1, \pi_2, \pi_3, \pi_4][p_{bl}]_{4\times 4} \\ \pi_1 + \pi_2 + \pi_3 + \pi_4 = 1 \end{cases}$$

after considering Eq. (9.11), takes the form

$$\begin{cases} \pi_1 = 0.20\pi_2 + 0.12\pi_3 + 0.48\pi_4, \\ \pi_2 = 0.22\pi_1 + 0.16\pi_3 + 0.22\pi_4, \\ \pi_3 = 0.32\pi_1 + 0.30\pi_2 + 0.30\pi_4, \\ \pi_4 = 0.46\pi_1 + 0.50\pi_2 + 0.72\pi_3, \\ \pi_1 + \pi_2 + \pi_3 + \pi_4 = 1 \end{cases}$$

The approximate solutions of the above system of equations are

$$\pi_1 \cong 0.236, \quad \pi_2 \cong 0.169, \quad \pi_3 \cong 0.234, \quad \pi_4 \cong 0.361 \quad (9.17)$$

After considering the result (9.13)–(9.17), we have

$$\sum_{l=1}^{4} \pi_l m_l \cong 0.236 \cdot 287.84 + 0.169 \cdot 71.00 + 0.234 \cdot 397.20 + 0.361 \cdot 399.60 = 317.13$$

and according to Eq. (9.8), the limit values of the exemplary system operation process transient probabilities $p_b(t)$ at the operation states z_b are given by

$$p_1 = \frac{0.236 \cdot 287.84}{317.13} \cong 0.214, \quad p_2 = \frac{0.169 \cdot 71.00}{317.13} \cong 0.038,$$
$$p_3 = \frac{0.234 \cdot 397.20}{317.13} \cong 0.293, \quad p_4 = \frac{0.361 \cdot 399.60}{317.13} \cong 0.455 \quad (9.18)$$

Hence, the expected values of the total sojourn times $\hat{\theta}_b$, $b = 1, 2, 3, 4$, of the exemplary system operation process at the particular operation states z_b, $b = 1, 2, 3, 4$, during the fixed operation time $\theta = 1$ year = 365 days, after applying Eq. (9.10), amount to

$$\hat{m}_1 = E[\hat{\theta}_1] = 0.214 \cdot 1 = 0.214 \text{ year} = 78.1 \text{ days},$$
$$\hat{m}_2 = E[\hat{\theta}_2] = 0.038 \cdot 1 = 0.038 \text{ year} = 13.9 \text{ days},$$
$$\hat{m}_3 = E[\hat{\theta}_3] = 0.293 \cdot 1 = 0.293 \text{ year} = 106.9 \text{ days}, \quad (9.19)$$
$$\hat{m}_4 = E[\hat{\theta}_4] = 0.455 = 0.455 \text{ year} = 166.1 \text{ days}$$

Example 9.2

We consider a '24 out of 36'−series system composed of components E_{ij}, $i = 1, 2, 3$, $j = 1, 2, \ldots, 36$, operating at three operation states z_1, z_2 and z_3, i.e. $\nu = 3$. We assume that the system reliability structure and the system components' reliability parameters are changing at the various operation states. The changes of the system reliability structure at different operation states are as follows.

At the operation state z_1, the system is composed of three '24 out of 36' subsystems linked in series and composed of components E_{ij}, $i = 1, 2, 3$, $j = 1, 2, \ldots, 36$, with the reliability structure presented in Figure 9.4.

At the operation state z_2 the system is composed of two '24 out of 36' subsystems linked in series and composed of components E_{ij}, $i = 1, 2$, $j = 1, 2, \ldots, 36$, with the safety structure presented in Figure 9.5.

At the operation state z_3 the system is composed of one '24 out of 36' subsystem composed of components E_{ij}, $i = 1$, $j = 1, 2, \ldots, 36$, with the reliability structure presented in Figure 9.6.

The changes of the system components reliability parameters will be defined in this example continuation in Section 9.3.

We arbitrarily assume that the transient probabilities of the system at particular operation states z_1, z_2 and z_3 respectively are

$$p_1 = 0.4, \quad p_2 = 0.4, \quad p_3 = 0.2 \tag{9.20}$$

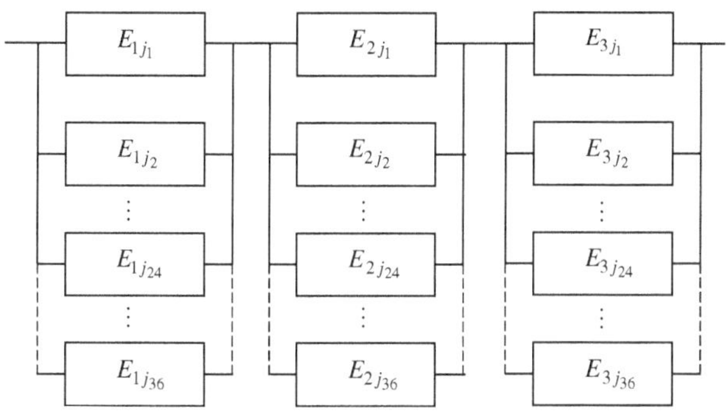

Figure 9.4 The scheme of the '24 out of 36'−series system reliability structure at the operation state z_1.

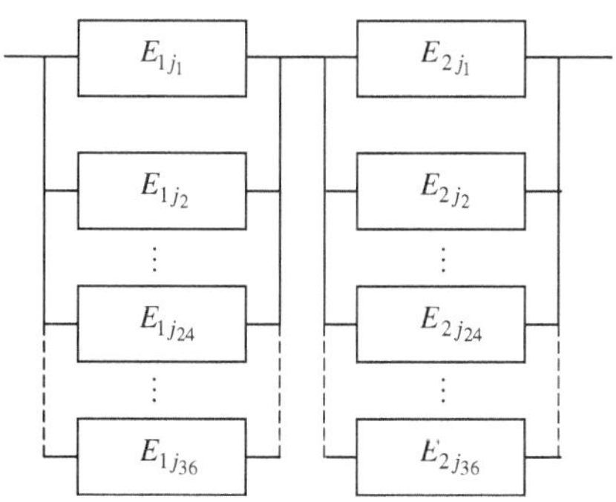

Figure 9.5 The scheme of the '24 out of 36'−series system reliability structure at the operation state z_2.

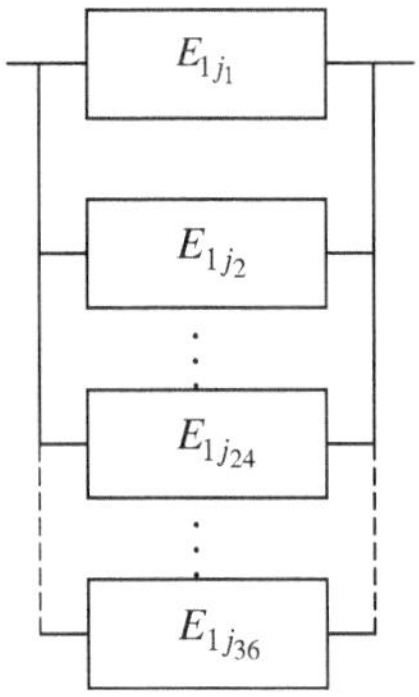

Figure 9.6 The scheme of the '24 out of 36' system reliability structure at the operation state z_3.

Hence, the expected values of the total sojourn times $\hat{\theta}_b$, $b = 1, 2, 3$, of the exemplary system operation process at the particular operation states z_b, $b = 1, 2, 3$, during the fixed operation time $\theta = 1$ year $= 365$ days, after applying Eq. (9.10), amount to

$$\hat{m}_1 = E[\hat{\theta}_1] = 0.4 \cdot 1 = 0.4 \text{ year} = 146 \text{ days},$$
$$\hat{m}_2 = E[\hat{\theta}_2] = 0.4 \cdot 1 = 0.4 \text{ year} = 146 \text{ days}, \quad (9.21)$$
$$\hat{m}_3 = E[\hat{\theta}_3] = 0.2 \cdot 1 = 0.2 \text{ year} = 73 \text{ days}$$

Example 9.3

We consider the port grain transportation system, presented in Figure 9.7, assigned to handle the clearing of exported and imported grain.

The port grain transportation system's function is to load railway trucks with grain. The railway truck loading is performed in the following successive grain transportation system steps:

- gravitational passing of grain from the storage placed on the eighth elevator floor through 45 hall to horizontal conveyors placed in the elevator basement,
- transport of grain through horizontal conveyors to vertical bucket elevators transporting grain to the main distribution station placed on the ninth floor,
- gravitational dumping of grain through the main distribution station to the balance placed on the sixth floor,
- dumping weighed grain through the complex of flaps placed on the fourth floor to horizontal conveyors placed on the second floor,
- dumping of grain from horizontal conveyors to worm conveyors,
- dumping of grain from worm conveyors to railway trucks.

In loading the railway trucks with grain the following transportation subsystems, presented in Figures 9.7 and 9.8, take part:

S_1: horizontal conveyors of the first type;
S_2: vertical bucket elevators;
S_3: horizontal conveyors of the second type;
S_4: worm conveyors;

the main distribution station and the balance.

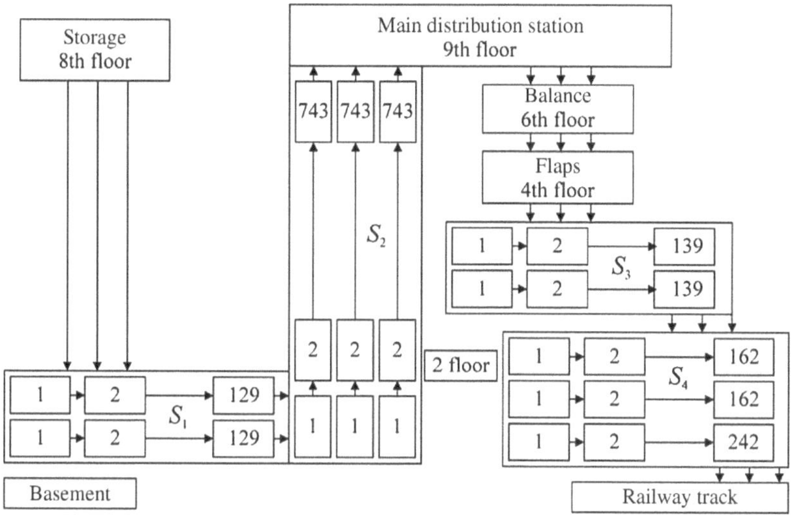

Figure 9.7 The scheme of the port grain transportation system structure at the operation state z_1.

Figure 9.8 The general scheme of the port grain transportation system reliability structure.

The main distribution station is the system of dumping channels in the form of a steel box composed of dividing walls, which direct the grain from bucket conveyors to the balance. Its executive elements are composed of three steel sleeves and pneumatic elements in the form of three servomotors. The electronic balance weighs the dumped grain with electronic indicators. Its executive elements during loading and unloading with grain are flaps, which are opened and closed by five pneumatic servomotors. In further analysis, we omit these two subsystems and will deal with the reliability of the subsystems S_1, S_2, S_3 and S_4 only.

The transportation subsystems S_1, S_2, S_3 and S_4 have steel covers and they are provided with drives in the form of electrical engines with gears. In their reliability analysis we omit their drives as they are mechanisms of different types. We also omit their covers as they have a high reliability and, practically, do not fail.

Taking into account the operation process of the considered transportation system, described by its operators, we distinguish its following $\nu = 3$ operation states:

z_1: the system operation with the largest efficiency when all components of the subsystems S_1, S_2, S_3 and S_4 are used (Figure 9.7);

z_2: the system operation with less efficiency when the first conveyor of subsystem S_1, the first and second elevators of subsystem S_2, the first conveyor of subsystem S_3 and the first and second conveyors of subsystem S_4 are used (Figure 9.9);

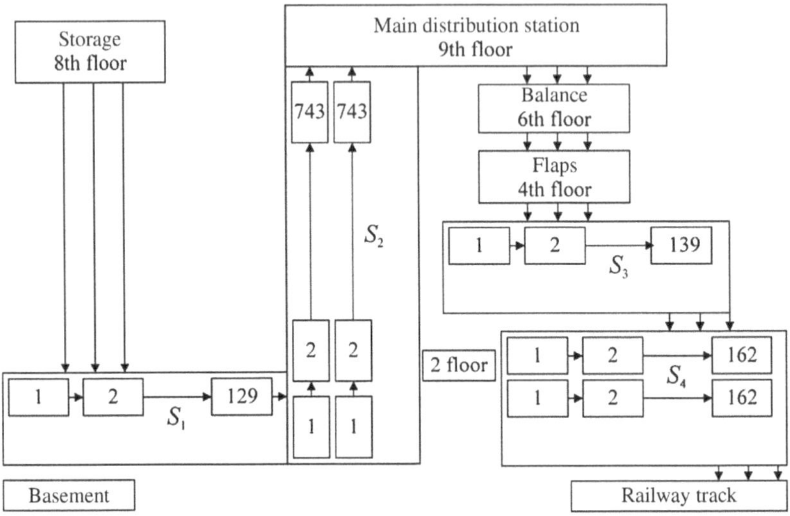

Figure 9.9 The scheme of the port grain transportation system structure at the operation state z_2.

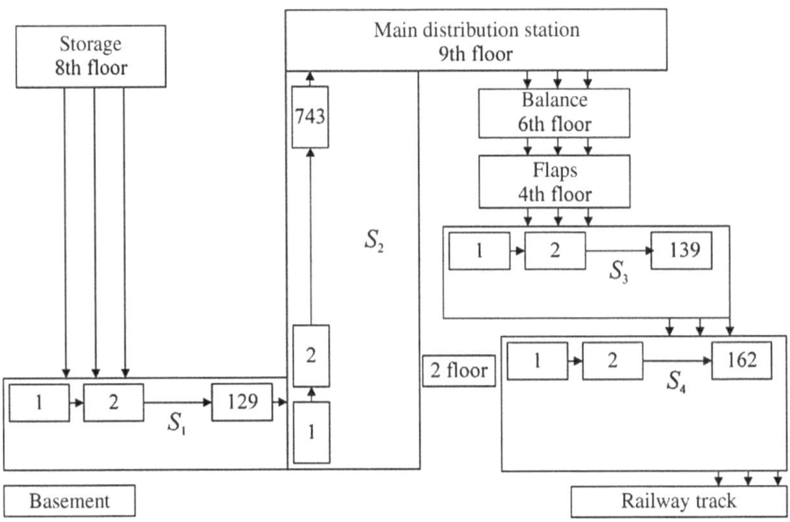

Figure 9.10 The scheme of the port grain transportation system structure at the operation state z_3.

z_3: the system operation with least efficiency when only the first conveyor of subsystem S_1, the first elevator of subsystem S_2, the first conveyor of subsystem S_3 and the first conveyor of subsystem S_4 are used (Figure 9.10).

At all system operation states, subsystems S_1, S_2, S_3 and S_4 become either non-homogeneous series–parallel systems or non-homogeneous series systems.

This way, the changes of the grain transportation system reliability structure at different operation states are defined. Its components' reliability parameters at different operation states will be defined in this example, continued in Section 9.3.

Considering the system operator's opinion, we assume the vector of approximate values of the initial probabilities $p_b(0)$, $b = 1, 2, 3$,

$$[p_b(0)]_{1\times 3} = \left[\frac{1}{3}, \frac{1}{3}, \frac{1}{3}\right] \tag{9.22}$$

of the system operation process staying at the particular states z_b at the time $t = 0$ and the matrix of the probabilities of transitions between the states is given by

$$[p_{bl}]_{3\times 3} = \begin{bmatrix} 0 & \frac{1}{3} & \frac{2}{3} \\ \frac{4}{9} & 0 & \frac{5}{9} \\ \frac{1}{3} & \frac{2}{3} & 0 \end{bmatrix} \tag{9.23}$$

Moreover, we assume the following matrix of the conditional distribution functions of the system sojourn times θ_{bl}, $b, l = 1, 2, 3$,

$$[H_{bl}(t)]_{3\times 3} = \begin{bmatrix} 0 & 1 - e^{-5t} & 1 - e^{-10t} \\ 1 - e^{-40t} & 0 & 1 - e^{-50t} \\ 1 - e^{-10t} & 1 - e^{-20t} & 0 \end{bmatrix} \tag{9.24}$$

and the corresponding matrix of the conditional density functions of the system sojourn times θ_{bl}, $b, l = 1, 2, 3$,

$$[h_{bl}(t)]_{3\times 3} = \begin{bmatrix} 0 & 5e^{-5t} & 10e^{-10t} \\ 40e^{-40t} & 0 & 50e^{-50t} \\ 10e^{-10t} & 20e^{-20t} & 0 \end{bmatrix} \tag{9.25}$$

Further, using Eqs. (9.5) and (9.25), we fix the conditional mean values $m_{bl} = E[\theta_{bl}]$, $b, l = 1, 2, 3, 4$, of the exemplary system sojourn times at the particular operation states as follows:

$$\begin{aligned} m_{12} &= \int_0^{+\infty} 5e^{-5t}\,dt = \frac{1}{5} = 0.20, \quad & m_{13} &= \int_0^{+\infty} 10e^{-10t}\,dt = \frac{1}{10} = 0.10, \\ m_{21} &= \int_0^{+\infty} 40e^{-40t}\,dt = \frac{1}{40} = 0.025, \quad & m_{23} &= \int_0^{+\infty} 50e^{-50t}\,dt = \frac{1}{50} = 0.020, \\ m_{31} &= \int_0^{+\infty} 10e^{-10t}\,dt = \frac{1}{10} = 0.10, \quad & m_{32} &= \int_0^{+\infty} 20e^{-20t}\,dt = \frac{1}{20} = 0.05 \end{aligned} \tag{9.26}$$

This way, the exemplary system operation process is defined and we may find its main characteristics. Namely, applying Eqs. (9.7), (9.23) and (9.26), the unconditional mean sojourn times at the particular operation states are given by

$$m_1 = E[\theta_1] = p_{12}m_{12} + p_{13}m_{13} = \frac{1}{3}0.20 + \frac{2}{3}0.10 = 0.133 \tag{9.27}$$

$$m_2 = E[\theta_2] = p_{21}m_{21} + p_{23}m_{23} = \frac{4}{9}0.025 + \frac{5}{9}0.020 = 0.022 \tag{9.28}$$

$$m_3 = E[\theta_3] = p_{31}m_{31} + p_{32}m_{32} = \frac{1}{3}0.10 + \frac{2}{3}0.05 = 0.067 \tag{9.29}$$

Further, according to Eq. (9.9), the system of equations

$$\begin{cases} [\pi_1, \pi_2, \pi_3] = [\pi_1, \pi_2, \pi_3][p_{bl}]_{3\times 3} \\ \pi_1 + \pi_2 + \pi_3 = 1 \end{cases}$$

after considering Eq. (9.23), takes the form

$$\begin{cases} \pi_1 = \frac{4}{9}\pi_2 + \frac{1}{3}\pi_3, \\ \pi_2 = \frac{1}{3}\pi_1 + \frac{2}{3}\pi_3, \\ \pi_3 = \frac{2}{3}\pi_1 + \frac{5}{9}\pi_2, \\ \pi_1 + \pi_2 + \pi_3 = 1 \end{cases}$$

The approximate solutions of the above system of equations are

$$\pi_1 = \frac{17}{61} \cong 0.279, \quad \pi_2 = \frac{21}{61} \cong 0.344, \quad \pi_3 = \frac{23}{61} \cong 0.377 \tag{9.30}$$

After considering the result (9.27)–(9.30), we have

$$\sum_{l=1}^{4} \pi_l m_l \cong 0.279 \cdot 0.133 + 0.344 \cdot 0.022 + 0.377 \cdot 0.067 = 0.070$$

and according to Eq. (9.8), the limit values of the exemplary system operation process transient probabilities $p_b(t)$ at the operation states z_b are given by

$$\begin{aligned} p_1 &= \frac{0.279 \cdot 0.133}{0.070} \cong 0.530, \quad p_2 = \frac{0.344 \cdot 0.022}{0.070} \cong 0.109 \\ p_3 &= \frac{0.377 \cdot 0.067}{0.070} \cong 0.361. \end{aligned} \tag{9.31}$$

Hence, the expected values of the total sojourn times $\hat{\theta}_b$, $b = 1, 2, 3, 4$, of the exemplary system operation process at the particular operation states z_b, $b = 1, 2, 3, 4$, during the fixed operation time $\theta = 1$ year = 365 days, after applying Eq. (9.10), amount to

$$\begin{aligned} \hat{m}_1 &= E[\hat{\theta}_1] = 0.530 \cdot 1 = 0.530 \text{ year} = 193.4 \text{ days}, \\ \hat{m}_2 &= E[\hat{\theta}_2] = 0.109 \cdot 1 = 0.109 \text{ year} = 39.8 \text{ days}, \\ \hat{m}_3 &= E[\hat{\theta}_3] = 0.361 \cdot 1 = 0.361 \text{ year} = 131.8 \text{ days} \end{aligned} \tag{9.32}$$

9.3 Complex System Reliability Modelling

We assume that the changes of the operation states of the system operation process $Z(t)$ have an influence on the system multi-state components' E_i, $i = 1, 2, \ldots, n$, reliability and the system reliability structure as well. Consequently, we denote the system multi-state component E_i, $i = 1, 2, \ldots, n$, conditional lifetime in the reliability state subset $\{u, u+1, \ldots, z\}$ while the system is at the operation state z_b, $b = 1, 2, \ldots, v$, by $T_i^{(b)}(u)$ and its conditional reliability function by the vector

$$[R_i(t, \cdot)]^{(b)} = \left[1, [R_i(t, 1)]^{(b)}, \ldots, [R_i(t, z)]^{(b)}\right] \tag{9.33}$$

with the coordinates defined by

$$[R_i(t, u)]^{(b)} = P(T_i^{(b)}(u) > t \,|\, Z(t) = z_b) \quad \text{for } t \in \langle 0, \infty), \ u = 1, 2, \ldots, z, \ b = 1, 2, \ldots, v \tag{9.34}$$

The reliability function $[R_i(t, u)]^{(b)}$ is the conditional probability that the component E_i lifetime $T_i^{(b)}(u)$ in the reliability state subset $\{u, u+1, \ldots, z\}$ is greater than t, while the system operation process $Z(t)$ is at the operation state z_b.

Similarly, we denote the system conditional lifetime in the reliability state subset $\{u, u+1, \ldots, z\}$ while the system is at the operation state z_b, $b = 1, 2, \ldots, v$, by $T^{(b)}(u)$ and the conditional reliability function of the system by the vector

$$[\boldsymbol{R}(t, \cdot)]^{(b)} = [1, [\boldsymbol{R}(t, 1)]^{(b)}, \ldots, [\boldsymbol{R}(t, z)]^{(b)}] \tag{9.35}$$

with the coordinates defined by

$$[\boldsymbol{R}(t, u)]^{(b)} = P(T^{(b)}(u) > t \,|\, Z(t) = z_b) \quad \text{for } t \in \langle 0, \infty), \ u = 1, 2, \ldots, z, \ b = 1, 2, \ldots, v \tag{9.36}$$

The reliability function $[\boldsymbol{R}(t, u)]^{(b)}$ is the conditional probability that the system lifetime $T^{(b)}(u)$ in the reliability state subset $\{u, u+1, \ldots, z\}$ is greater than t, while the system operation process $Z(t)$ is at the operation state z_b.

Thus, the system conditional lifetimes in the reliability states subset $\{u, u+1, \ldots, z\}$ at the operational state z_b

$$T^{(b)}(u) = T\left(T_1^{(b)}(u), T_2^{(b)}(u), \ldots, T_n^{(b)}(u)\right)$$

defined for $u = 1, 2, \ldots, z$, $b = 1, 2, \ldots, \nu$, $n \in N$, are dependent on the components' conditional lifetimes $T_1^{(b)}(u)$, $T_2^{(b)}(u)$, $\ldots$, $T_n^{(b)}(u)$, in the reliability states subset $\{u, u+1, \ldots, z\}$ at the operation state z_b and the coordinates of the system conditional reliability function at the operation state z_b

$$[\boldsymbol{R}(t,u)]^{(b)} = \boldsymbol{R}\left([R_1(t,u)]^{(b)}, [R_2(t,u)]^{(b)}, \ldots, [R_n(t,u)]^{(b)}\right)$$

defined for $t \in \langle 0, \infty)$, $u = 1, 2, \ldots, z$, $b = 1, 2, \ldots, \nu$, $n \in N$, are dependent on the coordinates $[R_1(t,u)]^{(b)}, [R_2(t,u)]^{(b)}, \ldots, [R_n(t,u)]^{(b)}$, of the components conditional reliability functions at the operation state z_b.

Further, we denote the system unconditional lifetime in the reliability state subset $\{u, u+1, \ldots, z\}$ by $T(u)$ and the unconditional reliability function of the system by the vector

$$\boldsymbol{R}(t, \cdot) = [1, \boldsymbol{R}(t,1), \ldots, \boldsymbol{R}(t,z)] \tag{9.37}$$

with the coordinates defined by

$$\boldsymbol{R}(t,u) = P(T(u) > t) \quad \text{for } t \in \langle 0, \infty), \quad u = 1, 2, \ldots, z$$

In cases when the system operation time θ is large enough, the coordinates of the unconditional reliability function of the system defined by Eq. (9.37) are given by

$$\boldsymbol{R}(t,u) \cong \sum_{b=1}^{\nu} p_b [\boldsymbol{R}(t,u)]^{(b)} \quad \text{for } t \geq 0, \quad u = 1, 2, \ldots, z \tag{9.38}$$

where $[\boldsymbol{R}(t,u)]^{(b)}$, $u = 1, 2, \ldots, z$, $b = 1, 2, \ldots, \nu$, are the coordinates of the system conditional reliability functions defined by Eq. (9.36) and p_b, $b = 1, 2, \ldots, \nu$, are the system operation process limit transient probabilities given by Eq. (9.9).

Thus, the mean value of the system unconditional lifetime $T(u)$ in the reliability state subset $\{u, u+1, \ldots, z\}$ is given by Refs. [97,105,159,160]

$$M(u) \cong \sum_{b=1}^{\nu} p_b M_b(u), \quad u = 1, 2, \ldots, z \tag{9.39}$$

where $M_b(u)$ are the mean values of the system conditional lifetimes $T^{(b)}(u)$ in the reliability state subset $\{u, u+1, \ldots, z\}$ at the operation state z_b, $b = 1, 2, \ldots, \nu$, and according to Eq. (3.13), given by

$$M_b(u) = \int_0^{\infty} [\boldsymbol{R}(t,u)]^{(b)} \, dt, \quad u = 1, 2, \ldots, z \tag{9.40}$$

$[\boldsymbol{R}(t,u)]^{(b)}$, $u = 1,2,\ldots,z$, $b = 1,2,\ldots,\nu$, are defined by Eq. (9.36) and p_b are given by Eq. (9.9). Whereas, the variances of the system unconditional lifetimes $T(u)$, according to Eqs. (3.14)–(3.15), are given by

$$\sigma^2(u) = 2\int_0^\infty t\,\boldsymbol{R}(t,u)\,\mathrm{d}t - [M(u)]^2, \quad u = 1,2,\ldots,z \tag{9.41}$$

where $\boldsymbol{R}(t,u)$, $u = 1,2,\ldots,z$, are given by Eq. (9.38) and $M(u)$, $u = 0,1,\ldots,z$, are given by Eq. (9.39).

Hence, according to Eq. (3.17), we get the following formulae for the mean values of the unconditional lifetimes of the system in particular reliability states:

$$\overline{M}(u) = M(u) - M(u+1), \quad u = 0,1,\ldots,z-1; \ \overline{M}(z) = M(z) \tag{9.42}$$

where $M(u)$, $u = 0,1,\ldots,z$, are given by Eq. (9.40).

Moreover, according Eqs. (3.18) and (3.19), if r is the system critical reliability state, then the system risk function is given by

$$\boldsymbol{r}(t) = 1 - \boldsymbol{R}(t,r), \quad t \in \langle 0,\infty) \tag{9.43}$$

where $\boldsymbol{R}(t,r)$ is the coordinate of the system unconditional reliability function given by Eq. (9.38) for $u = r$ and if τ is the moment when the system risk function exceeds a permitted level δ, then

$$\tau = \boldsymbol{r}^{-1}(\delta) \tag{9.44}$$

where $\boldsymbol{r}^{-1}(t)$, if it exists, is the inverse function of the risk function $\boldsymbol{r}(t)$ given by Eq. (9.43).

Further, we assume that the system components E_i, $i = 1,2,\ldots,n$, at the system operation states z_b, $b = 1,2,\ldots,v$, have the exponential reliability functions, i.e. their coordinates are given by

$$[R_i(t,u)]^{(b)} = P(T_i^{(b)}(u) > t \mid Z(t) = z_b) = \exp[-[\lambda_i(u)]^{(b)} t] \tag{9.45}$$

for $t \in \langle 0,\infty)$, $u = 1,2,\ldots,z$, $b = 1,2,\ldots,\nu$, and we have

$$\begin{aligned}[\boldsymbol{R}(t,u)]^{(b)} &= P(T^{(b)}(u) > t \mid Z(t) = z_b) \\ &= [\boldsymbol{R}([R_1(t,u)]^{(b)}, [R_2(t,u)]^{(b)}, \ldots, [R_n(t,u)]^{(b)})]^{(b)} \\ &= [\boldsymbol{R}(\exp[-[\lambda_1(u)]^{(b)} t], \exp[-[\lambda_2(u)]^{(b)} t], \ldots, \exp[-[\lambda_n(u)]^{(b)} t])]^{(b)}\end{aligned}$$

for $t \in \langle 0,\infty)$, $u = 1,2,\ldots,z$, $b = 1,2,\ldots,\nu$, $n \in N$.

The reason for this strong assumption about the system components is that the exponential distribution has 'no memory' expressed in the following property:

$$\begin{aligned}[R_i(t_0+t,u)]^{(b)} &= P(T_i^{(b)}(u) > t_0 + t / T_i^{(b)}(u) > t_0 \big| Z(t) = z_b) \\ &= P(T_i^{(b)}(u) > t_0 + t \cap T_i^{(b)}(u) > t_0 \big| Z(t) = z_b) / P(T_i^{(b)}(u) > t_0 \big| Z(t) = z_b) \\ &= P(T_i^{(b)}(u) > t_0 + t \big| Z(t) = z_b) / P(T_i^{(b)}(u) > t_0 \big| Z(t) = z_b) \\ &= \exp[-[\lambda_i(u)]^{(b)}(t+t_0) / \exp[-[\lambda_i(u)]^{(b)} t_o] = \exp[-[\lambda_i(u)]^{(b)} t] \\ &= P(T_i^{(b)}(u) > t \big| Z(t) = z_b) = [R_i(t,u)]^{(b)}\end{aligned}$$

for $t_0 \in <0, \infty)$ and $t \in <0, \infty)$, $u = 1, 2, \ldots, z$.

Both of them – the assumption about the exponential reliability functions of the system components and the above property – justify the following form of the formula (9.38)

$$\begin{aligned}\boldsymbol{R}(t,u) &\cong \sum_{b=1}^{\nu} p_b [\boldsymbol{R}(t,u)]^{(b)} \\ &= \sum_{b=1}^{\nu} p_b [\boldsymbol{R}(\exp[-[\lambda_1(u)]^{(b)} t], \exp[-[\lambda_2(u)]^{(b)} t], \ldots, \exp[-[\lambda_n(u)]^{(b)} t])]^{(b)}\end{aligned} \tag{9.46}$$

for $t \geq 0$, $u = 1, 2, \ldots, z$.

The application of the above formula and the results given in Chapter 3 yield the following results formulated in the form of the following proposition.

Proposition 9.1

If components of the multi-state system at the operation state z_b, $b = 1, 2, \ldots, \nu$, have the exponential reliability functions given by

$$[R_i(t,\cdot)]^{(b)} = \left[1, [R_i(t,1)]^{(b)}, \ldots, [R_i(t,z)]^{(b)}\right], \quad t \in (-\infty, \infty), \; b = 1, 2, \ldots, \nu \tag{9.47}$$

where

$$\begin{aligned}&[R_i(t,u)]^{(b)} = 1 \quad \text{for } t < 0 \\ &[R_i(t,u)]^{(b)} = \exp[-[\lambda_i(u)]^{(b)} t] \quad \text{for } t \geq 0, \; [\lambda_i(u)]^{(b)} > 0, \\ &i = 1, 2, \ldots, n, \quad u = 1, 2, \ldots, z, \quad b = 1, 2, \ldots, \nu\end{aligned} \tag{9.48}$$

in the case of series, parallel and 'm out of n' systems and respectively by

$$[R_{ij}(t,\cdot)]^{(b)} = \left[1, [R_{ij}(t,1)]^{(b)}, \ldots, [R_{ij}(t,z)]^{(b)}\right], \quad t \in (-\infty, \infty), \; b = 1, 2, \ldots, \nu \tag{9.49}$$

where

$$[R_{ij}(t,u)]^{(b)} = 1 \quad \text{for } t<0$$

$$[R_{ij}(t,u)]^{(b)} = \exp[-[\lambda_{ij}(u)]^{(b)}t] \quad \text{for } t \geq 0, \ [\lambda_{ij}(u)]^{(b)} > 0, \\ i=1,2,\ldots,k, \ j=1,2,\ldots,l_i, \ u=1,2,\ldots,z, \ b=1,2,\ldots,\nu \tag{9.50}$$

in the case of series–parallel, parallel–series, series–'m out of k' and 'm_i out of l_i'–series systems and the system operation time θ is large enough, then its multi-state unconditional reliability function is given by the vector

i. for a series system

$$\overline{\boldsymbol{R}}_n(t,\cdot) = [1, \overline{\boldsymbol{R}}_n(t,1), \ldots, \overline{\boldsymbol{R}}_n(t,z)] \tag{9.51}$$

where

$$\overline{\boldsymbol{R}}_n(t,u) = 1 \quad \text{for } t<0$$

$$\overline{\boldsymbol{R}}_n(t,u) \cong \sum_{b=1}^{\nu} p_b \exp\left[-\sum_{i=1}^{n}[\lambda_i(u)]^{(b)}t\right] \quad \text{for } t \geq 0, \ u=1,2,\ldots,z \tag{9.52}$$

ii. for a parallel system

$$\boldsymbol{R}_n(t,\cdot) = [1, \boldsymbol{R}_n(t,1), \ldots, \boldsymbol{R}_n(t,z)] \tag{9.53}$$

where

$$\overline{\boldsymbol{R}}_n(t,u) = 1 \quad \text{for } t<0$$

$$\boldsymbol{R}_n(t,u) \cong 1 - \sum_{b=1}^{\nu} p_b \prod_{i=1}^{n}[1-\exp[-[\lambda_i(u)]^{(b)}t]] \quad \text{for } t \geq 0, \ u=1,2,\ldots,z \tag{9.54}$$

iii. for an 'm out of n' system

$$\boldsymbol{R}_n^{m}(t,\cdot) = \left[1, \boldsymbol{R}_n^{m}(t,1), \ldots, \boldsymbol{R}_n^{m}(t,z)\right] \tag{9.55}$$

where

$$\boldsymbol{R}_n^{m}(t,u) = 1 \quad \text{for } t<0$$

$$\boldsymbol{R}_n^{m}(t,u) \cong 1 - \sum_{b=1}^{\nu} p_b \sum_{\substack{r_1,r_2,\ldots,r_n=0 \\ r_1+r_2+\cdots+r_n \leq m-1}}^{1} \prod_{i=1}^{n} \exp[-r_i[\lambda_i(u)]^{(b)}t][1-\exp[-[\lambda_i(u)]^{(b)}t]]^{1-r_i} \tag{9.56}$$

for $t \geq 0$, $u = 1,2,\ldots,z$, or

$$\overline{\boldsymbol{R}}_n^{\overline{m}}(t,\cdot) = [1, \boldsymbol{R}_n^{\overline{m}}(t,1), \ldots, \overline{\boldsymbol{R}}_n^{\overline{m}}(t,z)] \tag{9.57}$$

where

$$\overline{\boldsymbol{R}}_n^{\overline{m}}(t,u) = 1 \quad \text{for } t<0$$

$$\overline{\boldsymbol{R}}_n^{\overline{m}}(t,u) \cong \sum_{b=1}^{v} p_b \sum_{\substack{r_1,r_2,\ldots,r_n=0 \\ r_1+r_2+\cdots+r_n \le \overline{m}}}^{1} \prod_{i=1}^{n}[1-\exp[-[\lambda_i(u)]^{(b)}t]]^{r_i}\exp[-(1-r_i)[\lambda_i(u)]^{(b)}t] \tag{9.58}$$

for $t \ge 0$, $\overline{m} = n - m$, $u = 1,2,\ldots,z$

iv. for a series–parallel system

$$\boldsymbol{R}_{k;l_1,l_2,\ldots,l_k}(t,\cdot) = [1, \boldsymbol{R}_{k;l_1,l_2,\ldots,l_k}(t,1),\ldots,\boldsymbol{R}_{k;l_1,l_2,\ldots,l_k}(t,z)] \tag{9.59}$$

where

$$\boldsymbol{R}_{k;l_1,l_2,\ldots,l_k}(t,u) = 1 \quad \text{for } t<0$$

$$\boldsymbol{R}_{k;l_1,l_2,\ldots,l_k}(t,u) \cong 1 - \sum_{b=1}^{v} p_b \prod_{i=1}^{k}\left[1-\exp\left[-\sum_{j=1}^{l_i}[\lambda_{ij}(u)]^{(b)}t\right]\right] \quad \text{for } t \ge 0,\ u=1,2,\ldots,z \tag{9.60}$$

v. for a parallel–series system

$$\overline{\boldsymbol{R}}_{k;l_1,l_2,\ldots,l_k}(t,\cdot) = [1, \overline{\boldsymbol{R}}_{k;l_1,l_2,\ldots,l_k}(t,1),\ldots,\overline{\boldsymbol{R}}_{k;l_1,l_2,\ldots,l_k}(t,z)] \tag{9.61}$$

where

$$\overline{\boldsymbol{R}}_{k;l_1,l_2,\ldots,l_k}(t,u) = 1 \quad \text{for } t<0$$

$$\overline{\boldsymbol{R}}_{k;l_1,l_2,\ldots,l_k}(t,u) \cong \sum_{b=1}^{v} p_b \prod_{i=1}^{k}\left[1-\prod_{j=1}^{l_i}[1-\exp[-[\lambda_{ij}(u)]^{(b)}t]]\right] \quad \text{for } t \ge 0,\ u=1,2,\ldots,z \tag{9.62}$$

vi. for a series–'m out of k' system

$$\boldsymbol{R}^{m}_{k;l_1,l_2,\ldots,l_k}(t,\cdot) = [1, \boldsymbol{R}^{m}_{k;l_1,l_2,\ldots,l_k}(t,1),\ldots,\boldsymbol{R}^{m}_{k;l_1,l_2,\ldots,l_k}(t,z)] \tag{9.63}$$

where

$$\boldsymbol{R}^{m}_{k;l_1,l_2,\ldots,l_k}(t,u) = 1 \quad \text{for } t<0$$

$$\boldsymbol{R}^{m}_{k;l_1,l_2,\ldots,l_k}(t,u) \cong 1 - \sum_{b=1}^{v} p_b \sum_{\substack{r_1,r_2,\ldots,r_k=0 \\ r_1+r_2+\cdots+r_k \le m-1}}^{1} \prod_{i=1}^{k}\left[\prod_{j=1}^{l_i}\exp[-[\lambda_{ij}(u)]^{(b)}t]\right]^{r_i} \cdot \left[1-\prod_{j=1}^{l_i}\exp[-[\lambda_{ij}(u)]^{(b)}t]\right]^{1-r_i} \quad \text{for } t \ge 0,\ u=1,2,\ldots,z \tag{9.64}$$

or

$$\overline{\boldsymbol{R}}^{\overline{m}}_{k;l_1,l_2,\ldots,l_k}(t,\cdot)=\left[1,\overline{\boldsymbol{R}}^{\overline{m}}_{k;l_1,l_2,\ldots,l_k}(t,1),\ldots,\overline{\boldsymbol{R}}^{\overline{m}}_{k;l_1,l_2,\ldots,l_k}(t,z)\right] \tag{9.65}$$

where

$$\overline{\boldsymbol{R}}^{\overline{m}}_{k;l_1,l_2,\ldots,l_k}(t,u)=1 \quad \text{for } t<0$$

$$\overline{\boldsymbol{R}}^{\overline{m}}_{k;l_1,l_2,\ldots,l_k}(t,u)\cong\sum_{b=1}^{v}p_b\sum_{\substack{r_1,r_2,\ldots,r_k=0\\ r_1,r_2+\cdots+r_k\le\overline{m}}}^{1}\prod_{i=1}^{k}\left[1-\left[\prod_{i=1}^{l_i}\exp[-[\lambda_{ij}(u)]^{(b)}t]\right]^{r_i}\right.$$
$$\left.\cdot\left[\prod_{j=1}^{l_i}\exp[-[\lambda_{ij}(u)]^{(b)}t]\right]^{1-r_i}\right] \text{for } t\ge0,\ \overline{m}=k-m,\ u=1,2,\ldots,z \tag{9.66}$$

vii. for an 'm_i out of l_i'−series system

$$\overline{\boldsymbol{R}^{m_1,m_2,\ldots,m_k}_{k;l_1,l_2,\ldots,l_k}}(t,\cdot)=\left[1,\overline{\boldsymbol{R}^{m_1,m_2,\ldots,m_k}_{k;l_1,l_2,\ldots,l_k}}(t,1),\ldots,\overline{\boldsymbol{R}^{m_1,m_2,\ldots,m_k}_{k;l_1,l_2,\ldots,l_k}}(t,z)\right] \tag{9.67}$$

where

$$\overline{\boldsymbol{R}^{m_1,m_2,\ldots,m_k}_{k;l_1,l_2,\ldots,l_k}}(t,u)=1 \quad \text{for } t<0$$

$$\overline{\boldsymbol{R}^{m_1,m_2,\ldots,m_k}_{k;l_1,l_2,\ldots,l_k}}(t,u)\cong\sum_{b=1}^{v}p_b\prod_{i=1}^{k}\left[1-\sum_{\substack{r_1,r_2,\ldots,r_{l_i}=0\\ r_1+r_2+\ldots+r_{l_i}\le m_i-1}}^{1}\prod_{j=1}^{l_i}\exp[-r_j[\lambda_{ij}(u)]^{(b)}t]\right.$$
$$\left.\cdot[1-\exp[-[\lambda_{ij}(u)]^{(b)}t]]^{1-r_j}\right] \quad \text{for } t\ge0,\ u=1,2,\ldots,z \tag{9.68}$$

or

$$\overline{\overline{\boldsymbol{R}}^{\overline{m}_1,\overline{m}_2,\ldots,\overline{m}_k}_{k;l_1,l_2,\ldots,l_k}}(t,\cdot)=\left[1,\overline{\overline{\boldsymbol{R}}^{\overline{m}_1,\overline{m}_2,\ldots,\overline{m}_k}_{k;l_1,l_2,\ldots,l_k}}(t,1),\ldots,\overline{\overline{\boldsymbol{R}}^{\overline{m}_1,\overline{m}_2,\ldots,\overline{m}_k}_{k;l_1,l_2,\ldots,l_k}}(t,z)\right] \tag{9.69}$$

where

$$\overline{\overline{\boldsymbol{R}}^{\overline{m}_1,\overline{m}_2,\ldots,\overline{m}_k}_{k;l_1,l_2,\ldots,l_k}}(t,u)=1 \quad \text{for } t<0$$

$$\overline{\overline{\boldsymbol{R}}^{\overline{m}_1,\overline{m}_2,\ldots,\overline{m}_k}_{k;l_1,l_2,\ldots,l_k}}(t,u)$$

$$\cong\sum_{b=1}^{v}p_b\prod_{i=1}^{k}\left[\sum_{\substack{r_1,r_2,\ldots,r_{l_i}=0\\ r_1+r_2+\ldots+r_{l_i}\le\overline{m}_i}}^{1}\prod_{j=1}^{l_i}[1-\exp[-[\lambda_{ij}(u)]^{(b)}t]]^{r_j}\cdot\exp[-(1-r_j)[\lambda_{ij}(u)]^{(b)}t]\right] \tag{9.70}$$

for $t\ge0,\ \overline{m}_i=l_i-m_i,\ i=1,2,\ldots,k,\ u=1,2,\ldots,z$

Remark 9.1

The formulae for the reliability functions stated in Proposition 9.1 are valid for the considered systems under the assumption that they do not change their structures at different operation states z_b, $b = 1, 2, \ldots, \nu$. This limitation can be simply omitted by the replacement in these formulae of the system's structure shape constant parameters n, m, k, m_i, l_i respectively by their changing at different operation states z_b, $b = 1, 2, \ldots, \nu$, equivalent structure shape parameters $n^{(b)}$, $m^{(b)}$, $k^{(b)}$, $m_i^{(b)}$, $l_i^{(b)}$, $b = 1, 2, \ldots, \nu$.

For the exponential complex technical systems, considered in Proposition 9.1, we determine the mean values $M(u)$ and the standard deviations $\sigma(u)$ of the unconditional lifetimes of the system in the reliability state subsets $\{u, u+1, \ldots, z\}$, $u = 1, 2, \ldots, z$, the mean values $\overline{M}(u)$ of the unconditional lifetimes of the system in the particular reliability states u, $u = 1, 2, \ldots, z$, the system risk function $\boldsymbol{r}(t)$ and the moment τ when the system risk function exceeds a permitted level δ respectively defined by Eqs. (9.37)–(9.44), after substituting for $\boldsymbol{R}(t, u)$, $u = 1, 2, \ldots, z$, the coordinates of the unconditional reliability functions given respectively by Eqs. (9.47)–(9.70).

Example 9.1 (Continuation)

The considered here exemplary technical system reliability analysis at the constant operation conditions is performed in Section 5.6. Its variable at the time operation process is analysed in Section 9.2, where it is fixed that the exemplary system reliability structure and its subsystems and components reliability depend on its changing in time operation states. Considering the assumptions and agreements of these sections, we assume that its subsystems S_υ, $\upsilon = 1, 2$, are composed of four-state, i.e. $z = 3$, components $E_{ij}^{(\upsilon)}$, $\upsilon = 1, 2$, having the conditional reliability functions given by the vector

$$\left[R_{ij}^{(\upsilon)}(t,\cdot)\right]^{(b)} = \left[1, \left[R_{ij}^{(\upsilon)}(t,1)\right]^{(b)}, \left[R_{ij}^{(\upsilon)}(t,2)\right]^{(b)}, \left[R_{ij}^{(\upsilon)}(t,3)\right]^{(b)}\right], \quad b = 1, 2, 3, 4$$

with the exponential coordinates

$$[R_{ij}^{(\upsilon)}(t,1)]^{(b)} = \exp[-[\lambda_{ij}^{(\upsilon)}(1)]^{(b)}], [R_{ij}^{(\upsilon)}(t,2)]^{(b)} = \exp[-[\lambda_{ij}^{(\upsilon)}(2)]^{(b)}]$$
$$[R_{ij}^{(\upsilon)}(t,3)]^{(b)} = \exp[-[\lambda_{ij}^{(\upsilon)}(3)]^{(b)}]$$

different at various operation states z_b, $b = 1, 2, 3, 4$, and with the intensities of departure from the reliability state subsets $\{1, 2, 3\}$, $\{2, 3\}$, $\{3\}$, respectively

$$[\lambda_{ij}^{(\upsilon)}(1)]^{(b)}, [\lambda_{ij}^{(\upsilon)}(2)]^{(b)}, [\lambda_{ij}^{(\upsilon)}(3)]^{(b)}, \quad b = 1, 2, 3, 4$$

The influence of the system operation states changing on the changes of the system reliability structure and its components' reliability functions is as follows.

At the system operation state z_1, the system is composed of the series–parallel subsystem S_1 with the structure shown in Figure 9.2, containing two identical series subsystems ($k = 2$), each composed of 30 components ($l_1 = 30$, $l_2 = 30$) with the exponential reliability functions. In both series subsystems of the subsystem S_1 there are respectively

- the components $E_{ij}^{(1)}$, $i = 1, 2$, $j = 1, 2, \ldots, 20$, with the conditional reliability function coordinates,

$$[R_{ij}^{(1)}(t, 1)]^{(1)} = \exp[-0.0008t], \quad [R_{ij}^{(1)}(t, 2)]^{(1)} = \exp[-0.0009t],$$
$$[R_{ij}^{(1)}(t, 3)]^{(1)} = \exp[-0.0010t], \quad i = 1, 2, \ j = 1, 2, \ldots, 20$$

- the components $E_{ij}^{(1)}$, $i = 1, 2$, $j = 21, 22, \ldots, 30$, with the conditional reliability function coordinates

$$[R_{ij}^{(1)}(t, 1)]^{(1)} = \exp[-0.0011t], \quad [R_{ij}^{(1)}(t, 2)]^{(1)} = \exp[-0.0012t],$$
$$[R_{ij}^{(1)}(t, 3)]^{(1)} = \exp[-0.0013t], \quad i = 1, 2, \ j = 21, 22, \ldots, 30$$

Thus, at the operational state z_1, the system is identical with the subsystem S_1 that is a four-state series–parallel system with its structure shape parameters $k = 2$, $l_1 = 30$, $l_2 = 30$, and according to the formulae appearing after Definition 3.11 and Eqs. (9.59) and (9.60), its conditional reliability function is given by

$$[\boldsymbol{R}(t, \cdot)]^{(1)} = [1, [\boldsymbol{R}(t, 1)]^{(1)}, [\boldsymbol{R}(t, 2)]^{(1)}, [\boldsymbol{R}(t, 3)]^{(1)}], \quad t \geq 0 \tag{9.71}$$

where

$$\begin{aligned} [\boldsymbol{R}(t, 1)]^{(1)} &= \boldsymbol{R}_{2;30,30}(t, 1) = 1 - \prod_{i=1}^{2}\left[1 - \prod_{j=1}^{30}[R_{ij}^{(1)}(t, 1)]^{(1)}\right] \\ &= 1 - \prod_{i=1}^{2}\left[1 - \exp\left[-\sum_{j=1}^{30}[\lambda_{ij}^{(1)}(1)]^{(1)}t\right]\right] \\ &= 1 - [1 - \exp[-[0.0008 \cdot 20 + 0.0011 \cdot 10]t]]^2 \\ &= 1 - [1 - \exp[-0.027t]]^2 \\ &= 2\exp[-0.027t] - \exp[-0.054t] \end{aligned} \tag{9.72}$$

$$\begin{aligned} [\boldsymbol{R}(t, 2)]^{(1)} &= \boldsymbol{R}_{2;30,30}(t, 2) = 1 - \prod_{i=1}^{2}\left[1 - \prod_{j=1}^{30}[R_{ij}^{(1)}(t, 2)]^{(1)}\right] \\ &= 1 - \prod_{i=1}^{2}\left[1 - \exp\left[-\sum_{j=1}^{30}[\lambda_{ij}^{(1)}(2)]^{(1)}t\right]\right] \\ &= 1 - [1 - \exp[-[0.0009 \cdot 20 + 0.0012 \cdot 10]t]]^2 \\ &= 1 - [1 - \exp[-0.030t]]^2 \\ &= 2\exp[-0.030t] - \exp[-0.060t] \end{aligned} \tag{9.73}$$

$$[\boldsymbol{R}(t,3)]^{(1)} = \boldsymbol{R}_{2;30,30}(t,3) = 1 - \prod_{i=1}^{2}\left[1 - \prod_{j=1}^{30}[R_{ij}^{(1)}(t,3)]^{(1)}\right]$$
$$= 1 - \prod_{i=1}^{2}\left[1 - \exp\left[-\sum_{j=1}^{30}[\lambda_{ij}^{(1)}(3)]^{(1)}t\right]\right] \qquad (9.74)$$
$$= 1 - [1 - \exp[-[0.0010 \cdot 20 + 0.0013 \cdot 10]t]]^2$$
$$= 1 - [1 - \exp[-0.033t]]^2$$
$$= 2\exp[-0.033t] - \exp[-0.066t]$$

The expected values and standard deviations of the system conditional lifetimes in the reliability state subsets $\{1,2,3\}$, $\{2,3\}$, $\{3\}$ at the operation state z_1, calculated from the results given by Eqs. (9.71)–(9.74), according to Eq. (9.40), are respectively

$$M_1(1) = \int_0^{\infty} [\boldsymbol{R}(t,1)]^{(1)}\,dt = 2/0.027 - 1/0.054 \cong 55.56 \qquad (9.75)$$

$$M_1(2) = \int_0^{\infty} [\boldsymbol{R}(t,2)]^{(1)}\,dt = 2/0.030 - 1/0.060 = 50.00 \qquad (9.76)$$

$$M_1(3) = \int_0^{\infty} [\boldsymbol{R}(t,3)]^{(1)}\,dt = 2/0.033 - 1/0.066 \cong 45.45 \qquad (9.77)$$

At the system operation state z_2, the system is composed of the series–parallel subsystem S_2 with the structure shown in Figure 9.3, containing four identical series subsystems ($k = 4$), each composed of 40 components ($l_1 = 40$, $l_2 = 40$, $l_3 = 40$, $l_4 = 40$) with the exponential reliability functions. In all series subsystems of the subsystem S_2 there are respectively

- the components $E_{ij}^{(2)}$, $i = 1,2,3,4$, $j = 1,2,\ldots,20$, with the conditional reliability function coordinates

$$[R_{ij}^{(2)}(t,1)]^{(2)} = \exp[-0.0014t], \quad [R_{ij}^{(2)}(t,2)]^{(2)} = \exp[-0.0015t],$$
$$[R_{ij}^{(2)}(t,3)]^{(2)} = \exp[-0.0016t], \quad i = 1,2,3,4, \ j = 1,2,\ldots,20$$

- the components $E_{ij}^{(2)}$, $i = 1,2,3,4$, $j = 21,22,\ldots,40$, with the conditional reliability function coordinates

$$[R_{ij}^{(2)}(t,1)]^{(2)} = \exp[-0.0018t], \quad [R_{ij}^{(2)}(t,2)]^{(2)} = \exp[-0.0020t],$$
$$[R_{ij}^{(2)}(t,3)]^{(2)} = \exp[-0.0022t], \quad i = 1,2,3,4, \ j = 21,22,\ldots,40$$

Thus, at the operation state z_2, the system is identical with the subsystem S_2 that is a four-state series–parallel system with its structure shape parameters $k = 4$, $l_1 = 40$,

$l_2 = 40$, $l_3 = 40$, $l_4 = 40$, and according to the formulae appearing after Definition 3.11 and Eqs. (9.59) and (9.60), its conditional reliability function is given by

$$[\boldsymbol{R}(t,\cdot)]^{(2)} = [1, [\boldsymbol{R}(t,1)]^{(2)}, [\boldsymbol{R}(t,2)]^{(2)}, [\boldsymbol{R}(t,3)]^{(2)}], \quad t \geq 0 \tag{9.78}$$

where

$$\begin{aligned}
[\boldsymbol{R}(t,1)]^{(2)} &= \boldsymbol{R}_{4;40,40,40,40}(t,1) = 1 - \prod_{i=1}^{4}\left[1 - \prod_{j=1}^{40}[R_{ij}^{(2)}(t,1)]^{(2)}\right] \\
&= 1 - \prod_{i=1}^{4}\left[1 - \exp\left[-\sum_{j=1}^{40}[\lambda_{ij}^{(2)}(1)]^{(2)}t\right]\right] \\
&= 1 - [1 - \exp[-[0.0014 \cdot 20 + 0.0018 \cdot 20]t]^4 \\
&= 1 - [1 - \exp[-0.064t]]^4 \\
&= 4\exp[-0.064t] - 6\exp[-0.128t] \\
&\quad + 4\exp[-0.192t] - \exp[-0.256t]
\end{aligned} \tag{9.79}$$

$$\begin{aligned}
[\boldsymbol{R}(t,2)]^{(2)} &= \boldsymbol{R}_{4;40,40,40,40}(t,2) = 1 - \prod_{i=1}^{4}\left[1 - \prod_{j=1}^{40}[R_{ij}^{(2)}(t,2)]^{(2)}\right] \\
&= 1 - \prod_{i=1}^{4}\left[1 - \exp\left[-\sum_{j=1}^{40}[\lambda_{ij}^{(2)}(2)]^{(2)}t\right]\right] \\
&= 1 - [1 - \exp[-[0.0015 \cdot 20 + 0.0020 \cdot 20]t]^4 \\
&= 1 - [1 - \exp[-0.070t]]^4 \\
&= 4\exp[-0.070t] - 6\exp[-0.140t] \\
&\quad + 4\exp[-0.210t] - \exp[-0.280t]
\end{aligned} \tag{9.80}$$

$$\begin{aligned}
[\boldsymbol{R}(t,3)]^{(2)} &= \boldsymbol{R}_{4;40,40,40,40}(t,3) = 1 - \prod_{i=1}^{4}\left[1 - \prod_{j=1}^{40}[R_{ij}^{(2)}(t,3)]^{(2)}\right] \\
&= 1 - \prod_{i=1}^{4}\left[1 - \exp\left[-\sum_{j=1}^{40}[\lambda_{ij}^{(2)}(3)]^{(2)}t\right]\right] \\
&= 1 - [1 - \exp[-[0.0016 \cdot 20 + 0.0022 \cdot 20]t]^4 \\
&= 1 - [1 - \exp[-0.076t]]^4 \\
&= 4\exp[-0.076t] - 6\exp[-0.152t] \\
&\quad + 4\exp[-0.228t] - \exp[-0.304t]
\end{aligned} \tag{9.81}$$

The expected values and standard deviations of the system conditional lifetimes in the reliability state subsets $\{1,2,3\}$, $\{2,3\}$, $\{3\}$ at the operation state z_1, calculated from the results given by Eqs. (9.78)–(9.81), according to Eq. (9.40), are respectively

$$M_2(1) = \int_0^\infty [\boldsymbol{R}(t,1)]^{(2)}\mathrm{d}t = 4/0.064 - 6/0.128 + 4/0.192 - 1/0.256 \cong 32.55 \tag{9.82}$$

$$M_2(2) = \int_0^\infty [\boldsymbol{R}(t,2)]^{(2)}\mathrm{d}t = 4/0.070 - 6/0.140 + 4/0.210 - 1/0.280 \cong 29.76 \tag{9.83}$$

$$M_2(3) = \int_0^\infty [\boldsymbol{R}(t,3)]^{(2)}\mathrm{d}t = 4/0.076 - 6/0.152 + 4/0.228 - 1/0.304 \cong 27.41 \tag{9.84}$$

At the system operation state z_3, the system is a series system with the structure shown in Figure 9.1, composed of two series–parallel subsystems S_1 and S_2, illustrated respectively in Figures 9.2 and 9.3.

The subsystem S_1 with the structure shown in Figure 9.2, consists of two identical series subsystems ($k = 2$), each composed of 30 components ($l_1 = 30$, $l_2 = 30$) with the exponential reliability functions. In both series subsystems of the subsystem S_1 there are respectively

- the components $E_{ij}^{(1)}$, $i = 1,2$, $j = 1,2,\ldots,20$, with the conditional reliability function coordinates

$$[R_{ij}^{(1)}(t,1)]^{(3)} = \exp[-0.0009t],\ \ [R_{ij}^{(1)}(t,2)]^{(3)} = \exp[-0.0010t],$$
$$[R_{ij}^{(1)}(t,3)]^{(3)} = \exp[-0.0011t],\quad i = 1,2,\ \ j = 1,2,\ldots,20$$

- the components $E_{ij}^{(1)}$, $i = 1,2$, $j = 21,22,\ldots,30$, with the conditional reliability function coordinates

$$[R_{ij}^{(1)}(t,1)]^{(3)} = \exp[-0.0012t],\ \ [R_{ij}^{(1)}(t,2)]^{(3)} = \exp[-0.0014t],$$
$$[R_{ij}^{(1)}(t,3)]^{(3)} = \exp[-0.0016t],\quad i = 1,2,\ \ j = 21,22,\ldots,30$$

Thus, at the operation state z_3, the subsystem S_1 is a four-state series–parallel system with its structure shape parameters $k = 2$, $l_1 = 30$, $l_2 = 30$, and according to the formulae appearing after Definition 3.11 and Eqs. (9.59) and (9.60), its conditional reliability function is given by

$$[\boldsymbol{R}^{(1)}(t,\cdot)]^{(3)} = [1, [\boldsymbol{R}^{(1)}(t,1)]^{(3)}, [\boldsymbol{R}^{(1)}(t,2)]^{(3)}, [\boldsymbol{R}^{(1)}(t,3)]^{(3)}],\quad t \geq 0 \tag{9.82a}$$

where

$$[\boldsymbol{R}^{(1)}(t,1)]^{(3)} = \boldsymbol{R}_{2;30,30}(t,1) = 1 - \prod_{i=1}^{2}\left[1 - \prod_{j=1}^{30}[R_{ij}^{(1)}(t,1)]^{(3)}\right]$$
$$= 1 - \prod_{i=1}^{2}\left[1 - \exp\left[-\sum_{j=1}^{30}[\lambda_{ij}^{(1)}(1)]^{(3)}t\right]\right] \quad (9.83a)$$
$$= 1 - [1 - \exp[-[0.0009 \cdot 20 + 0.0012 \cdot 10]t]^2$$
$$= 1 - [1 - \exp[-0.030t]]^2$$
$$= 2\exp[-0.030t] - \exp[-0.060t]$$

$$[\boldsymbol{R}^{(1)}(t,2)]^{(3)} = \boldsymbol{R}_{2;30,30}(t,2) = 1 - \prod_{i=1}^{2}\left[1 - \prod_{j=1}^{30}[R_{ij}^{(1)}(t,2)]^{(3)}\right]$$
$$= 1 - \prod_{i=1}^{2}\left[1 - \exp\left[-\sum_{j=1}^{30}[\lambda_{ij}^{(1)}(2)]^{(3)}t\right]\right] \quad (9.84a)$$
$$= 1 - [1 - \exp[-[0.0010 \cdot 20 + 0.0014 \cdot 10]t]^2$$
$$= 1 - [1 - \exp[-0.034t]]^2$$
$$= 2\exp[-0.034t] - \exp[-0.068t]$$

$$[\boldsymbol{R}(t,3)]^{(1)} = \boldsymbol{R}_{2;30,30}(t,3) = 1 - \prod_{i=1}^{2}\left[1 - \prod_{j=1}^{30}[R_{ij}^{(1)}(t,3)]^{(1)}\right]$$
$$= 1 - \prod_{i=1}^{2}\left[1 - \exp\left[-\sum_{j=1}^{30}[\lambda_{ij}^{(1)}(3)]^{(1)}t\right]\right] \quad (9.85)$$
$$= 1 - [1 - \exp[-[0.0011 \cdot 20 + 0.0016 \cdot 10]t]^2$$
$$= 1 - [1 - \exp[-0.038t]]^2$$
$$= 2\exp[-0.038t] - \exp[-0.076t]$$

The subsystem S_2 with the structure shown in Figure 9.3, consists of four identical series subsystems ($k = 4$), each composed of 40 components ($l_1 = 40$, $l_2 = 40$, $l_3 = 40$, $l_4 = 40$) with the exponential reliability functions given below. In all series subsystems of the subsystem S_2 there are respectively

- the components $E_{ij}^{(2)}$, $i = 1,2,3,4$, $j = 1,2,\ldots,20$, with the conditional reliability function coordinates

$$[R_{ij}^{(2)}(t,1)]^{(3)} = \exp[-0.0010t], \quad [R_{ij}^{(2)}(t,2)]^{(3)} = \exp[-0.0011t],$$
$$[R_{ij}^{(2)}(t,3)]^{(3)} = \exp[-0.0012t], \quad i = 1,2,3,4, \quad j = 1,2,\ldots,20$$

- the components $E_{ij}^{(2)}$, $i = 1,2,3,4$, $j = 21,22,\ldots,40$, with the conditional reliability function coordinates

$$[R_{ij}^{(2)}(t,1)]^{(3)} = \exp[-0.0014t], \quad [R_{ij}^{(2)}(t,2)]^{(3)} = \exp[-0.0016t]$$
$$[R_{ij}^{(2)}(t,3)]^{(3)} = \exp[-0.0018t], \quad i = 1,2,3,4, \quad j = 21,22,\ldots,40$$

Thus, at the operation state z_3, the subsystem S_2 is a four-state series–parallel system with its structure shape parameters $k = 4$, $l_1 = 40$, $l_2 = 40$, $l_3 = 40$, $l_4 = 40$, and according to the formulae appearing after Definition 3.11 and Eqs. (9.59) and (9.60), its conditional reliability function is given by

$$[\boldsymbol{R}^{(2)}(t,\cdot)]^{(3)} = [1, [\boldsymbol{R}^{(2)}(t,1)]^{(3)}, [\boldsymbol{R}^{(2)}(t,2)]^{(3)}, [\boldsymbol{R}^{(2)}(t,3)]^{(3)}], \quad t \geq 0 \tag{9.86}$$

where

$$\begin{aligned}[\boldsymbol{R}^{(2)}(t,1)]^{(3)} &= \boldsymbol{R}_{4;40,40,40,40}(t,1) = 1 - \prod_{i=1}^{4}\left[1 - \prod_{j=1}^{40}[R_{ij}^{(2)}(t,1)]^{(3)}\right] \\ &= 1 - \prod_{i=1}^{4}\left[1 - \exp\left[-\sum_{j=1}^{40}[\lambda_{ij}^{(2)}(1)]^{(3)}t\right]\right] \\ &= 1 - [1 - \exp[-[0.0010 \cdot 20 + 0.0014 \cdot 20]t]^4 \\ &= 1 - [1 - \exp[-0.048t]]^4 \\ &= 4\exp[-0.048t] - 6\exp[-0.096t] \\ &\quad + 4\exp[-0.144t] - \exp[-0.192t]\end{aligned} \tag{9.87}$$

$$\begin{aligned}[\boldsymbol{R}^{(2)}(t,2)]^{(3)} &= \boldsymbol{R}_{4;40,40,40,40}(t,2) = 1 - \prod_{i=1}^{4}\left[1 - \prod_{j=1}^{40}[R_{ij}^{(2)}(t,2)]^{(3)}\right] \\ &= 1 - \prod_{i=1}^{4}\left[1 - \exp\left[-\sum_{j=1}^{40}[\lambda_{ij}^{(2)}(2)]^{(3)}t\right]\right] \\ &= 1 - [1 - \exp[-[0.0011 \cdot 20 + 0.0016 \cdot 20]t]^4 \\ &= 1 - [1 - \exp[-0.054t]]^4 \\ &= 4\exp[-0.054t] - 6\exp[-0.108t] \\ &\quad + 4\exp[-0.162t] - \exp[-0.216t]\end{aligned} \tag{9.88}$$

$$\begin{aligned}[\boldsymbol{R}^{(2)}(t,3)]^{(3)} &= \boldsymbol{R}_{4;40,40,40,40}(t,3) = 1 - \prod_{i=1}^{4}\left[1 - \prod_{j=1}^{40}[R_{ij}^{(2)}(t,3)]^{(3)}\right] \\ &= 1 - \prod_{i=1}^{4}\left[1 - \exp\left[-\sum_{j=1}^{40}[\lambda_{ij}^{(2)}(3)]^{(3)}t\right]\right] \\ &= 1 - [1 - \exp[-[0.0012 \cdot 20 + 0.0018 \cdot 20]t]^4 \\ &= 1 - [1 - \exp[-0.060t]]^4 \\ &= 4\exp[-0.060t] - 6\exp[-0.120t] \\ &\quad + 4\exp[-0.180t] - \exp[-0.240t]\end{aligned} \tag{9.89}$$

Considering that the system at the operation state z_3 is a four-state series system composed of subsystems S_1 and S_2, after applying the formulae appearing after Definition 3.4 and Eqs. (9.82a)–(9.84a), (9.85) and (9.86)–(9.88), its conditional reliability function is given by

$$[\boldsymbol{R}(t,\cdot)]^{(3)} = [1, [\boldsymbol{R}(t,1)]^{(3)}, [\boldsymbol{R}(t,2)]^{(3)}, [\boldsymbol{R}(t,3)]^{(3)}], \quad t \geq 0 \tag{9.90}$$

where

$$\begin{aligned}[\boldsymbol{R}(t,1)]^{(3)} &= \overline{\boldsymbol{R}}_2(t,1) = [\boldsymbol{R}^{(1)}(t,1)]^{(3)}[\boldsymbol{R}^{(2)}(t,1)]^{(3)} \\ &= 8\exp[-0.078t] - 12\exp[-0.126t] + 8\exp[-0.174t] \\ &\quad - 2\exp[-0.212t] - 4\exp[-0.108t] + 6\exp[-0.156t] \\ &\quad - 4\exp[-0.204t] + \exp[-0.252t]\end{aligned} \tag{9.91}$$

$$\begin{aligned}[\boldsymbol{R}(t,2)]^{(3)} &= \overline{\boldsymbol{R}}_2(t,2) = [\boldsymbol{R}^{(1)}(t,2)]^{(3)}[\boldsymbol{R}^{(2)}(t,2)]^{(3)} \\ &= 8\exp[-0.088t] - 12\exp[-0.142t] + 8\exp[-0.196t] \\ &\quad - 2\exp[-0.250t] - 4\exp[-0.118t] + 6\exp[-0.172t] \\ &\quad - 4\exp[-0.226t] + \exp[-0.280t]\end{aligned} \tag{9.92}$$

$$\begin{aligned}[\boldsymbol{R}(t,3)]^{(3)} &= \overline{\boldsymbol{R}}_2(t,3) = [\boldsymbol{R}^{(1)}(t,3)]^{(3)}[\boldsymbol{R}^{(2)}(t,3)]^{(3)} \\ &= 8\exp[-0.098t] - 12\exp[-0.158t] + 8\exp[-0.218t] \\ &\quad - 2\exp[-0.278t] - 4\exp[-0.128t] + 6\exp[-0.188t] \\ &\quad - 4\exp[-0.248t] + \exp[-0.308t]\end{aligned} \tag{9.93}$$

The expected values and standard deviations of the system conditional lifetimes in the reliability state subsets $\{1,2,3\}$, $\{2,3\}$, $\{3\}$ at the operation state z_3, calculated from the results given by Eqs. (9.90)–(9.93), according to Eq. (9.40), are respectively

$$\begin{aligned}M_3(1) = \int_0^\infty [\boldsymbol{R}(t,1)]^{(3)}\,dt &= 8/0.078 - 12/0.126 + 8/0.174 - 2/0.212 - 4/0.108 \\ &+ 6/0.156 - 4/0.204 + 1/0.252 \cong 29.65\end{aligned} \tag{9.94}$$

$$\begin{aligned}M_3(2) = \int_0^\infty [\boldsymbol{R}(t,2)]^{(3)}\,dt &= 8/0.088 - 12/0.142 + 8/0.196 - 2/0.250 - 4/0.118 \\ &+ 6/0.172 - 4/0.226 + 1/0.280 \cong 26.08\end{aligned} \tag{9.95}$$

$$\begin{aligned}M_3(3) = \int_0^\infty [\boldsymbol{R}(t,3)]^{(3)}\,dt &= 8/0.098 - 12/0.158 + 8/0.218 - 2/0.278 - 4/0.128 \\ &+ 6/0.188 - 4/0.248 + 1/0.308 \cong 22.97\end{aligned} \tag{9.96}$$

At the system operation state z_4, the system is a series system with the scheme shown in Figure 9.1, composed of the subsystem S_1 and S_2, illustrated respectively in Figures 9.2 and 9.3, whereas the subsystem S_1 is a series–parallel system and the subsystem S_2 is a series–'2 out of 4' system.

The subsystem S_1 consists of two identical series subsystems ($k = 2$), each composed of 30 components ($l_1 = 30$, $l_2 = 30$), with the exponential reliability functions the same as at the operation state z_1. Thus, according to Eqs. (9.82a)–(9.84a) and (9.85), the subsystem S_1 conditional reliability function at the operation state z_4 is given by

$$[\boldsymbol{R}^{(1)}(t,\cdot)]^{(4)} = [1, [\boldsymbol{R}^{(1)}(t,1)]^{(4)}, [\boldsymbol{R}^{(1)}(t,2)]^{(4)}, [\boldsymbol{R}^{(1)}(t,3)]^{(4)}], \quad t \geq 0 \tag{9.97}$$

where

$$[\boldsymbol{R}^{(1)}(t,1)]^{(4)} = 2\exp[-0.027t] - \exp[-0.054t] \tag{9.98}$$

$$[\boldsymbol{R}^{(1)}(t,2)]^{(4)} = 2\exp[-0.030t] - \exp[-0.060t] \tag{9.99}$$

$$[\boldsymbol{R}(t,3)]^{(4)} = 2\exp[-0.033t] - \exp[-0.066t] \tag{9.100}$$

The subsystem S_2 consists of four identical series subsystems ($k = 4$), each composed of 40 components ($l_1 = 40$, $l_2 = 40$, $l_3 = 40$, $l_4 = 40$) with the exponential reliability functions the same as at the operation state z_2, and is a series–'2 out of 4' system ($m = 2$). Thus, at the operation state z_4, the subsystem S_2 is a four-state series–'2 out of 4' system, with its structure shape parameters $k = 4$, $m = 2$, $l_1 = 40$, $l_2 = 40$, $l_3 = 40$, $l_4 = 40$, and according to the formulae appearing after Definition 8.1 and Eqs. (9.63)–(9.65), its conditional reliability function is given by

$$[\boldsymbol{R}^{(2)}(t,\cdot)]^{(4)} = [1, [\boldsymbol{R}^{(2)}(t,1)]^{(4)}, [\boldsymbol{R}^{(2)}(t,2)]^{(4)}, [\boldsymbol{R}^{(2)}(t,3)]^{(4)}], \quad t \geq 0 \tag{9.101}$$

where

$$\begin{aligned}
&[\boldsymbol{R}^{(2)}(t,1)]^{(4)} \\
&= \boldsymbol{R}^{2}_{4;40,40,40,40}(t,1) \\
&= 1 - \sum_{\substack{r_1,r_2,r_3,r_4=0 \\ r_1+r_2+r_3+r_4 \leq 1}}^{1} \prod_{i=1}^{4} \left[\prod_{j=1}^{40} [R_{ij}^{(2)}(t,1)]^{(4)}\right]^{r_i} \left[1 - \prod_{j=1}^{40} [R_{ij}^{(2)}(t,1)]^{(4)}\right]^{1-r_i} \\
&= 1 - \sum_{\substack{r_1,r_2,r_3,r_4=0 \\ r_1+r_2+r_3+r_4 \leq 1}}^{1} \prod_{i=1}^{4} \exp\left[-r_i \sum_{j=1}^{4} [\lambda_{ij}^{(2)}(1)]^{(4)} t\right] \left[1 - \exp\left[-\sum_{j=1}^{40} [\lambda_{ij}^{(2)}(1)]^{(4)} t\right]\right]^{1-r_i} \\
&= 1 - \sum_{\substack{r_1,r_2,r_3,r_4=0 \\ r_1+r_2+r_3+r_4 \leq 1}}^{1} \prod_{i=1}^{4} \exp[-r_i[0.0014 \cdot 20 + 0.0018 \cdot 20]t]
\end{aligned}$$

$$
\begin{aligned}
&\cdot [1-\exp[-[0.0014\cdot 20+1.0018\cdot 20]t]]^{1-r_i}\\
&=1-\sum_{\substack{r_1,r_2,r_3,r_4=0\\ r_1+r_2+r_3+r_4\le 1}}^{1}\prod_{i=1}^{4}\exp[-r_i 0.064t][1-\exp[-0.064t]]^{1-r_i}\\
&=1-\sum_{i=0}^{1}\binom{4}{i}\exp[-i\cdot 0.064t][1-\exp[-0.064t]]^{4-i}\\
&=1-\exp[-0\cdot 0.064t][1-\exp[-0.0064t]^{4}\\
&\quad-4\exp[-1\cdot 0.064t][1-\exp[-0.0064t]^{3}\\
&=6\exp[-0.128t]-8\exp[-0.192t]+3\exp[-0.256t]
\end{aligned}
\tag{9.102}
$$

$$
\begin{aligned}
[\boldsymbol{R}^{(2)}(t,2)]^{(4)}&=\boldsymbol{R}^{2}_{4;40,40,40,40}(t,2)\\
&=1-\sum_{\substack{r_1,r_2,r_3,r_4=0\\ r_1+r_2+r_3+r_4\le 1}}^{1}\prod_{i=1}^{4}\left[\prod_{j=1}^{40}[R_{ij}^{(2)}(t,2)]^{(4)}\right]^{r_i}\left[1-\prod_{j=1}^{40}[R_{ij}^{(2)}(t,2)]^{(4)}\right]^{1-r_i}\\
&=1-\sum_{\substack{r_1,r_2,r_3,r_4=0\\ r_1+r_2+r_3+r_4\le 1}}^{1}\prod_{i=1}^{4}\exp\left[-r_i\sum_{j=1}^{4}[\lambda_{ij}^{(2)}(2)]^{(4)}t\right]\left[1-\exp\left[-\sum_{j=1}^{40}[\lambda_{ij}^{(2)}(2)]^{(4)}t\right]\right]^{1-r_i}\\
&=1-\sum_{\substack{r_1,r_2,r_3,r_4=0\\ r_1+r_2+r_3+r_4\le 1}}^{1}\prod_{i=1}^{4}\exp[-r_i[0.0015\cdot 20+0.0020\cdot 20]t]\\
&\quad\cdot[1-\exp[-[0.0015\cdot 20+1.0020\cdot 20]t]]^{1-r_i}\\
&=1-\sum_{\substack{r_1,r_2,r_3,r_4=0\\ r_1+r_2+r_3+r_4\le 1}}^{1}\prod_{i=1}^{4}\exp[-r_i 0.070t][1-\exp[-0.070t]]^{1-r_i}\\
&=1-\sum_{i=0}^{1}\binom{4}{i}\exp[-i\cdot 0.070t][1-\exp[-0.070t]]^{4-i}\\
&=1-\exp[-0\cdot 0.070t][1-\exp[-0.0070t]^{4}\\
&\quad-4\exp[-1\cdot 0.070t][1-\exp[-0.0070t]^{3}\\
&=6\exp[-0.140t]-8\exp[-0.210t]+3\exp[-0.280t]
\end{aligned}
\tag{9.103}
$$

$$
\begin{aligned}
[\boldsymbol{R}^{(2)}(t,3)]^{(4)} &= \boldsymbol{R}^{2}_{4;40,40,40,40}(t,3)\\
&= 1 - \sum_{\substack{r_1,r_2,r_3,r_4=0\\ r_1+r_2+r_3+r_4\le 1}}^{1} \prod_{i=1}^{4}\left[\prod_{j=1}^{40}[R_{ij}^{(2)}(t,3)]^{(4)}\right]^{r_i}\left[1-\prod_{j=1}^{40}[R_{ij}^{(2)}(t,3)]^{(4)}\right]^{1-r_i}\\
&= 1 - \sum_{\substack{r_1,r_2,r_3,r_4=0\\ r_1+r_2+r_3+r_4\le 1}}^{1} \prod_{i=1}^{4}\exp\left[-r_i\sum_{j=1}^{4}[\lambda_{ij}^{(2)}(3)]^{(4)}t\right]\left[1-\exp\left[-\sum_{j=1}^{40}[\lambda_{ij}^{(2)}(3)]^{(4)}t\right]\right]^{1-r_i}\\
&= 1 - \sum_{\substack{r_1,r_2,r_3,r_4=0\\ r_1+r_2+r_3+r_4\le 1}}^{1} \prod_{i=1}^{4}\exp[-r_i[0.0016\cdot 20+0.0022\cdot 20]t]\\
&\quad\cdot[1-\exp[-[0.0016\cdot 20+1.0022\cdot 20]t]]^{1-r_i}\\
&= 1 - \sum_{\substack{r_1,r_2,r_3,r_4=0\\ r_1+r_2+r_3+r_4\le 1}}^{1} \prod_{i=1}^{4}\exp[-r_i 0.076t][1-\exp[-0.076t]]^{1-r_i}\\
&= 1-\sum_{i=0}^{1}\binom{4}{i}\exp[-i\cdot 0.076t][1-\exp[-0.076t]]^{4-i}\\
&= 1-\exp[-0\cdot 0.076t][1-\exp[-0.0076t]^{4}\\
&\quad -4\exp[-1\cdot 0.076t][1-\exp[-0.0076t]^{3}\\
&= 6\exp[-0.152t]-8\exp[-0.228t]+3\exp-0.304t]
\end{aligned}
\tag{9.104}
$$

Considering that the system at the operation state z_4 is a four-state series system composed of subsystems S_1 and S_2, after applying the formulae appearing after Definition 3.4 and Eqs. (9.97)–(9.104), its conditional reliability function is given by

$$[\boldsymbol{R}(t,\cdot)]^{(4)} = [1,[\boldsymbol{R}(t,1)]^{(4)},[\boldsymbol{R}(t,2)]^{(4)},[\boldsymbol{R}(t,3)]^{(4)}],\quad t\ge 0 \tag{9.105}$$

where

$$
\begin{aligned}
[\boldsymbol{R}(t,1)]^{(4)} &= \overline{\boldsymbol{R}}_2(t,1) = [\boldsymbol{R}^{(1)}(t,1)]^{(4)}[\boldsymbol{R}^{(2)}(t,1)]^{(4)}\\
&= 12\exp[-0.155t]-6\exp[-0.182t]-16\exp[-0.219t]\\
&\quad +8\exp[-0.246t]+6\exp[-0.283t]-3\exp[-0.310t]
\end{aligned}
\tag{9.106}
$$

$$
\begin{aligned}
[\boldsymbol{R}(t,2)]^{(4)} &= \overline{\boldsymbol{R}}_2(t,2) = [\boldsymbol{R}^{(1)}(t,2)]^{(4)}[\boldsymbol{R}^{(2)}(t,2)]^{(4)}\\
&= 12\exp[-0.170t]-6\exp[-0.200t]-16\exp[-0.240t]\\
&\quad +8\exp[-0.270t]+6\exp[-0.310t]-3\exp[-0.340t]
\end{aligned}
\tag{9.107}
$$

$$[\boldsymbol{R}(t,3)]^{(4)} = \overline{\boldsymbol{R}}_2(t,3) = [\boldsymbol{R}^{(1)}(t,3)]^{(4)}[\boldsymbol{R}^{(2)}(t,3)]^{(4)}$$
$$= 12\exp[-0.185t] - 6\exp[-0.218t] - 16\exp[-0.261t] + 8\exp[-0.294t] + 6\exp[-0.337t] - 3\exp[-0.370t] \quad (9.108)$$

The mean values of the system sojourn times $T(u)$ in the reliability state subsets, after applying the formula (9.40) and (9.105)–(9.108), are

$$M_4(1) = \int_0^\infty [\boldsymbol{R}(t,1)]^{(4)}\,dt = 12/0.155 - 6/0.182 - 16/0.219 + 8/0.246 + 6/0.283 - 3/0.310 \cong 15.44 \quad (9.109)$$

$$M_4(2) = \int_0^\infty [\boldsymbol{R}(t,2)]^{(4)}\,dt = 12/0.170 - 6/0.200 - 16/0.240 + 8/0.270 + 6/0.310 - 3/0.340 \cong 14.08 \quad (9.110)$$

$$M_4(3) = \int_0^\infty [\boldsymbol{R}(t,3)]^{(4)}\,dt = 12/0.185 - 6/0.218 - 16/0.261 + 8/0.294 + 6/0.337 - 3/0.370 \cong 12.95 \quad (9.111)$$

In the case when the system operation time is large enough, its unconditional four-state reliability function is given by the vector

$$\boldsymbol{R}(t,\cdot) = [1, \boldsymbol{R}(t,1), \boldsymbol{R}(t,2), \boldsymbol{R}(t,3)], \quad t \geq 0 \quad (9.112)$$

where according to Eq. (3.6) and considering the exemplary system operation process transient probabilities at the operation states determined by Eq. (9.18), the vector coordinates are given respectively by

$$\boldsymbol{R}(t,1) = p_1[\boldsymbol{R}(t,1)]^{(1)} + p_2[\boldsymbol{R}(t,1)]^{(2)} + p_3[\boldsymbol{R}(t,1)]^{(3)} + p_4[\boldsymbol{R}(t,1)]^{(4)}$$
$$= 0.214 \cdot [\boldsymbol{R}(t,1)]^{(1)} + 0.038 \cdot [\boldsymbol{R}(t,1)]^{(2)} + 0.293 \cdot [\boldsymbol{R}(t,1)]^{(3)} + 0.455 \cdot [\boldsymbol{R}(t,1)]^{(4)} \quad \text{for } t \geq 0 \quad (9.113)$$

$$\boldsymbol{R}(t,2) = p_1[\boldsymbol{R}(t,2)]^{(1)} + p_2[\boldsymbol{R}(t,2)]^{(2)} + p_3[\boldsymbol{R}(t,2)]^{(3)} + p_4[\boldsymbol{R}(t,2)]^{(4)}$$
$$= 0.214 \cdot [\boldsymbol{R}(t,2)]^{(1)} + 0.038 \cdot [\boldsymbol{R}(t,2)]^{(2)} + 0.293 \cdot [\boldsymbol{R}(t,2)]^{(3)} + 0.455 \cdot [\boldsymbol{R}(t,2)]^{(4)} \quad \text{for } t \geq 0 \quad (9.114)$$

$$\boldsymbol{R}(t,3) = p_1[\boldsymbol{R}(t,3)]^{(1)} + p_2[\boldsymbol{R}(t,3)]^{(2)} + p_3[\boldsymbol{R}(t,3)]^{(3)} + p_4[\boldsymbol{R}(t,3)]^{(4)}$$
$$= 0.214 \cdot [\boldsymbol{R}(t,3)]^{(1)} + 0.038 \cdot [\boldsymbol{R}(t,3)]^{(2)} + 0.293 \cdot [\boldsymbol{R}(t,3)]^{(3)} + 0.455 \cdot [\boldsymbol{R}(t,3)]^{(4)} \quad \text{for } t \geq 0 \quad (9.115)$$

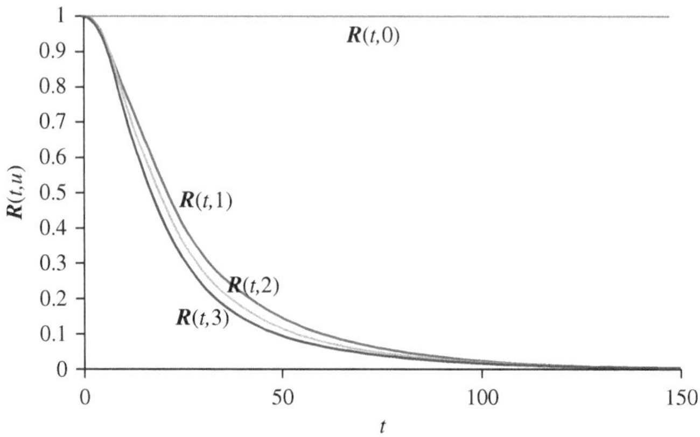

Figure 9.11 The graph of the exemplary system reliability function $\boldsymbol{R}(t,\cdot)$ coordinates.

where coordinates $[\boldsymbol{R}(t,1)]^{(1)}$, $[\boldsymbol{R}(t,1)]^{(2)}$, $[\boldsymbol{R}(t,1)]^{(3)}$, $[\boldsymbol{R}(t,1)]^{(4)}$ are given by Eqs. (9.72), (9.79), (9.90), (9.106), $[\boldsymbol{R}(t,2)]^{(1)}$, $[\boldsymbol{R}(t,2)]^{(2)}$, $[\boldsymbol{R}(t,2)]^{(3)}$, $[\boldsymbol{R}(t,2)]^{(4)}$ are given by Eqs. (9.73), (9.80), (9.91), (9.107) and $[\boldsymbol{R}(t,3)]^{(1)}$, $[\boldsymbol{R}(t,3)]^{(2)}$, $[\boldsymbol{R}(t,3)]^{(3)}$, $[\boldsymbol{R}(t,3)]^{(4)}$ are given by Eqs. (9.74), (9.81), (9.92), (9.108).

The graph of the four-state exemplary system reliability function is illustrated in Figure 9.11.

The expected values and standard deviations of the system unconditional lifetimes in the reliability state subsets {1, 2, 3}, {2, 3}, {3}, calculated from the results given by Eqs. (9.112)–(9.115), according to Eqs. (9.39) and (9.40) and considering Eqs. (9.18), (9.75)–(9.77), (9.82)–(9.84), (9.94)–(9.96) and (9.109)–(9.111), are respectively

$$\begin{aligned} M(1) &= p_1M_1(1)+p_2M_2(1)+p_3M_3(1)+p_4M_4(1) \\ &= 0.214\cdot 55.56+0.038\cdot 32.55+0.293\cdot 29.65+0.455\cdot 15.44 \\ &\cong 28.84 \end{aligned} \tag{9.116}$$

$$\begin{aligned} M(2) &= p_1M_1(2)+p_2M_2(2)+p_3M_3(2)+p_4M_4(2) \\ &= 0.214\cdot 50.00+0.038\cdot 29.76+0.293\cdot 26.08+0.455\cdot 14.08 \\ &\cong 25.88 \end{aligned} \tag{9.117}$$

$$\begin{aligned} M(3) &= p_1M_1(3)+p_2M_2(3)+p_3M_3(3)+p_4M_4(3) \\ &= 0.214\cdot 45.45+0.038\cdot 27.41+0.293\cdot 22.97+0.455\cdot 12.95 \\ &\cong 23.39 \end{aligned} \tag{9.118}$$

Further, considering Eqs. (9.42) and (9.116)–(9.118), the mean values of the system unconditional lifetimes in the particular reliability states 1, 2, 3, are respectively

$$\overline{M}(1)=M(1)-M(2)=2.96,\ \overline{M}(2)=M(2)-M(3)=2.49,\ \overline{M}(3)=M(3)=23.39 \tag{9.119}$$

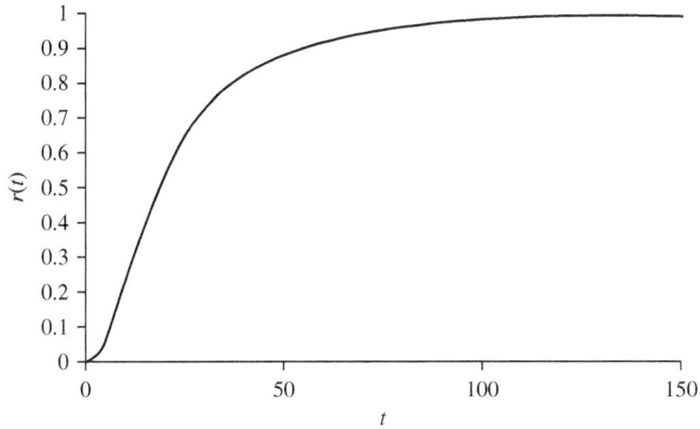

Figure 9.12 The graph of the exemplary system risk function $r(t)$.

Since the critical reliability state is $r = 2$, then the system risk function, according to Eq. (9.43), is given by

$$\boldsymbol{r}(t) = 1 - \boldsymbol{R}(t, 2) \quad \text{for } t \geq 0 \tag{9.120}$$

where $\boldsymbol{R}(t, 2)$ is given by Eq. (9.114).

Hence, by Eq. (9.44), the moment when the system risk function exceeds a permitted level, for instance $\delta = 0.05$, is

$$\tau = \boldsymbol{r}^{-1}(\delta) \cong 4.524 \tag{9.121}$$

The graph of the risk function $\boldsymbol{r}(t)$ of the exemplary four-state system operating at the variable conditions is shown in Figure 9.12.

The reliability characteristics of the exemplary system operating at the variable conditions predicted in this section are different from those determined in Section 5.6 for this system operating at constant conditions. This fact justifies the sensibility of considering the real systems at the variable operation conditions that is appearing out in a natural way from practice. This approach, based on the accuracy of the systems' operation processes identification, makes their reliability prediction more precise.

Example 9.2 (Continuation)

We distinguish four reliability states 0, 1, 2, 3, i.e. $z = 3$, of the components of the '24 out of 36'–series system considered in Section 9.2 and we fix that the critical reliability state is $r = 2$.

Consequently, we define the four-state reliability functions of the system components E_{ij}, $i = 1, 2, 3, j = 1, 2, \ldots, 36$, at the operation state z_1 in the form of the vector

$$[R_{ij}(t,\cdot)]^{(1)} = [1, [R_{ij}(t,1)]^{(1)}, [R_{ij}(t,2)]^{(1)}, [R_{ij}(t,3)]^{(1)}], \quad i = 1,2,3, \;\; j = 1,2,\ldots,36$$

with the exponential coordinates

$$[R_{ij}(t,u)]^{(1)} = \exp[-3ut], \quad u = 1,2,3, \;\; i = 1,2,3, \;\; j = 1,2,\ldots,36 \tag{9.122}$$

and the four-state reliability functions of the system components E_{ij}, $i = 1, 2$, $j = 1, 2, \ldots, 36$, at the operation state z_2 in the form of the vector

$$[R_{ij}(t,\cdot)]^{(2)} = [1, [R_{ij}(t,1)]^{(2)}, [R_{ij}(t,2)]^{(2)}, [R_{ij}(t,3)]^{(2)}], \quad i = 1,2, \;\; j = 1,2,\ldots,36$$

with the exponential coordinates

$$[R_{ij}(t,u)]^{(2)} = \exp[-2ut], \quad u = 1,2,3, \;\; i = 1,2, \;\; j = 1,2,\ldots,36 \tag{9.123}$$

and the four-state reliability functions of the system components E_{ij}, $i = 1$, $j = 1, 2, \ldots, 36$, at the operation state z_3 in the form of the vector

$$[R_{ij}(t,\cdot)]^{(3)} = [1, [R_{ij}(t,1)]^{(3)}, [R_{ij}(t,2)]^{(3)}, [R_{ij}(t,3)]^{(3)}], \quad i = 1, \;\; j = 1,2,\ldots,36$$

with the exponential coordinates

$$[R_{ij}(t,u)]^{(3)} = \exp[-ut], \quad u = 1,2,3, \;\; i = 1, \;\; j = 1,2,\ldots,36 \tag{9.124}$$

Since the shape parameters of the considered '24 out of 36'−series system are

- $k^{(1)} = 3$, $l^{(1)} = 36$, $m^{(1)} = 24$, at the operation state z_1,
- $k^{(2)} = 2$, $l^{(2)} = 36$, $m^{(2)} = 24$, at the operation state z_2,
- $k^{(3)} = 1$, $l^{(3)} = 36$, $m^{(3)} = 24$, at the operation state z_3,

then applying directly the formulae (9.68), taking for the case of a regular and homogeneous system the following form

$$\overline{\boldsymbol{R}^{m^{(b)}}_{k^{(b)};l^{(b)}}}(t,u) \cong \sum_{b=1}^{v} p_b \left[1 - \sum_{i=0}^{m^{(b)}-1} \binom{l^{(b)}}{i} \exp[-i[\lambda(u)]^{(b)} t]\right. \\ \left. \cdot [1 - \exp[-[\lambda(u)]^{(b)} t]]^{l^{(b)}-i}\right]^{k^{(b)}} \quad \text{for } t \geq 0, \;\; u = 1,2,\ldots,z \tag{9.125}$$

we get the system reliability function

$$\boldsymbol{R}(t,\cdot) = [1, \boldsymbol{R}(t,1), \boldsymbol{R}(t,2), \boldsymbol{R}(t,3)], \quad t \geq 0 \tag{9.126}$$

where

$$\boldsymbol{R}(t,u) \cong \sum_{b=1}^{3} p_b[\boldsymbol{R}(t,u)]^{(b)}$$

and

$$\begin{aligned}[\boldsymbol{R}(t,u)]^{(b)} &= \overline{\boldsymbol{R}^{24}_{k^{(b)};36}}(t,u)\\ &= \sum_{b=1}^{3} p_b\left[1-\sum_{i=0}^{23}\binom{36}{i}\exp[-i[\lambda(u)]^{(b)}t]\right.\\ &\quad\cdot[1-\exp[-[\lambda(u)]^{(b)}t]^{36-i}]^{k^{(b)}} \quad \text{for } t \geq 0, \;\; u=1,2,3, \;\; b=1,2,3\end{aligned} \tag{9.127}$$

Hence, particularly, considering Eqs. (9.20) and (9.122)–(9.124), we have

$$\begin{aligned}\boldsymbol{R}(t,1) \cong\, & 0.4\left[1-\sum_{i=0}^{23}\binom{36}{i}\exp[-i3t][1-\exp[-3t]]^{36-i}\right]^3\\ &+0.4\left[1-\sum_{i=0}^{23}\binom{36}{i}\exp[-i2t][1-\exp[-2t]]^{36-i}\right]^2\\ &+0.2\left[1-\sum_{i=0}^{23}\binom{36}{i}\exp[-it][1-\exp[-t]]^{36-i}\right] \quad \text{for } t \geq 0\end{aligned} \tag{9.128}$$

$$\begin{aligned}\boldsymbol{R}(t,2) \cong\, & 0.4\left[1-\sum_{i=0}^{23}\binom{36}{i}\exp[-i6t][1-\exp[-6t]]^{36-i}\right]^3\\ &+0.4\left[1-\sum_{i=0}^{23}\binom{36}{i}\exp[-i4t][1-\exp[-4t]]^{36-i}\right]^2\\ &+0.2\left[1-\sum_{i=0}^{23}\binom{36}{i}\exp[-i2t][1-\exp[-2t]]^{36-i}\right] \quad \text{for } t \geq 0\end{aligned} \tag{9.129}$$

$$\begin{aligned}\boldsymbol{R}(t,3) \cong\, & 0.4\left[1-\sum_{i=0}^{23}\binom{36}{i}\exp[-i9t][1-\exp[-9t]]^{36-i}\right]^3\\ &+0.4\left[1-\sum_{i=0}^{23}\binom{36}{i}\exp[-i6t][1-\exp[-6t]]^{36-i}\right]^2\\ &+0.2\left[1-\sum_{i=0}^{23}\binom{36}{i}\exp[-i3t][1-\exp[-3t]]^{36-i}\right] \quad \text{for } t \geq 0\end{aligned} \tag{9.130}$$

Table 9.1 The Values of the '24 Out of 36'–Series System Unconditional Reliability Function Coordinates

t	$\boldsymbol{R}(t, 1)$	t	$\boldsymbol{R}(t, 2)$	t	$\boldsymbol{R}(t, 3)$
0.00	1.0000	0.000	1.0000	0.00	1.0000
0.03	1.0000	0.015	1.0000	0.01	1.0000
0.06	0.9957	0.030	0.9957	0.02	0.9957
0.09	0.9254	0.045	0.9254	0.03	0.9254
0.12	0.7294	0.060	0.7294	0.04	0.7294
0.15	0.5446	0.750	0.5446	0.05	0.5446
0.18	0.4134	0.900	0.4134	0.06	0.4134
0.21	0.3098	0.105	0.3098	0.07	0.3098
0.24	0.2407	0.120	0.2407	0.08	0.2407
0.27	0.2029	0.135	0.2029	0.09	0.2029
0.30	0.1811	0.150	0.1811	0.10	0.1811
0.33	0.1636	0.165	0.1636	0.11	0.1636
0.36	0.1455	0.180	0.1455	0.12	0.1455
0.39	0.1259	0.195	0.1259	0.13	0.1259
0.42	0.1058	0.210	0.1058	0.14	0.1058
0.45	0.0863	0.225	0.0863	0.15	0.0863
0.48	0.0684	0.240	0.0684	0.16	0.0684
0.51	0.0527	0.255	0.0527	0.17	0.0527
0.54	0.0396	0.270	0.0396	0.18	0.0396
0.57	0.0290	0.285	0.0290	0.19	0.0290
0.60	0.0208	0.300	0.0208	0.20	0.0208
0.63	0.0146	0.315	0.0146	0.21	0.0146
0.66	0.0101	0.330	0.0101	0.22	0.0101
0.69	0.0068	0.345	0.0068	0.23	0.0068
0.72	0.0045	0.360	0.0045	0.24	0.0045
0.75	0.0030	0.375	0.0030	0.25	0.0030
0.78	0.0019	0.390	0.0019	0.26	0.0019
0.81	0.0012	0.405	0.0012	0.27	0.0012
0.84	0.0008	0.420	0.0008	0.28	0.0008
0.87	0.0005	0.435	0.0005	0.29	0.0005
0.90	0.0003	0.450	0.0003	0.30	0.0003
0.93	0.0002	0.465	0.0002	0.31	0.0002
0.96	0.0001	0.480	0.0001	0.32	0.0001
0.99	0.0001	0.495	0.0001	0.33	0.0001
1.02	0.0000	0.510	0.0000	0.34	0.0000

In Table 9.1, the values of the coordinates of the system unconditional reliability function $\boldsymbol{R}(t, \cdot)$ given by Eqs. (9.128)–(9.130) are presented.

The approximate graphs of the coordinates of the system for $t \geq 0$, $u = 1, 2, 3$, function are presented in Figure 9.13.

The system conditional reliability function $[\boldsymbol{R}(t, \cdot)]^{(1)}$ at operation state z_1, determined from Eq. (9.127), takes the following form:

$$[\boldsymbol{R}(t, \cdot)]^{(1)} = [[\boldsymbol{R}(t, 1)]^{(1)}, [\boldsymbol{R}(t, 2)]^{(1)}, [\boldsymbol{R}(t, 3)]^{(1)}] \tag{9.131}$$

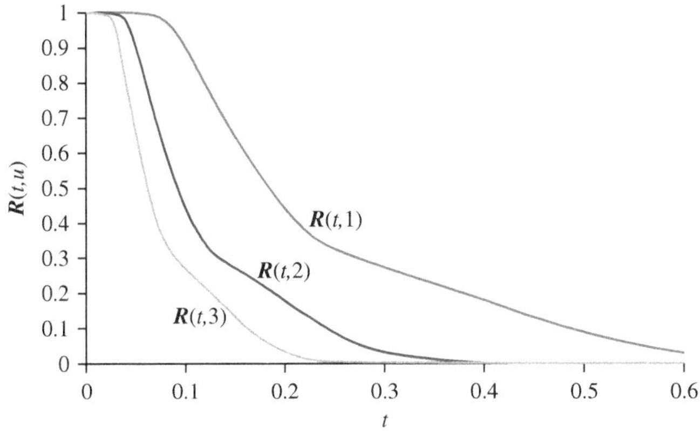

Figure 9.13 The graph of the '24 out of 36'−series reliability function $\boldsymbol{R}(t,\cdot)$ coordinates.

where

$$[\boldsymbol{R}(t,1)]^{(1)} = \overline{\boldsymbol{R}_{3;36}^{24}}(t,u) = \left[1 - \sum_{i=0}^{23} \binom{36}{i} \exp[-i3t][1-\exp[-3t]]^{36-i}\right]^3 \quad (9.132)$$

$$[\boldsymbol{R}(t,2)]^{(1)} = \overline{\boldsymbol{R}_{3;36}^{24}}(t,2) = \left[1 - \sum_{i=0}^{23} \binom{36}{i} \exp[-i6t][1-\exp[-6t]]^{36-i}\right]^3 \quad (9.133)$$

$$[\boldsymbol{R}(t,3)]^{(3)} = \overline{\boldsymbol{R}_{3;36}^{24}}(t,3) = \left[1 - \sum_{i=0}^{23} \binom{36}{i} \exp[-i9t][1-\exp[-9t]]^{36-i}\right]^3 \quad \text{for } t \geq 0 \quad (9.134)$$

In Table 9.2, there are presented the values of the coordinates of the system conditional reliability function $[\boldsymbol{R}(t,\cdot)]^{(1)}$ at the operation state z_1 determined according to Eqs. (9.132)−(9.134).

The expected values of the system conditional lifetimes at the operation state z_1 in the reliability state subsets $\{1,2,3\}$, $\{2,3\}$, $\{3\}$ calculated according to Eq. (9.40) from the results given by Eqs. (9.132)−(9.134), using the trapezium method of integration and suitable data given in Table 9.2, are

$$\begin{aligned} M_1(1) = \int_0^\infty [\boldsymbol{R}(t,1)]^{(1)}\, dt &\cong \frac{0.01}{2}[[\boldsymbol{R}(0.00,1)]^{(1)} + 2[[\boldsymbol{R}(0.01,1)]^{(1)} + [\boldsymbol{R}(0.02,1)]^{(1)} \\ &+ \cdots + [\boldsymbol{R}(0.22,1)]^{(1)}] + [\boldsymbol{R}(0.23,1)]^{(1)}] \\ &= \frac{0.01}{2}[1.0000 + 2[1.0000 + 1.0000 + \cdots + 0.0001] + 0.0000] \\ &\cong 0.1134 \end{aligned} \quad (9.135)$$

Table 9.2 The Values of the '24 Out of 36'–Series System Conditional Reliability Function Coordinates at the Operation State z_1

t	$[R(t,1)]^{(1)}$	t	$[R(t,2)]^{(1)}$	t	$[R(t,3)]^{(1)}$
0.00	1.0000	0.000	1.0000	0.0000	1.0000
0.01	1.0000	0.005	1.0000	0.0034	1.0000
0.02	1.0000	0.010	1.0000	0.0068	1.0000
0.03	1.0000	0.015	1.0000	0.0102	1.0000
0.04	0.9997	0.020	0.9997	0.0136	0.9997
0.05	0.9978	0.025	0.9978	0.0170	0.9974
0.06	0.9895	0.030	0.9895	0.0204	0.9877
0.07	0.9653	0.035	0.9653	0.0238	0.9599
0.08	0.9123	0.040	0.9123	0.0272	0.9003
0.09	0.8206	0.045	0.8206	0.0306	0.7998
0.10	0.6909	0.050	0.6909	0.0340	0.6614
0.11	0.5374	0.055	0.5374	0.0374	0.5027
0.12	0.3831	0.060	0.3831	0.0408	0.3485
0.13	0.2492	0.065	0.2492	0.0442	0.2195
0.14	0.1479	0.07	0.1479	0.0476	0.1257
0.15	0.0802	0.075	0.0802	0.0510	0.0657
0.16	0.0399	0.080	0.0399	0.0544	0.0314
0.17	0.0183	0.085	0.0183	0.0578	0.0138
0.18	0.0078	0.090	0.0078	0.0612	0.0056
0.19	0.0031	0.095	0.0031	0.0646	0.0021
0.20	0.0011	0.100	0.0011	0.0680	0.0007
0.21	0.0004	0.105	0.0004	0.0714	0.0002
0.22	0.0001	0.110	0.0001	0.0748	0.0001
0.23	0.0000	0.115	0.0000	0.0782	0.0000

$$\begin{aligned} M_1(2) &= \int_0^\infty [\boldsymbol{R}(t,2)]^{(1)}\,\mathrm{d}t \cong \frac{0.005}{2}[[\boldsymbol{R}(0.00,2)]^{(1)} + 2[[\boldsymbol{R}(0.005,2)]^{(1)} \\ &\quad + [\boldsymbol{R}(0.010,2)]^{(1)} + \cdots + [\boldsymbol{R}(0.110,2)]^{(1)}] + [\boldsymbol{R}(0.115,2)]^{(1)}] \\ &= \frac{0.005}{2}[1.0000 + 2[1.0000 + 1.0000 + \cdots + 0.0001] + 0.0000] \\ &\cong 0.0567 \end{aligned} \tag{9.136}$$

$$\begin{aligned} M_1(3) &= \int_0^\infty [\boldsymbol{R}(t,3)]^{(1)}\,\mathrm{d}t \cong \frac{0.0034}{2}[[\boldsymbol{R}(0.000,3)]^{(1)} + 2[[\boldsymbol{R}(0.0034,3)]^{(1)} \\ &\quad + [\boldsymbol{R}(0.0068,3)]^{(1)} + \cdots + [\boldsymbol{R}(0.066,3)]^{(1)}] + [\boldsymbol{R}(0.0782,3)]^{(1)}] \\ &= \frac{0.0034}{2}[1.0000 + 2[1.0000 + 1.0000 + \cdots + 0.0001] + 0.0000] \\ &\cong 0.0378 \end{aligned} \tag{9.137}$$

The system conditional reliability function $[\boldsymbol{R}(t,\cdot)]^{(2)}$ at operation state z_2, determined from Eq. (9.127), takes the following form:

$$[\boldsymbol{R}(t,\cdot)]^{(2)} = [[\boldsymbol{R}(t,1)]^{(2)}, [\boldsymbol{R}(t,2)]^{(2)}, [\boldsymbol{R}(t,3)]^{(2)}] \tag{9.138}$$

where

$$[\boldsymbol{R}(t,1)]^{(2)} = \overline{\boldsymbol{R}_{2;36}^{24}}(t,1) = \left[1-\sum_{i=0}^{23}\binom{36}{i}\exp[-i2t][1-\exp[-2t]]^{36-i}\right]^2 \tag{9.139}$$

$$[\boldsymbol{R}(t,2)]^{(2)} = \overline{\boldsymbol{R}_{2;36}^{24}}(t,2) = \left[1-\sum_{i=0}^{23}\binom{36}{i}\exp[-i4t][1-\exp[-4t]]^{36-i}\right]^2 \tag{9.140}$$

$$[\boldsymbol{R}(t,3)]^{(2)} = \overline{\boldsymbol{R}_{2;36}^{24}}(t,3) = \left[1-\sum_{i=0}^{23}\binom{36}{i}\exp[-i6t][1-\exp[-6t]]^{36-i}\right]^2 \quad \text{for } t \geq 0 \tag{9.141}$$

In Table 9.3, there are presented the values of the coordinates of the system conditional reliability function $[\boldsymbol{R}(t,\cdot)]^{(2)}$ at the operation state z_2 determined according to Eqs. (9.139)–(9.141).

The expected values of the system conditional lifetimes at the operation state z_2 in the reliability state subsets $\{1,2,3\}$, $\{2,3\}$, $\{3\}$ calculated according to Eq. (9.40) from the results given by Eqs. (9.139)–(9.141), using the trapezium method of integration and suitable data given in Table 9.3, are

$$\begin{aligned} M_2(1) &= \int_0^\infty [\boldsymbol{R}(t,1)]^{(2)}\,\mathrm{d}t \cong \frac{0.015}{2}[[\boldsymbol{R}(0.00,1)]^{(2)} + 2[[\boldsymbol{R}(0.015,1)]^{(2)} \\ &\quad + [\boldsymbol{R}(0.030,1)]^{(2)} + \cdots + [\boldsymbol{R}(0.390,1)]^{(2)}] + [\boldsymbol{R}(0.405,1)]^{(2)}] \\ &= \frac{0.015}{2}[1.0000 + 2[1.0000 + 1.0000 + \cdots + 0.0001] + 0.0000] \\ &\cong 0.1858 \end{aligned} \tag{9.142}$$

$$\begin{aligned} M_2(2) &= \int_0^\infty [\boldsymbol{R}(t,2)]^{(2)}\,\mathrm{d}t \cong \frac{0.0075}{2}[[\boldsymbol{R}(0.00,2)]^{(2)} + 2[[\boldsymbol{R}(0.0075,2)]^{(2)} \\ &\quad + [\boldsymbol{R}(0.0150,2)]^{(2)} + \cdots + [\boldsymbol{R}(0.1950,2)]^{(2)}] + [\boldsymbol{R}(0.2025,2)]^{(2)}] \\ &= \frac{0.0075}{2}[1.0000 + 2[1.0000 + 1.0000 + \cdots + 0.0001] + 0.0000] \\ &\cong 0.0929 \end{aligned} \tag{9.143}$$

Table 9.3 The Values of the '24 Out of 36'−Series System Conditional Reliability Function Coordinates at the Operation State z_2

t	$[\boldsymbol{R}(t,1)]^{(2)}$	t	$[\boldsymbol{R}(t,2)]^{(2)}$	t	$[\boldsymbol{R}(t,3)]^{(2)}$
0.000	1.0000	0.0000	1.0000	0.000	1.0000
0.015	1.0000	0.0075	1.0000	0.005	1.0000
0.030	1.0000	0.0150	1.0000	0.010	1.0000
0.045	1.0000	0.0225	1.0000	0.015	1.0000
0.060	0.9998	0.0300	0.9998	0.020	0.9998
0.075	0.9985	0.0375	0.9985	0.025	0.9985
0.090	0.9930	0.0450	0.9930	0.030	0.9930
0.105	0.9767	0.0525	0.9767	0.035	0.9767
0.120	0.9406	0.0600	0.9406	0.040	0.9406
0.135	0.8765	0.0675	0.8765	0.045	0.8765
0.150	0.7815	0.0750	0.7815	0.050	0.7815
0.165	0.6610	0.0825	0.6610	0.055	0.6610
0.180	0.5275	0.0900	0.5275	0.060	0.5275
0.195	0.3960	0.0975	0.3960	0.065	0.3960
0.210	0.2797	0.1050	0.2797	0.070	0.2797
0.225	0.1860	0.1125	0.1860	0.075	0.1860
0.240	0.1168	0.1200	0.1168	0.080	0.1168
0.255	0.0694	0.1275	0.0694	0.085	0.0694
0.270	0.0392	0.1350	0.0392	0.090	0.0392
0.285	0.0211	0.1425	0.0211	0.095	0.0211
0.300	0.0108	0.1500	0.0108	0.100	0.0108
0.315	0.0053	0.1575	0.0053	0.105	0.0053
0.330	0.0025	0.1650	0.0025	0.110	0.0025
0.345	0.0012	0.1725	0.0012	0.115	0.0012
0.360	0.0005	0.1800	0.0005	0.120	0.0005
0.375	0.0002	0.1875	0.0002	0.125	0.0002
0.390	0.0001	0.1950	0.0001	0.130	0.0001
0.405	0.0000	0.2025	0.0000	0.135	0.0000

$$\begin{aligned} M_2(3) &= \int_0^\infty [\boldsymbol{R}(t,3)]^{(2)}\,\mathrm{d}t \cong \frac{0.005}{2}[[\boldsymbol{R}(0.00,3)]^{(2)} + 2[[\boldsymbol{R}(0.005,3)]^{(2)} \\ &\quad + [\boldsymbol{R}(0.010,3)]^{(2)} + \cdots + [\boldsymbol{R}(0.130,3)]^{(2)}] + [\boldsymbol{R}(0.135,2)]^{(2)}] \\ &= \frac{0.005}{2}[1.0000 + 2[1.0000 + 1.0000 + \cdots + 0.0001] + 0.0000] \\ &\cong 0.0619 \end{aligned} \tag{9.144}$$

The values of the coordinates of the system conditional reliability function $[\boldsymbol{R}(t,\cdot)]^{(3)}$ at operation state z_3, determined from Eq. (9.127), takes the following form:

$$[\boldsymbol{R}(t,\cdot)]^{(3)} = [[\boldsymbol{R}(t,1)]^{(3)}, [\boldsymbol{R}(t,2)]^{(3)}, [\boldsymbol{R}(t,3)]^{(3)}] \tag{9.145}$$

where

$$[\boldsymbol{R}(t,1)]^{(3)} = \overline{\boldsymbol{R}_{1;36}^{24}}(t,1) = \left[1 - \sum_{i=0}^{23}\binom{36}{i}\exp[-it][1-\exp[-t]]^{36-i}\right] \tag{9.146}$$

$$[\boldsymbol{R}(t,2)]^{(3)} = \overline{\boldsymbol{R}_{1;36}^{24}}(t,2) = \left[1 - \sum_{i=0}^{23}\binom{36}{i}\exp[-i2t][1-\exp[-2t]]^{36-i}\right] \tag{9.147}$$

$$[\boldsymbol{R}(t,3)]^{(3)} = \overline{\boldsymbol{R}_{1;36}^{24}}(t,3) = \left[1 - \sum_{i=0}^{23}\binom{36}{i}\exp[-i3t][1-\exp[-3t]]^{36-i}\right] \quad \text{for } t \geq 0 \tag{9.148}$$

In Table 9.4, there are presented the values of the coordinates of the system conditional reliability function $[\boldsymbol{R}(t,\cdot)]^{(3)}$ at the operation state z_3 determined according to Eqs. (9.146)–(9.148).

The expected values of the system conditional lifetimes at the operation state z_3 in the reliability state subsets $\{1,2,3\}$, $\{2,3\}$, $\{3\}$ calculated according to Eq. (9.40) from the results given by Eqs. (9.146)–(9.148), using the trapezium method of integration and suitable data given in Table 9.3, are

$$\begin{aligned} M_3(1) &= \int_0^\infty [\boldsymbol{R}(t,1)]^{(3)}\,\mathrm{d}t \cong \frac{0.04}{2}[[\boldsymbol{R}(0.00,1)]^{(3)} + 2[[\boldsymbol{R}(0.04,1)]^{(3)} + [\boldsymbol{R}(0.08,1)]^{(3)} \\ &\quad + \cdots + [\boldsymbol{R}(1.08,1)]^{(3)}] + [\boldsymbol{R}(1.12,1)]^{(3)}] \\ &= \frac{0.04}{2}[1.0000 + 2[1.0000 + 1.0000 + \cdots + 0.0001] + 0.0000] \\ &\cong 0.4403 \end{aligned} \tag{9.149}$$

$$\begin{aligned} M_3(2) &= \int_0^\infty [R(t,2)]^{(3)}\,\mathrm{d}t \cong \frac{0.02}{2}[[\boldsymbol{R}(0.00,2)]^{(3)} + 2[[\boldsymbol{R}(0.02,2)]^{(1)} + [\boldsymbol{R}(0.04,2)]^{(1)} \\ &\quad + \cdots + [\boldsymbol{R}(0.54,2)]^{(1)}] + [\boldsymbol{R}(0.56,2)]^{(1)}] \\ &= \frac{0.02}{2}[1.0000 + 2[1.0000 + 1.0000 + \cdots + 0.0001] + 0.0000] \\ &\cong 0.2201 \end{aligned} \tag{9.150}$$

Table 9.4 The Values of the '24 Out of 36'–Series System Conditional Reliability Function Coordinates at the Operation State z_3

t	$[R(t,1)]^{(3)}$	t	$[R(t,2)]^{(3)}$	t	$[R(t,3)]^{(3)}$
0.00	1.0000	0.00	1.0000	0.0000	1.0000
0.04	1.0000	0.02	1.0000	0.0132	1.0000
0.08	1.0000	0.04	1.0000	0.0264	1.0000
0.12	0.9999	0.06	0.9999	0.0396	0.9999
0.16	0.9987	0.08	0.9987	0.0528	0.9988
0.20	0.9919	0.10	0.9919	0.0660	0.9925
0.24	0.9699	0.12	0.9699	0.0792	0.9718
0.28	0.9210	0.14	0.9210	0.0924	0.9254
0.32	0.8387	0.16	0.8387	0.1056	0.8465
0.36	0.7263	0.18	0.7263	0.1188	0.7374
0.40	0.5959	0.20	0.5959	0.1320	0.6093
0.44	0.4631	0.22	0.4631	0.1452	0.4774
0.48	0.3418	0.24	0.3418	0.1584	0.3554
0.52	0.2402	0.26	0.2402	0.1716	0.2521
0.56	0.1614	0.28	0.1614	0.1848	0.1711
0.60	0.1041	0.30	0.1041	0.1980	0.1115
0.64	0.0647	0.32	0.0647	0.2112	0.0700
0.68	0.0388	0.34	0.0388	0.2244	0.0425
0.72	0.0226	0.36	0.0226	0.2376	0.0250
0.76	0.0128	0.38	0.0128	0.2508	0.0143
0.80	0.0071	0.40	0.0071	0.2640	0.0080
0.84	0.0038	0.42	0.0038	0.2772	0.0043
0.88	0.0020	0.44	0.0020	0.2904	0.0023
0.92	0.0010	0.46	0.0010	0.3036	0.0012
0.96	0.0005	0.48	0.0005	0.3168	0.0006
1.00	0.0003	0.50	0.0003	0.3300	0.0003
1.04	0.0001	0.52	0.0001	0.3432	0.0002
1.08	0.0001	0.54	0.0001	0.3564	0.0001
1.12	0.0000	0.56	0.0000	0.3696	0.0000

$$
\begin{aligned}
M_3(3) = \int_0^{\infty} [\boldsymbol{R}(t,3)]^{(3)}\,\mathrm{d}t &\cong \frac{0.0132}{2}[[\boldsymbol{R}(0.00,3)]^{(3)} + 2[[\boldsymbol{R}(0.0132,3)]^{(3)} \\
&+ [\boldsymbol{R}(0.0264,3)]^{(3)} + \cdots + [\boldsymbol{R}(0.3564,3)]^{(3)}] + [\boldsymbol{R}(0.3696,3)]^{(3)}] \\
&= \frac{0.0132}{2}[1.0000 + 2[1.0000 + 1.0000 + \cdots + 0.0001] + 0.0000] \\
&\cong 0.1468
\end{aligned}
\tag{9.151}
$$

The expected value of the system unconditional lifetime in the reliability state subset $\{1,2,3\}$, calculated according to Eqs. (9.39) and (9.40) from the results given by Eqs. (9.20), (9.135), (9.142) and (9.149), is

$$\begin{aligned} M(1) &= p_1 M_1(1) + p_2 M_2(1) + p_3 M_3(1) \\ &\cong 0.4 \cdot 0.1134 + 0.4 \cdot 0.1858 + 0.2 \cdot 0.4403 \cong 0.2077 \end{aligned} \tag{9.152}$$

The expected value of the system unconditional lifetime in the reliability state subset $\{2,3\}$, calculated according to Eqs. (9.39) and (9.40) from the results given by Eqs. (9.20), (9.136), (9.143) and (9.150), is

$$\begin{aligned} M(2) &= p_1 M_1(2) + p_2 M_2(2) + p_3 M_3(2) \\ &\cong 0.4 \cdot 0.0567 + 0.4 \cdot 0.0929 + 0.2 \cdot 0.2201 \cong 0.1039 \end{aligned} \tag{9.153}$$

The expected value of the system unconditional lifetime in the reliability state subset $\{3\}$, calculated according to Eqs. (9.39) and (9.40) from the results given by Eqs. (9.20), (9.137), (9.144) and (9.151), is

$$\begin{aligned} M(3) &= p_1 M_1(3) + p_2 M_2(3) + p_3 M_3(3) \\ &\cong 0.4 \cdot 0.0378 + 0.4 \cdot 0.0619 + 0.2 \cdot 0.1468 \cong 0.0692 \end{aligned} \tag{9.154}$$

Further, considering Eqs. (9.152)–(9.154) and applying Eq. (9.42), the mean values of the unconditional lifetimes in the particular reliability states 1, 2, 3, are respectively

$$\begin{aligned} &\overline{M}(1) = M(1) - M(2) = 0.1038, \quad \overline{M}(2) = M(2) - M(3) = 0.0347, \\ &\overline{M}(3) = M(3) = 0.0692 \end{aligned} \tag{9.155}$$

Since the critical reliability state is $r = 2$, then the system risk function, according to Eqs. (9.43) and (9.127), is given by (Figure 9.14)

$$\begin{aligned} \boldsymbol{r}(t) &= 1 - \boldsymbol{R}(t,2) \\ &\cong 1 - 0.2\left[1 - \sum_{i=0}^{23}\binom{36}{i}\exp[-i2t][1-\exp[-2t]]^{36-i}\right] \\ &\quad - 0.4\left[1 - \sum_{i=0}^{23}\binom{36}{i}\exp[-i4t][1-\exp[-4t]]^{36-i}\right]^2 \\ &\quad - 0.4\left[1 - \sum_{i=0}^{23}\binom{36}{i}\exp[-i6t][1-\exp[-6t]]^{36-i}\right]^3, \quad \text{for } t \geq 0 \end{aligned} \tag{9.156}$$

Hence, by Eq. (9.44), the moment when the system risk function exceeds a permitted level, for instance $\delta = 0.025$, is

$$\tau = \boldsymbol{r}^{-1}(\delta) \cong 0.0379 \tag{9.157}$$

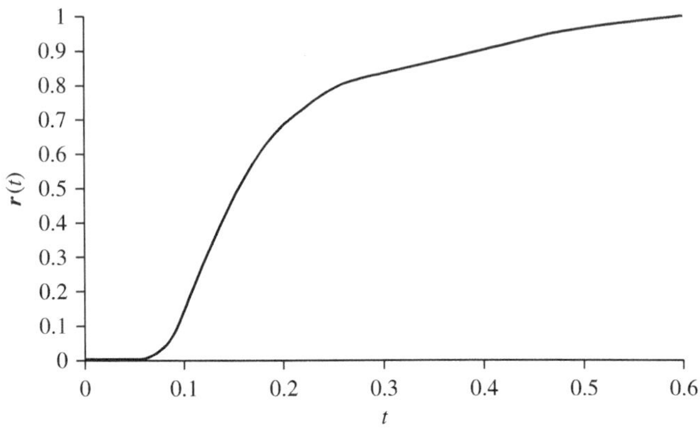

Figure 9.14 The graph of the '24 out of 36'−series system risk function $\boldsymbol{r}(t)$.

Example 9.3 (Continuation)

Taking into account the efficiency of the port grain transportation system considered in Section 9.2, we distinguish the following three reliability states of the systems and its components:

- state 2: the state ensuring the largest efficiency of the system and its conveyors;
- state 1: the state ensuring less efficiency of the system caused by throwing grain off the system conveyors;
- state 0: the state involving failure of the system.

Thus, the transportation system reliability analysis considered here at the constant operation conditions is performed in Section 6.2, whereas its varying in time operation process is analysed in Section 9.2. In Section 9.2, it is assumed that the system reliability structure and its subsystems and components' reliability depend on its changing in time operation states. Considering the assumptions and agreements of these sections, we assume that its subsystems S_υ, $\upsilon = 1, 2, 3, 4$, are composed of three-state, i.e. $z = 3$, components $E_{ij}^{(\upsilon)}$, $\upsilon = 1, 2, 3, 4$, having the conditional reliability functions given by the vector

$$[R_{ij}^{(\upsilon)}(t, \cdot)]^{(b)} = [1, [R_{ij}^{(\upsilon)}(t, 1)]^{(b)}, [R_{ij}^{(\upsilon)}(t, 2)]^{(b)}], \quad b = 1, 2, 3$$

with the exponential coordinates

$$[R_{ij}^{(\upsilon)}(t, 1)]^{(b)} = \exp[-[\lambda_{ij}^{(\upsilon)}(1)]^{(b)} t], [R_{ij}^{(\upsilon)}(t, 2)]^{(b)} = \exp[-[\lambda_{ij}^{(\upsilon)}(2)]^{(b)} t]$$

different at various operation states z_b, $b = 1, 2, 3$, and with the intensities of departure from the reliability state subsets $\{1, 2\}$, $\{2\}$ respectively

$$[\lambda_{ij}^{(\upsilon)}(1)]^{(b)}, [\lambda_{ij}^{(\upsilon)}(2)]^{(b)}, \quad b = 1, 2, 3$$

The influence of the system operation states changing on the changes of the system reliability structure and its components' reliability functions is as follows.

At the system operation state z_1, the system is a series system with the structure shown in Figure 9.8, composed of four series–parallel subsystems S_1, S_2, S_3 and S_4 illustrated respectively in Figure 9.7.

The subsystem S_1 with the structure shown in Figure 9.7, consists of two identical belt conveyors forming series subsystems ($k = 2$), each composed of 129 components ($l_1 = 129$, $l_2 = 129$) with the exponential reliability functions. In both series subsystems of the subsystem S_1 there are respectively

- 1 belt marked by $E_{ij}^{(1)}$, $i = 1,2$, $j = 1$, with reliability function coordinates

$$[R_{ij}^{(1)}(t,1)]^{(1)} = \exp[-0.0125t], \quad [R_{ij}^{(1)}(t,2)]^{(1)} = \exp[-0.0150t], \quad t \geq 0, \quad i = 1,2, \quad j = 1$$

- 2 drums marked by $E_{ij}^{(1)}$, $i = 1,2$, $j = 2,3$, with reliability function coordinates

$$[R_{ij}^{(1)}(t,1)]^{(1)} = \exp[-0.0015t], \quad [R_{ij}^{(1)}(t,2)]^{(1)} = \exp[-0.0020t], \quad t \geq 0, \quad i = 1,2, \quad j = 2,3$$

- 117 channelled rollers marked by $E_{ij}^{(1)}$, $i = 1,2$, $j = 4,5,\ldots,120$, with reliability function coordinates

$$[R_{ij}^{(1)}(t,1)]^{(1)} = \exp[-0.005t], \quad [R_{ij}^{(1)}(t,2)]^{(1)} = \exp[-0.010t], \quad t \geq 0, \quad i = 1,2, \quad j = 4,5,\ldots,120$$

- 9 supporting rollers marked by $E_{ij}^{(1)}$, $i = 1,2$, $j = 121,122,\ldots,129$, with reliability function coordinates

$$[R_{ij}^{(1)}(t,1)]^{(1)} = \exp[-0.004t], \quad [R_{ij}^{(1)}(t,2)]^{(1)} = \exp[-0.008t], \quad t \geq 0, \quad i = 1,2,$$
$$j = 121,122,\ldots,129$$

Thus, at the operation state z_1, the subsystem S_1 is a three-state series–parallel system with its structure shape parameters $k = 2$, $l_1 = 129$, $l_2 = 129$, and according to the formulae appearing after Definition 3.11 and Eqs. (9.59) and (9.60), its conditional reliability function is given by

$$[\boldsymbol{R}^{(1)}(t,\cdot)]^{(1)} = [1, [\boldsymbol{R}^{(1)}(t,1)]^{(1)}, [\boldsymbol{R}^{(1)}(t,2)]^{(1)}], \quad t \geq 0 \tag{9.158}$$

where

$$\begin{aligned}
[\boldsymbol{R}^{(1)}(t,1)]^{(1)} &= \boldsymbol{R}_{2;129,129}(t,1) = 1 - \prod_{i=1}^{2}\left[1 - \prod_{j=1}^{129}[R_{ij}^{(1)}(t,1)]^{(1)}\right] \\
&= 1 - \prod_{i=1}^{2}\left[1 - \exp\left[-\sum_{j=1}^{129}[\lambda_{ij}^{(1)}(1)]^{(1)}t\right]\right] \\
&= 1 - [1 - \exp[-[0.0125 \cdot 1 + 0.0015 \cdot 2 + 0.005 \cdot 117 + 0.004 \cdot 9]t]^2 \\
&= 1 - [1 - \exp[-0.6365t]]^2 \\
&= 2\exp[-0.6365t] - \exp[-1.2730t]
\end{aligned} \tag{9.159}$$

$$[\boldsymbol{R}^{(1)}(t,2)]^{(1)} = \boldsymbol{R}_{2;129,129}(t,2) = 1 - \prod_{i=1}^{2}\left[1 - \prod_{j=1}^{129}[R_{ij}^{(1)}(t,2)]^{(1)}\right]$$
$$= 1 - \prod_{i=1}^{2}\left[1 - \exp\left[-\sum_{j=1}^{129}[\lambda_{ij}^{(1)}(2)]^{(1)}t\right]\right]$$
$$= 1 - [1 - \exp[-[0.0150 \cdot 1 + 0.0020 \cdot 2 + 0.010 \cdot 117 + 0.008 \cdot 9]t]^2$$
$$= 1 - [1 - \exp[-1.2610t]]^2$$
$$= 2\exp[-1.2610t] - \exp[-2.5220t] \quad (9.160)$$

The subsystem S_2 with the structure shown in Figure 9.9, consists of three identical bucket elevators forming series subsystems ($k = 3$), each composed of 743 components ($l_1 = 743,\ l_2 = 743,\ l_3 = 743$) with the exponential reliability functions. In each series subsystems of the subsystem S_2 there are respectively

- 1 belt marked by $E_{ij}^{(2)}$, $i = 1, 2, 3, j = 1$, with reliability function coordinates

$$[R_{ij}^{(2)}(t,1)]^{(1)} = \exp[-0.025t],\ [R_{ij}^{(2)}(t,2)]^{(1)} = \exp[-0.030t], \quad t \geq 0,\ i = 1,2,3,\ j = 1$$

- 2 drums marked by $E_{ij}^{(2)}$, $i = 1, 2, 3, j = 2, 3$, with reliability function coordinates

$$[R_{ij}^{(2)}(t,1)]^{(1)} = \exp[-0.0015t],\ [R_{ij}^{(2)}(t,2)]^{(1)} = \exp[-0.0020t], \quad t \geq 0,\ i = 1,2,3,\ j = 2,3$$

- 740 buckets marked by $E_{ij}^{(2)}$, $i = 1, 2, 3, j = 4, 5, \ldots, 743$, with reliability function coordinates

$$[R_{ij}^{(2)}(t,1)]^{(1)} = \exp[-0.030t],\ [R_{ij}^{(2)}(t,2)]^{(1)} = \exp[-0.035t],\ t \geq 0,\ i = 1,2,3,\ j = 4,5,\ldots,743$$

Thus, at the operation state z_1, the subsystem S_2 is a three-state series–parallel system with its structure shape parameters $k = 3,\ l_1 = 743,\ l_2 = 743,\ l_3 = 743$, and according to the formulae appearing after Definition 3.11 and Eqs. (9.59) and (9.60), its conditional reliability function is given by

$$[\boldsymbol{R}^{(2)}(t,\cdot)]^{(1)} = [1, [\boldsymbol{R}^{(2)}(t,1)]^{(1)}, [\boldsymbol{R}^{(2)}(t,2)]^{(1)}], \quad t \geq 0 \quad (9.161)$$

where

$$[\boldsymbol{R}^{(2)}(t,1)]^{(1)} = \boldsymbol{R}_{3;743,743,743}(t,1) = 1 - \prod_{i=1}^{3}\left[1 - \prod_{j=1}^{743}[R_{ij}^{(2)}(t,1)]^{(1)}\right]$$
$$= 1 - \prod_{i=1}^{3}\left[1 - \exp\left[-\sum_{j=1}^{743}[\lambda_{ij}^{(2)}(1)]^{(1)}t\right]\right] \quad (9.162)$$
$$= 1 - [1 - \exp[-[0.025 \cdot 1 + 0.0015 \cdot 2 + 0.03 \cdot 740]t]^3$$
$$= 1 - [1 - \exp[-22.2280t]]^3$$
$$= 3\exp[-22.2280t] - 3\exp[-44.4560t] + \exp[-66.6840t]$$

$$[\boldsymbol{R}^{(2)}(t,2)]^{(1)} = \boldsymbol{R}_{3;743,743,743}(t,2) = 1 - \prod_{i=1}^{3}\left[1 - \prod_{j=1}^{743}[R_{ij}^{(2)}(t,2)]^{(1)}\right]$$

$$= 1 - \prod_{i=1}^{3}\left[1 - \exp\left[-\sum_{j=1}^{743}[\lambda_{ij}^{(2)}(2)]^{(1)}t\right]\right] \tag{9.163}$$

$$= 1 - [1 - \exp[-[0.030 \cdot 1 + 0.0020 \cdot 2 + 0.035 \cdot 740]t]^3$$

$$= 1 - [1 - \exp[-25.9340t]]^3$$

$$= 3\exp[-25.9340t] - 3\exp[-51.8680t] + \exp[-77.8020t]$$

The subsystem S_3 with the structure shown in Figure 9.9 consists of two identical belt conveyors forming series subsystems ($k = 2$), each composed of 139 components ($l_1 = 139$, $l_2 = 139$) with the exponential reliability functions. In both series subsystems of the subsystem S_3 there are respectively

- 1 belt marked by $E_{ij}^{(3)}$, $i = 1,2, j = 1$, with reliability function coordinates

$$[R_{ij}^{(3)}(t,1)]^{(1)} = \exp[-0.0125t],\ [R_{ij}^{(3)}(t,2)]^{(1)} = \exp[-0.0150t], \quad t \geq 0,\ i = 1,2,\ j = 1$$

- 2 drums marked by $E_{ij}^{(3)}$, $i = 1,2, j = 2,3$, with reliability function coordinates

$$[R_{ij}^{(3)}(t,1)]^{(1)} = \exp[-0.0015t],\ [R_{ij}^{(3)}(t,2)]^{(3)} = \exp[-0.0020t], \quad t \geq 0,\ i = 1,2,\ j = 2,3$$

- 117 channelled rollers marked by $E_{ij}^{(3)}$, $i = 1,2, j = 4,5,\ldots,120$, with reliability function coordinates

$$[R_{ij}^{(3)}(t,1)]^{(1)} = \exp[-0.005t],\ [R_{ij}^{(3)}(t,2)]^{(1)} = \exp[-0.010t], \quad t \geq 0,\ i = 1,2,\ j = 4,5,\ldots,120$$

- 19 supporting rollers marked by $E_{ij}^{(1)}$, $i = 1,2, j = 121,122,\ldots,139$, with reliability function coordinates

$$[R_{ij}^{(3)}(t,1)]^{(1)} = \exp[-0.004t],\ [R_{ij}^{(3)}(t,2)]^{(1)} = \exp[-0.008t], \quad t \geq 0,$$
$$i = 1,2,\ j = 121,122,\ldots,139$$

Thus, at the operation state z_1, the subsystem S_3 is a three-state series–parallel system with its structure shape parameters $k = 2$, $l_1 = 139$, $l_2 = 139$, and according to the formulae appearing after Definition 3.11 and Eqs. (9.59) and (9.60), its conditional reliability function is given by

$$[\boldsymbol{R}^{(3)}(t,\cdot)]^{(1)} = [1, [\boldsymbol{R}^{(3)}(t,1)]^{(1)}, [\boldsymbol{R}^{(3)}(t,2)]^{(1)}], \quad t \geq 0 \tag{9.164}$$

where

$$\begin{aligned}
[\boldsymbol{R}^{(3)}(t,1)]^{(1)} &= \boldsymbol{R}_{2;139,139}(t,1) = 1-\prod_{i=1}^{2}\left[1-\prod_{j=1}^{139}[R_{ij}^{(3)}(t,1)]^{(1)}\right]\\
&= 1-\prod_{i=1}^{2}\left[1-\exp\left[-\sum_{j=1}^{139}[\lambda_{ij}^{(3)}(1)]^{(1)}t\right]\right]\\
&= 1-[1-\exp[-[0.0125\cdot 1+0.0015\cdot 2+0.005\cdot 117+0.004\cdot 19]t]^2\\
&= 1-[1-\exp[-0.6765t]]^2\\
&= 2\exp[-0.6765t]-\exp[-1.3530t]
\end{aligned} \tag{9.165}$$

$$\begin{aligned}
[\boldsymbol{R}^{(3)}(t,2)]^{(1)} &= \boldsymbol{R}_{2;139,139}(t,2) = 1-\prod_{i=1}^{2}\left[1-\prod_{j=1}^{139}[R_{ij}^{(3)}(t,2)]^{(1)}\right]\\
&= 1-\prod_{i=1}^{2}\left[1-\exp\left[-\sum_{j=1}^{139}[\lambda_{ij}^{(3)}(2)]^{(1)}t\right]\right]\\
&= 1-[1-\exp[-[0.0150\cdot 1+0.0020\cdot 2+0.010\cdot 117+0.008\cdot 19]t]^2\\
&= 1-[1-\exp[-1.3410t]]^2\\
&= 2\exp[-1.3410t]-\exp[-2.6820t]
\end{aligned} \tag{9.166}$$

The subsystem S_4 with the structure shown in Figure 9.9 consists of three chain conveyors forming series subsystems ($k=3$), each composed of a wheel driving the belt, a reversible driving wheel and 160, 160 and 240 links respectively. Thus, two conveyors have 162 components and the remaining one has 242 components ($l_1=162$, $l_2=162$, $l_3=242$), which means that the subsystem is a non-homogeneous non-regular three-state series–parallel system with the exponential reliability functions. In two series subsystems of the subsystem S_4 there are respectively

- 2 two driving wheels marked by $E_{ij}^{(4)}$, $i=1,2$, $j=2$, with reliability function coordinates

$$[R_{ij}^{(4)}(t,1)]^{(1)}=\exp[-0.005t],\ [R_{ij}^{(4)}(t,2)]^{(1)}=\exp[-0.006t],\quad t\geq 0,\ i=1,2,\ j=2$$

- 160 links marked by $E_{ij}^{(4)}$, $i=1,2$, $j=3,4,\ldots,162$, with reliability function coordinates

$$[R_{ij}^{(4)}(t,1)]^{(1)}=\exp[-0.012t],\ [R_{ij}^{(4)}(t,2)]^{(1)}=\exp[-0.014t],\quad t\geq 0,\ i=1,2,\ j=3,4,\ldots,162$$

In the third series subsystems of the subsystem S_4 there are respectively

- 2 two driving wheels marked by $E_{ij}^{(4)}$, $i=3$, $j=2$, with reliability function coordinates

$$[R_{ij}^{(4)}(t,1)]^{(1)}=\exp[-0.022t],\ [R_{ij}^{(4)}(t,2)]^{(1)}=\exp[-0.024t],\quad t\geq 0,\ i=3,\ j=2$$

- 240 links marked by $E_{ij}^{(4)}$, $i=3, j=3,4,\ldots,242$, with reliability function coordinates

$$[R_{ij}^{(4)}(t,1)]^{(1)} = \exp[-0.034t],\ [R_{ij}^{(4)}(t,2)]^{(1)} = \exp[-0.040t], \quad t \geq 0,\ i=3,\ j=3,4,\ldots,242$$

Thus, at the operation state z_1, the subsystem S_4 is a three-state series–parallel system with its structure shape parameters $k=3$, $l_1=162$, $l_2=162$, $l_3=242$, and according to the formulae appearing after Definition 3.11 and Eqs. (9.59) and (9.60), its conditional reliability function is given by

$$[\boldsymbol{R}^{(4)}(t,\cdot)]^{(1)} = [1, [\boldsymbol{R}^{(4)}(t,\ 1)]^{(1)}, [\boldsymbol{R}^{(4)}(t,\ 2)]^{(1)}], \quad t \geq 0 \tag{9.167}$$

where

$$\begin{aligned}
[\boldsymbol{R}^{(4)}(t,1)]^{(1)} &= \boldsymbol{R}_{3;162,162,242}(t,1) = 1 - \prod_{i=1}^{2}\left[1 - \prod_{j=1}^{162}[R_{ij}^{(4)}(t,1)]^{(1)}\right]\left[1 - \prod_{j=1}^{242}[R_{3j}^{(4)}(t,1)]^{(1)}\right] \\
&= 1 - \prod_{i=1}^{2}\left[1 - \exp\left[-\sum_{j=1}^{162}[\lambda_{ij}^{(4)}(1)]^{(1)}t\right]\right]\left[1 - \exp\left[-\sum_{j=1}^{242}\lambda_{3j}^{(4)}(1)\right]\right] \\
&= 1 - [1 - \exp[-[0.005\cdot 2 + 0.012\cdot 160]t]^2 \\
&\quad \cdot [1 - \exp[-[0.022\cdot 2 + 0.034\cdot 240]t] \\
&= 1 - [1 - \exp[-1.930t]]^2 \cdot [1 - \exp[-8.204t] \\
&= 2\exp[-1.930t] - 2\exp[-10.134t] - \exp[-3.860t] \\
&\quad + \exp[-12.064t] + \exp[-8.204t]
\end{aligned} \tag{9.168}$$

$$\begin{aligned}
[\boldsymbol{R}^{(4)}(t,2)]^{(1)} &= \boldsymbol{R}_{3;162,162,242}(t,2) = 1 - \prod_{i=1}^{2}\left[1 - \prod_{j=1}^{162}[R_{ij}^{(4)}(t,2)]^{(1)}\right]\left[1 - \prod_{j=1}^{242}[R_{3j}^{(4)}(t,2)]^{(1)}\right] \\
&= 1 - \prod_{i=1}^{2}\left[1 - \exp\left[-\sum_{j=1}^{162}[\lambda_{ij}^{(4)}(2)]^{(1)}t\right]\right]\left[1 - \exp\left[-\sum_{j=1}^{242}\lambda_{3j}^{(4)}(2)\right]\right] \\
&= 1 - [1 - \exp[-[0.006\cdot 2 + 0.014\cdot 160]t]^2 \\
&\quad \cdot [1 - \exp[-[0.024\cdot 2 + 0.040\cdot 240]t] \\
&= 1 - [1 - \exp[-2.252t]]^2 \cdot [1 - \exp[-9.648t] \\
&= 2\exp[-2.252t] - 2\exp[-11.900t] - \exp[-4.504t] \\
&\quad + \exp[-14.152t] + \exp[-9.648t]
\end{aligned} \tag{9.169}$$

Considering that the system at the operation state z_1 is a three-state series system composed of subsystems S_1, S_2, S_3 and S_4, after applying the formulae appearing after Definition 3.4 and the results (9.158)–(9.169), its conditional reliability function is given by

$$[\boldsymbol{R}(t,\cdot)]^{(1)} = [1, [\boldsymbol{R}(t,1)]^{(1)}, [\boldsymbol{R}(t,2)]^{(1)}], \quad t \geq 0 \tag{9.170}$$

where

$$\begin{aligned}
[\boldsymbol{R}(t,1)]^{(1)} = \overline{\boldsymbol{R}}_4(t,1) = {} & [\boldsymbol{R}^{(1)}(t,1)]^{(1)}[\boldsymbol{R}^{(2)}(t,1)]^{(1)}[\boldsymbol{R}^{(3)}(t,1)]^{(1)}[\boldsymbol{R}^{(4)}(t,1)]^{(1)} \\
= {} & 24\exp[-25.4710t] - 24\exp[-47.6990t] + 8\exp[-69.9270t] \\
& -12\exp[-26.1075t] + 12\exp[-48.3355t] - 4\exp[-70.5635t] \\
& -12\exp[-26.1475t] + 12\exp[-48.3755t] - 4\exp[-70.6035t] \\
& +6\exp[-26.7840t] - 6\exp[-49.0120t] + 2\exp[-71.2400t] \\
& -24\exp[-33.6750t] + 24\exp[-55.9030t] - 8\exp[-78.1310t] \\
& +12\exp[-34.3115t] - 12\exp[-56.5395t] + 4\exp[-78.7675t] \\
& +12\exp[-34.3515t] - 12\exp[-56.5795t] + 4\exp[-78.8075t] \\
& -6\exp[-34.9880t] + 6\exp[-57.2160t] - 2\exp[-79.4440t] \\
& -12\exp[-27.4010t] + 12\exp[-49.6290t] - 4\exp[-71.8570t] \\
& +6\exp[-28.0375t] - 6\exp[-50.2655t] + 2\exp[-72.4935t] \\
& +6\exp[-28.0775t] - 6\exp[-50.3055t] + 2\exp[-72.5335t] \\
& -3\exp[-28.7140t] + 3\exp[-50.9420t] - \exp[-73.1700t] \\
& +12\exp[-35.6050t] - 12\exp[-57.8330t] + 4\exp[-80.0610t] \\
& -6\exp[-36.2415t] + 6\exp[-58.4695t] - 2\exp[-80.9750t] \\
& -6\exp[-36.2815t] + 6\exp[-58.5095t] - 2\exp[80.7375t] \\
& +3\exp[-36.9180t] - 3\exp[-59.1460t] + \exp[-81.3740t] \\
& +12\exp[-31.7450t] - 6\exp[-32.4215t] - 12\exp[-53.9730t] \\
& +6\exp[-54.6495t] + 4\exp[-76.2010t] - 2\exp[-76.8775t] \\
& -6\exp[-32.3815t] + 3\exp[-33.0580t] + 6\exp[-54.6095t] \\
& -3\exp[-55.2860t] - 2\exp[-76.8375t] + \exp[-77.5140t]
\end{aligned} \tag{9.171}$$

$$
\begin{aligned}
[\boldsymbol{R}(t,2)]^{(1)} = \overline{\boldsymbol{R}}_4(t,2) &= [\boldsymbol{R}^{(1)}(t,2)]^{(1)}[\boldsymbol{R}^{(2)}(t,2)]^{(1)}[\boldsymbol{R}^{(3)}(t,2)]^{(1)}[\boldsymbol{R}^{(4)}(t,2)]^{(1)} \\
&= 24\exp[-30.7880t] - 24\exp[-56.7220t] + 8\exp[-82.6560t] \\
&\quad - 12\exp[-32.0490t] + 12\exp[-57.9830t] - 4\exp[-83.9170t] \\
&\quad - 12\exp[-32.1290t] + 12\exp[-58.0630t] - 4\exp[-83.9970t] \\
&\quad + 6\exp[-33.3900t] - 6\exp[-59.3240t] + 2\exp[-86.1690t] \\
&\quad - 24\exp[-40.4360t] + 24\exp[-66.3700t] - 8\exp[-92.3040t] \\
&\quad + 12\exp[-41.6970t] - 12\exp[-67.6310t] + 4\exp[-93.5650t] \\
&\quad + 12\exp[-41.7770t] - 12\exp[-67.7110t] + 4\exp[-93.6450t] \\
&\quad - 6\exp[-43.0380t] + 6\exp[-68.9720t] - 2\exp[-94.9060t] \\
&\quad - 12\exp[-33.0400t] + 12\exp[-58.9740t] - 4\exp[-84.9080t] \\
&\quad + 6\exp[-34.3010t] - 6\exp[-60.2350t] + 2\exp[-85.2580t] \\
&\quad + 6\exp[-34.3810t] - 6\exp[-60.3150t] + 2\exp[-86.2490t] \\
&\quad - 3\exp[-35.6420t] + 3\exp[-61.5760t] - \exp[-87.5100t] \\
&\quad + 12\exp[-42.6880t] - 12\exp[-68.6220t] + 4\exp[-94.5560t] \\
&\quad - 6\exp[-43.9490t] + 6\exp[-69.8830t] - 2\exp[-95.8170t] \\
&\quad - 6\exp[-44.0290t] + 6\exp[-69.9630t] - 2\exp[-95.8970t] \\
&\quad + 3\exp[-45.2900t] - 3\exp[-71.2240t] + \exp[-97.1580t] \\
&\quad + 12\exp[-38.1840t] - 6\exp[-39.5250t] - 12\exp[-64.1180t] \\
&\quad + 6\exp[-65.4590t] + 4\exp[-90.0520t] - 2\exp[-91.3930t] \\
&\quad - 6\exp[-39.4450t] + 3\exp[-40.7860t] + 6\exp[-65.3790t] \\
&\quad - 3\exp[-66.7200t] - 2\exp[-91.3130t] + \exp[-92.6540t]
\end{aligned}
\tag{9.172}
$$

The expected values and standard deviations of the system conditional lifetimes in the reliability state subsets {1,2}, {2} at the operation state z_1, calculated from the results given by Eqs. (9.170)–(9.172), according to Eq. (9.40), are respectively

$$
\begin{aligned}
M_1(1) = \int_0^{\infty} [\boldsymbol{R}(t,1)]^{(1)}\,dt &= 24/25.4710 - 24/47.6990 + 8/69.9270 - 12/26.1075 \\
&+ 12/48.3355 - 4/70.5635 - 12/26.1475 + 12/48.3755 - 4/70.6035 \\
&+ 6/26.7840 - 6/49.0120 + 2/71.2400 - 24/33.6750 + 24/55.9030 \\
&- 8/78.1310 + 12/34.3115 - 12/56.5395 + 4/78.7675 + 12/34.3515 \\
&- 12/56.5795 + 4/78.8075 - 6/34.9880 + 6/57.2160 - 2/79.4440 \\
&- 12/27.4010 + 12/49.6290 - 4/71.8570 + 6/28.0375 - 6/50.2655 \\
&+ 2/72.4935 + 6/28.0775 - 6/50.3055 + 2/72.5335 - 3/28.7140 \\
&+ 3/50.9420 - 1/73.1700 + 12/35.6050 - 12/57.8330 + 4/80.0610 \\
&- 6/36.2415 + 6/58.4695 - 2/80.9750 - 6/36.2815 + 6/58.5095 \\
&- 2/80.7375 + 3/36.9180 - 3/59.1460 + 1/81.3740 + 12/31.7450 \\
&- 6/32.4215 - 12/53.9730 + 6/54.6495 + 4/76.2010 - 2/76.8775 \\
&- 6/32.3815 + 3/33.0580 + 6/54.6095 - 3/55.2860 - 2/76.8375 \\
&+ 1/77.5140 \cong 0.0813 \text{ year}
\end{aligned}
\tag{9.173}
$$

$$
\begin{aligned}
M_1(2) = \int_0^{\infty} [\boldsymbol{R}(t,2)]^{(1)}\,dt &= 24/30.7880 - 24/56.7220 + 8/82.6560 - 12/32.0490 \\
&+ 12/57.9830 - 4/83.9170 - 12/32.1290 + 12/58.0630 - 4/83.9970 \\
&+ 6/33.3900 - 6/59.3240 + 2/86.1690 - 24/40.4360 + 24/66.3700 \\
&- 8/92.3040 + 12/41.6970 - 12/67.6310 + 4/93.5650 + 12/41.7770 \\
&- 12/67.7110 + 4/93.6450 - 6/43.0380 + 6/68.9720 - 2/94.9060 \\
&- 12/33.0400 + 12/58.9740 - 4/84.9080 + 6/34.3010 - 6/60.2350 \\
&+ 2/85.2580 + 6/34.3810 - 6/60.3150 + 2/86.2490 - 3/35.6420 \\
&+ 3/61.5760 - 1/87.5100 + 12/42.6880 - 12/68.6220 + 4/94.5560 \\
&- 6/43.9490 + 6/69.8830 - 2/95.8170 - 6/44.0290 + 6/69.9630 \\
&- 2/95.8970 + 3/45.2900 - 3/71.2240 + 1/97.1580 + 12/38.1840 \\
&- 6/39.5250 - 12/64.1180 + 6/65.4590 + 4/90.0520 - 2/91.3930 \\
&- 6/39.4450 + 3/40.7860 + 6/65.3790 - 3/66.7200 - 2/91.3130 \\
&+ 1/92.6540 \cong 0.0691 \text{ year}
\end{aligned}
\tag{9.174}
$$

At the system operation state z_2, the system is a series system with the structure shown in Figure 9.8, composed of four subsystems S_1, S_2, S_3 and S_4 illustrated respectively in Figure 9.9.

The subsystem S_1 with the structure shown in Figure 9.9 consists of 1 belt conveyor forming a series systems ($k = 1$) composed of 129 components ($l_1 = 129$) with the exponential reliability functions. In the series system of the subsystem S_1 there are respectively

- 1 belt marked by $E_{ij}^{(1)}$, $i = 1, j = 1$, with reliability function coordinates

$$[R_{ij}^{(1)}(t,1)]^{(2)} = \exp[-0.0110t],\ [R_{ij}^{(1)}(t,2)]^{(2)} = \exp[-0.0120t],\quad t \geq 0,\ i = 1,\ j = 1$$

- 2 drums marked by $E_{ij}^{(1)}$, $i = 1, j = 2, 3$, with reliability function coordinates

$$[R_{ij}^{(1)}(t,1)]^{(2)} = \exp[-0.0010t],\ [R_{ij}^{(1)}(t,2)]^{(2)} = \exp[-0.0012t],\quad t \geq 0,\ i = 1,2,\ j = 2,3$$

- 117 channelled rollers marked by $E_{ij}^{(1)}$, $i = 1$, $j = 4, 5, \ldots, 120$, with reliability function coordinates

$$[R_{ij}^{(1)}(t,1)]^{(2)} = \exp[-0.003t],\ [R_{ij}^{(1)}(t,2)]^{(2)} = \exp[-0.006t],\quad t \geq 0,\ i = 1,\ j = 4,5,\ldots,120$$

- 9 supporting rollers marked by $E_{ij}^{(1)}$, $i = 1$, $j = 121, 122, \ldots, 129$, with reliability function coordinates

$$[R_{ij}^{(1)}(t,1)]^{(2)} = \exp[-0.002t],\ [R_{ij}^{(1)}(t,2)]^{(2)} = \exp[-0.003t],\quad t \geq 0,\ i = 1,$$
$$j = 121, 122, \ldots, 129$$

Thus, at the operation state z_2, the subsystem S_1 is a three-state series–parallel system (a series system) with its structure shape parameters $k = 1$, $l_1 = 129$, and according to the formulae appearing after Definition 3.11 and Eqs. (9.59) and (9.60), its conditional reliability function is given by

$$[\boldsymbol{R}^{(1)}(t,\cdot)]^{(2)} = [1, [\boldsymbol{R}^{(1)}(t,1)]^{(2)}, [\boldsymbol{R}^{(1)}(t,2)]^{(2)}],\quad t \geq 0 \qquad 9.175)$$

where

$$[\boldsymbol{R}^{(1)}(t,1)]^{(2)} = \boldsymbol{R}_{1;129}(t,1) = 1 - \prod_{i=1}^{1}\left[1 - \prod_{j=1}^{129}[R_{ij}^{(1)}(t,1)]^{(2)}\right]$$
$$= \prod_{j=1}^{129}[R_{1j}^{(1)}(t,1)]^{(2)}] = \exp\left[-\sum_{j=1}^{129}\lambda_{1j}^{(1)}(1)t\right] \qquad (9.176)$$
$$= \exp[-[0.0110 \cdot 1 + 0.0010 \cdot 2 + 0.003 \cdot 117 + 0.002 \cdot 9]t]$$
$$= \exp[-0.3710t]$$

$$[\boldsymbol{R}^{(1)}(t,2)]^{(2)} = \boldsymbol{R}_{1;129}(t,1) = 1 - \prod_{i=1}^{1}\left[1 - \prod_{j=1}^{129}[R_{ij}^{(1)}(t,2)]^{(2)}\right]$$
$$= \prod_{j=1}^{129}[R_{1j}^{(1)}(t,2)]^{(2)}] = \exp\left[-\sum_{j=1}^{129}\lambda_{1j}^{(1)}(2)t\right] \tag{9.177}$$
$$= \exp[-[0.0120\cdot 1 + 0.0012\cdot 2 + 0.006\cdot 117 + 0.003\cdot 9]t]$$
$$= \exp[-0.7434t]$$

The subsystem S_2 with the structure shown in Figure 9.9 consists of two identical bucket elevators forming series subsystems ($k = 2$), each composed of 743 components ($l_1 = 743$, $l_2 = 743$) with the exponential reliability functions. In each series subsystems of the subsystem S_2 there are respectively

- 1 belt marked by $E_{ij}^{(2)}$, $i = 1, 2$, $j = 1$, with reliability function coordinates

$$[R_{ij}^{(2)}(t,1)]^{(2)} = \exp[-0.020t],\ [R_{ij}^{(2)}(t,2)]^{(2)} = \exp[-0.024t], \quad t \ge 0,\ i = 1,2,\ j = 1$$

- 2 drums marked by $E_{ij}^{(2)}$, $i = 1, 2$, $j = 2, 3$, with reliability function coordinates

$$[R_{ij}^{(2)}(t,1)]^{(2)} = \exp[-0.0012t],\ [R_{ij}^{(2)}(t,2)]^{(2)} = \exp[-0.0014t], \quad t \ge 0,\ i = 1,2,\ j = 2,3$$

- 740 buckets marked by $E_{ij}^{(2)}$, $i = 1, 2$, $j = 4, 5, \ldots, 743$, with reliability function coordinates

$$[R_{ij}^{(2)}(t,1)]^{(2)} = \exp[-0.022t],\ [R_{ij}^{(2)}(t,2)]^{(2)} = \exp[-0.025t], \quad t \ge 0,\ i = 1,2,\ j = 4,5,\ldots,743$$

Thus, at the operation state z_2, the subsystem S_2 is a three-state series–parallel system with its structure shape parameters $k = 2$, $l_1 = 743$, $l_2 = 743$, and according to the formulae appearing after Definition 3.11 and Eqs. (9.59) and (9.60), its conditional reliability function is given by

$$[\boldsymbol{R}^{(2)}(t,\cdot)]^{(2)} = [1, [\boldsymbol{R}^{(2)}(t,1)]^{(2)}, [\boldsymbol{R}^{(2)}(t,2)]^{(2)}], \quad t \ge 0 \tag{9.178}$$

where

$$[\boldsymbol{R}^{(2)}(t,1)]^{(2)} = \boldsymbol{R}_{2;743,743}(t,1) = 1 - \prod_{i=1}^{2}\left[1 - \prod_{j=1}^{743}[R_{ij}^{(2)}(t,1)]^{(2)}\right]$$
$$= 1 - \prod_{i=1}^{2}\left[1 - \exp\left[-\sum_{j=1}^{743}[\lambda_{ij}^{(2)}(1)]^{(2)}t\right]\right] \tag{9.179}$$
$$= 1 - [1 - \exp[-[0.020\cdot 1 + 0.012\cdot 2 + 0.022\cdot 740]t]^2$$
$$= 1 - [1 - \exp[-16.324t]]^2$$
$$= 2\exp[-16.324t] - \exp[-32.648t]$$

$$[\boldsymbol{R}^{(2)}(t,2)]^{(2)} = \boldsymbol{R}_{2;743,743}(t,2) = 1 - \prod_{i=1}^{2}\left[1 - \prod_{j=1}^{743}[R_{ij}^{(2)}(t,2)]^{(2)}\right]$$

$$= 1 - \prod_{i=1}^{2}\left[1 - \exp\left[-\sum_{j=1}^{743}[\lambda_{ij}^{(2)}(2)]^{(2)}t\right]\right] \tag{9.180}$$

$$= 1 - [1 - \exp[-[0.024 \cdot 1 + 0.014 \cdot 2 + 0.025 \cdot 740]t]^2$$

$$= 1 - [1 - \exp[-18.552t]]^2$$

$$= 2\exp[-18.552t] - \exp[-37.104t]$$

The subsystem S_3 with the structure shown in Figure 9.9 consists of 1 belt conveyors forming a series system ($k = 1$) composed of 139 components ($l_1 = 139$) with the exponential reliability functions. In the series system of the subsystem S_3 there are respectively

- 1 belt marked by $E_{ij}^{(1)}$, $i = 1, j = 1$, with reliability function coordinates

$$[R_{ij}^{(3)}(t,1)]^{(2)} = \exp[-0.0110t],\ \ [R_{ij}^{(2)}(t,2)]^{(2)} = \exp[-0.0120t], \quad t \geq 0,\ \ i = 1,\ \ j = 1$$

- 2 drums marked by $E_{ij}^{(1)}$, $i = 1, j = 2, 3$, with reliability function coordinates

$$[R_{ij}^{(3)}(t,1)]^{(2)} = \exp[-0.0010t],\ \ [R_{ij}^{(3)}(t,2)]^{(2)} = \exp[-0.0012t], \quad t \geq 0,\ \ i = 1, 21,\ \ j = 2, 3$$

- 117 channelled rollers marked by $E_{ij}^{(1)}$, $i = 1$, $j = 4, 5, \ldots, 120$, with reliability function coordinates

$$[R_{ij}^{(3)}(t,1)]^{(2)} = \exp[-0.003t],\ [R_{ij}^{(3)}(t,2)]^{(2)} = \exp[-0.006t], \quad t \geq 0,\ i = 1,\ j = 4, 5, \ldots, 120$$

- 19 supporting rollers marked by $E_{ij}^{(1)}$, $i = 1, j = 121, 122, \ldots, 139$, with reliability function coordinates

$$[R_{ij}^{(3)}(t,1)]^{(2)} = \exp[-0.002t],\ \ [R_{ij}^{(3)}(t,2)]^{(2)} = \exp[-0.003t], \quad t \geq 0,$$

$$i = 1,\ \ j = 121, 122, \ldots, 129$$

Thus, at the operation state z_2, the subsystem S_3 is a three-state series−parallel system (a series system) with its structure shape parameters $k = 1$, $l_1 = 139$, and according to the formulae appearing after Definition 3.11 and Eqs. (9.59) and (9.60), its conditional reliability function is given by

$$[\boldsymbol{R}^{(3)}(t,\cdot)]^{(2)} = [1, [\boldsymbol{R}^{(3)}(t,1)]^{(2)}, [\boldsymbol{R}^{(3)}(t,2)]^{(2)}], \quad t \geq 0 \tag{9.181}$$

where

$$[\boldsymbol{R}^{(3)}(t,1)]^{(2)} = \boldsymbol{R}_{1;139}(t,1) = 1 - \prod_{i=1}^{1}\left[1 - \prod_{j=1}^{139}[R_{ij}^{(3)}(t,1)]^{(2)}\right]$$
$$= \prod_{j=1}^{139}[R_{1j}^{(3)}(t,1)]^{(2)}] = \exp\left[-\sum_{j=1}^{139}\lambda_{1j}^{(3)}(1)t\right]$$
$$= \exp[-[0.0110\cdot 1 + 0.0010\cdot 2 + 0.003\cdot 117 + 0.002\cdot 19]t]$$
$$= \exp[-0.3910t] \tag{9.182}$$

$$[\boldsymbol{R}^{(3)}(t,2)]^{(2)} = \boldsymbol{R}_{1;139}(t,1) = 1 - \prod_{i=1}^{1}\left[1 - \prod_{j=1}^{139}[R_{ij}^{(3)}(t,2)]^{(2)}\right]$$
$$= \prod_{j=1}^{139}[R_{1j}^{(3)}(t,2)]^{(2)}] = \exp\left[-\sum_{j=1}^{139}\lambda_{1j}^{(3)}(2)t\right]$$
$$= \exp[-[0.0120\cdot 1 + 0.0012\cdot 2 + 0.006\cdot 117 + 0.003\cdot 19]t]$$
$$= \exp[-0.7734t] \tag{9.183}$$

The subsystem S_4 with the structure shown in Figure 9.9 consists of three identical chain conveyors forming series subsystems ($k = 2$), each composed of a wheel driving the belt, a reversible driving wheel and 160 links ($l_1 = 162$, $l_2 = 162$), which means that the subsystem is a non-homogeneous non-regular three-state series–parallel system with the exponential reliability functions. In the series subsystems of the subsystem S_4 there are respectively

- 2 two driving wheels marked by $E_{ij}^{(4)}$, $i = 1, 2, j = 2$, with reliability function coordinates

$$[R_{ij}^{(4)}(t,1)]^{(2)} = \exp[-0.002t],\ [R_{ij}^{(4)}(t,2)]^{(2)} = \exp[-0.004t],\quad t \geq 0,\ \ i = 1,2,\ \ j = 2$$

- 160 links marked by $E_{ij}^{(4)}$, $i = 1, 2, j = 3, 4, \ldots, 162$, with reliability function coordinates

$$[R_{ij}^{(4)}(t,1)]^{(2)} = \exp[-0.008t],\ [R_{ij}^{(4)}(t,2)]^{(2)} = \exp[-0.010t],\quad t \geq 0,$$
$$i = 1,2,\ \ j = 3,4,\ldots,162$$

Thus, at the operation state z_2, the subsystem S_4 is a three-state series–parallel system with its structure shape parameters $k = 2$, $l_1 = 162$, $l_2 = 162$, and according to the formulae appearing after Definition 3.11 and Eqs. (9.59) and (9.60), its conditional reliability function is given by

$$[\boldsymbol{R}^{(4)}(t,\cdot)]^{(2)} = [1, [\boldsymbol{R}^{(4)}(t,\ 1)]^{(2)}, [\boldsymbol{R}^{(4)}(t,\ 2)]^{(2)}],\quad t \geq 0 \tag{9.184}$$

where

$$[\boldsymbol{R}^{(4)}(t,1)]^{(2)} = \boldsymbol{R}_{2;162,162}(t,1) = 1 - \prod_{i=1}^{2}\left[1 - \prod_{j=1}^{162}[R_{ij}^{(4)}(t,1)]^{(2)}\right]$$
$$= 1 - \prod_{i=1}^{2}\left[1 - \exp\left[-\sum_{j=1}^{162}[\lambda_{ij}^{(4)}(1)]^{(2)}t\right]\right]$$
$$= 1 - [1 - \exp[-[0.002 \cdot 2 + 0.008 \cdot 160]t]^2$$
$$= 1 - [1 - \exp[-1.284t]]^2$$
$$= 2\exp[-1.284t] - \exp[-2.5680t] \quad (9.185)$$

$$[\boldsymbol{R}^{(4)}(t,2)]^{(2)} = \boldsymbol{R}_{2;162,162}(t,2) = 1 - \prod_{i=1}^{2}\left[1 - \prod_{j=1}^{162}[R_{ij}^{(4)}(t,2)]^{(2)}\right]$$
$$= 1 - \prod_{i=1}^{2}\left[1 - \exp\left[-\sum_{j=1}^{162}[\lambda_{ij}^{(4)}(2)]^{(2)}t\right]\right]$$
$$= 1 - [1 - \exp[-[0.004 \cdot 2 + 0.010 \cdot 160]t]^2$$
$$= 1 - [1 - \exp[-1.608t]]^2$$
$$= 2\exp[-1.608t] - \exp[-3.2160t] \quad (9.186)$$

Considering that the system at the operation state z_2 is a three-state series system composed of subsystems S_1, S_2, S_3 and S_4, after applying the formulae appearing after Definition 3.4 and the results (9.175)–(9.186), its conditional reliability function is given by

$$[\boldsymbol{R}(t,\cdot)]^{(2)} = [1,\ [\boldsymbol{R}(t,1)]^{(2)}, [\boldsymbol{R}(t,2)]^{(2)}], \quad t \geq 0 \quad (9.187)$$

where

$$[\boldsymbol{R}(t,1)]^{(2)} = \overline{\boldsymbol{R}}_4(t,1) = [\boldsymbol{R}^{(1)}(t,1)]^{(2)}[\boldsymbol{R}^{(2)}(t,1)]^{(2)}[\boldsymbol{R}^{(3)}(t,1)]^{(2)}[\boldsymbol{R}^{(4)}(t,1)]^{(2)}$$
$$= 4\exp[-18.380t] - 2\exp[-34.694t] - 2\exp[-19.664t] + \exp[-35.978t] \quad (9.188)$$

$$[\boldsymbol{R}(t,2)]^{(2)} = \overline{\boldsymbol{R}}_4(t,2) = [\boldsymbol{R}^{(1)}(t,2)]^{(2)}[\boldsymbol{R}^{(2)}(t,2)]^{(2)}[\boldsymbol{R}^{(3)}(t,2)]^{(2)}[\boldsymbol{R}^{(4)}(t,2)]^{(2)}$$
$$= 4\exp[-21.6768t] - 2\exp[-40.2288t] - 2\exp[-23.2848t] + \exp[-41.8368t] \quad (9.189)$$

The expected values and standard deviations of the system conditional lifetimes in the reliability state subsets $\{1,2\}$, $\{2\}$ at the operation state z_2, calculated from the results given by Eqs. (9.187)–(9.189), according to Eq. (9.40), are respectively

$$M_2(1) = \int_0^\infty [R(t,1)]^{(2)}\,dt$$
$$= 4/18.380 - 2/34.694 - 2/19.664 + 1/35.978 \cong 0.0861 \text{ year} \quad (9.190)$$

$$M_2(2) = \int_0^\infty [R(t,2)]^{(2)}\,dt$$
$$= 4/21.6768 - 2/40.2288 - 2/23.2848 + 1/41.8368 \cong 0.0728 \text{ year} \quad (9.191)$$

At the system operational state z_3, the system is a series system with the structure shown in Figure 9.8, composed of four subsystems S_1, S_2, S_3 and S_4 illustrated respectively in Figure 9.10.

The subsystem S_1 with the structure shown in Figure 9.9 consists of 1 belt conveyor forming a series systems ($k = 1$) composed of 129 components ($l_1 = 129$) with the exponential reliability functions. In the series system of the subsystem S_1 there are respectively

- 1 belt marked by $E_{ij}^{(1)}$, $i = 1, j = 1$, with reliability function coordinates

$$[R_{ij}^{(1)}(t,1)]^{(3)} = \exp[-0.0100t],\ [R_{ij}^{(1)}(t,2)]^{(3)} = \exp[-0.0105t], \quad t \geq 0,\ i = 1,\ j = 1$$

- 2 drums marked by $E_{ij}^{(1)}$, $i = 1, j = 2, 3$, with reliability function coordinates

$$[R_{ij}^{(1)}(t,1)]^{(3)} = \exp[-0.0009t],\ [R_{ij}^{(1)}(t,2)]^{(3)} = \exp[-0.0011t], \quad t \geq 0,\ i = 1,21,\ j = 2,3$$

- 117 channelled rollers marked by $E_{ij}^{(1)}$, $i = 1$, $j = 4, 5, \ldots, 120$, with reliability function coordinates

$$[R_{ij}^{(1)}(t,1)]^{(3)} = \exp[-0.002t],\ [R_{ij}^{(1)}(t,2)]^{(3)} = \exp[-0.004t], \quad t \geq 0,\ i = 1,\ j = 4,5,\ldots,120$$

- 9 supporting rollers marked by $E_{ij}^{(1)}$, $i = 1$, $j = 121, 122, \ldots, 129$, with reliability function coordinates

$$[R_{ij}^{(1)}(t,1)]^{(3)} = \exp[-0.001t],\ [R_{ij}^{(1)}(t,2)]^{(3)} = \exp[-0.002t], \quad t \geq 0,\ i = 1,$$
$$j = 121, 122, \ldots, 129$$

Thus, at the operation state z_3, the subsystem S_1 is a three-state series–parallel system (a series system) with its structure shape parameters $k = 1$, $l_1 = 129$, and according to the formulae appearing after Definition 3.11 and Eqs. (9.59) and (9.60), its conditional reliability function is given by

$$[\boldsymbol{R}^{(1)}(t,\cdot)]^{(3)} = [1, [\boldsymbol{R}^{(1)}(t,\ 1)]^{(3)}, [\boldsymbol{R}^{(1)}(t,\ 2)]^{(3)}], \quad t \geq 0 \quad (9.192)$$

where

$$
\begin{aligned}
[\boldsymbol{R}^{(1)}(t,1)]^{(3)} &= \boldsymbol{R}_{1;129}(t,1) = 1 - \prod_{i=1}^{1}\left[1 - \prod_{j=1}^{129}[R_{ij}^{(1)}(t,1)]^{(3)}\right] \\
&= \prod_{j=1}^{129}[R_{1j}^{(1)}(t,1)]^{(3)}] = \exp\left[-\sum_{j=1}^{129}\lambda_{1j}^{(1)}(1)t\right] \\
&= \exp[-[0.0100\cdot 1 + 0.0009\cdot 2 + 0.002\cdot 117 + 0.001\cdot 9]t] \\
&= \exp[-0.2548t]
\end{aligned}
\tag{9.193}
$$

$$
\begin{aligned}
[\boldsymbol{R}^{(1)}(t,2)]^{(3)} &= \boldsymbol{R}_{1;129}(t,2) = 1 - \prod_{i=1}^{1}\left[1 - \prod_{j=1}^{129}[R_{ij}^{(1)}(t,2)]^{(3)}\right] \\
&= \prod_{j=1}^{129}[R_{1j}^{(1)}(t,2)]^{(3)}] = \exp\left[1-\sum_{j=1}^{129}\lambda_{1j}^{(1)}(2)t\right] \\
&= \exp[-[0.0105\cdot 1 + 0.0011\cdot 2 + 0.004\cdot 117 + 0.002\cdot 9]t] \\
&= \exp[-0.4987t]
\end{aligned}
\tag{9.194}
$$

The subsystem S_2 with the structure shown in Figure 9.10 consists of 1 bucket elevator forming a series system ($k = 1$) composed of 743 components ($l_1 = 743$) with the exponential reliability functions. In the series stem of the subsystem S_2 there are respectively

- 1 belt marked by $E_{ij}^{(2)}$, $i = 1, j = 1$, with reliability function coordinates

$$[R_{ij}^{(2)}(t,1)]^{(3)} = \exp[-0.018t],\ [R_{ij}^{(2)}(t,2)]^{(2)} = \exp[-0.022t],\quad t \geq 0,\ i = 1,\ j = 1$$

- 2 drums marked by $E_{ij}^{(2)}$, $i = 1, j = 2, 3$, with reliability function coordinates

$$[R_{ij}^{(2)}(t,1)]^{(3)} = \exp[-0.0011t],\ [R_{ij}^{(2)}(t,2)]^{(3)} = \exp[-0.0013t],\quad t \geq 0,\ i = 1,\ j = 2,3$$

- 740 buckets marked by $E_{ij}^{(2)}$, $i = 1, j = 4, 5, \ldots, 743$, with reliability function coordinates

$$[R_{ij}^{(2)}(t,1)]^{(3)} = \exp[-0.020t],\ [R_{ij}^{(2)}(t,2)]^{(2)} = \exp[-0.024t],\quad t \geq 0,\ i = 1,\ j = 4,5,\ldots,743$$

Thus, at the operation state z_3, the subsystem S_2 is a three-state series–parallel system (a series system) with its structure shape parameters $k = 1$, $l_1 = 743$, and according to the formulae appearing after Definition 3.11 and Eqs. (9.59) and (9.60), its conditional reliability function is given by

$$[\boldsymbol{R}^{(2)}(t,\cdot)]^{(3)} = [1, [\boldsymbol{R}^{(2)}(t,\ 1)]^{(3)}, [\boldsymbol{R}^{(2)}(t,\ 2)]^{(3)}],\quad t \geq 0 \tag{9.195}$$

where

$$[\boldsymbol{R}^{(2)}(t,1)]^{(3)} = \boldsymbol{R}_{1;743}(t,1) = 1 - \prod_{i=1}^{1}\left[1 - \prod_{j=1}^{743}[R_{ij}^{(2)}(t,1)]^{(3)}\right]$$
$$= \prod_{j=1}^{743}[[R_{1j}^{(2)}(t,1)]^{(3)}] = \exp\left[-\sum_{j=1}^{743}\lambda_{1j}^{(3)}(1)t\right] \tag{9.196}$$
$$= \exp[-[0.018 \cdot 1 + 0.011 \cdot 2 + 0.020 \cdot 740]t]$$
$$= \exp[-14.840t]$$

$$[\boldsymbol{R}^{(2)}(t,2)]^{(3)} = \boldsymbol{R}_{1;743}(t,2) = 1 - \prod_{i=1}^{1}\left[1 - \prod_{j=1}^{743}[R_{ij}^{(2)}(t,2)]^{(3)}\right]$$
$$= \prod_{j=1}^{743}[[R_{1j}^{(2)}(t,2)]^{(3)}] = \exp\left[-\sum_{j=1}^{743}\lambda_{1j}^{(3)}(2)t\right] \tag{9.197}$$
$$= \exp[-[0.022 \cdot 1 + 0.013 \cdot 2 + 0.024 \cdot 740]t]$$
$$= \exp[-17.808t]$$

The subsystem S_3 with the structure shown in Figure 9.10 consists of one belt conveyors forming a series system ($k = 1$) composed of 139 components ($l_1 = 139$) with the exponential reliability functions. In the series system of the subsystem S_3 there are respectively

- 1 belt marked by $E_{ij}^{(3)}$, $i = 1, j = 1$, with reliability function coordinates

$$[R_{ij}^{(3)}(t,1)]^{(3)} = \exp[-0.0100t],\ [R_{ij}^{(3)}(t,2)]^{(3)} = \exp[-0.0105t],\quad t \geq 0,\ i = 1,\ j = 1$$

- 2 drums marked by $E_{ij}^{(3)}$, $i = 1, j = 2, 3$, with reliability function coordinates

$$[R_{ij}^{(3)}(t,1)]^{(3)} = \exp[-0.0009t],\ [R_{ij}^{(3)}(t,2)]^{(3)} = \exp[-0.0011t],\quad t \geq 0,\ i = 1, 2,\ j = 2, 3$$

- 117 channelled rollers marked by $E_{ij}^{(3)}$, $i = 1$, $j = 4, 5, \ldots, 120$, with reliability function coordinates

$$[R_{ij}^{(3)}(t,1)]^{(3)} = \exp[-0.002t],\ [R_{ij}^{(3)}(t,2)]^{(3)} = \exp[-0.004t],\quad t \geq 0,\ i = 1,\ j = 4, 5, \ldots, 120$$

- 19 supporting rollers marked by $E_{ij}^{(3)}$, $i = 1, j = 121, 122, \ldots, 129$, with reliability function coordinates

$$[R_{ij}^{(3)}(t,1)]^{(3)} = \exp[-0.001t],\ [R_{ij}^{(3)}(t,2)]^{(3)} = \exp[-0.002t],\quad t \geq 0,\ i = 1,$$
$$j = 121, 122, \ldots, 129$$

Thus, at the operation state z_3, the subsystem S_3 is a three-state series−parallel system (a series system) with its structure shape parameters $k = 1$, $l_1 = 139$, and according to the formulae appearing after Definition 3.11 and Eqs. (9.59) and (9.60), its conditional reliability function is given by

$$[\boldsymbol{R}^{(3)}(t,\cdot)]^{(3)} = [1, [\boldsymbol{R}^{(3)}(t,\ 1)]^{(3)}, [\boldsymbol{R}^{(3)}(t,\ 2)]^{(3)}], \quad t \geq 0 \tag{9.198}$$

where

$$\begin{aligned}
[\boldsymbol{R}^{(3)}(t,1)]^{(3)} &= \boldsymbol{R}_{1;139}(t,1) = 1 - \prod_{i=1}^{1}\left[1 - \prod_{j=1}^{139}[R_{ij}^{(3)}(t,1)]^{(3)}\right] \\
&= \prod_{j=1}^{139}[R_{1j}^{(3)}(t,1)]^{(3)}] = \exp\left[-\sum_{j=1}^{139}\lambda_{1j}^{(3)}(1)t\right] \\
&= \exp[-[0.0100 \cdot 1 + 0.0009 \cdot 2 + 0.002 \cdot 117 + 0.001 \cdot 19]t] \\
&= \exp[-0.2648t]
\end{aligned} \tag{9.199}$$

$$\begin{aligned}
[\boldsymbol{R}^{(3)}(t,2)]^{(3)} &= \boldsymbol{R}_{1;139}(t,2) = 1 - \prod_{i=1}^{1}\left[1 - \prod_{j=1}^{139}[R_{ij}^{(3)}(t,2)]^{(3)}\right] \\
&= \prod_{j=1}^{139}[R_{1j}^{(3)}(t,2)]^{(3)}] = \exp\left[-\sum_{j=1}^{139}\lambda_{1j}^{(3)}(2)t\right] \\
&= \exp[-[0.0105 \cdot 1 + 0.0011 \cdot 2 + 0.004 \cdot 117 + 0.002 \cdot 19]t] \\
&= \exp[-0.5187t]
\end{aligned} \tag{9.200}$$

The subsystem S_4 with the structure shown in Figure 9.10 consists of 1 chain conveyor forming a series system ($k = 1$), composed of a wheel driving the belt, a reversible driving wheel and 160 links ($l_1 = 162$) with the exponential reliability functions. In the series system of the subsystem S_4 there are respectively

- 2 two driving wheels marked by $E_{ij}^{(4)}$, $i = 1, j = 2$, with reliability function coordinates

$$[R_{ij}^{(4)}(t,\ 1)]^{(3)} = \exp[-0.001t],\ [R_{ij}^{(4)}(t,2)]^{(3)} = \exp[-0.003t], \quad t \geq 0,\ i = 1,\ j = 2$$

- 160 links marked by $E_{ij}^{(4)}$, $i = 1, j = 3, 4, \ldots, 162$, with reliability function coordinates

$$[R_{ij}^{(4)}(t,\ 1)]^{(3)} = \exp[-0.007t],\ [R_{ij}^{(4)}(t,2)]^{(3)} = \exp[-0.009t], \quad t \geq 0,\ i = 1,$$
$$j = 3, 4, \ldots, 162$$

Thus, at the operation state z_3, the subsystem S_4 is a three-state series−parallel system (a series system) with its structure shape parameters $k = 1$, $l_1 = 162$, and according to the formulae appearing after Definition 3.11 and Eqs. (9.59) and (9.60), its conditional reliability function is given by

$$[\boldsymbol{R}^{(4)}(t,\cdot)]^{(3)} = [1, [\boldsymbol{R}^{(4)}(t,1)]^{(3)}, [\boldsymbol{R}^{(4)}(t,2)]^{(3)}], \quad t \geq 0 \tag{9.201}$$

where

$$\begin{aligned}[\boldsymbol{R}^{(4)}(t,1)]^{(3)} &= \boldsymbol{R}_{1;162}(t,1) = 1 - \prod_{i=1}^{1}\left[1 - \prod_{j=1}^{162}[R_{ij}^{(4)}(t,1)]^{(3)}\right] \\ &= \prod_{j=1}^{162}[R_{1j}^{(4)}(t,1)]^{(3)} = \exp\left[-\sum_{j=1}^{162}[\lambda_{1j}^{(3)}(1)]^{(3)}t\right] \\ &= \exp[-[0.001 \cdot 2 + 0.007 \cdot 160]t] \\ &= \exp[-1.122t]\end{aligned} \tag{9.202}$$

$$\begin{aligned}[\boldsymbol{R}^{(4)}(t,2)]^{(3)} &= \boldsymbol{R}_{1;162}(t,2) = 1 - \prod_{i=1}^{1}\left[1 - \prod_{j=1}^{162}[R_{ij}^{(4)}(t,2)]^{(3)}\right] \\ &= \prod_{j=1}^{162}[R_{1j}^{(4)}(t,2)]^{(3)} = \exp\left[-\sum_{j=1}^{162}[\lambda_{1j}^{(3)}(2)]^{(3)}t\right] \\ &= \exp[-[0.003 \cdot 2 + 0.009 \cdot 160]t] \\ &= \exp[-1.446t]\end{aligned} \tag{9.203}$$

Considering that the system at the operation state z_3 is a three-state series system composed of subsystems S_1, S_2, S_3 and S_4, after applying the formulae appearing after Definition 3.4 and the results (9.192)–(9.203), its conditional reliability function is given by

$$[\boldsymbol{R}(t,\cdot)]^{(3)} = [1, [\boldsymbol{R}(t,1)]^{(3)}, [\boldsymbol{R}(t,2)]^{(3)}], \quad t \geq 0 \tag{9.204}$$

where

$$\begin{aligned}[\boldsymbol{R}(t,1)]^{(3)} &= \overline{\boldsymbol{R}}_4(t,1) = [\boldsymbol{R}^{(1)}(t,1)]^{(3)}[\boldsymbol{R}^{(2)}(t,1)]^{(3)}[\boldsymbol{R}^{(3)}(t,1)]^{(3)}[\boldsymbol{R}^{(4)}(t,1)]^{(3)} \\ &= \exp[-16.4816t]\end{aligned} \tag{9.205}$$

$$\begin{aligned}[\boldsymbol{R}(t,2)]^{(3)} &= \overline{\boldsymbol{R}}_4(t,2) = [\boldsymbol{R}^{(1)}(t,2)]^{(3)}[\boldsymbol{R}^{(2)}(t,2)]^{(3)}[\boldsymbol{R}^{(3)}(t,2)]^{(3)}[\boldsymbol{R}^{(4)}(t,2)]^{(3)} \\ &= \exp[-20.2714t]\end{aligned} \tag{9.206}$$

The expected values and standard deviations of the system conditional lifetimes in the reliability state subsets $\{1,2\}$, $\{2\}$ at the operation state z_3, calculated from the results given by Eqs. (9.204) and (9.205), according to (9.40), are respectively

$$M_3(1) = \int_0^{\infty} [\boldsymbol{R}(t,1)]^{(3)} dt = 1/16.4816 \cong 0.0607 \text{ year} \tag{9.207}$$

$$M_3(2) = \int_0^{\infty} [\boldsymbol{R}(t,2)]^{(3)} dt = 1/20.2714 \cong 0.0493 \text{ year} \tag{9.208}$$

In the case when the system operation time is large enough its unconditional four-state reliability function is given by the vector

$$\boldsymbol{R}(t,\cdot) = [1, \boldsymbol{R}(t,1), \boldsymbol{R}(t,2)] \quad t \geq 0 \tag{9.209}$$

where, according to Eq. (9.38) and considering the system operation process transient probabilities at the operation states determined by Eq. (9.31), the vector coordinates are given respectively by

$$\begin{aligned}\boldsymbol{R}(t,1) &= p_1[\boldsymbol{R}(t,1)]^{(1)} + p_2[\boldsymbol{R}(t,1)]^{(2)} + p_3[\boldsymbol{R}(t,1)]^{(3)} \\ &= 0.530 \cdot [\boldsymbol{R}(t,1)]^{(1)} + 0.109 \cdot [\boldsymbol{R}(t,1)]^{(2)} + 0.361 \cdot [\boldsymbol{R}(t,1)]^{(3)} \quad \text{for } t \geq 0\end{aligned} \tag{9.210}$$

$$\begin{aligned}\boldsymbol{R}(t,2) &= p_1[\boldsymbol{R}(t,2)]^{(1)} + p_2[\boldsymbol{R}(t,2)]^{(2)} + p_3[\boldsymbol{R}(t,2)]^{(3)} \\ &= 0.530 \cdot [\boldsymbol{R}(t,2)]^{(1)} + 0.109 \cdot [\boldsymbol{R}(t,2)]^{(2)} + 0.361 \cdot [\boldsymbol{R}(t,2)]^{(3)} \quad \text{for } t \geq 0\end{aligned} \tag{9.211}$$

where the reliability functions $[\boldsymbol{R}(t,1)]^{(1)}$, $[\boldsymbol{R}(t,1)]^{(2)}$, $[\boldsymbol{R}(t,1)]^{(3)}$ are given by Eqs. (9.171), (9.188), (9.205) and $[\boldsymbol{R}(t,2)]^{(1)}$, $[\boldsymbol{R}(t,2)]^{(2)}$, $[\boldsymbol{R}(t,2)]^{(3)}$ are given by Eqs. (9.172), (9.189), (9.206).

In Table 9.5, the values of the coordinates of the port grain transportation system unconditional reliability function $\boldsymbol{R}(t,\cdot)$ given by Eqs. (9.210) and (9.211) are presented.

The graph of the three-state port grain transportation system unconditional reliability function is illustrated in Figure 9.15.

The expected value of the system unconditional lifetime in the reliability state subset $\{1,2\}$, calculated according to Eqs. (9.39) and (9.40) from the results given by Eqs. (9.31) and (9.210), is

$$\begin{aligned}M(1) &= p_1 M_1(1) + p_2 M_2(1) + p_3 M_3(1) \\ &= 0.530 \cdot 0.0813 + 0.109 \cdot 0.0861 + 0.361 \cdot 0.0607 \cong 0.0744 \text{ year}\end{aligned} \tag{9.212}$$

The expected value of the system unconditional lifetime in the reliability state subset $\{2\}$, calculated according to Eqs. (9.39) and (9.40) from the results given by Eqs. (9.31) and (9.211), is

$$\begin{aligned}M(2) &= p_1 M_1(2) + p_2 M_2(2) + p_3 M_3(2) \\ &= 0.530 \cdot 0.0691 + 0.109 \cdot 0.0728 + 0.361 \cdot 0.0493 \cong 0.0624 \text{ year}\end{aligned} \tag{9.213}$$

Table 9.5 The Values of the Port Grain Transportation System Unconditional Reliability Function Coordinates

t	$\boldsymbol{R}(t, 1)$	$\boldsymbol{R}(t, 2)$
0.00	1.0000	1.0000
0.02	0.8638	0.8302
0.04	0.6884	0.6191
0.06	0.5133	0.4259
0.08	0.3665	0.2793
0.10	0.2549	0.1782
0.12	0.1745	0.1121
0.14	0.1184	0.0699
0.16	0.0800	0.0435
0.18	0.0540	0.0271
0.20	0.0364	0.0169
0.22	0.0246	0.0106
0.24	0.0167	0.0066
0.26	0.0113	0.0042
0.28	0.0077	0.0027
0.30	0.0053	0.0017
0.32	0.0036	0.0011
0.34	0.0025	0.0007
0.36	0.0017	0.0004
0.38	0.0012	0.0003
0.40	0.0008	0.0002
0.42	0.0006	0.0001
0.43	0.0005	0.0001
0.44	0.0004	0.0001
0.45	0.0003	0.0001
0.46	0.0003	0.0001
0.48	0.0002	0.0000
0.50	0.0001	0.0000
0.52	0.0001	0.0000
0.54	0.0001	0.0000
0.56	0.0000	0.0000

Further, considering Eqs. (9.212) and (9.213) and applying Eq. (9.42), the mean values of the unconditional lifetimes in the particular reliability states 1, 2 respectively are

$$\overline{M}(1) = M(1) - M(2) = 0.0120 \text{ year}, \quad \overline{M}(2) = M(2) = 0.0624 \text{ year} \tag{9.214}$$

Since the critical reliability state is $r = 1$, then the system risk function, according to Eqs. (9.43) and (9.211), is given by (Figure 9.16)

$$\begin{aligned} \boldsymbol{r}(t) &= 1 - \boldsymbol{R}(t,1) \\ &= 0.530 \cdot [\boldsymbol{R}(t,1)]^{(1)} + 0.109 \cdot [\boldsymbol{R}(t,1)]^{(2)} + 0.361 \cdot [\boldsymbol{R}(t,1)]^{(3)} \quad \text{for } t \geq 0 \end{aligned} \tag{9.215}$$

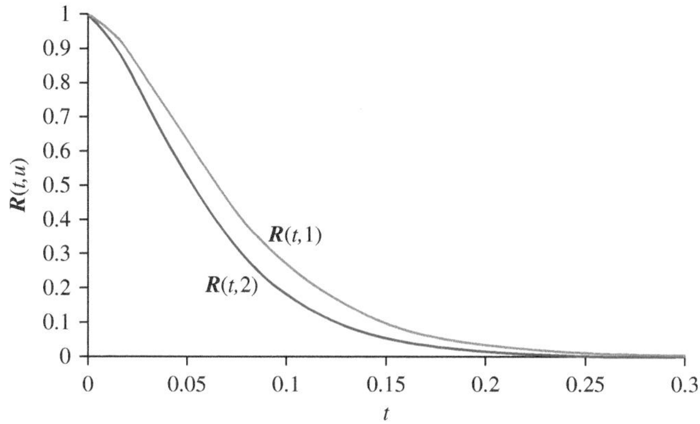

Figure 9.15 The graph of the port grain transportation system reliability function $\boldsymbol{R}(t,\cdot)$ coordinates.

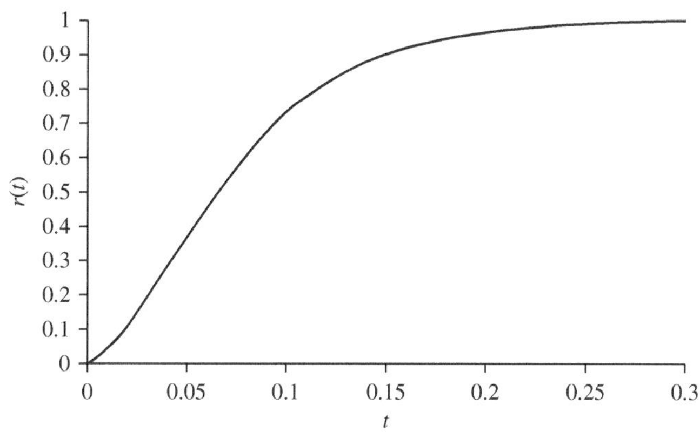

Figure 9.16 The graph of the port grain transportation system risk function $\boldsymbol{r}(t)$.

where $[\boldsymbol{R}(t,1)]^{(1)}$, $[\boldsymbol{R}(t,1)]^{(2)}$, $[\boldsymbol{R}(t,1)]^{(3)}$ are given by Eqs. (9.171), (9.188), (9.205).

Hence, by Eq. (9.44), the moment when the system risk function exceeds a permitted level, for instance $\delta = 0.05$, is

$$\tau = \boldsymbol{r}^{-1}(\delta) \cong 0.0100 \tag{9.216}$$

9.4 Complex System Reliability and Operation Optimization

Considering Eq. (9.38), it is natural to assume that the system operation process has a significant influence on the system reliability. This influence is also clearly expressed in Eq. (9.39) for the mean values of the system unconditional lifetimes in the reliability state subsets.

From the linear equation (9.39), we can see that the mean value of the system unconditional lifetime $M(u)$, $u = 1, 2, \ldots, z$, is determined by the limit values of transient probabilities p_b, $b = 1, 2, \ldots, \nu$, of the system operation process at the operation states given by Eq. (9.8) and the mean values $M_b(u)$, $b = 1, 2, \ldots, \nu$, $u = 1, 2, \ldots, z$, of the system conditional lifetimes in the reliability state subsets $\{u, u+1, \ldots, z\}$, $u = 1, 2, \ldots, z$, given by Eq. (9.40). Therefore, the system lifetime optimization approach based on the linear programming [61,100,105] can be proposed. Namely, we may look for the corresponding optimal values $\dot{p}_b$, $b = 1, 2, \ldots, \nu$, of the transient probabilities p_b, $b = 1, 2, \ldots, \nu$, of the system operation process at the operation states to maximize the mean value $M(u)$ of the unconditional system lifetimes in the reliability state subsets $\{u, u+1, \ldots, z\}$, $u = 1, 2, \ldots, z$, under the assumption that the mean values $M_b(u)$, $b = 1, 2, \ldots, \nu$, $u = 1, 2, \ldots, z$, of the system conditional lifetimes in the reliability state subsets are fixed. As a special and practically important case of the above formulated system lifetime optimization problem, if r, $r = 1, 2, \ldots, z$, is a system critical reliability state, we may look for the optimal values $\dot{p}_b$, $b = 1, 2, \ldots, \nu$, of the transient probabilities p_b, $b = 1, 2, \ldots, \nu$, of the system operation process at the system operation states to maximize the mean value $M(r)$ of the unconditional system lifetime in the reliability state subset $\{r, r+1, \ldots, z\}$, $r = 1, 2, \ldots, z$, under the assumption that the mean values $M_b(r)$, $b = 1, 2, \ldots, \nu$, $r = 1, 2, \ldots, z$, of the system conditional lifetimes in this reliability state subset are fixed. More exactly, we may formulate the optimization problem as a linear programming model with the objective function of the following form:

$$M(r) = \sum_{b=1}^{\nu} p_b M_b(r) \tag{9.217}$$

for a fixed $r \in \{1, 2, \ldots, z\}$ and with the following bound constraints:

$$\breve{p}_b \le p_b \le \hat{p}_b, \quad b = 1, 2, \ldots, \nu \tag{9.218}$$

$$\sum_{b=1}^{\nu} p_b = 1 \tag{9.219}$$

where

$$M_b(r), \ M_b(r) \ge 0, \quad b = 1, 2, \ldots, \nu \tag{9.220}$$

are fixed mean values of the system conditional lifetimes in the reliability state subset $\{r, r+1, \ldots, z\}$ and

$$\breve{p}_b, \ 0 \le \breve{p}_b \le 1 \text{ and } \hat{p}_b, \ 0 \le \hat{p}_b \le 1, \ \breve{p}_b \le \hat{p}_b, \quad b = 1, 2, \ldots, \nu \tag{9.221}$$

are lower and upper bounds of the unknown transient probabilities p_b, $b = 1, 2, \ldots, \nu$, respectively.

Now, we may look for the optimal solution of the linear programming problem formulated by (9.217)–(9.221), i.e. we can find the optimal values $\dot{p}_b$ of the transient probabilities p_b, $b = 1, 2, \ldots, \nu$, that maximize the objective function given by Eq. (9.217).

First, we arrange the system conditional lifetime mean values $M_b(r)$, $b = 1, 2, \ldots, \nu$, in non-increasing order

$$M_{b_1}(r) \geq M_{b_2}(r) \geq \cdots \geq M_{b_\nu}(r), \quad \text{where } b_i \in \{1, 2, \ldots, \nu\} \text{ for } i = 1, 2, \ldots, \nu$$

Next, we substitute

$$x_i = p_{b_i}, \ \breve{x}_i = \breve{p}_{b_i}, \ \hat{x}_i = \hat{p}_{b_i} \quad \text{for } i = 1, 2, \ldots, \nu \tag{9.222}$$

and we maximize with respect to x_i, $i = 1, 2, \ldots, \nu$, the linear form (9.217) that, after this transformation, takes the form

$$M(r) = \sum_{i=1}^{\nu} x_i M_{b_i}(r) \tag{9.223}$$

for a fixed $r \in \{1, 2, \ldots, z\}$ with the following bound constraints

$$\breve{x}_i \leq x_i \leq \hat{x}_i, \quad i = 1, 2, \ldots, \nu \tag{9.224}$$

$$\sum_{i=1}^{\nu} x_i = 1 \tag{9.225}$$

where

$$M_{b_i}(r), \ M_{b_i}(r) \geq 0, \quad i = 1, 2, \ldots, \nu$$

are fixed mean values of the system conditional lifetimes in the reliability state subset $\{r, r+1, \ldots, z\}$ arranged in non-increasing order and

$$\breve{x}_i, \ 0 \leq \breve{x}_i \leq 1 \text{ and } \hat{x}_i, \ 0 \leq \hat{x}_i \leq 1, \ \breve{x}_i \leq \hat{x}_i, \quad i = 1, 2, \ldots, \nu \tag{9.226}$$

are lower and upper bounds of the unknown probabilities x_i, $i = 1, 2, \ldots, \nu$, respectively.

To find the optimal values of x_i, $i = 1, 2, \ldots, \nu$, we define

$$\breve{x} = \sum_{i=1}^{\nu} \breve{x}_i, \ \hat{y} = 1 - \breve{x} \tag{9.227}$$

and

$$\breve{x}^0 = 0, \ \hat{x}^0 = 0 \text{ and } \breve{x}^I = \sum_{i=1}^{I} \breve{x}_i, \ \hat{x}^I = \sum_{i=1}^{I} \hat{x}_i \quad \text{for } I = 1, 2, \ldots, \nu \tag{9.228}$$

Next, we find the largest value $I \in \{0, 1, \ldots, \nu\}$ such that

$$\widehat{x}^{I} - \breve{x}^{I} < \hat{y} \tag{9.229}$$

and we fix the optimal solution that maximizes Eq. (9.223) in the following way:

i. if $I = 0$, the optimal solution is

$$\dot{x}_1 = \hat{y} + \breve{x}_1 \text{ and } \dot{x}_i = \breve{x}_i \quad \text{for } i = 2, 3, \ldots, \nu \tag{9.230}$$

ii. if $0 < I < \nu$, the optimal solution is

$$\dot{x}_i = \widehat{x}_i \quad \text{for } i = 1, 2, \ldots, I, \ \dot{x}_{I+1} = \hat{y} - \widehat{x}^{I} + \breve{x}^{I} + \breve{x}_{I+1} \text{ and } \dot{x}_i = \breve{x}_i \quad \text{for } i = I+2, I+3, \ldots, \nu \tag{9.231}$$

iii. if $I = \nu$, the optimal solution is

$$\dot{x}_i = \widehat{x}_i, \quad \text{for } i = 1, 2, \ldots, \nu \tag{9.232}$$

Finally, after making the inverse to Eq. (9.222) substitution, we get the optimal limit transient probabilities

$$\dot{p}_{b_i} = \dot{x}_i, \quad \text{for } i = 1, 2, \ldots, \nu \tag{9.233}$$

that maximize the system mean lifetime in the reliability state subset $\{r, r+1, \ldots, z\}$, defined by the linear form (9.217), giving its maximum value in the following form:

$$\dot{M}(r) = \sum_{b=1}^{\nu} \dot{p}_b M_b(r) \tag{9.234}$$

for a fixed $r \in \{1, 2, \ldots, z\}$.

From the expression (9.234) for the maximum mean value $\dot{M}(r)$ of the system unconditional lifetime in the reliability state subset $\{r, r+1, \ldots, z\}$, replacing in it the critical reliability state r with the reliability state u, $u = 1, 2, \ldots, z$, we obtain the corresponding optimal solutions for the mean values of the system unconditional lifetimes in the reliability state subsets $\{u, u+1, \ldots, z\}$ of the form

$$\dot{M}(u) = \sum_{b=1}^{\nu} \dot{p}_b M_b(u) \quad \text{for } u = 1, 2, \ldots, z \tag{9.235}$$

Further, according to Eqs. (9.37) and (9.38), the corresponding optimal unconditional multi-state reliability function of the system is the vector

$$\dot{\boldsymbol{R}}(t, \cdot) = [1, \dot{\boldsymbol{R}}(t, 1), \ldots, \dot{\boldsymbol{R}}(t, z)] \tag{9.236}$$

with the coordinates given by

$$\dot{\boldsymbol{R}}(t,u) \cong \sum_{b=1}^{v} \dot{p}_b[\boldsymbol{R}(t,u)]^{(b)} \quad \text{for } t \geq 0, \quad u = 1,2,\ldots,z \tag{9.237}$$

By applying Eq. (9.41), the corresponding optimal values of the variances of the system unconditional lifetimes in the system reliability state subsets are

$$\dot{\sigma}^2(u) = 2\int_0^\infty t\dot{\boldsymbol{R}}(t,u)\mathrm{d}t - [\dot{M}(u)]^2, \quad u = 1,2,\ldots,z \tag{9.238}$$

where $\dot{M}(u)$ is given by Eq. (9.235) and $\dot{\boldsymbol{R}}(t,u)$ is given by Eq. (9.237).

And, by Eq. (9.42), the optimal solutions for the mean values of the system unconditional lifetimes in the particular reliability states are

$$\dot{\overline{M}}(u) = \dot{M}(u) - \dot{M}(u+1), \quad u = 1,\ldots,z-1; \; \dot{\overline{M}}(z) = \dot{M}(z) \tag{9.239}$$

Moreover, considering Eqs. (9.43) and (9.44), the corresponding optimal system risk function and the optimal moment when the risk exceeds a permitted level δ are respectively given by

$$\dot{\boldsymbol{r}}(t) = 1 - \dot{\boldsymbol{R}}(t,r), \quad t \geq 0 \tag{9.240}$$

and

$$\dot{\tau} = \dot{\boldsymbol{r}}^{-1}(\delta) \tag{9.241}$$

where $\dot{\boldsymbol{R}}(t,r)$ is given by Eq. (9.238) for $u = r$ and $\dot{\boldsymbol{r}}^{-1}(t)$, if it exists, is the inverse function of the optimal risk function $\dot{\boldsymbol{r}}(t)$.

Replacing in Eq. (9.8) the limit transient probabilities p_b of the system operation process at the operation states with their optimal values $\dot{p}_b$, maximizing the mean value $M_b(r)$ of the system lifetime in the reliability states subset $\{r, r+1,\ldots,z\}$ defined by Eq. (9.217) and the mean values m_b of the unconditional sojourn times at the operation states by their corresponding unknown optimal values $\dot{m}_b$, we get the system of equations

$$\dot{p}_b = \frac{\pi_b \dot{m}_b}{\sum_{l=1}^{v} \pi_l \dot{m}_l}, \quad b = 1,2,\ldots,v \tag{9.242}$$

After simple transformations, the above system takes the form

$$\begin{aligned}
&(\dot{p}_1 - 1)\pi_1\dot{m}_1 + \dot{p}_1\pi_2\dot{m}_2 + \cdots + \dot{p}_1\pi_\nu\dot{m}_\nu = 0,\\
&\dot{p}_2\pi_1\dot{m}_1 + (\dot{p}_2 - 1)\pi_2\dot{m}_2 + \cdots + \dot{p}_2\pi_\nu\dot{m}_\nu = 0,\\
&\qquad\vdots\\
&\dot{p}_\nu\pi_1\dot{m}_1 + \dot{p}_\nu\pi_2\dot{m}_2 + \cdots + (\dot{p}_\nu - 1)\pi_\nu\dot{m}_\nu = 0
\end{aligned} \tag{9.243}$$

where $\dot{m}_b$ are unknown variables we want to find, $\dot{p}_b$ are optimal transient probabilities determined by Eq. (9.233) and π_b are steady probabilities determined by Eq. (9.9).

Since the system of equations (9.243) is homogeneous and it can be proved that the determinant of its main matrix is equal to zero, then it has non-zero solutions and moreover, these solutions are ambiguous. Thus, if we fix some of the optimal values $\dot{m}_b$ of the mean values m_b of the unconditional sojourn times at the operation states, for instance by arbitrarily fixing one or a few of them, we may find the values of the remaining once and this way get the solution of this equation.

Having this solution, it is also possible to look for the optimal values $\dot{m}_{bl}$ of the mean values m_{bl} of the conditional sojourn times at the operation states using the following system of equations:

$$\sum_{l=1}^{v} p_{bl}\dot{m}_{bl} = \dot{m}_b, \quad b = 1, 2, \ldots, v \tag{9.244}$$

obtained from Eq. (9.7) by replacing m_b by $\dot{m}_b$ and m_{bl} by $\dot{m}_{bl}$, where p_{bl} are known probabilities of the system operation process transitions between the operation states z_b and z_l, $b, l = 1, 2, \ldots, v$, $b \neq l$, defined by Eq. (9.2).

Another very useful set of tools, which are much more easily applied in practice and which can help in planning the operation processes of the complex technical systems, are the system operation process optimal mean values of the total system operation process sojourn times $\hat{\theta}_b$ at the particular operation states z_b, $b = 1, 2, \ldots, v$, during the fixed system operation time θ, that can be obtained by the replacement in the formula (9.10) of the transient probabilities p_b at the operation states z_b by their optimal values $\dot{p}_b$ and resulting in the following expression:

$$\dot{\hat{m}}_b = \dot{E}[\hat{\theta}_b] = \dot{p}_b\theta, \quad b = 1, 2, \ldots, v \tag{9.245}$$

The knowledge of the optimal values $\dot{m}_b$ of the mean values of the unconditional sojourn times and the optimal values $\dot{m}_{bl}$ of the mean values of the conditional sojourn times at the operation states and the optimal mean values $\dot{\hat{m}}_b$ of the total sojourn times at the particular operation states during the fixed system operation time may be the basis for changing the complex technical systems operation processes in order to ensure these systems operate more reliably.

Example 9.1 (Continuation)

The objective function defined by Eq. (9.217), in this case, as the exemplary system critical state is $r = 2$, according to Eq. (9.117), takes the form

$$M(2) = p_1 \cdot 50.00 + p_2 \cdot 29.76 + p_3 \cdot 26.08 + p_4 \cdot 14.08 \tag{9.246}$$

Arbitrarily assumed, the lower $\breve{p}_b$ and upper $\widehat{p}_b$ bounds of the unknown optimal values of transient probabilities p_b, $b = 1, 2, 3, 4$, are respectively

$$\breve{p}_1 = 0.201, \ \breve{p}_2 = 0.03, \ \breve{p}_3 = 0.245, \ \breve{p}_4 = 0.309,$$
$$\widehat{p}_1 = 0.351, \ \widehat{p}_2 = 0.105, \ \widehat{p}_3 = 0.395, \ \widehat{p}_4 = 0.459$$

Therefore, according to Eqs. (9.218) and (9.219), we assume the following bound constraints:

$$0.201 \le p_1 \le 0.351, 0.030 \le p_2 \le 0.105, 0.245 \le p_3 \le 0.395, 0.309 \le p_4 \le 0.459 \tag{9.247}$$

$$\sum_{b=1}^{4} p_b = 1 \tag{9.248}$$

Now, before we find optimal values $\dot{p}_b$ of the transient probabilities p_b, $b = 1, 2, 3, 4$, that maximize the objective function (9.246), we arrange the system conditional lifetime mean values $M_b(2)$, $b = 1, 2, 3, 4$, in non-increasing order

$$M_1(2) \ge M_2(2) \ge M_3(2) \ge M_4(2)$$

Further, according to Eq. (9.222), we substitute

$$x_1 = p_1, \quad x_2 = p_2, \quad x_3 = p_3, \quad x_4 = p_4 \tag{9.249}$$

and

$$\breve{x}_1 = \breve{p}_1 = 0.201, \quad \breve{x}_2 = \breve{p}_2 = 0.030, \quad \breve{x}_3 = \breve{p}_3 = 0.245, \quad \breve{x}_4 = \breve{p}_4 = 0.309 \tag{9.250}$$

$$\widehat{x}_1 = \widehat{p}_1 = 0.351, \quad \widehat{x}_2 = \widehat{p}_2 = 0.105, \quad \widehat{x}_3 = \widehat{p}_3 = 0.395, \quad \widehat{x}_4 = \widehat{p}_4 = 0.459 \tag{9.251}$$

and we maximize with respect to x_i, $i = 1, 2, 3, 4$, the linear form (9.246) that, according to Eqs. (9.223)–(9.225), takes the form

$$M(2) = x_1 \cdot 50.00 + x_2 \cdot 29.76 + x_3 \cdot 26.08 + x_4 \cdot 14.08 \tag{9.252}$$

with the following bound constraints

$$0.201 \le x_1 \le 0.351, \quad 0.030 \le x_2 \le 0.105, \quad 0.245 \le x_3 \le 0.395, \quad 0.309 \le x_4 \le 0.459 \tag{9.253}$$

$$\sum_{i=1}^{4} x_i = 1 \tag{9.254}$$

According to Eq. (9.227), we calculate

$$\breve{x} = \sum_{i=1}^{4} \breve{x}_i = 0.785, \quad \hat{y} = 1 - \breve{x} = 1 - 0.785 = 0.215 \tag{9.255}$$

and according to Eq. (9.228), we determine

$$
\begin{aligned}
&\overset{\smile}{x}^{0} = 0,\ \overset{\frown}{x}^{0} = 0,\ \overset{\frown}{x}^{0} - \overset{\smile}{x}^{0} = 0,\\
&\overset{\smile}{x}^{1} = 0.201,\ \overset{\frown}{x}^{1} = 0.351,\ \overset{\frown}{x}^{1} - \overset{\smile}{x}^{1} = 0.150,\\
&\overset{\smile}{x}^{2} = 0.231,\ \overset{\frown}{x}^{2} = 0.456,\ \overset{\frown}{x}^{2} - \overset{\smile}{x}^{2} = 0.225,\\
&\overset{\smile}{x}^{3} = 0.476,\ \overset{\frown}{x}^{3} = 0.851,\ \overset{\frown}{x}^{3} - \overset{\smile}{x}^{3} = 0.375,\\
&\overset{\smile}{x}^{4} = 0.785,\ \overset{\frown}{x}^{4} = 1.31,\ \overset{\frown}{x}^{4} - \overset{\smile}{x}^{4} = 0.525
\end{aligned}
\tag{9.256}
$$

From the above, as according to Eq. (9.255), the inequality (9.229) takes the form

$$\overset{\frown}{x}^{I} - \overset{\smile}{x}^{I} < 0.215 \tag{9.257}$$

which follows that the largest value $I \in \{0, 1, \ldots, 4\}$ such that this inequality holds is $I = 1$.

Therefore, we fix the optimal solution that maximizes linear function (9.252) according to the rule (9.231). Namely, we get

$$
\begin{aligned}
&\dot{x}_1 = \overset{\frown}{x}_1 = 0.351,\\
&\dot{x}_2 = \hat{y} - \overset{\frown}{x}^{1} + \overset{\smile}{x}^{1} + \overset{\smile}{x}_2 = 0.215 - 0.351 + 0.201 + 0.030 = 0.095,\\
&\dot{x}_3 = \overset{\smile}{x}_3 = 0.245, \quad \dot{x}_4 = \overset{\smile}{x}_4 = 0.309
\end{aligned}
\tag{9.258}
$$

Finally, after making the inverse to Eq. (9.249) substitution, we get the optimal transient probabilities

$$\dot{p}_1 = \dot{x}_1 = 0.351,\ \dot{p}_2 = \dot{x}_2 = 0.095,\ \dot{p}_3 = \dot{x}_3 = 0.245,\ \dot{p}_4 = \dot{x}_4 = 0.309 \tag{9.259}$$

that maximize the exemplary system mean lifetime $M(2)$ in the reliability state subset $\{2, 3\}$ expressed by the linear form (9.246) giving, according to Eqs. (9.234) and (9.259), its optimal value

$$
\begin{aligned}
\dot{M}(2) &= \dot{p}_1 \cdot 50.00 + \dot{p}_2 \cdot 29.76 + \dot{p}_3 \cdot 26.08 + \dot{p}_4 \cdot 14.08\\
&= 0.351 \cdot 50.00 + 0.095 \cdot 29.76 + 0.245 \cdot 26.08 + 0.309 \cdot 14.08 \cong 31.12
\end{aligned}
\tag{9.260}
$$

Substituting the optimal solution (9.259) into the formula (9.235), we obtain the optimal solution for the mean values of the exemplary system unconditional lifetimes in the reliability state subsets $\{1, 2, 3\}$ and $\{3\}$,that are as follows:

$$
\begin{aligned}
\dot{M}(1) &= \dot{p}_1 \cdot 55.56 + \dot{p}_2 \cdot 32.55 + \dot{p}_3 \cdot 29.65 + \dot{p}_4 \cdot 15.04\\
&= 0.351 \cdot 55.56 + 0.095 \cdot 32.55 + 0.245 \cdot 29.65 + 0.309 \cdot 15.44 \cong 34.63
\end{aligned}
\tag{9.261}
$$

$$\dot{M}(3) = \dot{p}_1 \cdot 45.45 + \dot{p}_2 \cdot 27.41 + \dot{p}_3 \cdot 22.97 + \dot{p}_4 \cdot 12.95$$
$$= 0.351 \cdot 45.45 + 0.095 \cdot 27.41 + 0.245 \cdot 22.97 + 0.309 \cdot 12.95 \cong 28.19 \tag{9.262}$$

and according to Eq. (9.239), the optimal values of the mean values of the system unconditional lifetimes in the particular reliability states 1, 2 and 3 are respectively

$$\dot{\bar{M}}(1) = \dot{M}(1) - \dot{M}(2) = 3.51,\ \dot{\bar{M}}(2) = \dot{M}(2) - \dot{M}(3) = 2.93,\ \dot{\bar{M}}(3) = \dot{M}(3) = 28.19 \tag{9.263}$$

Moreover, according to Eqs. (9.236) and (9.237), the corresponding optimal unconditional multi-state reliability function of the system is of the form

$$\dot{\boldsymbol{R}}(t,\cdot) = [1 - \dot{\boldsymbol{R}}(t,1), \dot{\boldsymbol{R}}(t,2), \dot{\boldsymbol{R}}(t,3)], \quad t \geq 0 \tag{9.264}$$

with the coordinates given by

$$\dot{\boldsymbol{R}}(t,1) = 0.351 \cdot [\boldsymbol{R}(t,1)]^{(1)} + 0.095 \cdot [\boldsymbol{R}(t,1)]^{(2)} + 0.245 \cdot [\boldsymbol{R}(t,1)]^{(3)} + 0.309 \cdot [\boldsymbol{R}(t,1)]^{(4)} \quad \text{for } t \geq 0 \tag{9.265}$$

$$\dot{\boldsymbol{R}}(t,2) = 0.351 \cdot [\boldsymbol{R}(t,1)]^{(1)} + 0.095 \cdot [\boldsymbol{R}(t,2)]^{(2)} + 0.245 \cdot [\boldsymbol{R}(t,2)]^{(3)} + 0.309 \cdot [\boldsymbol{R}(t,2)]^{(4)} \quad \text{for } t \geq 0 \tag{9.266}$$

$$\dot{\boldsymbol{R}}(t,3) = 0.351 \cdot [\boldsymbol{R}(t,3)]^{(1)} + 0.0095 \cdot [\boldsymbol{R}(t,3)]^{(2)} + 0.245 \cdot [\boldsymbol{R}(t,3)]^{(3)} + 0.309 \cdot [\boldsymbol{R}(t,3)]^{(4)} \quad \text{for } t \geq 0 \tag{9.267}$$

where $[\boldsymbol{R}(t,1)]^{(b)}$, $[\boldsymbol{R}(t,2)]^{(b)}$, $[\boldsymbol{R}(t,3)]^{(b)}$, $b = 1,2,3,4$, are fixed in Section 9.3.

The graph of the exemplary system optimal reliability function $\dot{\boldsymbol{R}}(t,\cdot)$ given by Eqs. (9.264)–(9.267) is presented in Figure 9.17.

As the critical reliability state is $r = 2$, then the exemplary system optimal system risk function, according to Eq. (9.240), is given by (Figure 9.18)

$$\dot{\boldsymbol{r}}(t) = 1 - \dot{\boldsymbol{R}}(t,2) \quad \text{for } t \geq 0 \tag{9.268}$$

where $\dot{\boldsymbol{R}}(t,2)$ is given by Eq. (9.266).

Hence, and considering Eq. (9.241), the moment when the optimal system risk function exceeds a permitted level, for instance $\delta = 0.025$, is

$$\dot{\tau} = \dot{\boldsymbol{r}}^{-1}(\delta) \cong 5.099 \tag{9.269}$$

Substituting the exemplary operation process optimal transient probabilities at operation states

$$\dot{p}_1 = 0.351,\ \dot{p}_2 = 0.095,\ \dot{p}_3 = 0.245,\ \dot{p}_4 = 0.309$$

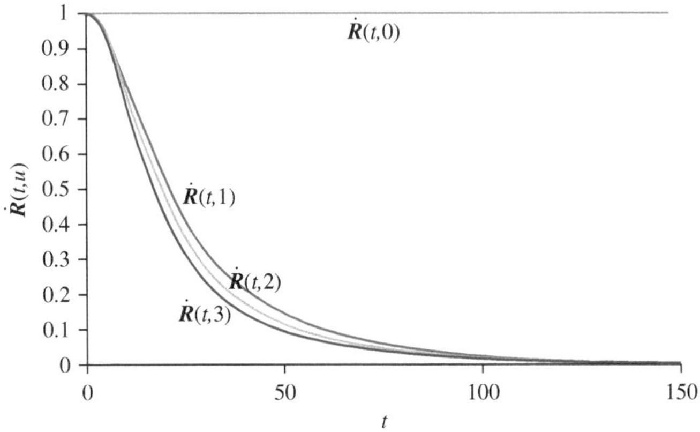

Figure 9.17 The graph of the exemplary system optimal reliability function $\dot{\boldsymbol{R}}(t,\cdot)$ coordinates.

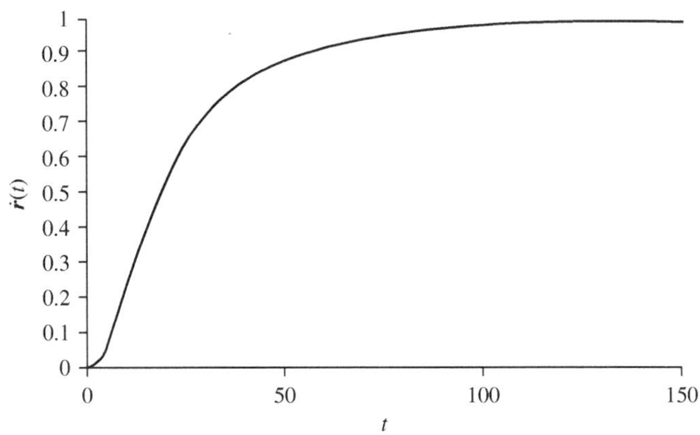

Figure 9.18 The graph of the exemplary system optimal risk function $\dot{\boldsymbol{r}}(t)$.

determined by Eq. (9.259) and the steady probabilities

$$\pi_1 \cong 0.236, \quad \pi_2 \cong 0.169, \quad \pi_3 \cong 0.234, \quad \pi_4 \cong 0.361$$

determined by Eq. (9.17) into Eq. (9.243), we get the following system of equations with the unknown optimal mean values $\dot{m}_b$ of the exemplary system operation process unconditional sojourn times at the operation states we are looking for:

$$\begin{aligned} -0.153164\dot{m}_1 + 0.059319\dot{m}_2 + 0.082134\dot{m}_3 + 0.126711\dot{m}_4 &= 0, \\ 0.02242\dot{m}_1 - 0.152945\dot{m}_2 + 0.02223\dot{m}_3 + 0.034295\dot{m}_4 &= 0, \\ 0.05782\dot{m}_1 + 0.041405\dot{m}_2 - 0.17667\dot{m}_3 + 0.088445\dot{m}_4 &= 0, \\ 0.072924\dot{m}_1 + 0.052221\dot{m}_2 + 0.072306\dot{m}_3 - 0.249451\dot{m}_4 &= 0 \end{aligned} \tag{9.270}$$

The determinant of the main matrix of the above homogeneous system of equations is equal to zero and therefore there are non-zero solutions of this system of equations that are ambiguous and dependent on one or more parameters. Thus, we may fix some of them and determine the remaining ones. To show the way of solving this system of equations, we may suppose that we are arbitrarily interested in fixing the value of $\dot{m}_4$ and we put

$$\dot{m}_4 = 400$$

Substituting the above value into the system of equations (9.270), we get

$$\begin{aligned}
-0.153164\dot{m}_1 + 0.059319\dot{m}_2 + 0.082134\dot{m}_3 &= -50.6844,\\
0.02242\dot{m}_1 - 0.152945\dot{m}_2 + 0.02223\dot{m}_3 &= -13.7180,\\
0.05782\dot{m}_1 + 0.041405\dot{m}_2 - 0.17667\dot{m}_3 &= -35.3780,\\
0.072924\dot{m}_1 + 0.052221\dot{m}_2 + 0.072306\dot{m}_3 &= 99.7804
\end{aligned}$$

and we solve it with respect to $\dot{m}_1$, $\dot{m}_2$ and $\dot{m}_3$, after omitting its last equation. This way, we obtain the following solutions that satisfy Eq. (9.270)

$$\dot{m}_1 \cong 689, \quad \dot{m}_2 \cong 261, \quad \dot{m}_3 \cong 487, \quad \dot{m}_4 = 400 \tag{9.271}$$

It can be seen that these solutions differ greatly from the values m_1, m_2, m_3 and m_4 estimated in Section 9.2 and given by Eq. (9.13)–(9.16).

Having these solutions, it is also possible to look for the optimal values $\dot{m}_{bl}$ of the mean values m_{bl} of the exemplary system operation process conditional sojourn times at operation states. Namely, substituting the values $\dot{m}_b$ instead of m_b, the probabilities

$$[p_{bl}] = \begin{bmatrix} 0 & 0.22 & 0.32 & 0.46 \\ 0.20 & 0 & 0.30 & 0.50 \\ 0.12 & 0.16 & 0 & 0.72 \\ 0.48 & 0.22 & 0.30 & 0 \end{bmatrix}$$

of the exemplary system operation process transitions between the operation states given by Eq. (9.11) and replacing m_{bl} by $\dot{m}_{bl}$ in Eq. (9.244), we get the following system of equations:

$$\begin{aligned}
0.22\dot{m}_{12} + 0.32\dot{m}_{13} + 0.46\dot{m}_{14} &= 689,\\
0.20\dot{m}_{21} + 0.30\dot{m}_{23} + 0.50\dot{m}_{24} &= 261,\\
0.12\dot{m}_{31} + 0.16\dot{m}_{32} + 0.72\dot{m}_{34} &= 487,\\
0.48\dot{m}_{41} + 0.2\dot{2}_{42} + 0.30\dot{m}_{14} &= 400
\end{aligned} \tag{9.272}$$

with the unknown optimal values $\dot{m}_{bl}$ we want to find.

As the solutions of the above system of equations are ambiguous, then we fix some of them, say that because of practically important reasons, and we find the remaining ones. For instance,

$$\begin{aligned}
&\text{we fix in the first equation } \dot{m}_{12} = 200,\ \ \dot{m}_{13} = 500 \text{ and we find } \dot{m}_{14} \cong 1054\\
&\text{we fix in the second equation } \dot{m}_{21} = 100,\ \dot{m}_{23} = 100 \text{ and we find } \dot{m}_{24} \cong 422\\
&\text{we fix in the third equation } \dot{m}_{31} = 900,\ \dot{m}_{32} = 500 \text{ and we find } \dot{m}_{34} \cong 415\\
&\text{we fix in the fourth equation } \dot{m}_{41} = 300,\ \dot{m}_{42} = 500 \text{ and we find } \dot{m}_{43} \cong 487
\end{aligned} \tag{9.273}$$

It can be seen that these solutions differ greatly from the mean values of the exemplary system conditional sojourn times at the particular operation states before its operation process optimization given by Eq. (9.12).

Another very useful set of tools that is much easier to apply in practice and which can help in planning the operation process of the exemplary system are the system operation process optimal mean values of the total sojourn times at the particular operation states during the system operation time that, by the same assumption as in Section 9.2, are equal to $\theta = 1$ year $= 365$ days. Under this assumption, after applying Eq. (9.245), we get the optimal values of the exemplary system operation process total sojourn times at the particular operation states during 1 year:

$$\begin{aligned}
\dot{\hat{m}}_1 &= \dot{E}[\hat{\theta}_1] = \dot{p}_1\theta = 0.341 \cdot 365 \cong 124.5,\\
\dot{\hat{m}}_2 &= \dot{E}[\hat{\theta}_2] = \dot{p}_2\theta = 0.105 \cdot 365 \cong 38.3,\\
\dot{\hat{m}}_3 &= \dot{E}[\hat{\theta}_3] = \dot{p}_3\theta = 0.245 \cdot 365 \cong 89.4,\\
\dot{\hat{m}}_4 &= \dot{E}[\hat{\theta}_4] = \dot{p}_4\theta = 0.309 \cdot 365 \cong 112.8
\end{aligned} \tag{9.274}$$

that differ greatly from the values of $\hat{m}_1, \hat{m}_2, \hat{m}_3, \hat{m}_4$, determined by Eq. (9.19).

In practice, the knowledge of the optimal values of $\dot{m}_b$, $\dot{m}_{bl}$ and $\dot{\hat{m}}_b$ given respectively by Eqs. (9.271), (9.273) and (9.274) can be very important and helpful in planning and improving the system operation process in order to make the system operation more reliable.

Example 9.2 (Continuation)

The objective function defined by Eq. (9.217), in this case, as the '24 out of 36'−series system critical state is $r = 2$, according to Eq. (9.153), takes the form

$$M(2) = p_1 \cdot 0.0567 + p_2 \cdot 0.0929 + p_3 \cdot 0.2201 \tag{9.275}$$

Arbitrarily assumed, the lower $\breve{p}_b$ and upper $\frown{p}_b$ bounds of the unknown optimal values of transient probabilities p_b, $b = 1, 2, 3$, are respectively

$$\begin{aligned}
&\breve{p}_1 = 0.3,\ \ \breve{p}_2 = 0.3,\ \ \breve{p}_3 = 0.1,\\
&\frown{p}_1 = 0.5,\ \ \frown{p}_2 = 0.5,\ \ \frown{p}_3 = 0.3
\end{aligned}$$

Therefore, according to Eqs. (9.218) and (9.219), we assume the following bound constraints

$$0.3 \le p_1 \le 0.4,\ \ 0.3 \le p_2 \le 0.4,\ \ 0.1 \le p_3 \le 0.3 \tag{9.276}$$

$$\sum_{b=1}^{3} p_b = 1 \tag{9.277}$$

Now, before we find optimal values $\dot{p}_b$ of the transient probabilities p_b, $b = 1, 2, 3$, that maximize the objective function (9.275), we arrange the system conditional lifetime mean values $M_b(2)$, $b = 1, 2, 3$, in non-increasing order

$$M_3(2) \ge M_2(2) \ge M_1(2)$$

Further, according to Eq. (9.222), we substitute

$$x_1 = p_1,\ \ x_2 = p_2,\ \ x_3 = p_1 \tag{9.278}$$

and

$$\breve{x}_1 = \breve{p}_3 = 0.1,\ \ \breve{x}_2 = \breve{p}_2 = 0.3,\ \ \breve{x}_3 = \breve{p}_1 = 0.3 \tag{9.279}$$

$$\widehat{x}_1 = \widehat{p}_3 = 0.3,\ \ \widehat{x}_2 = \widehat{p}_2 = 0.5,\ \ \widehat{x}_3 = \widehat{p}_3 = 0.5 \tag{9.280}$$

and we maximize with respect to x_i, $i = 1, 2, 3$, the linear form (9.275) that, according to Eqs. (9.223)–(9.225), takes the form

$$M(2) = x_1 \cdot 0.2201 + x_2 \cdot 0.0929 + x_3 \cdot 0.0567 \tag{9.281}$$

with the following bound constraints

$$0.1 \le x_1 \le 0.3,\ \ 0.3 \le x_2 \le 0.5,\ \ 0.3 \le x_3 \le 0.5 \tag{9.282}$$

$$\sum_{i=1}^{3} x_i = 1 \tag{9.283}$$

According to Eq. (9.227), we calculate

$$\breve{x} = \sum_{i=1} \breve{x}_i = 0.7,\ \ \hat{y} = 1 - \breve{x} = 1 - 0.7 = 0.3 \tag{9.284}$$

and according to Eq. (9.228), we determine

$$\begin{aligned}
&\breve{x}^0 = 0,\ \widehat{x}^0 = 0,\ \widehat{x}^0 - \breve{x}^0 = 0,\\
&\breve{x}^1 = 0.1,\ \widehat{x}^1 = 0.3,\ \widehat{x}^1 - \breve{x}^1 = 0.2,\\
&\breve{x}^2 = 0.4,\ \widehat{x}^2 = 0.8,\ \widehat{x}^2 - \breve{x}^2 = 0.4,\\
&\breve{x}^3 = 0.7,\ \widehat{x}^3 = 1.3,\ \widehat{x}^3 - \breve{x}^3 = 0.6
\end{aligned} \tag{9.285}$$

From the above, as according to Eq. (9.284), the inequality (9.229) takes the form

$$\widehat{x}^I - \breve{x}^I < 0.3 \tag{9.286}$$

which follows that the largest value $I \in \{0, 1, 2, 3\}$ such that this inequality holds is $I = 1$.

Therefore, we fix the optimal solution that maximizes linear function (9.281) according to the rule (9.231). Namely, we get

$$\begin{aligned}
&\dot{x}_1 = \widehat{x}_1 = 0.3,\\
&\dot{x}_2 = \hat{y} - \widehat{x}^1 + \breve{x}^1 + \breve{x}_2 = 0.3 - 0.3 + 0.1 + 0.3 = 0.4,\\
&\dot{x}_3 = \breve{x}_3 = 0.3
\end{aligned} \tag{9.287}$$

Finally, after making the inverse to Eq. (9.278) substitution, we get the optimal transient probabilities

$$\dot{p}_1 = \dot{x}_3 = 0.3,\ \ \dot{p}_2 = \dot{x}_2 = 0.4,\ \ \dot{p}_3 = \dot{x}_1 = 0.3 \tag{9.288}$$

that maximize the exemplary system mean lifetime $M(2)$ in the reliability state subset $\{2, 3\}$ expressed by the linear form (9.275) giving, according to Eqs. (9.234) and (9.288), its optimal value

$$\begin{aligned}
\dot{M}(2) &= \dot{p}_1 \cdot 0.0567 + \dot{p}_2 \cdot 0.0929 + \dot{p}_3 \cdot 0.2201\\
&= 0.3 \cdot 0.0567 + 0.4 \cdot 0.0929 + 0.3 \cdot 0.2201 \cong 0.1202
\end{aligned} \tag{9.289}$$

Substituting the optimal solution (9.288) into the formula (9.235), we obtain the optimal solution for the mean values of the exemplary system unconditional lifetimes in the reliability state subsets $\{1, 2, 3\}$ and $\{3\}$, that are as follows:

$$\begin{aligned}
\dot{M}(1) &= \dot{p}_1 \cdot 0.1134 + \dot{p}_2 \cdot 0.1858 + \dot{p}_3 \cdot 0.4403\\
&= 0.3 \cdot 0.1134 + 4 \cdot 0.1858 + 0.3 \cdot 0.4403 \cong 0.2404
\end{aligned} \tag{9.290}$$

$$\begin{aligned}
\dot{M}(3) &= \dot{p}_1 \cdot 0.0378 + \dot{p}_2 \cdot 0.0619 + \dot{p}_3 \cdot 0.1134\\
&= 0.3 \cdot 0.0378 + 0.4 \cdot 0.0619 + 0.3 \cdot 0.1468 \cong 0.0801
\end{aligned} \tag{9.291}$$

and according to Eq. (9.239), the optimal values of the mean values of the system unconditional lifetimes in the particular reliability states 1, 2 and 3, are respectively

$$\dot{\overline{M}}(1) = \dot{M}(1) - \dot{M}(2) = 0.1202, \quad \dot{\overline{M}}(2) = \dot{M}(2) - \dot{M}(3) = 0.0401, \quad \dot{\overline{M}}(3) = \dot{M}(3) = 0.0801 \tag{9.292}$$

Moreover, according to Eqs. (9.236) and (9.237), the corresponding optimal unconditional multi-state reliability function of the system is of the form

$$\dot{\boldsymbol{R}}(t,\cdot) = [1 - \dot{\boldsymbol{R}}(t,1), \dot{\boldsymbol{R}}(t,2), \dot{\boldsymbol{R}}(t,3)], \quad t \geq 0 \tag{9.293}$$

with the coordinates given by

$$\dot{\boldsymbol{R}}(t,1) = 0.3 \cdot [\boldsymbol{R}(t,1)]^{(1)} + 0.4 \cdot [\boldsymbol{R}(t,1)]^{(2)} + 0.3 \cdot [\boldsymbol{R}(t,1)]^{(3)} \quad \text{for } t \geq 0 \tag{9.294}$$

$$\dot{\boldsymbol{R}}(t,2) = 0.3 \cdot [\boldsymbol{R}(t,2)]^{(1)} + 0.4 \cdot [\boldsymbol{R}(t,2)]^{(2)} + 0.3 \cdot [\boldsymbol{R}(t,2)]^{(3)} \quad \text{for } t \geq 0 \tag{9.295}$$

$$\dot{\boldsymbol{R}}(t,3) = 0.3 \cdot [\boldsymbol{R}(t,3)]^{(1)} + 0.4 \cdot [\boldsymbol{R}(t,3)]^{(2)} + 0.3 \cdot [\boldsymbol{R}(t,3)]^{(3)} \quad \text{for } t \geq 0 \tag{9.296}$$

where $[\boldsymbol{R}(t,1)]^{(b)}$, $[\boldsymbol{R}(t,2)]^{(b)}$, $[\boldsymbol{R}(t,3)]^{(b)}$, $b = 1,2,3$, are fixed in Section 9.3.

The graph of the '24 out of 36'–series system optimal reliability function $\dot{\boldsymbol{R}}(t,\cdot)$ given by Eqs. (9.293)–(9.296) is presented in Figure 9.19.

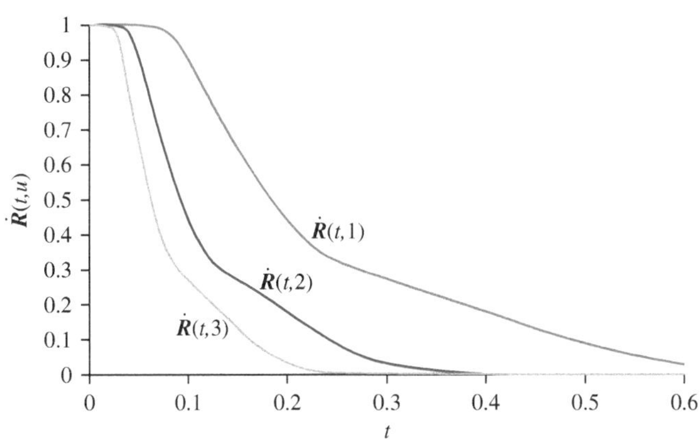

Figure 9.19 The graph of the '24 out of 36'–series system optimal reliability function $\dot{\boldsymbol{R}}(t,\cdot)$ coordinates.

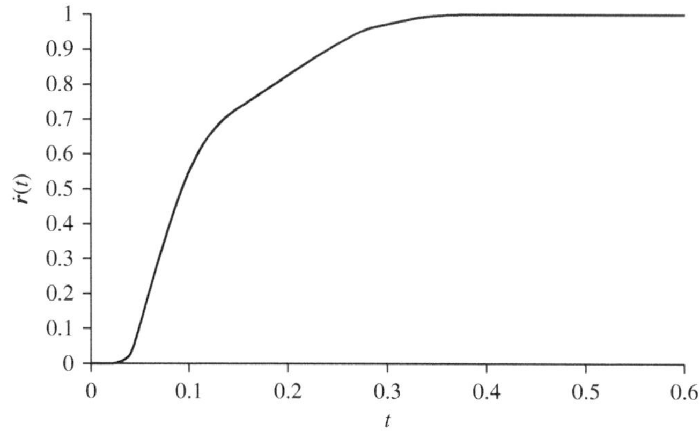

Figure 9.20 The graph of the '24 out of 36'—series system optimal risk function $\dot{\boldsymbol{r}}(t)$.

As the critical reliability state is $r=2$, then the '24 out of 36'—series system optimal system risk function, according to Eq. (9.240), is given by (Figure 9.20)

$$\dot{\boldsymbol{r}}(t) = 1 - \dot{\boldsymbol{R}}(t,2) \quad \text{for } t \geq 0 \tag{9.297}$$

where $\dot{\boldsymbol{R}}(t,2)$ is given by Eq. (9.295).

Hence, and considering Eq. (9.241), the moment when the optimal system risk function exceeds a permitted level, for instance $\delta = 0.025$, is

$$\dot{\tau} = \dot{\boldsymbol{r}}^{-1}(\delta) \cong 0.0395 \tag{9.298}$$

Substituting the exemplary operation process optimal transient probabilities at operation states

$$\dot{p}_1 = 0.3, \ \ \dot{p}_2 = 0.4, \ \ \dot{p}_3 = 0.3$$

determined by Eq. (9.288) into the formulae given by Eq. (9.245), we get the optimal values of the '24 out of 36'—series system operation process total sojourn times at the particular operation states during 1 year

$$\begin{aligned} \dot{\hat{m}}_1 &= \dot{E}[\hat{\theta}_1] = \dot{p}_1\theta = 0.3 \cdot 365 \cong 109.5 \text{ days}, \\ \dot{\hat{m}}_2 &= \dot{E}[\hat{\theta}_2] = \dot{p}_2\theta = 0.4 \cdot 365 \cong 146.0 \text{ days}, \\ \dot{\hat{m}}_3 &= \dot{E}[\hat{\theta}_3] = \dot{p}_3\theta = 0.3 \cdot 365 \cong 109.5 \text{ days} \end{aligned} \tag{9.299}$$

that differ greatly from the values of $\hat{m}_1$, $\hat{m}_2$, $\hat{m}_3$, determined by Eq. (9.21).

In practice, the knowledge of the optimal values of $\dot{\hat{m}}_b$ given by Eq. (9.299) can be important and helpful for planning and improving the system operation process in order to make the system operation more reliable.

Example 9.3 (Continuation)
The objective function defined by Eq. (9.217), in this case, as the port grain transportation system critical state is $r=1$, according to Eq. (9.212) takes the form

$$M(1)=p_1\cdot 0.0813+p_2\cdot 0.0861+p_3\cdot 0.0607 \tag{9.300}$$

The lower $\breve{p}_b$ and upper $\widehat{p}_b$ bounds of the unknown optimal values of transient probabilities p_b, $b=1,2,3$, are respectively

$$\breve{p}_1=0.460,\ \ \breve{p}_2=0.080,\ \ \breve{p}_3=0.310,$$
$$\widehat{p}_1=0.600,\ \ \widehat{p}_2=0.140,\ \ \widehat{p}_3=0.410$$

Therefore, according to Eqs. (9.218) and (9.219), we assume the following bound constraints:

$$0.460\le p_1\le 0.600,\ \ 0.080\le p_2\le 0.140,\ \ 0.310\le p_3\le 0.410 \tag{9.301}$$

$$\sum_{b=1}^{3} p_b=1 \tag{9.302}$$

Now, before we find optimal values $\dot{p}_b$ of the transient probabilities p_b, $b=1,2,3$, that maximize the objective function (9.300), we arrange the system conditional lifetime mean values $M_b(1)$, $b=1,2,3$, in non-increasing order

$$M_2(1)\ge M_1(1)\ge M_3(1)$$

Further, according to Eq. (9.222), we substitute

$$x_1=p_2,\ \ x_2=p_1,\ \ x_3=p_3 \tag{9.303}$$

and

$$\breve{x}_1=\breve{p}_2=0.080,\ \ \breve{x}_2=\breve{p}_1=0.460,\ \ \breve{x}_3=\breve{p}_3=0.310 \tag{9.304}$$

$$\widehat{x}_1=\widehat{p}_2=0.140,\ \ \widehat{x}_2=\widehat{p}_1=0.600,\ \ \widehat{x}_3=\widehat{p}_3=0.410 \tag{9.305}$$

and we maximize with respect to x_i, $i=1,2,3$, the linear form (9.300) that according to Eqs. (9.223)–(9.225) takes the form

$$M(2)=x_1\cdot 0.0873+x_2\cdot 0.0812+x_3\cdot 0.0605 \tag{9.306}$$

with the following bound constraints:

$$0.080\le x_1\le 0.140,\ \ 0.460\le x_2\le 0.600,\ \ 0.310\le x_3\le 0.410 \tag{9.307}$$

$$\sum_{i=1}^{3} x_i = 1 \tag{9.308}$$

According to Eq. (9.227), we calculate

$$\breve{x} = \sum_{i=1} \breve{x}_i = 0.850 \quad \hat{y} = 1 - \breve{x} = 1 - 0.850 = 0.150 \tag{9.309}$$

and according to Eq. (9.228), we determine

$$\begin{aligned} &\breve{x}^{0} = 0, \ \widehat{x}^{0} = 0, \ \widehat{x}^{0} - \breve{x}^{0} = 0, \\ &\breve{x}^{1} = 0.080, \ \widehat{x}^{1} = 0.140, \ \widehat{x}^{1} - \breve{x}^{1} = 0.060, \\ &\breve{x}^{2} = 0.540, \ \widehat{x}^{2} = 0.740, \ \widehat{x}^{2} - \breve{x}^{2} = 0.200, \\ &\breve{x}^{3} = 0.850, \ \widehat{x}^{3} = 1.150, \ \widehat{x}^{3} - \breve{x}^{3} = 0.300 \end{aligned} \tag{9.310}$$

From the above, as according to Eq. (9.309), the inequality (9.229) takes the form

$$\widehat{x}^{I} - \breve{x}^{I} < 0.150 \tag{9.311}$$

which follows that the largest value $I \in \{0, 1, 2, 3\}$ such that this inequality holds is $I = 1$.

Therefore, we fix the optimal solution that maximizes linear function (9.306) according to the rule (9.231). Namely, we get

$$\begin{aligned} &\dot{x}_1 = \widehat{x}_1 = 0.140, \\ &\dot{x}_2 = \hat{y} - \widehat{x}^{1} + \breve{x}^{1} + \breve{x}_2 = 0.150 - 0.140 + 0.080 + 0.460 = 0.550, \\ &\dot{x}_3 = \breve{x}_3 = 0.310 \end{aligned} \tag{9.312}$$

Finally, after making the inverse to Eq. (9.303) substitution, we get the optimal transient probabilities

$$\dot{p}_1 = \dot{x}_2 = 0.550, \ \dot{p}_2 = \dot{x}_1 = 0.140, \ \dot{p}_3 = \dot{x}_3 = 0.310 \tag{9.313}$$

that maximize the exemplary system mean lifetime $M(1)$ in the reliability state subset $\{1, 2\}$ expressed by the linear form (9.300) giving, according to Eqs. (9.234) and (9.313), its optimal value

$$\begin{aligned} \dot{M}(1) &= \dot{p}_1 \cdot 0.0813 + \dot{p}_2 \cdot 0.0861 + \dot{p}_3 \cdot 0.0607 \\ &= 0.55 \cdot 0.0813 + 0.14 \cdot 0.0861 + 0.31 \cdot 0.0607 = 0.0756 \text{ year} \end{aligned} \tag{9.314}$$

Substituting the optimal solution (9.313) into the formula (9.235), we obtain the optimal solution for the mean values of the exemplary system unconditional lifetimes in the reliability state subset {2}, which is as follows:

$$\begin{aligned}\dot{M}(2) &= \dot{p}_1 \cdot 0.0691 + \dot{p}_2 \cdot 0.0728 + \dot{p}_3 \cdot 0.0493 \\ &= 0.55 \cdot 0.0691 + 0.14 \cdot 0.0728 + 0.31 \cdot 0.0493 = 0.0635 \text{ year}\end{aligned} \tag{9.315}$$

and according to Eq. (9.239), the optimal values of the mean values of the system unconditional lifetimes in the particular reliability states 1 and 2, are respectively

$$\dot{\overline{M}}(1) = \dot{M}(1) - \dot{M}(2) = 0.0121 \text{ year}, \quad \dot{\overline{M}}(2) = \dot{M}(2) = 0.0635 \text{ year} \tag{9.316}$$

Moreover, according to Eqs. (9.236) and (9.237), the corresponding optimal unconditional multi-state reliability function of the port grain transportation system is of the form

$$\dot{\boldsymbol{R}}(t,\cdot) = [1, \dot{\boldsymbol{R}}(t,1), \dot{\boldsymbol{R}}(t,2)], \quad t \geq 0 \tag{9.317}$$

with the coordinates given by

$$\dot{\boldsymbol{R}}(t,1) = 0.55 \cdot [\boldsymbol{R}(t,1)]^{(1)} + 0.14 \cdot [\boldsymbol{R}(t,1)]^{(2)} + 0.31 \cdot [\boldsymbol{R}(t,1)]^{(3)} \text{ for } t \geq 0 \tag{9.318}$$

$$\dot{\boldsymbol{R}}(t,2) = 0.55 \cdot [\boldsymbol{R}(t,2)]^{(1)} + 0.14 \cdot [\boldsymbol{R}(t,2)]^{(2)} + 0.31 \cdot [\boldsymbol{R}(t,2)]^{(3)} \text{ for } t \geq 0 \tag{9.319}$$

where $[\boldsymbol{R}(t,1)]^{(b)}$, $[\boldsymbol{R}(t,2)]^{(b)}$, $b = 1, 2, 3$, are fixed in Section 9.3.

The graph of the port grain transportation system optimal reliability function $\dot{\boldsymbol{R}}(t,\cdot)$ given by Eqs. (9.317)–(9.319) is presented in Figure 9.21.

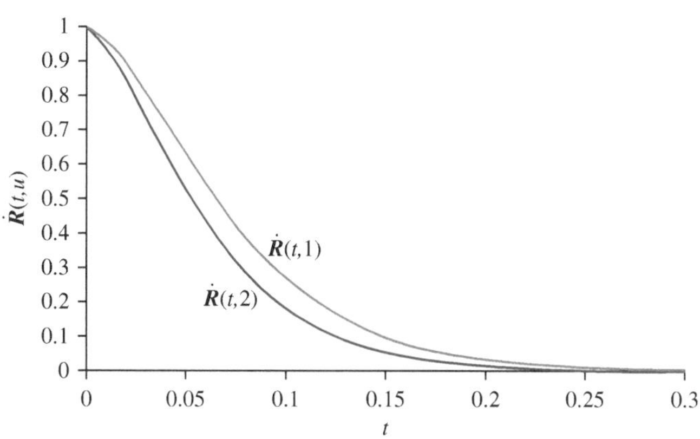

Figure 9.21 The graph of the port grain transportation system optimal reliability function $\dot{\boldsymbol{R}}(t,\cdot)$ coordinates.

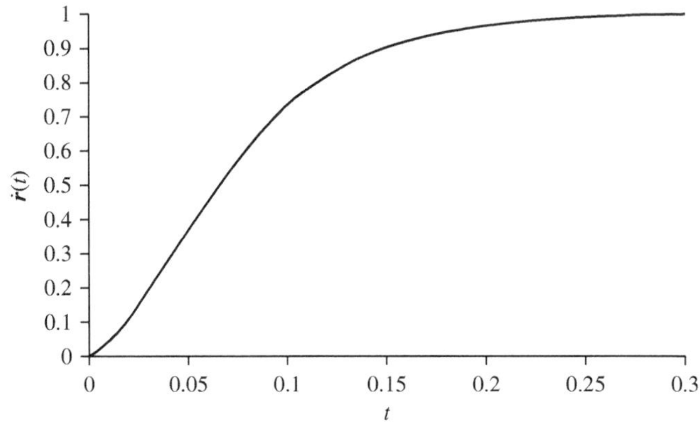

Figure 9.22 The graph of the port grain transportation system optimal risk function $\dot{r}(t)$.

As the critical reliability state is $r = 1$, then the exemplary system optimal system risk function, according to Eq. (9.240), is given by (Figure 9.22)

$$\dot{r}(t) = 1 - \dot{\boldsymbol{R}}(t, 1) \quad \text{for } t \geq 0 \tag{9.320}$$

where $\dot{\boldsymbol{R}}(t, 1)$ is given by Eq. (9.318).

Hence, and considering Eq. (9.241), the moment when the optimal system risk function exceeds a permitted level, for instance $\delta = 0.05$, is

$$\dot{\tau} = \dot{\boldsymbol{r}}^{-1}(\delta) \cong 0.0114 \text{ year} \tag{9.321}$$

Substituting the exemplary operation process optimal transient probabilities at operation states

$$\dot{p}_1 = 0.5, \ \ \dot{p}_2 = 0.14, \ \ \dot{p}_3 = 0.31$$

determined by Eq. (9.313) and the steady probabilities

$$\pi_1 \cong 0.279, \ \ \pi_2 \cong 0.344, \ \ \pi_3 \cong 0.377$$

determined by Eq. (9.30) in Section 9.2 into (9.243), we get the following system of equations with the unknown optimal mean values $\dot{m}_b$ of the port grain transportation system operation process unconditional sojourn times at the operation states we are looking for

$$\begin{aligned} -0.12555\dot{m}_1 + 0.18920\dot{m}_2 + 0.20735\dot{m}_3 &= 0, \\ 0.03906\dot{m}_1 - 0.29584\dot{m}_2 + 0.05278\dot{m}_3 &= 0, \\ 0.08649\dot{m}_1 + 0.10664\dot{m}_2 - 0.26013\dot{m}_3 &= 0 \end{aligned} \tag{9.322}$$

The determinant of the main matrix of the above homogeneous system of equations is equal to zero and therefore there are non-zero solutions of this system of equations that are ambiguous and dependent on one or more parameters. Thus, we may fix some of them and determine the remaining ones. To show the method of solving this system of equations, we may suppose that we are arbitrarily interested in fixing the value of $\dot{m}_3$ and we put

$$\dot{m}_3 = 0.070$$

Substituting the above value into the system of equations (9.322), we get

$$\begin{aligned} -0.12555\dot{m}_1 + 0.18920\dot{m}_2 &= -0.0145145, \\ 0.03906\dot{m}_1 - 0.29584\dot{m}_2 &= -0.0036946, \\ 0.08649\dot{m}_1 + 0.10664\dot{m}_2 &= 0.0182091 \end{aligned}$$

and we solve it with respect to $\dot{m}_1$ and $\dot{m}_2$, after omitting its last equation. This way, we obtain the following solutions that satisfy Eq. (9.322)

$$\dot{m}_1 \cong 0.1678, \quad \dot{m}_2 \cong 0.0346, \quad \dot{m}_3 = 0.0700 \tag{9.323}$$

It can be seen that these solutions are different from the values m_1, m_2 and m_3 estimated in Section 9.2 and given by Eqs. (9.27)–(9.29).

Having these solutions, it is also possible to look for the optimal values $\dot{m}_{bl}$ of the mean values m_{bl} of the port grain transportation system operation process conditional sojourn times at operation states. Namely, substituting the values $\dot{m}_b$ instead of m_b, the probabilities

$$[p_{bl}]_{3\times 3} = \begin{bmatrix} 0 & \frac{1}{3} & \frac{2}{3} \\ \frac{4}{9} & 0 & \frac{5}{9} \\ \frac{1}{3} & \frac{2}{3} & 0 \end{bmatrix}$$

of the port grain transportation system operation process transitions between the operation states given by Eq. (9.23) and replacing m_{bl} by $\dot{m}_{bl}$ in Eq. (9.244), we get the following system of equations:

$$\begin{aligned} \frac{1}{3}\dot{m}_{12} + \frac{2}{3}\dot{m}_{13} &= 0.1678, \\ \frac{4}{9}\dot{m}_{21} + \frac{5}{9}\dot{m}_{23} &= 0.0346, \\ \frac{1}{3}\dot{m}_{31} + \frac{2}{3}\dot{m}_{32} &= 0.0700 \end{aligned} \tag{9.324}$$

with the unknown optimal values $\dot{m}_{bl}$ we want to find.

As the solutions of the above system of equations are ambiguous, then we fix some of them, say that because of practically important reasons, and we find the remaining ones. For instance,

$$\begin{aligned}&\text{we fix in the first equation } \dot{m}_{13} = 0.050 \text{ and we find } \dot{m}_{12} = 0.403,\\&\text{we fix in the second equation } \dot{m}_{23} = 0.010 \text{ and we find } \dot{m}_{21} = 0.065,\\&\text{we fix in the third equation } \dot{m}_{32} = 0.100 \text{ and we find } \dot{m}_{31} = 0.010\end{aligned} \tag{9.325}$$

It can be seen that these solutions differ greatly from the mean values of the exemplary system conditional sojourn times at the particular operation states before its operation process optimization, given by Eq. (9.26).

Another very useful set of tools, which are much more easily applied in practice and which can help in planning the operation process of the port grain transportation system, are the system operation process optimal mean values of the total sojourn times at the particular operation states during the system operation time that, by the same assumption as in Section 9.2, is equal to $\theta = 1$ year $= 365$ days. Under this assumption, after applying Eq. (9.245), we get the optimal values of the exemplary system operation process total sojourn times at the particular operation states during 1 year

$$\begin{aligned}\dot{\hat{m}}_1 &= \dot{E}[\hat{\theta}_1] = \dot{p}_1\theta = 0.55 \cdot 365 \cong 200.75,\\\dot{\hat{m}}_2 &= \dot{E}[\hat{\theta}_2] = \dot{p}_2\theta = 0.14 \cdot 365 \cong 51.1,\\\dot{\hat{m}}_3 &= \dot{E}[\hat{\theta}_3] = \dot{p}_3\theta = 0.31 \cdot 365 = 113.15\end{aligned} \tag{9.326}$$

that are different from the values of $\hat{m}_1$, $\hat{m}_2$, $\hat{m}_3$, determined by Eq. (9.32).

In practice, the knowledge of the optimal values of $\dot{m}_b$, $\dot{m}_{bl}$ and $\dot{\hat{m}}_b$ given respectively by Eqs. (9.323), (9.325), (9.326) can be very important and helpful in planning and improving the system operation process in order to make the system operation more reliable.

The comparison of the values of the selected port grain transportation system reliability characteristics before the system operation process optimization given by Eqs. (9.212)–(9.214) and (9.216) with their values after the system operation process optimization respectively given by Eqs. (9.314)–(9.316) and (9.321) justifies the sensibility of the performed system operation process optimization.

From the analysis of the results of the port grain transportation system operation process optimization, it can be suggested to organize the system operation process in a way that causes the replacement of (or the approaching/convergence to) the conditional mean sojourn times m_{bl} of the system at the particular operation states before the optimization given by Eq. (9.26) by their optimal values $\dot{m}_{bl}$ after the optimization given by Eq. (9.325). However, the suggested change of the parameters of the system operation process very often is not easy to perform in practice.

An easier way might be to change the operation process characteristics that results in replacement of (or the approaching/convergence to) the unconditional mean sojourn times m_b of the port grain transportation system at the particular operation states before the optimization given by Eqs. (9.27)–(9.29) by their optimal values $\dot{m}_b$ after the optimization given by Eq. (9.323).

Practically, the easiest way of the system operation process reorganizing might be to replace (or to approach/converge to) the total sojourn times $\hat{m}_b$ of the port grain transportation system operation process at the particular operation states during the operation time $\theta = 1$ year before the optimization given by Eq. (9.32) by their optimal values $\dot{\hat{m}}_b$ after the optimization given by Eq. (9.326).

The evaluation and prediction of the port grain transportation system operating at the varying in time operation conditions reliability are approximate because of insufficiently exact input data. Therefore, the achieved results may only be considered as an illustration of the possibilities of applications of the proposed methods and procedures of this system operation and reliability analysis, prediction and optimization. However, the obtained evaluation may be a very useful example in the prediction of reliability characteristics of other port grain transportation systems and other complex technical systems, especially during the design and when planning and improving the effectiveness of their operation processes.

10 Large Complex Systems

10.1 Asymptotic Approach to Reliability of Large Complex Systems

In the case of large complex systems, the determination of their exact reliability functions and their risk functions leads us to very complicated formulae that are often useless for reliability practitioners. One of the important techniques that can be useful in this situation is the asymptotic approach [85,90,154–156,162] to that system reliability evaluation. In this approach, instead of the preliminary complex formula for the large complex system reliability function, after assuming that the number of system components tends to infinity and finding the limit reliability of the system, we can obtain its simplified form. Moreover, in the case of large complex systems, the possibility of combining the results of the reliability joint models of complex systems and the results concerning the limit reliability functions of the considered systems is possible [85,90,154–156,162]. This way, the results concerned with the asymptotic approach to the estimation of multi-state systems at variable operation conditions may be obtained. The main results concerning asymptotic approach to multi-state large system reliability with ageing components in the constant operation conditions are comprehensively presented in the monograph [85] and some of these results' extensions to the systems operating at the variable conditions can be found in Refs. [115,154–156,160,162].

In order to combine the results on the reliability of multi-state systems related to their operation processes and the results concerning the limit reliability functions of the multi-state systems, and to obtain the results on the asymptotic approach to the evaluation of the large complex systems reliability, we assume the following definition [160].

Definition 10.1
A reliability function

$$\mathfrak{R}(t,\cdot) = [1, \mathfrak{R}(t,1), \ldots, \mathfrak{R}(t,z)], \quad t \in (-\infty, \infty) \tag{10.1}$$

where

$$\mathfrak{R}(t,u) = \sum_{b-1}^{v} p_b [\mathfrak{R}(t,u)]^{(b)}, \quad u = 1, 2, \ldots, z \tag{10.2}$$

Reliability of Large Systems. DOI: http://dx.doi.org/10.1016/B978-0-08-099949-4.00010-6

is called a limit reliability function of a large complex multi-state system with the reliability function sequence

$$\boldsymbol{R}_n(t,\cdot)=[1,\boldsymbol{R}_n(t,1),\ldots,\boldsymbol{R}_n(t,z)],\quad t\in(-\infty,\infty),\ n\in N \tag{10.3}$$

where

$$\boldsymbol{R}_n(t,u)\cong\sum_{b=1}^{v}p_b[\boldsymbol{R}_n(t,u)]^{(b)},\quad u=1,2,\ldots,z \tag{10.4}$$

if there exist normalizing constants

$$a_n^{(b)}(u)>0,\ \ b_n^{(b)}(u)\in(-\infty,\infty),\quad u=1,2,\ldots,z,\ \ b=1,2,\ldots,v$$

such that

$$\lim_{n\to\infty}[\boldsymbol{R}_n(a_n^{(b)}(u)t+b_n^{(b)}(u),u)]^{(b)}=[\Re(t,u)]^{(b)}$$

for all t from the sets of continuity points $C_{[\Re(u)]^{(b)}}$ of the functions $[\Re(t,u)]^{(b)}$, $u=1,2,\ldots,z$, $b=1,2,\ldots,v$.

Hence, for sufficiently large n, the following approximate formulae are valid:

$$\boldsymbol{R}_n(t,\cdot)=[1,\boldsymbol{R}_n(t,1),\ldots,\boldsymbol{R}_n(t,z)],\quad t\in(-\infty,\infty) \tag{10.5}$$

where

$$\boldsymbol{R}_n(t,u)\cong\sum_{b=1}^{v}p_b\left[\Re\left(\frac{t-b_n^{(b)}(u)}{a_n^{(b)}(u)},u\right)\right]^{(b)},\quad t\in(-\infty,\infty),\ u=1,2,\ldots,z \tag{10.6}$$

10.2 Limit Reliability Functions of Large Complex Systems

The selected propositions concerned with the reliability evaluation of large series–parallel, series–'m out of l' and 'm out of l'–series exponential systems operating at the variable operation states are exemplary results that can be worked out on the basis of the results included in Chapter 7 for the large systems considered there.

From Proposition 9.1, Proposition 7.4 (Case 2) and Corollary 5.12, we get the following auxiliary theorem.

Proposition 10.1

If components of the non-homogeneous, regular multi-state series–parallel complex system with the structure shape parameters

$$k=k^{(b)},\ \ l=l^{(b)},\ \ a=a^{(b)},\ \ e_i=e_i^{(b)},\quad i=1,2,\ldots,a^{(b)},\ \ b=1,2,\ldots,\nu \tag{10.7}$$

and

$$q_i = q_i^{(b)}, \ \ p_{ij} = p_{ij}^{(b)}, \quad i = 1,2,\ldots,a^{(b)}, \ j = 1,2,\ldots,e_i^{(b)}, \ b = 1,2,\ldots,\nu \quad (10.8)$$

at the operation state z_b, $b = 1,2,\ldots,\nu$, have exponential reliability functions

$$[R^{(i,j)}(t,\cdot)]^{(b)} = [1,[R^{(i,j)}(t,1)]^{(b)},\ldots,[R^{(i,j)}(t,z)]^{(b)}, \quad t\in(-\infty,\infty), \ b = 1,2,\ldots,\nu$$

where

$$[R^{(i,j)}(t,u)]^{(b)} = 1 \ \text{ for } t<0, \ [R^{(i,j)}(t,u)]^{(b)} = \exp[-[\lambda_{ij}(u)]^{(b)}t] \ \text{ for } t\geq 0, \ \lambda_{ij}(u)>0,$$
$$u = 1,2,\ldots,z, \ i = 1,2,\ldots,a^{(b)}, \ j = 1,2,\ldots,e_i^{(b)}, \ b = 1,2,\ldots,\nu$$

and

$$k^{(b)} = \text{constant}, \ \ l^{(b)} \to \infty \ \text{ as } n \to \infty, \ \ b = 1,2,\ldots,v \quad (10.9)$$

$$a_n^{(b)}(u) = \frac{1}{[\lambda(u)]^{(b)} l^{(b)}}, \ \ b_n^{(b)}(u) = 0, \quad u = 1,2,\ldots,z, \ \ b = 1,2,\ldots,v \quad (10.10)$$

where

$$[\lambda_i(u)]^{(b)} = \sum_{j=1}^{e_i^{(b)}} p_{ij}^{(b)}[\lambda_{ij}(u)]^{(b)}, \quad u = 1,2,\ldots,z, \ \ i = 1,2,\ldots,a^{(b)}, \\ b = 1,2,\ldots,\nu \quad (10.11)$$

$$[\lambda(u)]^{(b)} = \min_{1\leq i\leq a^{(b)}} \{[\lambda_i(u)]^{(b)}\}, \quad u = 1,2,\ldots,z, \ \ b = 1,2,\ldots,\nu \quad (10.12)$$

then

$$\Re(t,\cdot) = [1,\Re(t,1),\ldots,\Re(t,z)], \quad t\in(-\infty,\infty) \quad (10.13)$$

where

$$\Re(t,u) = \sum_{b=1}^{v} p_b[\Re(t,u)]^{(b)}, \quad u = 1,2,\ldots,z \quad (10.14)$$

$$[\Re(t,u)]^{(b)} = \Re'_9(t,u) = 1 \quad \text{for } \ t<0$$

$$[\Re(t,u)]^{(b)} = \Re'_9(t,u) = 1 - \prod_{i=1}^{a^{(b)}} \left[1-\exp\left[-\frac{[\lambda_i(u)]^{(b)}}{[\lambda(u)]^{(b)}}t\right]\right]^{q_i^{(b)}k^{(b)}} \\ \text{for } t\geq 0 \ \ u = 1,2,\ldots,z, \ \ b = 1,2,\ldots,\nu \quad (10.15)$$

is its limit reliability function.

From Proposition 9.1, Proposition 7.1 and Corollary 5.13, we get the following result.

Proposition 10.2
If components of the non-homogeneous, multi-state series complex system with the structure shape parameters

$$n = n^{(b)}, \quad a = a^{(b)}, \quad b = 1, 2, \ldots, \nu \tag{10.16}$$

and

$$q_i = q_i^{(b)}, \quad i = 1, 2, \ldots, a^{(b)}, \quad b = 1, 2, \ldots, \nu \tag{10.17}$$

at the operation state z_b, $b = 1, 2, ..., \nu$, have exponential reliability functions

$$[R^{(i)}(t, \cdot)]^{(b)} = [1, [R^{(i)}(t, 1)]^{(b)}, \ldots, [R^{(i)}(t, z)]^{(b)}], \quad t \in (-\infty, \infty), \ b = 1, 2, \ldots, \nu$$

where

$$[R^{(i}(t,u)]^{(b)} = 1 \ \text{ for } t < 0, \ [R^{(i)}(t,u)]^{(b)} = \exp[-[\lambda_i(u)]^{(b)} t] \ \text{ for } t \geq 0, \ \lambda_i(u) > 0,$$
$$u = 1, 2, \ldots, z, \ i = 1, 2, \ldots, a^{(b)}, \ b = 1, 2, \ldots, \nu$$

and

$$n^{(b)} \to \infty, \quad b = 1, 2, \ldots, \nu \tag{10.18}$$

$$a_n^{(b)}(u) = \frac{1}{[\lambda(u)]^{(b)} n^{(b)}}, \quad b_n^{(b)}(u) = 0, \quad u = 1, 2, \ldots, z, \ b = 1, 2, \ldots, \nu \tag{10.19}$$

where

$$[\lambda(u)]^{(b)} = \max_{1 \leq i \leq a^{(b)}} \{[\lambda_i(u)]^{(b)}\}, \quad u = 1, 2, \ldots, z, \ b = 1, 2, \ldots, \nu \tag{10.20}$$

then

$$\mathfrak{R}(t, \cdot) = [1, \mathfrak{R}(t, 1), \ldots, \mathfrak{R}(t, z)], \quad t \in (-\infty, \infty) \tag{10.21}$$

where

$$\mathfrak{R}(t, u) = \sum_{b=1}^{\nu} p_b [\mathfrak{R}(t, u)]^{(b)}, \quad u = 1, 2, \ldots, z \tag{10.22}$$

$$[\mathfrak{R}(t, u)]^{(b)} = \overline{\mathfrak{R}}_2'(t, u) = 1 \quad \text{for } t < 0$$

$$[\mathfrak{R}(t,u)]^{(b)} = \overline{\mathfrak{R}}'_2(t,u) = \exp\left[-\sum_{i=1}^{a^{(b)}} q_i^{(b)} \frac{[\lambda_i(u)]^{(b)}}{[\lambda(u)]^{(b)}} t\right] \tag{10.23}$$

for $t \geq 0,\ \ u = 1,2,\ldots,z,\ \ b = 1,2,\ldots,\nu$

is its limit reliability function.

Application of Proposition 9.1 and slight modification of Proposition 7.3 (Case 2) and Corollary 5.14 result in the auxiliary theorem of the following form.

Proposition 10.3

If components of the homogeneous, regular multi-state 'm out of l'−series complex system with the structure shape parameters

$$k = k^{(b)},\ \ l = l^{(b)},\ \ m = m^{(b)},\quad b = 1,2,\ldots,\nu \tag{10.24}$$

at the operation state z_b, $b = 1,2,\ldots,\nu$, have exponential reliability functions

$$[R_{ij}(t,\cdot)]^{(b)} = [1,[R_{ij}(t,1)]^{(b)},[R_{ij}(t,2)]^{(b)},\ldots,[R_{ij}(t,z)]^{(b)}],$$
$$i = 1,2,\ldots,k^{(b)},\ \ j = 1,2,\ldots,l^{(b)}$$

with the coordinates

$$[R_{ij}(t,u)]^{(b)} = 1 \quad \text{for } t<0,\ \ [R_{ij}(t,u)]^{(b)} = \exp[-[\lambda(u)]^{(b)}t],\quad \text{for } t \geq 0,\ \ \lambda(u)>0,$$
$$u = 1,2,\ldots,z,\ \ i = 1,2,\ldots,k^{(b)},\ \ j = 1,2,\ldots,l^{(b)}$$

$$k^{(b)} = \text{constant},\ \ l^{(b)} \to \infty \ \text{ as } n \to \infty,\ \ b = 1,2,\ldots,\nu \tag{10.25}$$

$$m^{(b)}/l^{(b)} \to \mu^{(b)},\ \ 0<\mu^{(b)}<1,\ \text{ as } n \to \infty,\ \ b = 1,2,\ldots,\nu \tag{10.26}$$

$$\begin{aligned} a_n^{(b)}(u) &= \frac{1}{[\lambda(u)]^{(b)}\mu^{(b)}\sqrt{l^{(b)}+1}}\sqrt{\mu^{(b)}(1-\mu^{(b)})},\ \ b_n^{(b)}(u) \\ &= -\frac{1}{\lambda}\log \mu^{(b)},\ \ u = 1,2,\ldots,z,\ \ b = 1,2,\ldots,\nu \end{aligned} \tag{10.27}$$

then

$$\mathfrak{R}(t,\cdot) = [1,\mathfrak{R}(t,1),\ldots,\mathfrak{R}(t,z)],\quad t \in (-\infty,\infty) \tag{10.28}$$

where

$$\mathfrak{R}(t,u) = \sum_{b=1}^{\nu} p_b[\mathfrak{R}(t,u)]^{(b)},\quad u = 1,2,\ldots,z \tag{10.29}$$

$$[\mathfrak{R}(t,u)]^{(b)} = \mathfrak{R}^{(\mu_b)}(t,u) = \left[1 - \frac{1}{\sqrt{2\pi}} \int_{-\infty}^{t} e^{-x^2/2}\, dx\right]^{k^{(b)}} \quad (10.30)$$

$$\text{for } t \in (-\infty, \infty), \quad u = 1, 2, \ldots, z, \quad b = 1, 2, \ldots, \nu$$

is the limit multi-state reliability function of this system.

10.3 Examples of Large Complex Systems Reliability Evaluation

Example 10.1

We consider a large multi-state non-homogeneous system composed of a series–parallel and a large series–'m out of l' subsystems linked in series changing its reliability structures and its components reliability parameters at variable operation conditions, analysed in Example 9.1.

At the operation state z_1, the considered system is identical with a series–parallel subsystem S_1 composed of two identical series systems, then we may apply Proposition 10.1. Namely, since the system structure shape parameters are

$$k^{(1)} = 2, \quad l^{(1)} = 30, \quad a^{(1)} = 1, \quad e_1^{(1)} = 2$$

and

$$q_1^{(1)} = 1, \quad p_{11}^{(1)} = \frac{2}{3}, \quad p_{12}^{(1)} = \frac{1}{3}$$

the number of reliability states is $z = 3$ and considering the values of component' intensities of departure from the reliability state subsets defined in Example 9.1, by Eqs. (10.9)–(10.12), we have

$$[\lambda_i(1)]^{(1)} = \frac{2}{3}0.0008 + \frac{1}{3}0.0011 = 0.0009,$$

$$[\lambda_i(2)]^{(1)} = \frac{2}{3}0.0009 + \frac{1}{3}0.0012 = 0.0010,$$

$$[\lambda_i(3)]^{(1)} = \frac{2}{3}0.0010 + \frac{1}{3}0.0013 = 0.0011 \quad \text{for } i = 1, 2$$

$$[\lambda(1)]^{(1)} = \min\{0.0009\} = 0.0009,$$

$$[\lambda(2)]^{(1)} = \min\{0.0010\} = 0.0010,$$

$$[\lambda(3)]^{(1)} = \min\{0.0011\} = 0.0011$$

$$a_n^{(1)}(1) = \frac{1}{0.0009 \cdot 30} = \frac{1}{0.027}, \quad b_n^{(1)}(1) = 0,$$

$$a_n^{(1)}(2) = \frac{1}{0.0010 \cdot 30} = \frac{1}{0.030}, \quad b_n^{(1)}(2) = 0,$$

$$a_n^{(1)}(3) = \frac{1}{0.0011 \cdot 30} = \frac{1}{0.033}, \quad b_n^{(1)}(3) = 0$$

then according to Eqs. (10.13)–(10.15), after considering Eq. (10.6), the approximate formula for the system reliability function takes the form

$$\begin{aligned}[\boldsymbol{R}(t,\cdot)]^{(1)} &\cong \Re((t - b_n^{(1)}(u))/a_n^{(1)}(u),\cdot) \\ &= [[\Re(t,1)]^{(1)}, [\Re(t,2)]^{(1)}, [\Re(t,1)]^{(1)}] \quad \text{for } t \geq 0\end{aligned} \tag{10.31}$$

where

$$\begin{aligned}[\Re(t,1)]^{(1)} &= \Re_9'\left(\frac{t - b_n^{(1)}(1)}{a_n^{(1)}(1)}, 1\right) = 1 - [1 - \exp[-0.027t]]^2 \\ &= 2\exp[-0.027t] - \exp[-0.054t]\end{aligned} \tag{10.32}$$

$$\begin{aligned}[\Re(t,2)]^{(1)} &= \Re_9'\left(\frac{t - b_n^{(1)}(2)}{a_n^{(1)}(2)}, 2\right) = 1 - [1 - \exp[-0.030t]]^2 \\ &= 2\exp[-0.030t] - \exp[-0.060t]\end{aligned} \tag{10.33}$$

$$\begin{aligned}[\Re(t,3)]^{(1)} &= \Re_9'\left(\frac{t - b_n^{(1)}(3)}{a_n^{(1)}(3)}, 3\right) = 1 - [1 - \exp[-0.033t]]^2] \\ &= 2\exp[-0.033t] - \exp[-0.066t] \quad \text{for } t \geq 0\end{aligned} \tag{10.34}$$

At the operation state z_2, the considered system is identical with a series–parallel subsystem S_2 composed of four identical series systems, then we may apply Proposition 10.1. Namely, since the system structure shape parameters are

$$k^{(2)} = 4, \quad l^{(2)} = 40, \quad a^{(2)} = 1, \quad e_1^{(2)} = 2$$

and

$$q_1^{(1)} = 1, \quad p_{11}^{(2)} = 0.5, \quad p_{12}^{(2)} = 0.5$$

the number of reliability states is $z = 3$ and considering the values of component' intensities of departure from the reliability state subsets defined in Example 9.1, by Eqs. (10.9)–(10.12), we have

$$[\lambda_i(1)]^{(2)} = 0.5 \cdot 0.0014 + 0.5 \cdot 0.0018 = 0.0016,$$

$$[\lambda_i(2)]^{(2)} = 0.5 \cdot 0.0015 + 0.5 \cdot 0.0020 = 0.00175,$$

$$[\lambda_i(3)]^{(2)} = 0.5 \cdot 0.0016 + 0.5 \cdot 0.0022 = 0.0019, \quad \text{for } i = 1, 2, 3, 4$$

$$[\lambda(1)]^{(2)} = \min\{0.0016\} = 0.0016,$$

$$[\lambda(2)]^{(2)} = \min\{0.00175\} = 0.00175,$$

$$[\lambda(3)]^{(2)} = \min\{0.0019\} = 0.0019$$

$$a_n^{(2)}(1) = \frac{1}{0.0016 \cdot 40} = \frac{1}{0.064}, \quad b_n^{(2)}(1) = 0,$$

$$a_n^{(2)}(2) = \frac{1}{0.00175 \cdot 40} = \frac{1}{0.070}, \quad b_n^{(2)}(2) = 0,$$

$$a_n^{(2)}(3) = \frac{1}{0.0019 \cdot 40} = \frac{1}{0.076}, \quad b_n^{(1)}(3) = 0$$

then according to Eqs. (10.13)–(10.15), after considering Eq. (10.6), the approximate formula for the system reliability function takes the form

$$\begin{aligned}[\boldsymbol{R}(t,\cdot)]^{(2)} &\cong \mathfrak{R}((t - b_n^{(2)}(u))/a_n^{(2)}(u), \cdot) \\ &= [[\mathfrak{R}(t,1)]^{(2)}, [\mathfrak{R}(t,2)]^{(2)}, [\mathfrak{R}(t,1)]^{(2)}] \quad \text{for } t \geq 0\end{aligned} \tag{10.35}$$

where

$$\begin{aligned}[\mathfrak{R}(t,1)]^{(2)} &= \mathfrak{R}'_9\left(\frac{t - b_n^{(2)}(1)}{a_n^{(2)}(1)}, 1\right) = 1 - [1 - \exp[-0.064t]]^4 \\ &= 4\exp[-0.064t] - 6\exp[-0.128t] \\ &\quad + 4\exp[-0.192t] - \exp[-0.256t]\end{aligned} \tag{10.36}$$

$$\begin{aligned}[\mathfrak{R}(t,2)]^{(2)} &= \mathfrak{R}'_9\left(\frac{t - b_n^{(2)}(2)}{a_n^{(2)}(2)}, 2\right) = 1 - [1 - \exp[-0.070t]]^4 \\ &= 4\exp[-0.070t] - 6\exp[-0.140t] \\ &\quad + 4\exp[-0.210t] - \exp[-0.280t]\end{aligned} \tag{10.37}$$

$$[\Re(t,3)]^{(2)} = \Re'_9\left(\frac{t-b_n^{(2)}(3)}{a_n^{(2)}(3)},3\right) = 1-[1-\exp[-0.076t]]^4$$
$$= 4\exp[-0.076t]-6\exp[-0.152t]$$
$$+4\exp[-0.228t]-\exp[-0.304t] \quad \text{for } t \geq 0 \tag{10.38}$$

At the operation state z_3, the considered system is a series system composed of two series–parallel subsystems S_1 and S_2. The subsystem S_1, at the operation state z_3 is composed of two identical series systems; then we may apply Proposition 10.1. Namely, since the system structure shape parameters are

$$k^{(3)} = 2,\ \ l^{(3)} = 30,\ \ a^{(3)} = 1,\ \ e_1^{(3)} = 2$$

and

$$q_1^{(3)} = 1,\ \ p_{11}^{(3)} = \frac{2}{3},\ \ p_{12}^{(3)} = \frac{1}{3}$$

the number of reliability states is $z = 3$ and considering the values of component' intensities of departure from the reliability state subsets defined in Example 9.1, by Eqs. (10.9)–(10.12), we have

$$[\lambda_i(1)]^{(3)} = \frac{2}{3}0.0009 + \frac{1}{3}0.0012 = 0.0010,$$

$$[\lambda_i(2)]^{(3)} = \frac{2}{3}0.0010 + \frac{1}{3}0.0014 = 0.0011(3),$$

$$[\lambda_i(3)]^{(3)} = \frac{2}{3}0.0011 + \frac{1}{3}0.0016 = 0.0012(6) \quad \text{for } i = 1,2$$

$$[\lambda(1)]^{(3)} = \min\{0.0010\} = 0.0010,$$

$$[\lambda(2)]^{(3)} = \min\{0.0011(3)\} = 0.0011(3),$$

$$[\lambda(3)]^{(3)} = \min\{0.0012(6)\} = 0.0012(6)$$

$$a_n^{(3)}(1) = \frac{1}{0.0010\cdot 30} = \frac{1}{0.030},\ \ b_n^{(3)}(1) = 0,$$

$$a_n^{(3)}(2) = \frac{1}{0.0011(3)\cdot 30} = \frac{1}{0.034},\ \ b_n^{(3)}(2) = 0,$$

$$a_n^{(3)}(3) = \frac{1}{0.0012(6)\cdot 30} = \frac{1}{0.038},\ \ b_n^{(3)}(3) = 0$$

then according to Eqs. (10.13)–(10.15), after considering Eq. (10.6), the approximate formula for the subsystem S_1 reliability function takes the form

$$[\boldsymbol{R}^{(1)}(t,\cdot)]^{(3)} \cong \mathfrak{R}^{(1)}((t-b_n^{(3)}(u))/a_n^{(3)}(u),\cdot) \\ = [[\mathfrak{R}^{(1)}(t,1)]^{(3)},[\mathfrak{R}^{(1)}(t,2)]^{(3)},[\mathfrak{R}^{(1)}(t,3)]^{(3)}] \quad \text{for } t \geq 0 \tag{10.39}$$

where

$$[\mathfrak{R}^{(1)}(t,1)]^{(3)} = \mathfrak{R}'_9\left(\frac{t-b_n^{(3)}(1)}{a_n^{(3)}(1)},1\right) = 1-[1-\exp[-0.030t]]^2 \\ = 2\exp[-0.030t]-\exp[-0.060t] \tag{10.40}$$

$$[\mathfrak{R}^{(1)}(t,2)]^{(3)} = \mathfrak{R}'_9\left(\frac{t-b_n^{(3)}(2)}{a_n^{(3)}(2)},2\right) = 1-[1-\exp[-0.034t]]^2 \\ = 2\exp[-0.034t]-\exp[-0.068t] \tag{10.41}$$

$$[\mathfrak{R}^{(1)}(t,3)]^{(3)} = \mathfrak{R}'_9\left(\frac{t-b_n^{(3)}(3)}{a_n^{(3)}(3)},3\right) = 1-[1-\exp[-0.038t]]^2 \\ 2\exp[-0.038t]-\exp[-0.076t] \quad \text{for } t \geq 0 \tag{10.42}$$

The subsystem S_2 at the operation state z_3 is composed of four identical series systems; then we may apply Proposition 10.1. Namely, since the system structure shape parameters are

$$k^{(3)}=4,\ \ l^{(3)}=40,\ \ a^{(3)}=1,\ \ e_1^{(3)}=2$$

and

$$q_1^{(3)}=1,\ \ p_{11}^{(3)}=0.5,\ \ p_{12}^{(3)}=0.5$$

the number of reliability states is $z=3$ and considering the values of component' intensities of departure from the reliability state subsets defined in Example 9.1, by Eqs. (10.9)–(10.12), we have

$$[\lambda_i(1)]^{(3)} = 0.5\cdot 0.0010 + 0.5\cdot 0.0014 = 0.0012,$$

$$[\lambda_i(2)]^{(3)} = 0.5\cdot 0.0011 + 0.5\cdot 0.0016 = 0.00135,$$

$$[\lambda_i(3)]^{(3)} = 0.5\cdot 0.0012 + 0.5\cdot 0.0018 = 0.0015, \quad \text{for } i=1,2,3,4$$

$$[\lambda(1)]^{(3)} = \min\{0.0012\} = 0.0012,$$

$$[\lambda(2)]^{(3)} = \min\{0.00135\} = 0.00135,$$

$$[\lambda(3)]^{(3)} = \min\{0.0015\} = 0.0015$$

$$a_n^{(3)}(1) = \frac{1}{0.0012 \cdot 40} = \frac{1}{0.048}, \quad b_n^{(3)}(1) = 0,$$

$$a_n^{(3)}(2) = \frac{1}{0.00135 \cdot 40} = \frac{1}{0.054}, \quad b_n^{(3)}(2) = 0,$$

$$a_n^{(3)}(3) = \frac{1}{0.0015 \cdot 40} = \frac{1}{0.060}, \quad b_n^{(3)}(3) = 0$$

then according to Eqs. (10.13)–(10.15), after considering Eq. (10.6), the approximate formula for the subsystem system S_2 reliability function takes the form

$$\begin{aligned}[\boldsymbol{R}^{(2)}(t,\cdot)]^{(3)} &\cong \mathfrak{R}^{(2)}((t - b_n^{(3)}(u))/a_n^{(3)}(u),\cdot) \\ &= [[\mathfrak{R}^{(2)}(t,1)]^{(3)}, [\mathfrak{R}^{(2)}(t,2)]^{(3)}, [\mathfrak{R}^{(2)}(t,3)]^{(3)}] \quad \text{for } t \geq 0\end{aligned} \tag{10.43}$$

where

$$\begin{aligned}[\mathfrak{R}^{(2)}(t,1)]^{(3)} &= \mathfrak{R}'_9\left(\frac{t - b_n^{(3)}(1)}{a_n^{(3)}(1)}, 1\right) = 1 - [1 - \exp[-0.048t]]^4 \\ &= 4\exp[-0.048t] - 6\exp[-0.096t] \\ &\quad + 4\exp[-0.144t] - \exp[-0.192t]\end{aligned} \tag{10.44}$$

$$\begin{aligned}[\mathfrak{R}^{(2)}(t,2)]^{(3)} &= \mathfrak{R}'_9\left(\frac{t - b_n^{(3)}(2)}{a_n^{(3)}(2)}, 2\right) = 1 - [1 - \exp[-0.054t]]^4 \\ &= 4\exp[-0.054t] - 6\exp[-0.108t] \\ &\quad + 4\exp[-0.162t] - \exp[-0.216t]\end{aligned} \tag{10.45}$$

$$\begin{aligned}[\mathfrak{R}^{(2)}(t,3)]^{(3)} &= \mathfrak{R}'_9\left(\frac{t - b_n^{(3)}(3)}{a_n^{(3)}(3)}, 3\right) = 1 - [1 - \exp[-0.060t]]^4 \\ &= 4\exp[-0.060t] - 6\exp[-0.120t] \\ &\quad + 4\exp[-0.180t] - \exp[-0.240t] \quad \text{for } t \geq 0\end{aligned} \tag{10.46}$$

Considering that the system at the operation state z_3 is a four-state series system composed of subsystems S_1 and S_2, after applying the formulae appearing after Definition 3.4 and Eqs. (10.39)–(10.46), its conditional reliability function is given by

$$[\boldsymbol{R}(t,\cdot)]^{(3)} = [[\mathfrak{R}(t,1)]^{(3)}, [\mathfrak{R}(t,2)]^{(3)}, [\mathfrak{R}(t,3)]^{(3)}] \quad \text{for } t \geq 0 \tag{10.47}$$

where

$$\begin{aligned}[\mathfrak{R}(t,1)]^{(3)} &= [\mathfrak{R}^{(1)}(t,1)]^{(3)}[\mathfrak{R}^{(2)}(t,1)]^{(3)} \\ &= 8\exp[-0.078t] - 12\exp[-0.126t] + 8\exp[-0.174t] \\ &\quad - 2\exp[-0.212t] - 4\exp[-0.108t] + 6\exp[-0.156t] \\ &\quad - 4\exp[-0.204t] + \exp[-0.252t]\end{aligned} \tag{10.48}$$

$$\begin{aligned}[\mathfrak{R}(t,2)]^{(3)} &= [\mathfrak{R}^{(1)}(t,2)]^{(3)}[\mathfrak{R}^{(2)}(t,2)]^{(3)} \\ &= 8\exp[-0.088t] - 12\exp[-0.142t] + 8\exp[-0.196t] \\ &\quad - 2\exp[-0.250t] - 4\exp[-0.118t] + 6\exp[-0.172t] \\ &\quad - 4\exp[-0.226t] + \exp[-0.280t]\end{aligned} \tag{10.49}$$

$$\begin{aligned}[\mathfrak{R}(t,3)]^{(3)} &= [\mathfrak{R}^{(1)}(t,3)]^{(3)}[\mathfrak{R}^{(2)}(t,3)]^{(3)} \\ &= 8\exp[-0.098t] - 12\exp[-0.158t] + 8\exp[-0.218t] \\ &\quad - 2\exp[-0.278t] - 4\exp[-0.128t] + 6\exp[-0.188t] \\ &\quad - 4\exp[-0.248t] + \exp[-0.308t] \quad \text{for } t \geq 0\end{aligned} \tag{10.50}$$

At the operation state z_4, the considered system is a series system composed of a series–parallel subsystem S_1 and a series–'2 out of 4' subsystem S_2. The subsystem S_1 at the operation state z_4 is composed of two identical series systems and has the structure shape parameters and the intensities of its components' departure from the reliability state subsets the same as at the operation state z_1. Thus, applying Proposition 10.1 and proceeding in analogous way as in this case, we get the result identical with (10.31)–(10.35) what means that the approximate formula for the subsystem S_1 reliability function at the operation state z_4 takes the form

$$[\boldsymbol{R}^{(1)}(t,\cdot)]^{(4)} = [\mathfrak{R}^{(1)}(t,1)]^{(4)}, [\mathfrak{R}^{(1)}(t,2)]^{(4)}, [\mathfrak{R}^{(1)}(t,3)]^{(4)}] \quad \text{for } t \geq 0 \tag{10.51}$$

where

$$[\mathfrak{R}^{(1)}(t,1)]^{(4)} = 2\exp[-0.027t] - \exp[-0.054t] \tag{10.52}$$

$$[\mathfrak{R}^{(1)}(t,2)]^{(4)} = 2\exp[-0.030t] - \exp[-0.060t] \tag{10.53}$$

$$[\mathfrak{R}^{(1)}(t,3)]^{(4)} = 2\exp[-0.033t] - \exp[-0.066t] \quad \text{for } t \geq 0 \tag{10.54}$$

The series–'2 out of 4' subsystem S_2 at the operation state z_4 is composed of four identical series systems with the intensities of its components' departure from the reliability state subsets the same as at the operation state z_2. Thus, we may apply Proposition 10.2 to each of the series systems. Namely, since the system structure shape parameters are

$$n^{(4)} = 40, \quad a^{(4)} = 2$$

and

$$q_1^{(4)} = 0.5, \quad q_2^{(4)} = 0.5$$

the number of reliability states is $z = 3$ and considering the values of component' intensities of departure from the reliability state subsets defined in Example 9.1, by Eqs. (10.18)–(10.20), we have

$$[\lambda(1)]^{(4)} = \max\{0.0014, 0.0018\} = 0.0018,$$

$$[\lambda(2)]^{(4)} = \max\{0.0015, 0.0020\} = 0.0020,$$

$$[\lambda(3)]^{(4)} = \max\{0.0016, 0.0022\} = 0.0022$$

$$a_n^{(4)}(1) = \frac{1}{0.0018 \cdot 40} = \frac{1}{0.072}, \quad b_n^{(3)}(1) = 0,$$

$$a_n^{(4)}(2) = \frac{1}{0.0020 \cdot 40} = \frac{1}{0.080}, \quad b_n^{(3)}(2) = 0,$$

$$a_n^{(4)}(3) = \frac{1}{0.0022 \cdot 40} = \frac{1}{0.088}, \quad b_n^{(3)}(3) = 0$$

then according to Eqs. (10.22)–(10.23), after considering Eq. (10.6), the approximate formula for each of the series systems of the subsystem system S_2 reliability function takes the form

$$\begin{aligned}[\boldsymbol{R}_i^{(2)}(t,\cdot)]^{(4)} &\cong \mathfrak{R}_i^{(2)}((t - b_n^{(4)}(u))/a_n^{(4)}(u),\cdot) \\ &= [[\mathfrak{R}_i^{(2)}(t,1)]^{(4)}, [\mathfrak{R}_i^{(2)}(t,2)]^{(4)}, [\mathfrak{R}_i^{(2)}(t,3)]^{(4)}] \\ &\quad \text{for } t \geq 0, \ i = 1,2,3,4\end{aligned}$$

where

$$[\mathfrak{R}_i^{(2)}(t,1)]^{(4)} = \overline{\mathfrak{R}}_2'\left(\frac{t - b_n^{(4)}(1)}{a_n^{(4)}(1)}, 1\right) = \exp[-0.064t]$$

$$[\mathfrak{R}_i^{(2)}(t,2)]^{(4)} = \overline{\mathfrak{R}}_2'\left(\frac{t - b_n^{(4)}(2)}{a_n^{(4)}(2)}, 2\right) = \exp[-0.070t]$$

$$[\mathfrak{R}_i^{(2)}(t,3)]^{(4)} = \overline{\mathfrak{R}}_2'\left(\frac{t-b_n^{(4)}(3)}{a_n^{(4)}(3)},3\right) = \exp[-0.076t] \quad \text{for } t \ge 0, \quad i = 1,2,3,4$$

Further, using the above results, according to the formulae (3.24)–(3.25) for the homogeneous '2 out of 4' system reliability function, the reliability function of the series–'2 out of 4' subsystem S_2 at the operation state z_4 takes the form

$$[\boldsymbol{R}^{(2)}(t,\cdot)]^{(4)} = [[\mathfrak{R}^{(2)}(t,1)]^{(4)}, [\mathfrak{R}^{(2)}(t,2)]^{(4)}, [\mathfrak{R}^{(2)}(t,3)]^{(4)}] \quad \text{for } t \ge 0 \tag{10.55}$$

where

$$\begin{aligned}[\mathfrak{R}^{(2)}(t,1)]^{(4)} &= 1 - \sum_{i=0}^{1}\binom{4}{i}\exp[-i0.064t][1-\exp[-0.064t]]^{4-i} \\ &= 6\exp[-0.128t] - 8\exp[-0.192t] + 3\exp[-0.256t]\end{aligned} \tag{10.56}$$

$$\begin{aligned}[\mathfrak{R}^{(2)}(t,2)]^{(4)} &= 1 - \sum_{i=0}^{1}\binom{4}{i}\exp[-i0.070t][1-\exp[-0.070t]]^{4-i} \\ &= 6\exp[-0.140t] - 8\exp[-0.210t] + 3\exp[-0.280t]\end{aligned} \tag{10.57}$$

$$\begin{aligned}[\mathfrak{R}^{(2)}(t,3)]^{(4)} &= 1 - \sum_{i=0}^{1}\binom{4}{i}\exp[-i0.076t][1-\exp[-0.076t]]^{4-i} \\ &= 6\exp[-0.152t] - 8\exp[-0.228t] + 3\exp[-0.304t] \quad \text{for } t \ge 0\end{aligned} \tag{10.58}$$

Considering that the system at the operation state z_4 is a four-state series system composed of subsystems S_1 and S_2, after applying the formulae appearing after Definition 3.4 and Eqs. (10.51)–(10.58), its conditional reliability function is given by

$$[\boldsymbol{R}(t,\cdot)]^{(4)} = [[\mathfrak{R}(t,1)]^{(4)}, [\mathfrak{R}(t,2)]^{(4)}, [\mathfrak{R}(t,3)]^{(4)}] \quad \text{for } t \ge 0 \tag{10.59}$$

where

$$\begin{aligned}[\mathfrak{R}(t,1)]^{(4)} &= [\mathfrak{R}^{(1)}(t,1)]^{(4)}[\mathfrak{R}^{(2)}(t,1)]^{(4)} \\ &= 12\exp[-0.155t] - 6\exp[-0.182t] - 16\exp[-0.219t] \\ &\quad + 8\exp[-0.246t] + 6\exp[-0.283t] - 3\exp[-0.310t]\end{aligned} \tag{10.60}$$

$$[\mathfrak{R}(t,2)]^{(4)} = [\mathfrak{R}^{(1)}(t,2)]^{(4)}[\mathfrak{R}^{(2)}(t,2)]^{(4)}$$
$$= 12\exp[-0.170t] - 6\exp[-0.200t] - 16\exp[-0.240t] \quad (10.61)$$
$$+ 8\exp[-0.270t] + 6\exp[-0.310t] - 3\exp[-0.340t]$$

$$[\mathfrak{R}(t,3)]^{(4)} = [\mathfrak{R}^{(1)}(t,3)]^{(4)}[\mathfrak{R}^{(2)}(t,3)]^{(4)}$$
$$= 12\exp[-0.185t] - 6\exp[-0.218t] - 16\exp[-0.261t]$$
$$+ 8\exp[-0.294t] + 6\exp[-0.337t] - 3\exp[-0.370t] \quad \text{for } t \geq 0$$
$$(10.62)$$

Finally, considering Eqs. (10.6) and (9.18), the approximate formula for the system unconditional reliability function takes the form

$$[\boldsymbol{R}(t,\cdot)] = [[\mathfrak{R}(t,1)], [\mathfrak{R}(t,2)], [\mathfrak{R}(t,3)]] \quad \text{for } t \geq 0 \quad (10.63)$$

where

$$[\mathfrak{R}(t,1)] = 0.214[\mathfrak{R}(t,1)]^{(1)} + 0.038[\mathfrak{R}(t,1)]^{(2)} + 0.293[\mathfrak{R}(t,1)]^{(3)} + 0.455[\mathfrak{R}(t,1)]^{(4)} \quad (10.64)$$

$$[\mathfrak{R}(t,2)] = 0.214[\mathfrak{R}(t,2)]^{(1)} + 0.038[\mathfrak{R}(t,2)]^{(2)} + 0.293[\mathfrak{R}(t,2)]^{(3)} + 0.455[\mathfrak{R}(t,2)]^{(4)} \quad (10.65)$$

$$[\mathfrak{R}(t,3)] = 0.214[\mathfrak{R}(t,3)]^{(1)} + 0.038[\mathfrak{R}(t,3)]^{(2)} + 0.293[\mathfrak{R}(t,3)]^{(3)} + 0.455[\mathfrak{R}(t,3)]^{(4)} \quad \text{for } t \geq 0 \quad (10.66)$$

where $[\mathfrak{R}(t,1)]^{(1)}$, $[\mathfrak{R}(t,1)]^{(2)}$, $[\mathfrak{R}(t,1)]^{(3)}$, $[\mathfrak{R}(t,1)]^{(4)}$ are given by Eqs. (10.32), (10.36), (10.48), (10.60), $[\mathfrak{R}(t,2)]^{(1)}$, $[\mathfrak{R}(t,2)]^{(2)}$, $[\mathfrak{R}(t,2)]^{(3)}$, $[\mathfrak{R}(t,2)]^{(4)}$ are given by Eqs. (10.33), (10.37), (10.49), (10.61) and $[\mathfrak{R}(t,3)]^{(1)}$, $[\mathfrak{R}(t,3)]^{(2)}$, $[\mathfrak{R}(t,3)]^{(3)}$, $[\mathfrak{R}(t,3)]^{(4)}$ are given by Eqs. (10.34), (10.38), (10.50), (10.62).

The results given by Eqs. (10.63)–(10.66) are the same as the results (9.112)–(9.115), which means that the system approximate reliability function and the system exact reliability function are identical.

Example 10.2

We consider a large multi-state homogenous '24 out of 36'–series system changing its reliability structures and its components reliability parameters at variable operation conditions, analysed in Example 9.2.

Applying Proposition 10.3, since the shape parameters of the considered '24 out of 36'–series system at the operation states z_b, $b = 1, 2, 3$, respectively are

$$k^{(1)} = 3, \quad l^{(1)} = 36, \quad m^{(1)} = 24,$$
$$k^{(2)} = 2, \quad l^{(1)} = 36, \quad m^{(2)} = 24,$$
$$k^{(3)} = 1, \quad l^{(3)} = 36, \quad m^{(3)} = 24$$

the number of reliability states is $z=3$ and by Eqs. (10.25)–(10.27) and (9.126)–(9.124):

- at the operation state z_1

$$\mu^{(1)} = \frac{24}{36} = \frac{2}{3}$$

$$a_n^{(1)}(1) = \frac{1}{3\frac{2}{3}\sqrt{35}}\sqrt{\frac{2}{3}\left(1-\frac{2}{3}\right)} = \frac{1}{6\sqrt{17.5}}, \quad b_n^{(1)}(1) = -\frac{1}{3}\log\frac{2}{3},$$

$$a_n^{(1)}(2) = \frac{1}{6\frac{2}{3}\sqrt{35}}\sqrt{\frac{2}{3}\left(1-\frac{2}{3}\right)} = \frac{1}{12\sqrt{17.5}}, \quad b_n^{(1)}(2) = -\frac{1}{6}\log\frac{2}{3},$$

$$a_n^{(1)}(3) = \frac{1}{9\frac{2}{3}\sqrt{35}}\sqrt{\frac{2}{3}\left(1-\frac{2}{3}\right)} = \frac{1}{18\sqrt{17.5}}, \quad b_n^{(1)}(1) = -\frac{1}{9}\log\frac{2}{3}$$

- at the operation state z_2

$$\mu^{(2)} = \frac{24}{36} = \frac{2}{3}$$

$$a_n^{(2)}(1) = \frac{1}{2\frac{2}{3}\sqrt{35}}\sqrt{\frac{2}{3}\left(1-\frac{2}{3}\right)} = \frac{1}{4\sqrt{17.5}}, \quad b_n^{(2)}(1) = -\frac{1}{2}\log\frac{2}{3},$$

$$a_n^{(2)}(2) = \frac{1}{4\frac{2}{3}\sqrt{35}}\sqrt{\frac{2}{3}\left(1-\frac{2}{3}\right)} = \frac{1}{8\sqrt{17.5}}, \quad b_n^{(2)}(2) = -\frac{1}{4}\log\frac{2}{3},$$

$$a_n^{(2)}(3) = \frac{1}{6\frac{2}{3}\sqrt{35}}\sqrt{\frac{2}{3}\left(1-\frac{2}{3}\right)} = \frac{1}{12\sqrt{17.5}}, \quad b_n^{(2)}(1) = -\frac{1}{6}\log\frac{2}{3}$$

- at the operation state z_3

$$\mu^{(3)} = \frac{24}{36} = \frac{2}{3}$$

$$a_n^{(3)}(1) = \frac{1}{1\frac{2}{3}\sqrt{35}}\sqrt{\frac{2}{3}\left(1-\frac{2}{3}\right)} = \frac{1}{2\sqrt{17.5}},\quad b_n^{(3)}(1) = -\frac{1}{1}\log\frac{2}{3},$$

$$a_n^{(3)}(2) = \frac{1}{2\frac{2}{3}\sqrt{35}}\sqrt{\frac{2}{3}\left(1-\frac{2}{3}\right)} = \frac{1}{4\sqrt{17.5}},\quad b_n^{(3)}(2) = -\frac{1}{2}\log\frac{2}{3},$$

$$a_n^{(3)}(3) = \frac{1}{3\frac{2}{3}\sqrt{35}}\sqrt{\frac{2}{3}\left(1-\frac{2}{3}\right)} = \frac{1}{6\sqrt{17.5}},\quad b_n^{(3)}(1) = -\frac{1}{3}\log\frac{2}{3}$$

then according to Eqs. (10.28)–(10.30), after considering Eq. (10.6) and (9.20), the approximate formula for the system unconditional reliability function takes the form

$$\boldsymbol{R}(t,\cdot) \cong \mathfrak{R}((t-b_n(u))/a_n(u),\cdot) \cong [1, \mathfrak{R}(t,1), \mathfrak{R}(t,2), \mathfrak{R}(t,3)],\quad t\in(-\infty,\infty) \tag{10.67}$$

where

$$\begin{aligned}\mathfrak{R}(t,1) &\cong \sum_{b=1}^{v} p_b\left[\mathfrak{R}\left(\frac{t-b_n^{(b)}(1)}{a_n^{(b)}(1)},1\right)\right]^{(b)}\\
&= 0.4\left[1-\frac{1}{\sqrt{2\pi}}\int_{-\infty}^{6\sqrt{17.5}t+2\sqrt{17.5}\log(2/3)} e^{-x^2/2}\,dx\right]^3\\
&+ 0.4\left[1-\frac{1}{\sqrt{2\pi}}\int_{-\infty}^{4\sqrt{17.5}t+2\sqrt{17.5}\log(2/3)} e^{-x^2/2}\,dx\right]^2\\
&+ 0.2\left[1-\frac{1}{\sqrt{2\pi}}\int_{-\infty}^{2\sqrt{17.5}t+2\sqrt{17.5}\log(2/3)} e^{-x^2/2}\,dx\right]\end{aligned} \tag{10.68}$$

$$\mathfrak{R}(t,2) \cong \sum_{b=1}^{v} p_b \left[\mathfrak{R}\left(\frac{t-b_n^{(b)}(2)}{a_n^{(b)}(2)}, 2 \right) \right]^{(b)}$$
$$= 0.4 \left[1 - \frac{1}{\sqrt{2\pi}} \int_{-\infty}^{12\sqrt{17.5}t+2\sqrt{17.5}\log(2/3)} e^{-x^2/2}\, dx \right]^3$$
$$+ 0.4 \left[1 - \frac{1}{\sqrt{2\pi}} \int_{-\infty}^{8\sqrt{17.5}t+2\sqrt{17.5}\log(2/3)} e^{-x^2/2}\, dx \right]^2 \quad (10.69)$$
$$+ 0.2 \left[1 - \frac{1}{\sqrt{2\pi}} \int_{-\infty}^{4\sqrt{17.5}t+2\sqrt{17.5}\log(2/3)} e^{-x^2/2}\, dx \right]$$

$$\mathfrak{R}(t,3) \cong \sum_{b=1}^{v} p_b \left[\mathfrak{R}\left(\frac{t-b_n^{(b)}(3)}{a_n^{(b)}(3)}, 3 \right) \right]^{(b)}$$
$$= 0.4 \left[1 - \frac{1}{\sqrt{2\pi}} \int_{-\infty}^{18\sqrt{17.5}t+2\sqrt{17.5}\log(2/3)} e^{-x^2/2}\, dx \right]^3$$
$$+ 0.4 \left[1 - \frac{1}{\sqrt{2\pi}} \int_{-\infty}^{12\sqrt{17.5}t+2\sqrt{17.5}\log(2/3)} e^{-x^2/2}\, dx \right]^2 \quad (10.70)$$
$$+ 0.2 \left[1 - \frac{1}{\sqrt{2\pi}} \int_{-\infty}^{6\sqrt{17.5}t+2\sqrt{17.5}\log(2/3)} e^{-x^2/2}\, dx \right]$$

It is obvious that using the approximate formulae (10.68)–(10.70) is easier than using the exact formulae (9.128)–(9.130) for the considered system reliability function. In Table 10.1, the values of the coordinates of the system approximate reliability function (10.67) with the coordinates given by Eqs. (10.68)–(10.70) are presented.

The accuracy of this approximation is illustrated in Table 10.2, presenting the differences between the values of the system exact reliability function given in Table 9.1 and the values of the system approximate reliability function given in Table 10.1.

Table 10.1 The Values of the Coordinates of the '24 Out of 36'–Series System Approximate Reliability Function

t	$\mathfrak{R}(t, 1)$	t	$\mathfrak{R}(t, 2)$	t	$\mathfrak{R}(t, 3)$
0.00	0.9993	0.000	0.9993	0.00	0.9993
0.03	0.9934	0.015	0.9934	0.01	0.9934
0.06	0.9599	0.030	0.9599	0.02	0.9599
0.09	0.8629	0.045	0.8629	0.03	0.8629
0.12	0.6687	0.060	0.6687	0.04	0.6687
0.15	0.4890	0.750	0.4890	0.05	0.4890
0.18	0.3677	0.900	0.3677	0.06	0.3677
0.21	0.2790	0.105	0.2790	0.07	0.2790
0.24	0.2139	0.120	0.2139	0.08	0.2139
0.27	0.1824	0.135	0.1824	0.09	0.1824
0.30	0.1632	0.150	0.1632	0.10	0.1632
0.33	0.1484	0.165	0.1484	0.11	0.1484
0.36	0.1311	0.180	0.1311	0.12	0.1311
0.39	0.1119	0.195	0.1119	0.13	0.1119
0.42	0.0920	0.210	0.0920	0.14	0.0920
0.45	0.0726	0.225	0.0726	0.15	0.0726
0.48	0.0519	0.240	0.0519	0.16	0.0519
0.51	0.0395	0.255	0.0395	0.17	0.0395
0.54	0.0271	0.270	0.0271	0.18	0.0271
0.57	0.0177	0.285	0.0177	0.19	0.0177
0.60	0.0110	0.300	0.0110	0.20	0.0110
0.63	0.0064	0.315	0.0064	0.21	0.0064
0.66	0.0036	0.330	0.0036	0.22	0.0036
0.69	0.0019	0.345	0.0019	0.23	0.0019
0.72	0.0009	0.360	0.0009	0.24	0.0009
0.75	0.0004	0.375	0.0004	0.25	0.0004
0.78	0.0002	0.390	0.0002	0.26	0.0002
0.81	0.0001	0.405	0.0001	0.27	0.0001
0.84	0.0000	0.420	0.0000	0.28	0.0000

Example 10.3

We consider a port grain transportation system composed of three large multi-state non-homogeneous series–parallel subsystems linked in series changing its reliability structures and its components reliability parameters at variable operation conditions, analysed in Example 9.3.

At the operation state z_1, the considered system is a series system composed of four series–parallel subsystems S_1, S_2, S_3 and S_4. The subsystem S_1 at the operation state z_1 is composed of two identical belt conveyors forming series systems, then we may apply Proposition 10.1. Namely, since the system structure shape parameters are

$$k^{(1)} = 2, \quad l^{(1)} = 129, \quad a^{(1)} = 1, \quad e_1^{(1)} = 4$$

Table 10.2 The Differences Between the Values of the Coordinates of the '24 Out of 36'–Series System Exact and Approximate Reliability Function

t	$R(t,1)-\Re(t,1)$	t	$R(t,2)-\Re(t,2)$	t	$R(t,3)-\Re(t,3)$
0.00	0.0007	0.000	0.0007	0.00	0.0007
0.03	0.0066	0.015	0.0066	0.01	0.0066
0.06	0.0358	0.030	0.0358	0.02	0.0358
0.09	**0.0625**	0.045	**0.0625**	0.03	**0.0625**
0.12	0.0607	0.060	0.0607	0.04	0.0607
0.15	0.0556	0.750	0.0556	0.05	0.0556
0.18	0.0457	0.900	0.0457	0.06	0.0457
0.21	0.0308	0.105	0.0308	0.07	0.0308
0.24	0.0268	0.120	0.0268	0.08	0.0268
0.27	0.0205	0.135	0.0205	0.09	0.0205
0.30	0.0179	0.150	0.0179	0.10	0.0179
0.33	0.0152	0.165	0.0152	0.11	0.0152
0.36	0.0144	0.180	0.0144	0.12	0.0144
0.39	0.0140	0.195	0.0140	0.13	0.0140
0.42	0.0138	0.210	0.0138	0.14	0.0138
0.45	0.0137	0.225	0.0137	0.15	0.0137
0.48	0.0157	0.240	0.0157	0.16	0.0157
0.51	0.0132	0.255	0.0132	0.17	0.0132
0.54	0.0125	0.270	0.0125	0.18	0.0125
0.57	0.0113	0.285	0.0113	0.19	0.0113
0.60	0.0098	0.300	0.0098	0.20	0.0098
0.63	0.0082	0.315	0.0082	0.21	0.0082
0.66	0.0065	0.330	0.0065	0.22	0.0065
0.69	0.0049	0.345	0.0049	0.23	0.0049
0.72	0.0035	0.360	0.0035	0.24	0.0035
0.75	0.0026	0.375	0.0026	0.25	0.0026
0.78	0.0017	0.390	0.0017	0.26	0.0017
0.81	0.0011	0.405	0.0011	0.27	0.0011
0.84	0.0008	0.420	0.0008	0.28	0.0008
0.87	0.0005	0.435	0.0005	0.29	0.0005
0.90	0.0003	0.450	0.0003	0.30	0.0003
0.93	0.0002	0.465	0.0002	0.31	0.0002
0.96	0.0001	0.480	0.0001	0.32	0.0001
0.99	0.0001	0.495	0.0001	0.33	0.0001
1.02	0.0000	0.510	0.0000	0.34	0.0000

and

$$q_1^{(1)} = 1,\ p_{11}^{(1)} = \frac{1}{129},\ p_{12}^{(1)} = \frac{2}{129},\ p_{13}^{(1)} = \frac{117}{129},\ p_{14}^{(1)} = \frac{9}{129}$$

the number of reliability states is $z = 3$ and considering the values of component' intensities of departure from the reliability state subsets defined in Example 9.3, by Eqs. (10.9)–(10.12), we have

$$[\lambda_i(1)]^{(1)} = \frac{1}{129}0.0125 + \frac{2}{129}0.0015 + \frac{117}{129}0.005 + \frac{9}{129}0.004 = 0.6365/129$$

$$[\lambda_i(2)]^{(1)} = \frac{1}{129}0.0150 + \frac{2}{129}0.0020 + \frac{117}{129}0.010 + \frac{9}{129}0.008 = 1.2610/129,$$

$$i = 1, 2$$

$$[\lambda(1)]^{(1)} = \min\{0.6365/129\} = 0.6365/129,$$

$$[\lambda(2)]^{(1)} = \min\{1.2610/129\} = 1.2610/129$$

$$a_n^{(1)}(1) = \frac{1}{(0.6365/129) \cdot 129} = \frac{1}{0.6365}, \quad b_n^{(1)}(1) = 0,$$

$$a_n^{(1)}(2) = \frac{1}{(1.2610/129) \cdot 129} = \frac{1}{1.2610}, \quad b_n^{(1)}(2) = 0$$

then according to Eqs. (10.13)–(10.15), after considering Eq. (10.6), the approximate formula for the subsystem S_1 reliability function takes the form

$$\begin{aligned}[\boldsymbol{R}^{(1)}(t, \cdot)]^{(1)} &\cong \mathfrak{R}^{(1)}((t - b_n^{(1)}(u))/a_n^{(1)}(u), \cdot) \\ &= [[\mathfrak{R}^{(1)}(t, 1)]^{(1)}, [\mathfrak{R}^{(1)}(t, 2)]^{(1)}] \quad \text{for } t \geq 0\end{aligned} \tag{10.71}$$

where

$$\begin{aligned}[\mathfrak{R}^{(1)}(t, 1)]^{(1)} &= \mathfrak{R}'_9\left(\frac{t - b_n^{(1)}(1)}{a_n^{(1)}(1)}, 1\right) = 1 - [1 - \exp[-0.6365t]]^2 \\ &= 2\exp[-0.6365t] - \exp[-1.2730t]\end{aligned} \tag{10.72}$$

$$\begin{aligned}[\mathfrak{R}^{(1)}(t, 2)]^{(1)} &= \mathfrak{R}'_9\left(\frac{t - b_n^{(1)}(2)}{a_n^{(1)}(2)}, 2\right) = 1 - [1 - \exp[-1.2610t]]^2 \\ &= 2\exp[-1.2610t] - \exp[-2.5220t] \quad \text{for } t \geq 0\end{aligned} \tag{10.73}$$

The subsystem S_2 at the operation state z_1 is composed of three identical bucket elevators forming series systems, then we may apply Proposition 10.1. Namely, since the system structure shape parameters are

$$k^{(1)} = 3, \quad l^{(1)} = 743, \quad a^{(1)} = 1, \quad e_1^{(1)} = 3$$

and

$$q_1^{(1)} = 1, \quad p_{11}^{(1)} = \frac{1}{743}, \quad p_{12}^{(1)} = \frac{2}{743}, \quad p_{13}^{(1)} = \frac{740}{743}$$

the number of reliability states is $z = 3$ and considering the values of component' intensities of departure from the reliability state subsets defined in Example 9.3, by Eqs. (10.9)–(10.12), we have

$$[\lambda_i(1)]^{(1)} = \frac{1}{743}0.025 + \frac{2}{743}0.0015 + \frac{740}{743}0.030 = 22.2280/743,$$

$$[\lambda_i(2)]^{(1)} = \frac{1}{743}0.030 + \frac{2}{743}0.0020 + \frac{740}{743}0.035 = 25.9340/743, \quad i = 1, 2, 3$$

$$[\lambda(1)]^{(1)} = \min\{22.2280/743\} = 22.2280/743,$$

$$[\lambda(2)]^{(1)} = \min\{25.9340/743\} = 25.9340/743$$

$$a_n^{(1)}(1) = \frac{1}{(22.2280/743) \cdot 743} = \frac{1}{22.2280}, \quad b_n^{(1)}(1) = 0,$$

$$a_n^{(1)}(2) = \frac{1}{(25.9340/743) \cdot 743} = \frac{1}{25.9340}, \quad b_n^{(1)}(2) = 0$$

then according to Eqs. (10.13)–(10.15), after considering Eq. (10.6), the approximate formula for the subsystem S_2 reliability function takes the form

$$\begin{aligned}[\boldsymbol{R}^{(2)}(t, \cdot)]^{(1)} &\cong \Re^{(2)}((t - b_n^{(1)}(u))/a_n^{(1)}(u), \cdot) \\ &= [[\Re^{(2)}(t, 1)]^{(1)}, [\Re^{(2)}(t, 2)]^{(1)}] \quad \text{for } t \geq 0\end{aligned} \tag{10.74}$$

where

$$\begin{aligned}[\Re^{(2)}(t, 1)]^{(1)} &= \Re'_9\left(\frac{t - b_n^{(1)}(1)}{a_n^{(1)}(1)}, 1\right) = 1 - [1 - \exp[-22.2280t]]^3 \\ &= 3\exp[-22.2280t] - 3\exp[-44.4560t] \\ &\quad + \exp[-66.6840t]\end{aligned} \tag{10.75}$$

$$\begin{aligned}[\Re^{(2)}(t, 2)]^{(1)} &= \Re'_9\left(\frac{t - b_n^{(1)}(2)}{a_n^{(1)}(2)}, 2\right) = 1 - [1 - \exp[-25.9340t]]^3 \\ &= 3\exp[-25.9340t] - 3\exp[-51.8680t] \\ &\quad + \exp[-77.8020t]\end{aligned} \tag{10.76}$$

The subsystem S_3 at the operation state z_1 is composed of two identical belt conveyors forming series systems, then we may apply Proposition 10.1. Namely, since the system structure shape parameters are

$$k^{(1)} = 2,\ \ l^{(1)} = 139,\ \ l^{(1)} = 139,\ \ e_1^{(1)} = 4$$

and

$$q_1^{(1)} = 1,\ p_{11}^{(1)} = \frac{1}{139},\ p_{12}^{(1)} = \frac{2}{139},\ p_{13}^{(1)} = \frac{117}{139},\ p_{14}^{(1)} = \frac{19}{139}$$

the number of reliability states is $z = 3$ and considering the values of component' intensities of departure from the reliability state subsets defined in Example 9.3, by Eqs. (10.9)–(10.12), we have

$$[\lambda_i(1)]^{(1)} = \frac{1}{139}0.0125 + \frac{2}{139}0.0015 + \frac{117}{139}0.005 + \frac{19}{139}0.004 = 0.6765/139,$$

$$[\lambda_i(2)]^{(1)} = \frac{1}{139}0.0150 + \frac{2}{139}0.0020 + \frac{117}{139}0.010 + \frac{19}{139}0.008 = 1.3410/139,$$

$$i = 1, 2$$

$$[\lambda(1)]^{(1)} = \min\{0.6765/139\} = 0.6765/139,$$

$$[\lambda(2)]^{(1)} = \min\{1.3410/139\} = 1.3410/139$$

$$a_n^{(1)}(1) = \frac{1}{(0.6765/139)\cdot 139} = \frac{1}{0.6765},\ \ b_n^{(1)}(1) = 0,$$

$$a_n^{(1)}(2) = \frac{1}{(1.3410/139)\cdot 139} = \frac{1}{1.3410},\ \ b_n^{(1)}(2) = 0$$

then according to Eqs. (10.13)–(10.15), after considering Eq. (10.6), the approximate formula for the subsystem S_3 reliability function takes the form

$$\begin{aligned}[\boldsymbol{R}^{(3)}(t,\cdot)]^{(1)} &\cong \mathfrak{R}^{(3)}((t - b_n^{(1)}(u))/a_n^{(1)}(u),\cdot)\\ &= [[\mathfrak{R}^{(3)}(t,1)]^{(1)}, [\mathfrak{R}^{(3)}(t,2)]^{(1)}] \quad \text{for } t \geq 0\end{aligned} \tag{10.77}$$

where

$$\begin{aligned}[\mathfrak{R}^{(3)}(t,1)]^{(1)} &= \mathfrak{R}'_9\left(\frac{t - b_n^{(1)}(1)}{a_n^{(1)}(1)}, 1\right) = 1 - [1 - \exp[-0.6765t]]^2\\ &= 2\exp[-0.6765t] - \exp[-1.3530t]\end{aligned} \tag{10.78}$$

$$[\mathfrak{R}^{(3)}(t,2)]^{(1)} = \mathfrak{R}'_9\left(\frac{t-b_n^{(1)}(2)}{a_n^{(1)}(2)},2\right) = 1-[1-\exp[-1.3410t]]^2$$
$$= 2\exp[-1.3410t] - \exp[-2.6820t] \quad \text{for } t \geq 0 \tag{10.79}$$

The subsystem S_4 at the operation state z_1 is a non-regular series–parallel system composed of three chain conveyors forming series systems. To make it regular, we conventionally complete its two first conveyors that have 162 components with 80 components that do not fail, i.e. we assume that the additional components have 'reliability functions' with the following coordinates:

$$[R_{ij}^{(4)}(t,1)]^{(1)} = \exp[-[\lambda_{ij}^{(4)}(1)]^{(1)}t)], [R_{ij}^{(4)}(t,2)]^{(1)} = \exp[-[\lambda_{ij}^{(4)}(2)]^{(1)}t)] \quad \text{for } t \geq 0,$$
$$i = 1,2, \; j = 163, 164, \ldots, 242$$

where

$$[\lambda_{ij}^{(4)}(1)]^{(1)} = 0[\lambda_{ij}^{(4)}(2)]^{(1)} = 0 \quad \text{for } i = 1,2, \; j = 163, 164, \ldots, 242 \tag{10.80}$$

After that modification, we may apply Proposition 10.1. Namely, since the system structure shape parameters are

$$k^{(1)} = 3, \quad l^{(1)} = 242, \quad a^{(1)} = 2, \quad e_1^{(1)} = 3, \quad e_2^{(1)} = 2$$

and

$$q_1^{(1)} = \frac{2}{3}, \; p_{11}^{(1)} = \frac{2}{242}, \; p_{12}^{(1)} = \frac{16}{242}, \; p_{13}^{(1)} = \frac{80}{242}$$
$$q_2^{(1)} = \frac{1}{3}, \; p_{21}^{(1)} = \frac{2}{242}, \; p_{22}^{(1)} = \frac{240}{242}$$

the number of reliability states is $z = 3$ and considering the values of component' intensities of departure from the reliability state subsets defined in Example 9.3 and in Eq. (10.80), by Eqs. (10.9)–(10.12), we have

$$[\lambda_i(1)]^{(1)} = \frac{2}{242}0.005 + \frac{160}{242}0.012 + \frac{80}{242}0.0 = 1.93/242,$$
$$[\lambda_i(2)]^{(1)} = \frac{2}{242}0.006 + \frac{160}{242}0.0014 + \frac{80}{242}0.0 = 2.252/242, \quad i = 1,2$$

$$[\lambda_3(1)]^{(1)} = \frac{2}{242}0.022 + \frac{240}{242}0.034 = 8.204/242,$$
$$[\lambda_3(2)]^{(1)} = \frac{2}{242}0.024 + \frac{240}{242}0.040 = 9.648/242$$

$$[\lambda(1)]^{(1)} = \min\{1.93/242, 8.204/242\} = 1.93/242,$$
$$[\lambda(2)]^{(1)} = \min\{2.252/242, 9.648/242] = 2.252/242$$

$$a_n^{(1)}(1) = \frac{1}{(1.93/242)\cdot 242} = \frac{1}{1.93}, \quad b_n^{(1)}(1) = 0,$$
$$a_n^{(1)}(2) = \frac{1}{(2.252/242)\cdot 242} = \frac{1}{2.252}, \quad b_n^{(1)}(2) = 0$$

then according to Eqs. (10.13)–(10.15), after considering Eq. (10.6), the approximate formula for the subsystem S_4 reliability function takes the form

$$\begin{aligned}[\boldsymbol{R}^{(4)}(t,\cdot)]^{(1)} &\cong \mathfrak{R}^{(4)}((t - b_n^{(1)}(u))/a_n^{(1)}(u),\cdot) \\ &= [[\mathfrak{R}^{(4)}(t,1)]^{(1)}, [\mathfrak{R}^{(4)}(t,2)]^{(1)}] \quad \text{for } t \geq 0\end{aligned} \tag{10.81}$$

where

$$\begin{aligned}[\mathfrak{R}^{(4)}(t,1)]^{(1)} &\cong \mathfrak{R}'_9\left(\frac{t - b_n^{(1)}(1)}{a_n^{(1)}(1)}, 1\right) = 1 - [1 - \exp[-1.93t]]^2 \\ &= 2\exp[-1.93t] - \exp[-3.86t]\end{aligned} \tag{10.82}$$

$$\begin{aligned}[\mathfrak{R}^{(4)}(t,2)]^{(1)} &= \mathfrak{R}'_9\left(\frac{t - b_n^{(1)}(2)}{a_n^{(1)}(2)}, 2\right) = 1 - [1 - \exp[-2.252]]^2 \\ &= 2\exp[-2.252t] - \exp[-4.504t] \quad \text{for } t \geq 0\end{aligned} \tag{10.83}$$

Considering that the system at the operation state z_1 is a four-state series system composed of subsystems S_1, S_2, S_3 and S_4, after applying the formulae appearing after Definition 3.4 and Eqs. (10.71)–(10.79) and (10.81)–(10.83), its conditional reliability function is given by

$$[\boldsymbol{R}(t,\cdot)]^{(1)} = [[\mathfrak{R}(t,1)]^{(1)}, [\mathfrak{R}(t,2)]^{(1)}] \quad \text{for } t \geq 0 \tag{10.84}$$

where

$$\begin{aligned}[\mathfrak{R}(t,1)]^{(1)} &= [\mathfrak{R}^{(1)}(t,1)]^{(1)}[\mathfrak{R}^{(2)}(t,1)]^{(1)}[\mathfrak{R}^{(3)}(t,1)]^{(1)}[\mathfrak{R}^{(4)}(t,1)]^{(1)} \\ &= 24\exp[-25.4710t] - 24\exp[-47.6990t] + 8\exp[-69.9260t] \\ &\quad - 12\exp[-26.1075t] + 12\exp[-48.3355t] - 4\exp[-70.5635t] \\ &\quad - 12\exp[-27.4010t] + 12\exp[-49.6290t] - 4\exp[-71.8570t] \\ &\quad + 6\exp[-28.0375t] - 6\exp[-50.2655t] + 2\exp[-72.4935t] \\ &\quad - 12\exp[-26.1475t] + 12\exp[-48.3755t] - 4\exp[-70.6025t] \\ &\quad + 6\exp[-26.7840t] - 6\exp[-49.0120t] + 2\exp[-71.2400t] \\ &\quad + 6\exp[-28.0775t] - 6\exp[-50.3055t] + 2\exp[-72.5325t] \\ &\quad - 3\exp[-28.7140t] + 3\exp[-50.9420t] - \exp[-73.1700t]\end{aligned} \tag{10.85}$$

$$\begin{aligned}[\mathfrak{R}(t,2)]^{(1)} &= [\mathfrak{R}^{(1)}(t,2)]^{(1)}[\mathfrak{R}^{(2)}(t,2)]^{(1)}[\mathfrak{R}^{(3)}(t,2)]^{(1)}[\mathfrak{R}^{(4)}(t,2)]^{(1)}\\
&= 24\exp[-30.7880t]-24\exp[-56.7220t]+8\exp[-82.6560t]\\
&\quad -12\exp[-32.0490t]+12\exp[-57.9830t]-4\exp[-83.9170t]\\
&\quad -12\exp[-33.0400t]+12\exp[-58.9740t]-4\exp[-84.9080t]\\
&\quad +6\exp[-34.3010t]-6\exp[-60.2350t]+2\exp[-86.1690t]\\
&\quad -12\exp[-32.1290t]+12\exp[-58.0630t]-4\exp[-83.9970t]\\
&\quad +6\exp[-33.3900t]-6\exp[-59.3240t]+2\exp[-85.2580t]\\
&\quad +6\exp[-34.3810t]-6\exp[-60.3150t]+2\exp[-86.2490t]\\
&\quad -3\exp[-35.6420t]+3\exp[-61.5760t]-\exp[-87.5100t] \quad \text{for } t\geq 0
\end{aligned} \tag{10.86}$$

At the operation state z_2, the considered system is a series system composed of four either series or series–parallel subsystems S_1, S_2, S_3 and S_4. The subsystem S_1 at the operation state z_2 is composed of one belt conveyor forming series systems, then we may apply Proposition 10.1. Namely, since the system structure shape parameters are

$$k^{(2)} = 1,\ \ l^{(2)} = 129,\ \ a^{(2)} = 1,\ \ e_1^{(2)} = 4$$

and

$$q_1^{(2)} = 1,\ \ p_{11}^{(2)} = \frac{1}{129},\ \ p_{12}^{(2)} = \frac{2}{129},\ \ p_{13}^{(2)} = \frac{117}{129},\ \ p_{14}^{(2)} = \frac{9}{129}$$

the number of reliability states is $z = 3$ and considering the values of component' intensities of departure from the reliability state subsets defined in Example 9.3, by Eqs. (10.9)–(10.12), we have

$$[\lambda_i(1)]^{(2)} = \frac{1}{129}0.0110 + \frac{2}{129}0.0010 + \frac{117}{129}0.003 + \frac{9}{129}0.002 = 0.3710/129,$$

$$[\lambda_i(2)]^{(2)} = \frac{1}{129}0.0120 + \frac{2}{129}0.0012 + \frac{117}{129}0.006 + \frac{9}{129}0.003 = 0.7434/129,$$

$$i = 1$$

$$[\lambda(1)]^{(2)} = \min\{0.3710/129\} = 0.3710/129,$$

$$[\lambda(2)]^{(2)} = \min\{0.7434/129\} = 0.7434/129$$

$$a_n^{(2)}(1) = \frac{1}{(0.3710/129)\cdot 129} = \frac{1}{0.3710},\ \ b_n^{(2)}(1) = 0,$$

$$a_n^{(2)}(2) = \frac{1}{(1.2610/129)\cdot 129} = \frac{1}{1.2610},\ \ b_n^{(2)}(2) = 0$$

then according to Eqs. (10.13)–(10.15), after considering Eq. (10.6), the approximate formula for the subsystem S_1 reliability function takes the form

$$[\boldsymbol{R}^{(1)}(t,\cdot)]^{(2)} \cong \mathfrak{R}^{(1)}((t-b_n^{(2)}(u))/a_n^{(2)}(u),\cdot) = [[\mathfrak{R}^{(1)}(t,1)]^{(2)}, [\mathfrak{R}^{(1)}(t,2)]^{(2)}] \quad \text{for } t \geq 0 \tag{10.87}$$

where

$$[\mathfrak{R}^{(1)}(t,1)]^{(2)} = \mathfrak{R}'_9\left(\frac{t-b_n^{(2)}(1)}{a_n^{(2)}(1)},1\right) = 1-[1-\exp[-0.3710t]]^1 = \exp[-0.3710t] \tag{10.88}$$

$$[\mathfrak{R}^{(1)}(t,2)]^{(2)} = \mathfrak{R}'_9\left(\frac{t-b_n^{(2)}(2)}{a_n^{(2)}(2)},2\right) = 1-[1-\exp[-0.7434t]]^1 = \exp[-0.7434t] \quad \text{for } t \geq 0 \tag{10.89}$$

The subsystem S_2 at the operation state z_2 is composed of two identical bucket elevators forming series systems, then we may apply Proposition 10.1. Namely, since the system structure shape parameters are

$$k^{(2)} = 2, \ \ l^{(2)} = 743, \ \ a^{(2)} = 1, \ \ e_1^{(2)} = 3$$

and

$$q_1^{(2)} = 1, \ \ p_{11}^{(2)} = \frac{1}{743}, \ \ p_{12}^{(2)} = \frac{2}{743}, \ \ p_{13}^{(2)} = \frac{740}{743}$$

the number of reliability states is $z = 3$ and considering the values of component' intensities of departure from the reliability state subsets defined in Example 9.3, by Eqs. (10.9)–(10.12), we have

$$[\lambda_i(1)]^{(2)} = \frac{1}{743}0.020 + \frac{2}{743}0.0012 + \frac{740}{743}0.022 = 16.324/743,$$

$$[\lambda_i(2)]^{(2)} = \frac{1}{743}0.024 + \frac{2}{743}0.0014 + \frac{740}{743}0.025 = 18.552/743, \quad i = 1,2$$

$$[\lambda(1)]^{(2)} = \min\{16.324/743\} = 16.324/743,$$

$$[\lambda(2)]^{(2)} = \min\{18.552/743\} = 18.552/743$$

$$a_n^{(2)}(1) = \frac{1}{(16.324/743)\cdot 743} = \frac{1}{16.324}, \quad b_n^{(2)}(1) = 0,$$

$$a_n^{(2)}(2) = \frac{1}{(18.552/743)\cdot 743} = \frac{1}{18.552}, \quad b_n^{(2)}(2) = 0$$

then according to Eqs. (10.13)–(10.15), after considering Eq. (10.6), the approximate formula for the subsystem S_2 reliability function takes the form

$$\begin{aligned}[\boldsymbol{R}^{(2)}(t,\cdot)]^{(2)} &\cong \mathfrak{R}^{(2)}((t-b_n^{(2)}(u))/a_n^{(2)}(u),\cdot) \\ &= [[\mathfrak{R}^{(2)}(t,1)]^{(2)}, [\mathfrak{R}^{(2)}(t,2)]^{(2)}] \quad \text{for } t \geq 0\end{aligned} \tag{10.90}$$

where

$$\begin{aligned}[\mathfrak{R}^{(2)}(t,1)]^{(2)} &= \mathfrak{R}'_9\left(\frac{t-b_n^{(2)}(1)}{a_n^{(2)}(1)},1\right) 1-[1-\exp[-16.324t]]^2 \\ &= 2\exp[-16.324t]-\exp[-32.648t]\end{aligned} \tag{10.91}$$

$$\begin{aligned}[\mathfrak{R}^{(2)}(t,2)]^{(2)} &= \mathfrak{R}'_9\left(\frac{t-b_n^{(2)}(2)}{a_n^{(2)}(2)},2\right) = 1-[1-\exp[-18.552t]]^2 \\ &= 2\exp[-18.552t]-\exp[-37.104t] \quad \text{for } t \geq 0\end{aligned} \tag{10.92}$$

The subsystem S_3 at the operation state z_2 is composed of one belt conveyors forming series system, then we may apply Proposition 10.1. Namely, since the system structure shape parameters are

$$k^{(2)} = 1, \quad l^{(2)} = 139, \quad a^{(2)} = 1, \quad e_1^{(2)} = 4$$

and

$$q_1^{(2)} = 1, \quad p_{11}^{(2)} = \frac{1}{139}, \quad p_{12}^{(2)} = \frac{2}{139}, \quad p_{12}^{(2)} = \frac{117}{139}, \quad p_{14}^{(2)} = \frac{19}{139}$$

the number of reliability states is $z = 3$ and considering the values of component' intensities of departure from the reliability state subsets defined in Example 9.3, by Eqs. (10.9)–(10.12), we have

$$[\lambda_i(1)]^{(2)} = \frac{1}{139}0.0110 + \frac{2}{139}0.0010 + \frac{117}{139}0.003 + \frac{19}{139}0.002 = 0.3910/139$$

$$[\lambda_i(2)]^{(2)} = \frac{1}{139}0.0120 + \frac{2}{139}0.0012 + \frac{117}{139}0.006 + \frac{19}{139}0.003 = 0.7734/139,$$

$$i = 1$$

$$[\lambda(1)]^{(2)} = \min\{0.3910/139\} = 0.3910/139,$$

$$[\lambda(2)]^{(2)} = \min\{0.7734/139\} = 0.7734/139$$

$$a_n^{(2)}(1) = \frac{1}{(0.3910/139)\cdot 139} = \frac{1}{0.3910}, \quad b_n^{(1)}(1) = 0,$$

$$a_n^{(2)}(2) = \frac{1}{(0.7734/139)\cdot 139} = \frac{1}{0.7734}, \quad b_n^{(2)}(2) = 0$$

then according to Eqs. (10.13)–(10.15), after considering Eq. (10.6), the approximate formula for the subsystem S_3 reliability function takes the form

$$\begin{aligned}[\boldsymbol{R}^{(3)}(t,\cdot)]^{(2)} &\cong \mathfrak{R}^{(3)}((t-b_n^{(2)}(u))/a_n^{(2)}(u),\cdot)\\ &= [[\mathfrak{R}^{(3)}(t,1)]^{(2)},[\mathfrak{R}^{(3)}(t,2)]^{(2)}] \quad \text{for } t \geq 0\end{aligned} \tag{10.93}$$

where

$$\begin{aligned}[\mathfrak{R}^{(3)}(t,1)]^{(2)} &= \mathfrak{R}'_9\left(\frac{t-b_n^{(2)}(1)}{a_n^{(2)}(1)},1\right) = 1-[1-\exp[-0.3910t]]^1\\ &= \exp[-0.3910t]\end{aligned} \tag{10.94}$$

$$\begin{aligned}[\mathfrak{R}^{(3)}(t,2)]^{(2)} &= \mathfrak{R}'_9\left(\frac{t-b_n^{(2)}(2)}{a_n^{(2)}(2)},2\right) = 1-[1-\exp[-0.7734t]]^1\\ &= \exp[-0.7734t] \quad \text{for } t \geq 0\end{aligned} \tag{10.95}$$

The subsystem S_4 at the operation state z_2 is a regular series–parallel system composed of two identical chain conveyors forming series systems, then we may apply Proposition 10.1. Namely, since the system structure shape parameters are

$$k^{(2)} = 2, \quad l^{(2)} = 162, \quad a^{(2)} = 1, \quad e_1^{(2)} = 2$$

and

$$q_1^{(2)} = 1, \quad p_{11}^{(2)} = \frac{2}{162}, \quad p_{12}^{(2)} = \frac{160}{162}$$

the number of reliability states is $z = 3$ and considering the values of component' intensities of departure from the reliability state subsets defined in Example 9.3, by Eqs. (10.9)–(10.12), we have

$$[\lambda_i(1)]^{(?)} = \frac{2}{162}0.002 + \frac{160}{162}0.008 = 1.284/162,$$

$$[\lambda_i(2)]^{(2)} = \frac{2}{162}0.004 + \frac{160}{162}0.0010 = 1.608/162, \quad i = 1,2$$

$$[\lambda(1)]^{(2)} = \min\{1.284/162\} = 1.284/162,$$

$$[\lambda(2)]^{(2)} = \min\{1.608/162] = 1.608/162$$

$$a_n^{(2)}(1) = \frac{1}{(1.284/162)\cdot 162} = \frac{1}{1.284}, \quad b_n^{(2)}(1) = 0,$$

$$a_n^{(2)}(2) = \frac{1}{(1.608/162)\cdot 162} = \frac{1}{1.608}, \quad b_n^{(2)}(2) = 0$$

then according to Eqs. (10.13)–(10.15), after considering Eq. (10.6), the approximate formula for the subsystem S_4 reliability function takes the form

$$\begin{aligned}[\boldsymbol{R}^{(4)}(t,\cdot)]^{(1)} &\cong \mathfrak{R}^{(4)}((t - b_n^{(1)}(u))/a_n^{(1)}(u), \cdot) \\ &= [[\mathfrak{R}^{(4)}(t,1)]^{(1)}, [\mathfrak{R}^{(4)}(t,2)]^{(1)}] \quad \text{for } t \geq 0\end{aligned} \tag{10.96}$$

where

$$\begin{aligned}[\mathfrak{R}^{(4)}(t,1)]^{(2)} &\cong \mathfrak{R}'_9\left(\frac{t - b_n^{(2)}(1)}{a_n^{(2)}(1)}, 1\right) = 1 - [1 - \exp[-1.284t]]^2 \\ &= 2\exp[-1.284t] - \exp[-2.5680t]\end{aligned} \tag{10.97}$$

$$\begin{aligned}[\mathfrak{R}^{(4)}(t,2)]^{(2)} &= \mathfrak{R}'_9\left(\frac{t - b_n^{(2)}(2)}{a_n^{(2)}(2)}, 2\right) = 1 - [1 - \exp[-1.608t]]^2 \\ &= 2\exp[-1.608t] - \exp[-3.2160t] \quad \text{for } t \geq 0\end{aligned} \tag{10.98}$$

Considering that the system at the operation state z_2 is a four-state series system composed of subsystems S_1, S_2, S_3 and S_4, after applying the formulae appearing after Definition 3.4 and Eqs. (10.71)–(10.79) and (10.81)–(10.83), its conditional reliability function is given by

$$[\boldsymbol{R}(t,\cdot)]^{(2)} = [[\mathfrak{R}(t,1)]^{(2)}, [\mathfrak{R}(t,2)]^{(2)}] \quad \text{for } t \geq 0 \tag{10.99}$$

where

$$\begin{aligned}[\mathfrak{R}(t,1)]^{(2)} &= [\mathfrak{R}^{(1)}(t,1)]^{(2)}[\mathfrak{R}^{(2)}(t,1)]^{(2)}[\mathfrak{R}^{(3)}(t,1)]^{(2)}[\mathfrak{R}^{(4)}(t,1)]^{(2)} \\ &= 4\exp[-18.380t] - 2\exp[-34.694t] - 2\exp[-19.664t] \\ &\quad + \exp[-35.978]\end{aligned} \tag{10.100}$$

$$[\mathfrak{R}(t,2)]^{(2)} = [\mathfrak{R}^{(1)}(t,2)]^{(2)}[\mathfrak{R}^{(2)}(t,2)]^{(2)}[\mathfrak{R}^{(3)}(t,2)]^{(2)}[\mathfrak{R}^{(4)}(t,2)]^{(2)}$$
$$= 4\exp[-21.6768t] - 2\exp[-40.2288t] - 2\exp[-23.2848t]$$
$$+\exp[-41.8368] \quad \text{for } t \geq 0 \tag{10.101}$$

At the operation state z_3, the considered system is a series system composed of four series subsystems S_1, S_2, S_3 and S_4. The subsystem S_1 at the operation state z_3 is composed of one belt conveyor forming series systems, then we may apply Proposition 10.1. Namely, since the system structure shape parameters are

$$k^{(3)} = 1, \; l^{(3)} = 129, \; a^{(3)} = 1, \; e_1^{(3)} = 4$$

and

$$q_1^{(3)} = 1, \; p_{11}^{(3)} = \frac{1}{129}, \; p_{12}^{(3)} = \frac{2}{129}, \; p_{13}^{(3)} = \frac{117}{129}, \; p_{14}^{(3)} = \frac{9}{129}$$

the number of reliability states is $z = 3$ and considering the values of component' intensities of departure from the reliability state subsets defined in Example 9.3, by Eqs. (10.9)–(10.12), we have

$$[\lambda_i(1)]^{(3)} = \frac{1}{129}0.0100 + \frac{2}{129}0.0009 + \frac{117}{129}0.002 + \frac{9}{129}0.001 = 0.2548/129,$$

$$[\lambda_i(2)]^{(3)} = \frac{1}{129}0.0105 + \frac{2}{129}0.0011 + \frac{117}{129}0.004 + \frac{9}{129}0.002 = 0.4987/129,$$

$$i = 1$$

$$[\lambda(1)]^{(3)} = \min\{0.2548/129\} = 0.2548/129,$$

$$[\lambda(2)]^{(3)} = \min\{0.4987/129\} = 0.4987/129$$

$$a_n^{(3)}(1) = \frac{1}{(0.2548/129)\cdot 129} = \frac{1}{0.2548}, \; b_n^{(3)}(1) = 0,$$

$$a_n^{(3)}(2) = \frac{1}{(0.4987/129)\cdot 129} = \frac{1}{0.4987}, \; b_n^{(3)}(2) = 0$$

then according to Eqs. (10.13)–(10.15), after considering Eq. (10.6), the approximate formula for the subsystem S_1 reliability function takes the form

$$[\boldsymbol{R}^{(1)}(t,\cdot)]^{(3)} \cong \mathfrak{R}^{(1)}((t - b_n^{(3)}(u))/a_n^{(3)}(u),\cdot)$$
$$= [[\mathfrak{R}^{(1)}(t,1)]^{(3)}, [\mathfrak{R}^{(1)}(t,2)]^{(3)}] \quad \text{for } t \geq 0 \tag{10.102}$$

where

$$[\mathfrak{R}^{(1)}(t,1)]^{(3)} = \mathfrak{R}'_9\left(\frac{t-b_n^{(3)}(1)}{a_n^{(3)}(1)},1\right) = 1-[1-\exp[-0.2548t]]^1$$
$$= \exp[-0.2548t] \tag{10.103}$$

$$[\mathfrak{R}^{(1)}(t,2)]^{(3)} = \mathfrak{R}'_9\left(\frac{t-b_n^{(3)}(2)}{a_n^{(3)}(2)},2\right) = 1-[1-\exp[-0.4987t]]^1$$
$$= \exp[-0.4987t] \quad \text{for } t \geq 0 \tag{10.104}$$

The subsystem S_2 at the operation state z_3 is composed of one bucket elevator forming series system, then we may apply Proposition 10.1. Namely, since the system structure shape parameters are

$$k^{(3)} = 1,\ \ l^{(3)} = 743,\ \ a^{(3)} = 1,\ \ e_1^{(3)} = 3$$

and

$$q_1^{(3)} = 1,\ \ p_{11}^{(3)} = \frac{1}{743},\ \ p_{12}^{(3)} = \frac{2}{743},\ \ p_{13}^{(3)} = \frac{740}{743}$$

the number of reliability states is $z = 3$ and considering the values of component' intensities of departure from the reliability state subsets defined in Example 9.3, by Eqs. (10.9)–(10.12), we have

$$[\lambda_i(1)]^{(3)} = \frac{1}{743}0.018 + \frac{2}{743}0.0011 + \frac{740}{743}0.020 = 14.840/743,$$

$$[\lambda_i(2)]^{(3)} = \frac{1}{743}0.022 + \frac{2}{743}0.0013 + \frac{740}{743}0.024 = 17.808/743, \quad i = 1$$

$$[\lambda(1)]^{(3)} = \min\{14.840/743\} = 14.840/743,$$

$$[\lambda(2)]^{(3)} = \min\{17.808/743\} = 17.808/743$$

$$a_n^{(3)}(1) = \frac{1}{(14.840/743)\cdot 743} = \frac{1}{14.840},\ \ b_n^{(3)}(1) = 0,$$

$$a_n^{(3)}(2) = \frac{1}{(17.808/743)\cdot 743} = \frac{1}{17.808},\ \ b_n^{(3)}(2) = 0$$

then according to Eqs. (10.13)–(10.15), after considering Eq. (10.6), the approximate formula for the subsystem S_2 reliability function takes the form

$$\begin{aligned}[\boldsymbol{R}^{(2)}(t,\cdot)]^{(3)} &\cong \mathfrak{R}^{(2)}((t-b_n^{(3)}(u))/a_n^{(3)}(u),\cdot)\\ &= [[\mathfrak{R}^{(2)}(t,1)]^{(3)},[\mathfrak{R}^{(2)}(t,2)]^{(3)}] \quad \text{for } t \geq 0\end{aligned} \tag{10.105}$$

where

$$\begin{aligned}[\mathfrak{R}^{(2)}(t,1)]^{(3)} &= \mathfrak{R}'_9\left(\frac{t-b_n^{(3)}(1)}{a_n^{(3)}(1)},1\right) = 1-[1-\exp[-14.840t]]^1\\ &= \exp[-14.840t]\end{aligned} \tag{10.106}$$

$$\begin{aligned}[\mathfrak{R}^{(2)}(t,2)]^{(3)} &= \mathfrak{R}'_9\left(\frac{t-b_n^{(3)}(2)}{a_n^{(3)}(2)},2\right) = 1-[1-\exp[-17.808t]]^2\\ &= \exp[-17.808t] \quad \text{for } t \geq 0\end{aligned} \tag{10.107}$$

The subsystem S_3 at the operation state z_3 is composed of one belt conveyors forming series system, then we may apply Proposition 10.1. Namely, since the system structure shape parameters are

$$k^{(3)} = 1,\;\; l^{(3)} = 139,\;\; a^{(3)} = 1,\;\; e_1^{(3)} = 4$$

and

$$q_1^{(3)} = 1,\;\; p_{11}^{(3)} = \frac{1}{139},\;\; p_{12}^{(3)} = \frac{2}{139},\;\; p_{12}^{(3)} = \frac{117}{139},\;\; p_{14}^{(3)} = \frac{19}{139}$$

the number of reliability states is $z = 3$ and considering the values of component' intensities of departure from the reliability state subsets defined in Example 9.3, by Eqs. (10.9)–(10.12), we have

$$[\lambda_i(1)]^{(3)} = \frac{1}{139}0.0100 + \frac{2}{139}0.0009 + \frac{117}{139}0.002 + \frac{19}{139}0.001 = 0.2648/139,$$

$$[\lambda_i(2)]^{(3)} = \frac{1}{139}0.0105 + \frac{2}{139}0.0011 + \frac{117}{139}0.004 + \frac{19}{139}0.002 = 0.5187/139,$$

$$i = 1$$

$$[\lambda(1)]^{(3)} = \min\{0.2648/139\} = 0.2648/139,$$

$$[\lambda(2)]^{(3)} = \min\{0.5187/139\} = 0.5187/139$$

$$a_n^{(3)}(1) = \frac{1}{(0.2648/139)\cdot 139} = \frac{1}{0.2648}, \quad b_n^{(3)}(1) = 0,$$

$$a_n^{(3)}(2) = \frac{1}{(0.5187/139)\cdot 139} = \frac{1}{0.5187}, \quad b_n^{(3)}(2) = 0$$

then according to Eqs. (10.13)–(10.15), after considering Eq. (10.6), the approximate formula for the subsystem S_3 reliability function takes the form

$$\begin{aligned}[\boldsymbol{R}^{(3)}(t,\cdot)]^{(3)} &\cong \mathfrak{R}^{(3)}((t-b_n^{(3)}(u))/a_n^{(3)}(u),\cdot) \\ &= [[\mathfrak{R}^{(3)}(t,1)]^{(3)},[\mathfrak{R}^{(3)}(t,2)]^{(3)}] \quad \text{for } t \geq 0\end{aligned} \tag{10.108}$$

where

$$\begin{aligned}[\mathfrak{R}^{(3)}(t,1)]^{(3)} &= \mathfrak{R}'_9\left(\frac{t-b_n^{(3)}(1)}{a_n^{(3)}(1)},1\right) = 1-[1-\exp[-0.2648t]]^1 \\ &= \exp[-0.2648t]\end{aligned} \tag{10.109}$$

$$\begin{aligned}[\mathfrak{R}^{(3)}(t,2)]^{(3)} &= \mathfrak{R}'_9\left(\frac{t-b_n^{(3)}(2)}{a_n^{(3)}(2)},2\right) = 1-[1-\exp[-0.5187t]]^1 \\ &= \exp[-0.5187t] \quad \text{for } t \geq 0\end{aligned} \tag{10.110}$$

The subsystem S_4 at the operation state z_3 is a series system composed of one chain conveyor forming series system, then we may apply Proposition 10.1. Namely, since the system structure shape parameters are

$$k^{(3)} = 1, \quad l^{(3)} = 162, \quad a^{(3)} = 1, \quad e_1^{(3)} = 2$$

and

$$q_1^{(3)} = 1, \quad p_{11}^{(3)} = \frac{2}{162}, \quad p_{12}^{(3)} = \frac{160}{162}$$

the number of reliability states is $z = 3$ and considering the values of component' intensities of departure from the reliability state subsets defined in Example 9.3, by Eqs. (10.9)–(10.12), we have

$$[\lambda_i(1)]^{(3)} = \frac{2}{162}0.001 + \frac{160}{162}0.007 = 1.222/162,$$

$$[\lambda_i(2)]^{(3)} = \frac{2}{162}0.003 + \frac{160}{162}0.0009 = 1.446/162, \quad i = 1$$

$$[\lambda(1)]^{(3)} = \min\{1.222/162\} = 1.222/162,$$

$$[\lambda(2)]^{(3)} = \min\{1.446/162] = 1.446/162$$

$$a_n^{(3)}(1) = \frac{1}{(1.222/162)\cdot 162} = \frac{1}{1.222}, \quad b_n^{(3)}(1) = 0,$$

$$a_n^{(3)}(2) = \frac{1}{(1.446/162)\cdot 262} = \frac{1}{1.446}, \quad b_n^{(3)}(2) = 0$$

then according to Eqs. (10.13)–(10.15), after considering Eq. (10.6), the approximate formula for the subsystem S_4 reliability function takes the form

$$\begin{aligned}[\boldsymbol{R}^{(4)}(t,\cdot)]^{(3)} &\cong \mathfrak{R}^{(4)}((t - b_n^{(3)}(u))/a_n^{(3)}(u), \cdot) \\ &= [[\mathfrak{R}^{(4)}(t,1)]^{(3)}, [\mathfrak{R}^{(4)}(t,2)]^{(3)}] \quad \text{for } t \geq 0\end{aligned} \tag{10.111}$$

where

$$\begin{aligned}[\mathfrak{R}^{(4)}(t,1)]^{(3)} &\cong \mathfrak{R}'_9\left(\frac{t - b_n^{(3)}(1)}{a_n^{(3)}(1)}, 1\right) = 1 - [1 - \exp[-1.222t]]^1 \\ &= \exp[-1.222t]\end{aligned} \tag{10.112}$$

$$\begin{aligned}[\mathfrak{R}^{(4)}(t,2)]^{(3)} &= \mathfrak{R}'_9\left(\frac{t - b_n^{(3)}(2)}{a_n^{(3)}(2)}, 2\right) = 1 - [1 - \exp[-1.446t]]^2 \\ &= \exp[-1.446t] \quad \text{for } t \geq 0\end{aligned} \tag{10.113}$$

Considering that the system at the operation state z_3 is a four-state series system composed of subsystems S_1, S_2, S_3 and S_4, after applying the formulae appearing after Definition 3.4 and Eqs. (10.71)–(10.79) and (10.81)–(10.83), its conditional reliability function is given by

$$[\boldsymbol{R}(t,\cdot)]^{(3)} = [[\mathfrak{R}(t,1)]^{(3)}, [\mathfrak{R}(t,2)]^{(3)}] \quad \text{for } t \geq 0 \tag{10.114}$$

where

$$\begin{aligned}[\mathfrak{R}(t,1)]^{(3)} &= [\mathfrak{R}^{(1)}(t,1)]^{(3)}[\mathfrak{R}^{(2)}(t,1)]^{(3)}[\mathfrak{R}^{(3)}(t,1)]^{(3)}[\mathfrak{R}^{(4)}(t,1)]^{(3)} \\ &= \exp[-16.4816t]\end{aligned} \tag{10.115}$$

$$\begin{aligned}[\mathfrak{R}(t,2)]^{(3)} &= [\mathfrak{R}^{(1)}(t,2)]^{(3)}[\mathfrak{R}^{(2)}(t,2)]^{(3)}[\mathfrak{R}^{(3)}(t,2)]^{(3)}[\mathfrak{R}^{(4)}(t,2)]^{(3)} \\ &= \exp[-20.2714t] \quad \text{for } t \geq 0\end{aligned} \tag{10.116}$$

Table 10.3 The Values of the Coordinates of the Port Grain Transportation System Approximate Reliability Function

t	$\mathfrak{R}(t,1)$	$\mathfrak{R}(t,2)$
0.00	1.0000	1.0000
0.02	0.8632	0.8294
0.04	0.6867	0.6171
0.06	0.5109	0.4235
0.08	0.3641	0.2771
0.10	0.2528	0.1766
0.12	0.1729	0.1109
0.14	0.1172	0.0692
0.16	0.0792	0.0431
0.18	0.0534	0.0268
0.20	0.0361	0.0167
0.22	0.0244	0.0105
0.24	0.0165	0.0066
0.26	0.0112	0.0042
0.28	0.0077	0.0026
0.30	0.0053	0.0017
0.32	0.0036	0.0011
0.34	0.0025	0.0007
0.36	0.0017	0.0004
0.38	0.0012	0.0003
0.40	0.0008	0.0002
0.42	0.0006	0.0001
0.44	0.0004	0.0001
0.46	0.0003	0.0001
0.48	0.0002	0.0000
0.50	0.0001	0.0000
0.52	0.0001	0.0000
0.54	0.0001	0.0000
0.56	0.0000	0.0000

Finally, considering Eqs. (10.6) and (9.18), the approximate formula for the system unconditional reliability function takes the form

$$[\boldsymbol{R}(t,\cdot)] = [[\mathfrak{R}(t,1)],[\mathfrak{R}(t,2)]] \quad \text{for } t \geq 0 \tag{10.117}$$

where

$$[\mathfrak{R}(t,1)] = 0.530[\mathfrak{R}(t,1)]^{(1)} + 0.109[\mathfrak{R}(t,1)]^{(2)} + 0.361[\mathfrak{R}(t,1)]^{(3)} \tag{10.118}$$

$$[\mathfrak{R}(t,2)] = 0.530[\mathfrak{R}(t,2)]^{(1)} + 0.109[\mathfrak{R}(t,2)]^{(2)} + 0.361[\mathfrak{R}(t,2)]^{(3)} \quad \text{for } t \geq 0 \tag{10.119}$$

Table 10.4 The Differences Between the Values of the Coordinates of the Port Grain Transportation System Exact and Approximate Reliability Function

t	$\boldsymbol{R}(t,1)-\mathfrak{R}(t,1)$	$\boldsymbol{R}(t,2)-\mathfrak{R}(t,2)$
0.00	0.0000	0.0000
0.02	0.0006	0.0008
0.04	0.0017	0.0019
0.06	0.0023	**0.0024**
0.08	**0.0024**	0.0022
0.10	0.0021	0.0017
0.12	0.0016	0.0011
0.14	0.0012	0.0007
0.16	0.0008	0.0004
0.18	0.0005	0.0003
0.20	0.0004	0.0002
0.22	0.0002	0.0001
0.24	0.0001	0.0000
0.26	0.0001	0.0000
0.28	0.0001	0.0000
0.30	0.0000	0.0000
0.32	0.0000	0.0000
0.34	0.0000	0.0000
0.36	0.0000	0.0000
0.38	0.0000	0.0000
0.40	0.0000	0.0000
0.42	0.0000	0.0000
0.44	0.0000	0.0000
0.46	0.0000	0.0000
0.48	0.0000	0.0000
0.50	0.0000	0.0000
0.52	0.0000	0.0000
0.54	0.0000	0.0000
0.56	0.0000	0.0000

where $[\mathfrak{R}(t,1)]^{(1)}$, $[\mathfrak{R}(t,1)]^{(2)}$, $[\mathfrak{R}(t,1)]^{(3)}$ are given by Eqs. (10.85), (10.100), (10.115) and $[\mathfrak{R}(t,2)]^{(1)}$, $[\mathfrak{R}(t,2)]^{(2)}$, $[\mathfrak{R}(t,2)]^{(3)}$ are given by Eqs. (10.86), (10.101), (10.116).

In Table 10.3, the values of the coordinates of the port grain transportation system approximate unconditional reliability function $\boldsymbol{R}(t,\ \cdot)$ given by Eqs. (10.118) and (10.119) are presented.

The results given by Eqs. (10.117)–(10.119) are not much different from the results (9.209)–(9.211), which are illustrated in Table 10.4, presenting the differences between the values of the port grain transportation system exact reliability function the values of the system approximate reliability function.

The approximate expected value of the port grain transportation system unconditional lifetime in the reliability state subset {1,2}, calculated according to Eqs. (9.39) and (9.40) from the results given by Eqs. (10.118) and (9.31), is

$$M(1) = p_1 M_1(1) + p_2 M_2(1) + p_3 M_3(1)$$
$$= 0.530 \cdot 0.0807 + 0.109 \cdot 0.0861 + 0.361 \cdot 0.0607 \cong 0.0741 \text{ year} \tag{10.120}$$

The approximate expected value of the port grain transportation system unconditional lifetime in the reliability state subset {2}, calculated according to Eqs. (9.39) and (940) from the results given by Eqs. (10.119) and (9.31), is

$$M(2) = p_1 M_1(2) + p_2 M_2(2) + p_3 M_3(2)$$
$$= 0.530 \cdot 0.0687 + 0.109 \cdot 0.0728 + 0.361 \cdot 0.0493 \cong 0.0621 \text{ year} \tag{10.121}$$

Further, considering Eqs. (9.120) and (9.121) and applying Eq. (9.42), the approximate mean values of the system unconditional lifetimes in the particular reliability states 1, 2, respectively are

$$\overline{M}(1) = M(1) - M(2) = 0.0120 \text{ year}, \quad \overline{M}(2) = M(2) = 0.0621 \text{ year} \tag{10.122}$$

The approximate values of the unconditional system lifetimes given by Eqs. (10.120)–(10.122) are almost identical with their exact values, determined respectively by Eqs. (9.212)–(9.214). Thus, the accuracy of the approximation is very good.

Summary

In this book, the asymptotic approach to the reliability evaluation of homogeneous and non-homogeneous series and parallel systems, homogeneous '*m* out of *n*' systems and homogeneous and non-homogeneous regular series–parallel and parallel–series systems has been completely analysed. For these systems, in the case where their components are two-state as well in the case where they are multi-state, the classes of limit reliability functions have been fixed. Moreover, the auxiliary theorems useful for finding limit reliability functions of real technical systems composed of components having any reliability functions have been formulated and motivated. The series–parallel and parallel–series systems have been considered in the case where their reliability structures are regular. However, this fact does not restrict the completeness of the performed analysis, since by conventional joining of a suitable number of failed components in parallel subsystems of the non-regular parallel–series systems we get the regular non-homogeneous parallel–series systems considered in the book. Similarly, conventional joining of a suitable number of components which do not fail, in series subsystems of the non-regular series–parallel systems, leads us to the regular non-homogeneous series–parallel systems considered in the book. Thus the problem has been analysed exhaustively.

In addition to the general solutions, a practically important case when the components of the considered systems have exponential reliability functions has been considered separately. In this case, the class of limit reliability functions for the considered multi-state systems has been fixed and other practically useful theorems have been proposed. The results obtained in this case may play the role of an easy-to-use guide necessary in quick reliability evaluations of real large technical systems, as well as during their operations and when they are designed. There are proposed algorithms presented in the form of tables giving simple sequential steps in systems reliability evaluation. To make these algorithms an easy and useful tool for reliability practitioners, their usage is illustrated by various examples of their application to the evaluation of the real system reliability characteristics. Thinking about the practitioners using computers, on the basis of the algorithms the computer program has been elaborated and the results of its practical use have been illustrated in the book.

Theoretical results have been illustrated by many practical examples of their application in reliability evaluation of an extensive range of large systems. The evaluations of the real technical systems presented here have been obtained on the basis of non-precise component reliability data and therefore first of all they should be treated as an illustration of a wide possibility of applications of the proposed asymptotic approach in system reliability analysis. Reliability data come from

Reliability of Large Systems. DOI: http://dx.doi.org/10.1016/B978-0-08-099949-4.00018-0

experts and from the literature and are concerned with components' mean lifetimes and their expected reliability function types. These evaluations, despite not being precise may be a very useful, simple and quick tool in approximate reliability evaluation, especially during the design of large systems, and when planning and improving their safety and effectiveness operation processes. Optimization of the reliability structures [61,110,123,124] of large systems with respect to their safety and costs is complicated and often not possible for practitioners to perform because of the mathematical complexity of the exact methods.

The proposed method offers enough simplified formulae to allow significant simplifying of reliability optimizing calculations. This is testified by the recent partial and preliminary results presented in the last chapter. Especially important are results concerned with exponential systems. Their joining with non-precise data coming from experts concerned with the component mean lifetimes and expected reliability function only allow the constructors to evaluate and to optimize the system structures and their operation processes during designing and before including them in everyday practice [51,52,141].

The results presented in the book have become the basis of investigations on domains of attraction of system limit reliability functions and initiated the problem of the speed at which system reliability function sequences reach their limit reliability functions. The problem of the domains of attraction for fixed limit reliability functions, presented partly in the book, has been completely solved for two-state systems in Ref. [120] and generalized to multi-state systems in Ref. [93]. The solutions deliver the necessary and sufficient conditions that the reliability functions of the system particular components have to satisfy in order that the system limit reliability function is one of the reliability functions from the fixed class of possible limit reliability functions for this system. This way, on the basis of data about the types of system component reliability functions and the system shape, it is possible to predict which of the reliability functions is its limit reliability function.

Another significant problem in applying the proposed method to the reliability evaluation of real technical system is its accuracy. This problem has been completely solved for the considered two-state systems in Ref. [93]. Practical examples presented in the book testify that the mistakes in the approximation of the exact system reliability functions by their limiting forms are not often significant in practice for not-very-large systems. Moreover, in the asymptotic approach to system reliability evaluation it is possible to get lower and upper bounds of the exact system reliability characteristics, which is illustrated in the investigation of the ship-rope elevator. The complete solution of the proposed method, i.e. the evaluation of the speed of convergence of the system exact reliability sequences to their limiting forms – presented partly in the book for two-state systems – is easy to transfer to multi-state systems. The way of proceeding in this transfer is commented on in the book and solved in Ref. [93]. The asymptotic method accuracy evaluations are complicated, so it is probably not possible for reliability practitioners to use them in everyday practice. However, it seems to be practically possible to make the accuracy evaluation supported by computer calculations.

Additionally, the main results of the book have initiated and become the basis for further investigations on limit reliability functions. In particular, the investigations on limit reliability functions of practically important large series–'*m* out of *n*' and '*m* out of *n*'–series systems and hierarchical systems have been significantly developed [27,136,162]. Another problem concerned with the methods of the improving of large systems reliability, which are briefly presented in the book, has been completely solved for series systems in Ref. [122].

All the results of the investigations on limit reliability functions have initiated practically a very important aspect of joining them with the reliability models of the complex system subjected to their operation processes considered widely in Ref. [110]. This problem have been formulated in the book and the latest results on operation and reliability modelling, prediction and optimization of large complex multistate exponential systems changing their reliability structures and their component reliability parameters at variable operation conditions have been presented. Unfortunately, it is not possible to find analytically general solutions after omitting the assumption on multi-state system components' exponential reliability functions and the Monte Carlo simulation method [178] can be suggested as a sometimes useful tool in solving this complex problem [94,118].

The results presented in the book can be very helpful backgrounds for very important investigations of the safety of complex systems [106,110,158], the safety of critical infrastructures [92], and the safety of nanosystems [33], which currently are the main research areas in science around the world [112].

Description/Abstract

The book is concerned with the application of limit reliability functions to the reliability evaluation of large and complex systems. Two-state and multi-state large systems composed of independent components are considered. The main emphasis is on multi-state systems with degrading (ageing) components because of the importance of such an approach in safety analysis, assessment and prediction, and in analysing the effectiveness of operation processes of real technical systems.

The book contains complete current theoretical results of the asymptotic approach to reliability evaluation of large two-state and multi-state series, parallel, '*m* out of *n*,' series–parallel, and parallel–series systems. Some partial results on the asymptotic approach to reliability evaluation of '*m* out of *n*'–series, series–'*m* out of *n*' and hierarchical systems, and their application to large systems reliability improvement are presented in the book. Additionally, large complex systems reliability analysis in the variable operation conditions is performed and modelling, prediction and optimization of their operation and reliability is comprehensively presented.

Theoretical results are illustrated by many practical examples of their application in reliability evaluation of an extensive range of large and complex systems.

These evaluations may be a very useful, simple and quick tool in approximate reliability evaluation, especially during the design of large complex systems, and when planning and improving their safety and effectiveness of their operation processes.

The book also contains the evaluation of the presented results and the perspective of further investigations on the considered problems.

Bibliography

[1] A. Abouammoh, M. Al-Kadi, Component relevancy in multi-state reliability models, IEEE Trans. Reliab. 40 (1991) 370–375.
[2] S.V. Amari, R.B. Misra, Comment on: dynamic reliability analysis of coherent multistate systems, IEEE Trans. Reliab. 46 (1997) 460–461.
[3] C.W. Anderson, Extreme value theory for a class of discrete distributions with applications to some stochastic processes, J. Appl. Probab. 7 (1970) 99–113.
[4] T. Aven, Reliability evaluation of multi-state systems with multi-state components, IEEE Trans. Reliab. 34 (1985) 473–479.
[5] T. Aven, U. Jensen, Stochastic Models in Reliability, Springer-Verlag, New York, NY, 1999.
[6] T. Aven, On performance measures for multi-state monotone systems, Reliab. Eng. Syst. Saf. 41 (1993) 259–266.
[7] R.E. Barlow, F. Proschan, Statistical Theory of Reliability and Life Testing. Probability Models, Holt Rinehart and Winston, Inc., New York, NY, 1975.
[8] R.E. Barlow, A.S. Wu, Coherent systems with multi-state components, Math. Oper. Res. 4 (1978) 275–281.
[9] O. Barndorff-Nielsen, On the limit behaviour of extreme order statistics, Ann. Math. Stat. 34 (1963) 992–1002.
[10] S.M. Berman, Limiting distribution of the maximum term in sequences of dependent random variables, Ann. Math. Stat. 33 (1962) 894–908.
[11] S.M. Berman, Limit theorems for the maximum term in stationary sequences, Ann. Math. Stat. 35 (1964) 502–516.
[12] H.W. Block, T.H. Savitis, A decomposition for multi-state monotone systems, J. Appl. Probab. 19 (1982) 391–402.
[13] A. Blokus-Roszkowska, K. Kołowrocki, Xiuju Fu, Integrated software tools supporting decision making on operation, identification, prediction and optimization of complex technical systems reliability and safety. Parts 1–3, J. Pol. Saf. Reliab. Assoc. Summer Saf. Reliab. Semin. 2 (2) (2011) 267–296.
[14] A. Blokus-Roszkowska, K. Kołowrocki, M. Salahuddin Habibullah, Integrated package of tools supporting decision making on operation, identification, prediction and optimization of complex technical systems reliability and safety. Parts 4–6, J. Pol. Saf. Reliab. Assoc. Summer Saf. Reliab. Semin. 2 (2) (2011) 297–336.
[15] D. Bobrowski, Mathematical Models in Reliability Theory (in Polish), WNT, Warsaw, 1985.
[16] R. Boedigheimer, K. Kapur, Customer-driven reliability models for multi-state coherent systems, IEEE Trans. Reliab. 43 (1994) 45–50.
[17] A. Brausch, Calculation of critical importance for multi-state components, IEEE Trans. Reliab. 36 (1987) 247–249.
[18] R.D. Brunelle, K.C. Kapur, Review and classification of reliability measures for multistate and continuum models, IEEE Trans. 31 (1999) 1117–1180.
[19] D. Butler, A complete importance ranking for components of binary coherent systems with extension to multi state systems, Naval Res. Logistics 26 (1979) 556–578.

[20] D. Butler, Bounding the reliability of multi-state systems, Oper. Res. 30 (1982) 530–544.

[21] L. Cardalora, Coherent systems with multi-state components, Nucl. Eng. Des. 58 (1980) 127–139.

[22] E. Castillo, Extreme Value Theory in Engineering, Boston Academic Press, Boston, MA, 1988.

[23] H. Chernoff, H. Teicher, Limit distributions of the minimax of independent identically distributed random variables, Proc. Am. Math. Soc. 116 (1965) 474–491.

[24] Certificate for steel wire rope. Casar Drahtseilwerk Saar GmbH, Kirkel, 1996.

[25] A. Cichocki, D. Kurowicka, B. Milczek, On limit reliability functions of large systems (Chapter 12), in: D.C. Ionescu, N. Limnios (Eds.), Statistical and Probabilistic Models in Reliability, Birkhauser, Boston, MA, 1998, pp. 184–193.

[26] A. Cichocki, Limit reliability functions of some homogeneous regular series–parallel and parallel–series systems of higher order, Appl. Math. Comput. 120 (2001) 55–72.

[27] A. Cichocki, Determining of limit reliability functions of hierarchical systems under power standardisation (in Polish). PhD Thesis. Gdynia Maritime University, System Research Institute, Warsaw, 2003.

[28] J. Collet, Some remarks on rare-event approximation, IEEE Trans. Reliab. 45 (1996) 106–108.

[29] H.E. Daniels, The statistical theory of the strength of bundles of threads, J. Proc. R. Soc. 183 (1945) 404–435.

[30] L. De Haan, On Regular Variation and its Application to the Weak Convergence of Sample Extremes, Mathematical Centre Tracts, 32, Mathematics Centre, Amsterdam, 1970.

[31] W. Dziubdziela, Limit distributions of order statistics (in Polish), Appl. Math. 9 (1977) 45–71.

[32] N. Ebrahimi, Multistate reliability models, Naval Res. Logistics 31 (1984) 671–680.

[33] N. Ebrahimi, Assessing a linear nanosystem's limiting reliability from its components, J. Appl. Probab. 45 (2008) 879–887.

[34] E. El-Neweihi, F. Proschan, J. Setchuraman, Multi-state coherent systems, J. Appl. Probab. 15 (1978) 675–688.

[35] M.N. Fardis, C.A. Cornell, Analysis of coherent multistate systems, IEEE Trans. Reliab. 30 (1981) 117–122.

[36] F. Ferreira, A. Pacheco, Comparison of level-crossing times for Markov and semi-Markov processes, Stat. Probab. Lett. 77 (2) (2007) 151–157.

[37] R.A. Fisher, L.H.C. Tippett, Limiting forms of the frequency distribution of the largest and smallest member of a sample, Proc. Cambridge Philos. Soc. 24 (1928) 180–190.

[38] M. Frechet, Sur la loi de probabilite de l'ecart maximum, Annales de la Société Polonaise de Mathématique 6 (1927) 93–116.

[39] J. Galambos, Limit laws for mixtures with applications to asymptotic theory of extremes, Z. Wahrscheinlichkeitstheorie Verw. Gebiete 32 (1975) 197–207.

[40] P.W. Glynn, P.J. Haas, Laws of large numbers and functional central limit theorems for generalized semi-Markov processes, Stochastic Models 22 (2) (2006) 201–231.

[41] B.W. Gniedenko, Sur la distribution limite du terme maximum d'une serie aleatoire, Ann. Math. 44 (1943) 432–453.

[42] F. Grabski, Semi-Markov Models of Systems Reliability and Operations. Monograph. Analysis (in Polish), System Research Institute, Polish Academy of Science, Warsaw, 2002.

[43] W.S. Griffith, Multi-state reliability models, J. Appl. Probab. 17 (1980) 735–744.

[44] E.J. Gumbel, Les valeurs extremes des distributions statistiques, Annales de l'Institut Henri Poincaré 4 (1935) 115.

[45] E.J. Gumbel, Statistics of Extremes, New York, NY, 1962.

[46] S. Guze, K. Kołowrocki, J. Soszyńska, Modeling environment and infrastructure influence on reliability and operation processes of port transportation systems, Proc. Summer Saf. Reliab. Semin. 2 (1) (2008) 179–188.

[47] M.S. Habibullah, E. Lumanpauw, K. Kolowrocki, J. Soszynska, N.G. Ming, A computational tool for general model of industrial systems. Operation processes, Electron. J. Reliab. Risk Anal. Theor. Appl. 2 (4) (2009) 181–191.

[48] D.G. Harlow, S.L. Phoenix, Approximations for the strength distribution and size effects in an idealised lattice model of material breakdown, J. Mech. Phys. Solids 39 (1991) 173–200.

[49] D.G. Harlow, Statistical properties of hybrid composites: asymptotic distributions for strain, Reliab. Eng. Syst. Saf. 56 (1997) 197–208.

[50] R. Harris, An application of extreme value theory to reliability theory, Ann. Math. Stat. 41 (1970) 1456–1465.

[51] S. Helvacioglu, M. Insel, Expert system applications in marine technologies, Ocean Eng. 35 (11-12) (2008) 1067–1074.

[52] O. Hryniewicz, Lifetime tests for imprecise data and fuzzy reliability requirements, in: T. Onisawa, J. Kacprzyk (Eds.), Reliability and Safety Analyses under Fuzziness, Physica Verlag, Heidelberg, 1995, pp. 169–182.

[53] J. Huang, M.J. Zuo, Y. Wu, Generalized multi-state k-out-of-n:G systems, IEEE Trans. Reliab. 49 (2000) 105–111.

[54] J.C. Hudson, K.C. Kapur, Reliability theory for multistate systems with multistate components, Microelectron. Reliab. 22 (1982) 1–7.

[55] J.C. Hudson, K.C. Kapur, Reliability analysis of multistate systems with multistate components, Trans. Inst. Ind. Eng. 15 (1983) 127–135.

[56] J. Hudson, K.C. Kapur, Modules in coherent multistate systems, IEEE Trans. Reliab. 32 (1983) 183–185.

[57] J. Hudson, K. Kapur, Reliability bounds for multistate systems with multistate components, Oper. Res. 33 (1985) 735–744.

[58] J. Jaźwinski, K. Fiok-Ważyńska, Reliability of Technical Systems (in Polish), PWN, Warsaw, 1990.

[59] J. Karpiński, E. Korczak, Methods of Reliability Evaluation of Two-State Technical Systems (in Polish), System Research Institute, Polish Academy of Sciences, Warsaw, 1990.

[60] L.M. Kaufman, J.B. Dugan, B.W. Johnson, Using statistics of the extremes for software reliability analysis, IEEE Trans. Reliab. 48 (3) (1999) 292–299.

[61] D. Klabjan, D. Adelman, Existence of optimal policies for semi-Markov decision processes using duality for infinite linear programming, Siam J. Control Optim. 44 (6) (2006) 2104–2122.

[62] K. Kołowrocki, On a class of limit reliability functions of some regular homogeneous series–parallel systems, Reliab. Eng. Syst. Saf. 39 (1993) 11–23.

[63] K. Kołowrocki, On asymptotic reliability functions of series–parallel and parallel–series systems with identical components, Reliab. Eng. Syst. Saf. 41 (1993) 251–257.

[64] K. Kołowrocki, On a class of limit reliability functions of some regular homogeneous series–parallel systems, Appl. Math. 36 (1993) 55–69.

[65] K. Kołowrocki, On a Class of Limit Reliability Functions for Series Parallel and Parallel–Series Systems. Monograph, Maritime University Press, Gdynia, 1993.

[66] K. Kołowrocki, A remark on the class of limit reliability functions of series–parallel systems, Exploit. Prob. Machines 2 (98) (1994) 279–296.
[67] K. Kołowrocki, The classes of asymptotic reliability functions for series–parallel and parallel–series systems, Reliab. Eng. Syst. Saf. 46 (1994) 179–188.
[68] K. Kołowrocki, Limit reliability functions of some series–parallel and parallel–series systems, Appl. Math. Comput. 62 (1994) 129–151.
[69] K. Kołowrocki, Limit reliability functions of some non-homogeneous series–parallel and parallel–series systems, Reliab. Eng. Syst. Saf. 46 (1994) 171–177.
[70] K. Kołowrocki, On limiting forms of the reliability functions sequence of the series–parallel and parallel–series systems, Appl. Math. Comput. Sci. 4 (1994) 575–590.
[71] K. Kołowrocki, On a class of limit reliability functions for series–parallel and parallel–series systems, Int. J. Press. Vessels Pip. 61 (1995) 541–569.
[72] K. Kołowrocki, Asymptotic reliability functions of some non-homogeneous series–parallel and parallel–series systems, Appl. Math. Comput. 73 (1995) 133–151.
[73] K. Kołowrocki, On applications of asymptotic reliability functions to the reliability and risk evaluation of pipelines, Int. J. Press. Vessels Pip. 75 (1998) 545–558.
[74] K. Kołowrocki, On limit reliability functions of large systems (Chapter 11), in: D.C. Ionescu, N. Limnios (Eds.), Statistical and Probabilistic Models in Reliability, Birkhauser, Boston, MA, 1999, pp. 153–183.
[75] K. Kołowrocki, On reliability and risk of large multi-state systems with degrading components, Exploit. Prob. Machines (1999) 189–210.
[76] K. Kołowrocki, On asymptotic approach to multi-state systems reliability evaluation (Chapter 11), in: N. Limnios, M. Nikulin (Eds.), Recent Advances in Reliability Theory: Methodology, Practice and Inference, Birkhauser, Boston, MA, 2000, pp. 163–180.
[77] K. Kołowrocki, Weibull distribution applications to reliability evaluation of transportation systems, Arch. Trans. 12 (2) (2000) 17–31.
[78] K. Kołowrocki, Asymptotic approach to reliability evaluation of piping and rope transportation systems, Exploit. Prob. Machines 2 (122) (2000) 111–133.
[79] K. Kołowrocki, Reliability evaluation of port transportation systems (in Polish). Consultants: Z. Baranowski, J. Cichosz, D. Jewasiński. Report. Maritime Academy, Gdynia, 2000.
[80] K. Kołowrocki, Asymptotic approach to reliability evaluation of a rope transportation system, Reliab. Eng. Syst. Saf. 71 (1) (2001) 57–64.
[81] K. Kołowrocki, Asymptotic Approach to System Reliability Analysis (in Polish). Monograph, System Research Institute, Polish Academy of Science, Warsaw, 2001.
[82] K. Kołowrocki, On limit reliability functions of large multi-state systems with ageing components, Appl. Math. Comput. 121 (2001) 313–361.
[83] K. Kołowrocki, An asymptotic approach to reliability evaluation of large multi-state systems with applications to piping transportation systems, Int. J. Press. Vessels Pip. 80 (2003) 59–73.
[84] K. Kołowrocki, Asymptotic approach to reliability analysis of large systems with degrading components, Int. J. Reliab. Qual. Saf. Eng. 10 (3) (2003) 249–288.
[85] K. Kołowrocki, Reliability of Large Systems, Elsevier, Amsterdam, Boston, Heidelberg, London, New York, Oxford, Paris, San Diego, San Francisco, Singapore, Sydney, Tokyo, 2004.
[86] K. Kołowrocki, Reliability and risk evaluation of complex systems in their operation processes, Int. J. Mater. Struct. Reliab. 4 (2) (2006) 129–147.

[87] K. Kołowrocki, Reliability modeling of complex systems—Part 1, Electron. J. Reliab. Theor. Appl. 2 (3-4) (2007) 116–127.

[88] K. Kołowrocki, Reliability modeling of complex systems—Part 2, Electron. J. Reliab. Theor. Appl. 2 (3-4) (2007) 128–139.

[89] K. Kołowrocki, Reliability and risk analysis of multi-state systems with degrading components, Proc. Summer Saf. Reliab. Semin. 2 (2) (2008) 205–216.

[90] K. Kołowrocki, Reliability of large systems (ISBN 978-0-470-03549-8; pp. 1466–1471) Encyclopedia of Quantitative Risk Analysis and Assessment, 4, John Wiley & Sons, Hoboken, San Francisco, 2008.

[91] K. Kołowrocki, Reliability modelling of complex systems (Chapter I; pp. 3–54) in: H. Pham (Ed.), Safety and Risk Modelling and its Applications, Springer, London, Dordrecht, Heidelberg, New York, 2011.

[92] K. Kołowrocki, Safety of critical infrastructures, J. Pol. Saf. Reliab. Assoc. Summer Saf. Reliab. Semin. 4 (1) (2013) 51–72.

[93] K. Kołowrocki, A. Cichocki, D. Kurowicka, B. Milczek, B. Sarnecka, Limit reliability functions of large multistate systems and their applications to transportation and durability problems. Parts 1, 2, 3 (in Polish). Report. Project founded by the Polish Committee for Scientific Research. Maritime University, Gdynia, 2000, p. 567.

[94] K. Kołowrocki, E. Kuligowska, Monte Carlo simulation application to reliability evaluation of port grain transportation system operating at variable conditions, J. Pol. Saf. Reliab. Assoc. Summer Saf. Reliab. Semin. 4 (1) (2013) 73–82.

[95] K. Kołowrocki, D. Kurowicka, Limit reliability functions for non-homogeneous systems (Chapter 3), Techniques in Representing High Dimensional Distributions, Delft University, Delft, 2001, pp. 63–86.

[96] K. Kołowrocki, B. Kwiatuszewska-Sarnecka, Reliability and risk analysis of large systems with ageing components, Reliab. Eng. Syst. Saf. 93 (2008) 1821–1829.

[97] Kołowrocki K., Soszyńska J.: Reliability and availability of complex systems. Qual. Reliab. Eng. Int. 22(1) (2006) 79–99, John Wiley & Sons, Hoboken, San Francisco.

[98] K. Kołowrocki, J. Soszyńska, A general model of industrial systems operation processes related to their environment and infrastructure, Proc. Summer Saf. Reliab. Semin. 2 (2) (2008) 223–226.

[99] K. Kołowrocki, J. Soszyńska, Modeling environment and infrastructure influence on reliability and operation process of port oil transportation system, Electron. J. Reliab. Theor. Appl. 2 (3) (2009) 131–142.

[100] K. Kołowrocki, J. Soszyńska, Reliability, risk and availability based optimization of complex technical systems operation processes. Part 1. Theoretical backgrounds, Electron. J. Reliab. Theor. Appl. 2 (4) (2009) 141–152.

[101] K. Kołowrocki, J. Soszyńska, Reliability, risk and availability based optimization of complex technical systems operation processes. Part 2. Application in Port Transportation, Electron. J. Reliab. Theor. Appl. 2 (4) (2009) 153–167.

[102] K. Kołowrocki, J. Soszyńska, Statistical identification and prediction of the port oil pipeline system's operation process and its reliability and risk evaluation, Proc. Summer Saf. Reliab. Semin. 3 (2) (2009) 241–250.

[103] K. Kołowrocki, J. Soszyńska, Methods and algorithms for evaluating unknown parameters of components reliability of complex technical systems, Proc. Summer Saf. Reliab. Semin. 4 (2) (2009) 223–230.

[104] K. Kołowrocki, J. Soszyńska, Reliability modelling of a port oil transportation system's operation processes, Int. J. Perform. Eng. 6 (1) (2010) 77–87.

[105] K. Kołowrocki, J. Soszyńska, Reliability, availability and safety of complex technical systems: modeling—identification—prediction—optimization, J. Pol. Saf. Reliab. Assoc. Summer Saf. Reliab. Semin. 1 (1) (2010) 133–158.

[106] K. Kołowrocki, J. Soszyńska, Safety and risk evaluation of a ferry technical system, J. Pol. Saf. Reliab. Assoc. Summer Saf. Reliab. Semin. 1 (1) (2010) 159–172.

[107] K. Kołowrocki, J. Soszyńska, On safety analysis of complex maritime transportation system, J. Risk Reliab. 225 (3) (2011) 345–354.

[108] K. Kołowrocki, J. Soszyńska, M. Jurdziński, P. Dziula, On multi-state safety analysis in shipping, Int. J. Reliab. Qual. Saf. Eng. 14 (6) (2007) 547–567.

[109] K. Kołowrocki, J. Soszyńska, M. Xie, M. Kien, M. Salahudin, Safety and reliability of complex industrial systems and process, Proc. Summer Saf. Reliab. Semin. 4 (2) (2008) 227–234.

[110] K. Kołowrocki, J. Soszyńska-Budny, Reliability and Safety of Complex Technical Systems and Processes: Modeling—Identification—Prediction—Optimization, Springer, London, Dordrecht, Heidelberg, New York, 2011.

[111] K. Kołowrocki, J. Soszyńska-Budny, Preliminary approach to safety analysis of critical infrastructures, J. Pol. Saf. Reliab. Assoc. Summer Saf. Reliab. Semin. 3 (1) (2012) 73–88.

[112] K. Kołowrocki, J. Soszyńska-Budny, Current problems of technical systems reliability and safety, J. Pol. Saf. Reliab. Assoc. Summer Saf. Reliab. Semin. 2 (3) (2011) 1–9.

[113] K. Kołowrocki, J. Soszyńska-Budny, Ng Kien Ming, Integrated package of tools supporting decision making on operation, identification, prediction and optimization of complex technical systems reliability and safety. Parts 1–3, J. Pol. Saf. Reliab. Assoc. Summer Saf. Reliab. Semin. 2 (2) (2011) 361–398.

[114] K. Kołowrocki, J. Soszyńska-Budny, M. Xie, Integrated package of tools supporting decision making on operation, identification, prediction and optimization of complex technical systems reliability and safety. Parts 4–7, J. Pol. Saf. Reliab. Assoc. Summer Saf. Reliab. Semin. 2 (2) (2011) 399–451.

[115] K. Kołowrocki, J. Soszyńska-Budny, Asymptotic approach to reliability of large complex systems, Electron. J. Reliab. Theor. Appl. 2 (4) (2011) 103–128.

[116] A. Kossow, W. Preuss, Reliability of linear consecutively-connected systems with multistate components, IEEE Trans. Reliab. 44 (1995) 518–522.

[117] B. Krajewski, C. Pawluk, An opinion on reliability and exploitation of ship rope elevator (in Polish). Naval Shipyard, Gdynia, 1999.

[118] E. Kuligowska, Preliminary Monte Carlo approach to complex technical system reliability analysis, J. Pol. Saf. Reliab. Assoc. Summer Saf. Reliab. Semin. 3 (1) (2012) 59–71.

[119] D. Kurowicka, Domains of attraction of asymptotic reliability functions of some homogeneous series–parallel systems, Appl. Math. Comput. 98 (1998) 61–74.

[120] D. Kurowicka, Techniques in representing high dimensional distributions. PhD Thesis. Gdynia Maritime University-Delft University, 2001.

[121] B. Kwiatuszewska-Sarnecka, On a class of limit reliability functions of large series–parallel systems with assisting components, Appl. Math. Comput. 123 (2001) 155–177.

[122] B. Kwiatuszewska-Sarnecka, Reliability analysis of reservation effectiveness in series systems (in Polish). PhD Thesis. Gdynia Maritime University, System Research Institute, Warsaw, 2003.

[123] W. Kuo, V.R. Prasad, An annotated overview of system-reliability optimization, IEEE Trans. Reliab. 49 (2) (2000) 176–187.

[124] W. Kuo, M.J. Zuo, Optimal Reliability Modeling: Principles and Applications, John Wiley & Sons, Hoboken, NJ, 2003.
[125] M.R. Leadbetter, On extreme values in stationary sequences, Z. Wahrscheinlichkeitstheorie Verw. Gebiete 28 (1974) 289–303.
[126] G. Levitin, A. Lisnianski, H. Ben Haim, D. Elmakis, Redundancy optimisation for multistate series–parallel systems, IEEE Trans. Reliab. 47 (1998) 165–172.
[127] G. Levitin, A. Lisnianski, Joint redundancy and maintenance optimisation for series–parallel multistate systems, Reliab. Eng. Syst. Saf. 64 (1998) 33–42.
[128] G. Levitin, A. Lisnianski, Importance and sensitivity analysis of multi-state systems using universal generating functions method, Reliab. Eng. Syst. Saf. 65 (1999) 271–282.
[129] G. Levitin, A. Lisnianski, Optimisation of imperfect preventive maintenance for multistate systems, Reliab. Eng. Syst. Saf. 67 (2000) 193–203.
[130] G. Levitin, A. Lisnianski, Optimal replacement scheduling in multi-state series–parallel systems, Qual. Reliab. Eng. Int. 16 (2000) 157–162.
[131] G. Levitin, A. Lisnianski, Structure optimisation of multi-state system with two failure modes, Reliab. Eng. Syst. Saf. 72 (2001) 75–89.
[132] A. Lisnianski, G. Levitin, Multi-State System Reliability. Assessment, Optimisation and Applications, World Scientific Publishing Co., New Jersey, London, Singapore, Hong Kong, 2003.
[133] N. Limnios, G. Oprisan, Semi-Markov Processes and Reliability, Birkhauser, Boston, MA, 2001.
[134] F. Meng, Component-relevancy and characterisation in multi-state systems, IEEE Trans. Reliab. 42 (1993) 478–483.
[135] B. Milczek, On the class of limit reliability functions of homogeneous series-"k out of n" systems, Appl. Math. Comput. 137 (2001) 161–174.
[136] B. Milczek, Reliability of large series-"k out of n" systems (in Polish). PhD Thesis. Gdynia Maritime University, System Research Institute, Warsaw, 2004.
[137] B. Natvig, Two suggestions of how to define a multi-state coherent system, Adv. Appl. Probab. 14 (1982) 434–455.
[138] B. Natvig, A. Streller, The steady-state behaviour of multi-state monotone systems, J. Appl. Probab. 21 (1984) 826–835.
[139] B. Natvig, Multi-state coherent systems, Encyclopaedia of Statistical Sciences, Wiley and Sons, New York, NY, 1984.
[140] S. Mercier, Numerical bounds for semi-Markovian quantities and application to reliability, Method. Comput. Appl. Probab. 10 (2) (2008) 179–198.
[141] J.R.W. Merrick, R. van Dorp, Speaking the truth in maritime risk assessment, Risk Anal. 26 (1) (2006) 223–237.
[142] F. Ohio, T. Nishida, On multi-state coherent systems, IEEE Trans. Reliab. 33 (1984) 284–287.
[143] E. Pantcheva, Limit theorems for extreme order statistics under non-linear normalisation, Lecture Notes Math. 1155 (1984) 284–309.
[144] S. Piasecki, Elements of Reliability Theory and Multi-state Objects Exploitation (in Polish), System Research Institute, Polish Academy of Science, Warsaw, 1995.
[145] Polish Norm PN-68/M-80-200. Steel Ropes. Classification and Construction (in Polish).
[146] Polish Norm PN-81/M-46-650. Bucket Conveyors. Classification and Construction (in Polish).
[147] O. Pourret, J. Collet, J-L. Bon, Evaluation of the unavailability of a multistate component system using a binary model, IEEE Trans. Reliab. 64 (1999) 13–17.

[148] V. Prasad, W. Kuo, K. Kim, Optimal allocation of s-identical multi-functional spares in a series systems, IEEE Trans. Reliab. 48 (2) (1999) 118–126.

[149] S.I. Resnick, Limit laws for recorded values, Stochastic Processes Appl. 1 (1973) 67–82.

[150] S. Ross, Multi-valued state component systems, Ann. Probab. 7 (1979) 379–383.

[151] N.W. Smirnow, Predielnyje Zakony Raspredielenija dla Czlienow Wariacjonnogo Riada, Trudy Matem. Inst. im. W. A. Stieklowa (1949).

[152] R.L. Smith, The asymptotic distribution of the strength of a series–parallel system with equal load-sharing, Ann. Probab. 10 (1982) 137–171.

[153] R.L. Smith, Limit theorems and approximations for the reliability of load sharing systems, Adv. Appl. Probab. 15 (1983) 304–330.

[154] J. Soszyńska, Reliability of large series system in variable operation conditions, Joint Proceedings, No 17, Gdynia Maritime University Press, 2004, pp. 36–43.

[155] J. Soszyńska, Reliability of large parallel systems at variable operation conditions. (in Polish), Sci. Works Faculty Navig. 16 (2004) 168–180 Gdynia.

[156] J. Soszyńska, Reliability of large series–parallel system in variable operation conditions, Int. J. Autom. Comput. 3 (2) (2006) 199–206.

[157] J. Soszyńska, Reliability evaluation of a port oil transportation system in variable operation conditions, Int. J. Press. Vessels Pip 83 (4) (2006) 304–310.

[158] J. Soszyńska, Multistate systems safety analysis at variable operation conditions. (in Polish), Diagnostic 39 (3) (2006) 25–34.

[159] J. Soszyńska, Systems reliability analysis in variable operation conditions, Int. J. Reliab. Qual. Saf. Eng. 14 (6) (2007) 617–634.

[160] J. Soszyńska, Systems reliability analysis in variable operation conditions (in Polish). PhD Thesis, Gdynia Maritime University, System Research Institute, Warsaw, 2007.

[161] J. Soszyńska, Systems reliability analysis in variable operation conditions, Elect. J. Reliab. Theor. Appl. 2 (3-4) (2007) 186–197.

[162] J. Soszyńska, Asymptotic approach to reliability evaluation of large "m out of l"—series system in variable operation conditions, Proc. Summer Saf. Reliab. Semin. 2 (2) (2008) 323–346.

[163] J. Soszyńska, Reliability and risk evaluation of a port oil pipeline transportation system in variable operation conditions, Int. J. Press. Vessels Pip. 87 (2-3) (2010) 81–87.

[164] J. Soszyńska-Budny, Reliability, risk and availability analysis of a container gantry crane, Electron. J. Reliab. Theor. Appl. 1 (1) (2012) 126–141.

[165] J. Soszyńska, K. Kołowrocki, A. Blokus-Roszkowska, S. Guze, Prediction of complex technical systems operation processes, J. Pol. Saf. Reliab. Assoc. Summer Saf. Reliab. Semin. 1 (2) (2010) 379–510.

[166] Z.L. Sun, Ng K. Ming, J. Soszyńska-Budny, M.S. Habibullah, Application of the LP-ELM model on transportation system lifetime optimization, IEEE Trans. Intell. Trans. Syst. 12 (4) (2011) 1484–1494.

[167] L.S. Sutherland, C.G. Soares, Review of probabilistic models of the strength of composite materials, Reliab. Eng. Syst. Saf. 56 (1997) 183–196.

[168] H. Tang, B.Q. Yin, H.S. Xi, Error bounds of optimization algorithms for semi-Markov decision processes, Int. J. Syst. Sci. 38 (9) (2007) 725–736.

[169] M.N. Tata, On outstanding values in a sequence of random variables, Z. Wahrscheinlichkeitstheorie Verw. Gebiete 12 (1969) 9–20.

[170] Trade Norm BN-75/2118-01. Cranes. Instructions for Expenditure Evaluation of Steel Ropes (in Polish).

[171] S. Vercellis, Data Mining and Optimization for Decision Making, John Wiley & Sons, Hoboken, San Francisco, 2009.
[172] R. Von Mises, La distribution de la plus grande de n valeurs, Revue Mathematique de l'Union Interbalkanique 1 (1936) 141–160.
[173] R.C. Watherhold, Probabilistic aspects of the strength of short fibre composites with planar distributions, J. Mater. Sci. 22 (1987) 663–669.
[174] J. Xue, On multi-state system analysis, IEEE Trans. Reliab. 34 (1985) 329–337.
[175] J. Xue, K. Yang, Dynamic reliability analysis of coherent multi-state systems, IEEE Trans. Reliab. 4 (44) (1995) 683–688.
[176] J. Xue, K. Yang, Symmetric relations in multi-state systems, IEEE Trans. Reliab. 4 (44) (1995) 689–693.
[177] K. Yu, I. Koren, Y. Guo, Generalised multistate monotone coherent systems, IEEE Trans. Reliab. 43 (1994) 242–250.
[178] E. Zio, M. Marseguerra, Basics of the Monte Carlo Method with Application to System Reliability, LiLoLe, Verlag GmbH, 2002.

Kolibri 360 GmbH
Ferdinand-Jühlke-Straße 7
99095 Erfurt, Deutschland
produktsicherheit@kolibri360.de